Yanpei Liu

Algebraic Elements of Graphs

Also of Interest

Topological Theory of Graphs
Yanpei Liu, 2017
ISBN 978-3-11-047669-9, e-ISBN (PDF) 978-3-11-047949-2
e-ISBN (EPUB) 978-3-11-047922-5

Algebraic Graph Theory. Morphisms, Monoids and Matrices
Ulrich Knauer, 2011
ISBN 978-3-11-025408-2, e-ISBN 978-3-11-025509-6

Regular Graphs. A Spectral Approach
Zoran Stanić, 2017
ISBN 978-3-11-035128-6, e-ISBN (PDF) 978-3-11-035134-7,
e-ISBN (EPUB) 978-3-11-038336-2

Discrete Algebraic Methods.
Arithmetic, Cryptography, Automata and Groups
Volker Diekert, Manfred Kufleitner, Gerhard Rosenberger,
Ulrich Hertrampf, 2016
ISBN 978-3-11-041332-8, e-ISBN (PDF) 978-3-11-041333-5,
e-ISBN (EBUP) 978-3-11-041632-9

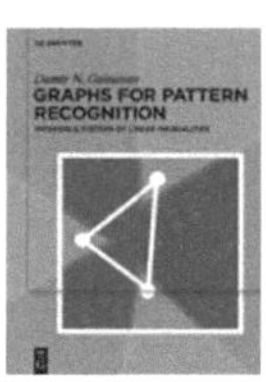

Graphs for Pattern Recognition.
Infeasible Systems of Linear Inequalities
Damir Gainanov, 2016
ISBN 978-3-11-048013-9, e-ISBN (PDF) 978-3-11-048106-8,
e-ISBN (EPUB) 978-3-11-048030-6

Yanpei Liu

Algebraic Elements of Graphs

DE GRUYTER

Author
Prof. Dr. Yanpei Liu
Beijing Jiaotong University
School of Science
Dept. of Mathematics
Haidian District
3 Shangyuancun
100044 Beijing
People's Republic of China
ypliu@bjtu.edu.cn

ISBN 978-3-11-048073-3
e-ISBN (PDF) 978-3-11-048184-6
e-ISBN (EPUB) 978-3-11-048075-7
Set-ISBN 978-3-11-048185-3

Library of Congress Cataloging-in-Publication Data
A CIP catalog record for this book has been applied for at the Library of Congress.

Bibliographic information published by the Deutsche Nationalbibliothek
The Deutsche Nationalbibliothek lists this publication in the Deutsche Nationalbibliografie; detailed bibliographic data are available on the Internet at http://dnb.dnb.de.

Typesetting: Integra Software Services Pvt. Ltd.
Printing and binding: CPI books GmbH, Leck
Cover image: rozdesign/iStock/Getty Images
♾ Printed on acid-free paper
Printed in Germany

www.degruyter.com

Preface (DG Edition)

The first edition of my monograph *Elements of Algebraic Graphs* was printed by USTC (University of Science and Technology of China) Press in 2013 and the book went out of print only a few years later. I then agreed to prepare a revised version of this book for the DG (De Gruyter) edition. The reason is that new relevant developments urge me to add a certain number of notices for updating and revising the version.

First, the latest advances in polyhedral theory enable us to prove that the Weak Embedding Conjecture (WEC) is true. From this, the Conjecture on Double Cover by Circuits (CDCC) is confirmatively derived and then the Strong Embedding Conjecture (SEC) is shown to be equivalent to the CDCC under a trivial condition. All those can be seen in LiuYP[80], [81], [82] and [83](2015). This enables us to add Notice 1.1 including Theorem I.1 and Notice 1.2 including Theorem I.2 in the Notes Section 1.5 of Chapter 1; Notice 2.1 including Theorem II.1, Notice 2.2 including Theorem II.2 and Notice 2.3 including Theorem II.3 in the Notes Section 2.6 of Chapter 2; from Notice 5.3 through Notice 5.6 in Notes Section 5.6 of Chapter 5; Notice 6.1 and Notice 6.2 in the Notes Section 6.6 of Chapter 6.

Second, the recent advances in combinatorial functional equations extracted from algebraic principles of suitably decomposing an infinite set, as shown in Liu [84–88] enable us to add Notice 13.1 through Notice 13.9, Notices 13.11 and 13.12 in the Notes Section 13.7 of Chapter 13.

Third, the latest advances in solving functional equations algebraically, as shown in Liu [97], enable us to add Notice 14.1, from Notice 14.6 through Notice 14.13 in the Notes Section 14.7 of Chapter 14; from Notice 15.8 through 15.10 and 12.12 in Notes Section 15.6 of Chapter 15.

Fourth, the advances (old and new) by employing joint tree model to represent graph embedding for carrying out up-embeddability, maximum genus, minimum genus (or genus), average genus, etc., enable us to add Notice 1.5, Notice 1.6, Notice 1.9 and Notice 1.12 in Notes Section 1.5 of Chapter 1; from Notice 5.7 through Notice 5.10 in Notes Section 5.6 of Chapter 5; from Notice 6.3 through Notice 6.7 in Notes Section 6.6 of Chapter 6; Notice 10.10 and Notice 10.11 in Notes Section 10.6 of Chapter 10; from Notice 16.1 through Notice 16.10 and Notice 16.12 in Notes Section 16.7 of Chapter 16. References are referred to as in Liu [64], Cai et al. [8], Cha Li and Liu[13], Dong et al. [25], Dong et al. [26], Li and Liu [49], Wang and Liu [161], Lv and Liu [118–120], Shao and Liu [141, 143, 145], Chen and Liu [15], Wei and Liu [164], etc.

Fifth, the advances in genus polynomials, map and embedding distributions, for a variety of graphs, based on joint trees, enable us to add Notice 1.3, Notice 1.4, Notice 1.10 and Notice 1.11 in Notes Section 1.5 of Chapter 1; Notice 9.17 and Notice 9.18 in Notes Section 9.5 of Chapter 9; Notice 12.2, Notice 12.3, from Notice 12.9 through Notice 12.12 in Notes Section 12.5 of Chapter 12; Notice 17.1 and Notice 17.2 in Notes Section 17.5 of Chapter 17. References are referred to as in Liu[67, 78], Yang and Liu[173, 175], Xu

DOI 10.1515/9783110481846-202

and Liu[169], Wan and Liu [159], Hao et al. [35], Shao and Liu [142] and Shao et al. [146], etc.

Sixth, the advances on the theory of combinatorial maps and their census enable us to add Notice 3.1 in Notes Section 3.5 of Chapter 3; Notice 4.8 and Notice 4.9 in Notes Section 4.5 of Chapter 4; from Notice 7.2 through 7.6 in Notes Section 7.6 of Chapter 7; from Notice 8.3 through Notice 8.6 in Notes Section 8.5 of Chapter 8; Notice 9.1 through Notice 9.16 in Notes Section 9.5 of Chapter 9; from Notice 10.1 through Notice 10.9, Notice 10.12 and Notice 10.13 in Notes Section 10.6 of Chapter 10; from Notice 11.1 through Notice 11.12 in Notes Section 11.5 of Chapter 11; Notice 15.2, Notice 15.4, Notice 15.5, Notice 15.7 and Notice 15.11 in Notes Section 15.6 of Chapter 15. References might be seen in Liu [67, 78], Cai et al. [9], Cai and Liu [10, 11], Dong and Liu [21], Li and Liu [52, 54], Liu and Liu [58, 59], Pan and Liu [130], Ren and Liu [132–134], Xu and Liu [167, 168], Yang and Liu [174–176], Zhang and Liu [180], etc.

The title of this edition is renamed from the previous one as *Algebraic Elements of Graphs* by considering the suggestions of the reviewers and the editor to reflect my initial idea with emphasis on fundamentals of an algebra of graphs.

The effective completion of the present version of the book depends on the help of many people. It is a pleasure for me to acknowledge that help for the DG edition. A number of readers pointed out to me errors and possible improvements in the version. They include E.L. Wei, W.L. He, Y.C. Chen, L.X. Wan, Y. Xu, W.Z. Liu, Z.L. Shao, Y. Yang, G.H. Dong, S.X. Lv, J.C. Zeng, Y.L. Zhang and L.Y. Pan. I am notably indebted to J.L. Cai, R.X. Hao and Z.X. Li for thoughtfully providing me with an especially large number of corrections and comments.

The author was supported in part by NNSFC grant 11371052.

Y.P. Liu
Beijing, China
December, 2016

Preface (USTC Edition)

Graphs as a combinatoric topic was formed from Euler for a solution of Konigsberg Seven Bridge problem dated in 1736. Maps as a mathematical topic arose probably from the four color problem, see in Birkhoff [7] and Ore [129] and the more general map coloring problem, see in HilbertD-CohnVossen [41], Ringel [137–139] and Liu [70] in the mid-nineteenth century. I could not list even main references on them because it is well known for a large range of readers and beyond the scope of this book. Here, I only intend to present a comprehensive theory of maps and graphs as algebraic structures, which has been developed mostly by myself only in recent few decades.

However, as described in Liu [74], maps can be seen as from polyhedra in origin to graphs in development via abstraction. This is why algebraic graphs are much concerned with in the present stage.

In the beginning, maps in mathematics were as a topological, or geometric object even with geographical consideration in Kempe [45]. The first formal definition of a map was done by Heffter [40] in the nineteenth century. However, mathematicians had not paid attention to this until 1960 when Edmonds published a note in the AMS Notices with the dual form of Heffter's in Edmonds [30] and Liu [62].

Although this concept was widely used in literature as Liu [60, 61, 63–65], Ringel [137–139], Stahl [147, 148] et al, its disadvantage for the nonorientable case involved does not bring with convenience for clarifying the related mathematical thinking.

Since Tutte described the nonorientability in a new way as in Tutte [150–152], a number of authors begin to develop it in combinatorization of continuous objects in Little [55], Liu [66–68], Vince [153, 154] et al.

The above representations are all with complication in constructing an embedding, or all distinct embeddings of a graph on a surface. However, the joint tree model of an embedding completed in recent years and initiated from the early articles at the end of seventies in the last century by the present author in Liu [60, 61] enables us to make the complication much simpler.

Because of the generality that in any asymmetric object there is some kind of local symmetry, the concepts of graphs and maps are just put in such a rule. In fact, the former is corresponding to that a group of two elements sticks on an edge and the latter is that a group of four elements sticks on an edge such that a graph without symmetry at all is in company with local symmetry. This treatment will bring more advantages for observing the structure of a graph. Of course, the latter is with restriction of the former because the latter is a permutation and the former a partition.

The joint tree representation of an embedding of a graph on 2-dimensional manifolds (or simply 2-manifolds), particularly surfaces (compact 2-manifolds without boundary in our case), is described in Liu [78] for simplifying a number of results, old and new.

This book contains the following subjects.

DOI 10.1515/9783110481846-203

In Chapter 1, an abstract graph and its embedding on surfaces are much concerned because they are motivated to building up the theory of abstract graphs.

The second chapter is for the formal definition of abstract maps. One can see that this matter is a natural generalization of graph embedding on surfaces.

The third chapter is on the duality not only for maps themselves but also for operations on maps from one surface to another. One can see how naturally the duality is deduced from the abstract maps described in the second chapter.

The fourth chapter is on the orientability. One can see how formally the orientability is designed as a combinatorial invariant. The fifth chapter concentrates on the classification of orientable maps. The sixth chapter is for the classification of nonorientable maps.

From Chapters 5 and 6, one can see how the procedure is simplified for these classifications.

The seventh chapter is on the isomorphisms of maps and provides an efficient algorithm for the justification and recognition of an isomorphism of two maps, which has been shown to be useful for determining the automorphism group of a map in the eighth chapter. Moreover, it enables us to access an automorphism of a graph much simply.

The ninth and the tenth chapters observe the number of distinct asymmetrized maps with the size as a parameter. In the former, only one vertex maps are counted by favorite formulas and in the latter general maps are counted from differential equations. More progresses about this kind of counting are referred to in Liu [67] and many further articles: Baxter [4], Bender et al. [5], Cai and Liu [10, 11], and Ren and Liu [132–134], etc.

Chapter 11 only presents some ideas for accessing the symmetric census of maps and further, of graphs. This topic can be done on the basis of the relationship between maps and embeddings.

Chapter 12 discusses in brief on genus polynomial of a graph and all its upper maps rooted and unrooted on the basis of the joint tree model. Recent progresses on this aspect are referred to in Chen and Liu [14, 15], Chen et al. [18], Hao and Liu [36, 37], Huang and Liu [42, 43], Li and Liu [49], Mao and Liu [124], Mao et al. [126], WanLX and Liu [157, 158], Zhao and Liu [184, 185], etc.

Chapter 13 is on the census of maps with vertex or face partitions. Although such census involves much complication and difficulty, because of the recent progress on a basic topic about trees via an elementary method first used by the author himself, we are able to do a number of types of such census in very simple way. This chapter reflects on such aspects around.

Chapter 14 is on functional equations discovered in the census of a variety of maps on sphere and general surfaces. All of them have not yet been solved up to now.

Chapters 15–17 mainly focus on to graphs via relationship among polyhedra, embeddings and maps.

Chapter 18 is on surface embeddability of graphs. Four approaches are described. More notably, one of them in the specific case of genus 0 turns out all the classic planarity theorems of Lefschetz (on double covering) in Lefschetz [48], Whitney (on duality) in Whitney [166] and MacLane (on cycle basis) in MacLane [123] (1937), which are all much generalized and much simplified at a time.

Each chapter has a section of Notes in all of which more than 200 research problems difficult and accessible to certain extent are mentioned with some historical remarks.

Three appendices are complement to the book. One provides the clarification of the concepts of polyhedra, surfaces, embeddings and maps and their relationship. The other two are for exhaustively calculating numerical results and listing all rooted and unrooted maps for small graphs.

Although I have tried to design this book to be self-contained as much as possible, some books such as Dixon and Mortimer [20], Massey [127] and Garey and Johnson [31] might be helpful to those not familiar with basic knowledge of permutation groups, topology and computing complexity as background.

Since the early nineties of the last century, a number of my former and present graduates were or are engaged with topics related to this book. Among them, I would like to mention Dr. Yi. Liu, Dr. Y.Q. Huang, Dr. J.L. Cai, Dr. D.M. Li, Dr. H. Ren, Dr. R.X. Hao, Dr. Z.X. Li, Dr. L.F. Mao, Dr. E.L. Wei, Dr. W.L. He, Dr. L.X. Wan, Dr. Y.C. Chen, Dr. Y. Xu, Dr. W.Z. Liu, Dr. Z.L. Shao, Dr. Y. Yang, Dr. G.H. Dong, Dr. J.C. Zeng, Dr. S.X. Lv, Ms. X.M. Zhao, Mr. L.F. Li, Ms. H.Y. Wang, Ms. Z. Chai, and Mr. Z.L. Zhu et al. for their successful work on this aspect.

On this occasion, I should express my heartiest appreciation of the financial support by KOSEF of Korea from the Com^{2}MaC (Combinatorial and Computational Mathematics Research Center) of the Pohang University of Science and Technology in the summer of 2001. In that period, the intention of this book was established. Moreover, I would also like to appreciated the Natural Science Foundation of China for the research development reflected in this book under its Grants (60373030, 10571013 and 10871021).

Y.P. Liu
Beijing, China
May, 2012

Contents

DOI 10.1515/9783110481846-204

1 Abstract Graphs

- A graph is considered to be a partition on the union of sets obtained from each element of a given set the binary group $B = \{0, 1\}$ sticks on.
- A surface, i.e., a compact two-manifold without boundary in topology, is seen as a polygon of even edges identified pairwise.
- An embedding of a graph on a surface is represented by a joint tree of the graph. A joint tree of a graph consists of a plane extended tree with labeled cotree semiedges. Two semiedges of a cotree edge have the same label as the cotree edge with a binary index. An extended tree is compounded of a spanning tree with cotree semiedges as articulate edges.
- Combinatorial properties of an embedding in abstraction are particularly discussed for the formal definition of a map.

1.1 Graphs and networks

Let X be a finite set. For any $x \in X$, the binary group $B = \{0, 1\}$ *sticks on* (seen as a type of group action!) x to obtain $Bx = \{x(0), x(1)\}$. $x(0)$ and $x(1)$ are called the *ends* of x, or Bx. If Bx is seen as an ordered set $\langle x(0), x(1)\rangle$, then $x(0)$ and $x(1)$ are, respectively, *initial* and *terminal* ends of x. Let

$$\mathscr{X} = \sum_{x \in X} Bx, \tag{1.1}$$

i.e., the disjoint union of all Bx, $x \in X$. $\mathscr{X}$ is called the *ground set*.

A (*directed*) *pregraph* is a *partition* $\mathrm{Par} = \{P_1, P_2, \ldots\}$ of the ground set $\mathscr{X}$, i.e.,

$$\mathscr{X} = \sum_{i \geq 1} P_i. \tag{1.2}$$

Bx (or $\langle x(0), x(1)\rangle$), or simply denoted by x itself, $x \in X$, is called an (*arc*) *edge* and P_i, $i \geq 1$, a *node* or *vertex*.

A (directed) pregraph is written as $G = (V, E)$, where $V = \mathrm{Par}$ and

$$E = B(X) = \{Bx | x \in X\}$$
$$(= \{\langle x(0), x(1)\rangle | x \in X\}).$$

If X is a finite set, the (directed) pregraph is called *finite*; otherwise, *infinite*. In this book, (directed) pregraphs are all finite.

If $X = \emptyset$, then the (directed) pregraph is said to be *empty* as well.

An edge (arc) is considered to have two semiedges, each of them being incident with only one end (semiarcs with directions of one from the end and the other

DOI 10.1515/9783110481846-001

to the end). An edge (arc) with two ends identified is called a selfloop (di-selfloop); otherwise, a link (di-link). If t edges (arcs) have same ends (same direction), they are called a multiedge (multiarc), or t-edge (t-arc).

Example 1.1 There are two directed pregraphs on $X = \{x\}$, i.e.,

$$\text{Par}_1 = \{\{x(0)\}, \{x(1)\}\};\ \text{Par}_2 = \{\{x(0), x(1)\}\}.$$

They are all distinct pregraphs, as shown in Figure 1.1.

Further, pregraphs of size 2 are also observed.

Example 1.2 On $X = \{x_1, x_2\}$, the 15 directed pregraphs are as follows:

$$\begin{aligned}
\text{Par}_1 &= \{\{x_1(0)\}, \{x_1(1)\}, \{x_2(0)\}, \{x_2(1)\}\};\\
\text{Par}_2 &= \{\{x_1(0), x_1(1)\}, \{x_2(0)\}, \{x_2(1)\}\};\\
\text{Par}_3 &= \{\{x_1(0), x_2(0)\}, \{x_1(1)\}, \{x_2(1)\}\};\\
\text{Par}_4 &= \{\{x_1(0), x_2(1)\}, \{x_1(1)\}, \{x_2(0)\}\};\\
\text{Par}_5 &= \{\{x_1(0)\}, \{x_1(1), x_2(0)\}, \{x_2(1)\}\};\\
\text{Par}_6 &= \{\{x_1(0)\}, \{x_1(1), x_2(1)\}, \{x_2(0)\}\};\\
\text{Par}_7 &= \{\{x_1(0)\}, \{x_1(1)\}, \{x_2(1), x_2(0)\}\};\\
\text{Par}_8 &= \{\{x_1(0), x_1(1), x_2(0)\}, \{x_2(1)\}\};\\
\text{Par}_9 &= \{\{x_1(0), x_1(1), x_2(1)\}, \{x_2(0)\}\};\\
\text{Par}_{10} &= \{\{x_1(0), x_2(0), x_2(1)\}, \{x_1(1)\}\};\\
\text{Par}_{11} &= \{\{x_1(0)\}, \{x_1(1), x_2(0), x_2(1)\}\};\\
\text{Par}_{12} &= \{\{x_1(0), x_1(1), x_2(0), x_2(1)\}\};\\
\text{Par}_{13} &= \{\{x_1(0), x_1(1)\}, \{x_2(0), x_2(1)\}\};\\
\text{Par}_{14} &= \{\{x_1(0), x_2(0)\}, \{x_1(1), x_2(1)\}\};\\
\text{Par}_{15} &= \{\{x_1(0), x_2(1)\}, \{x_1(1), x_2(0)\}\}.
\end{aligned}$$

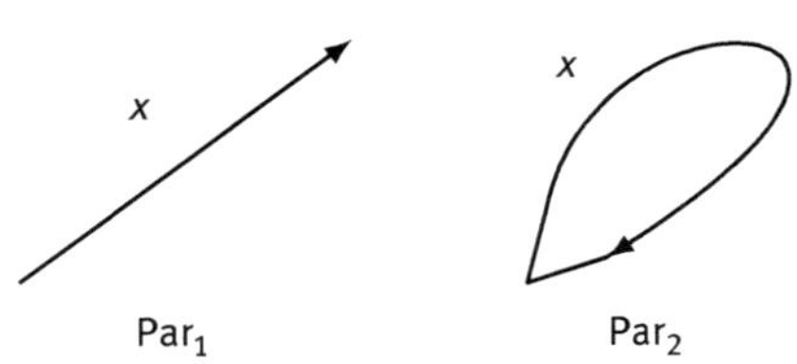

Figure 1.1: Directed pregraphs of 1 edge.

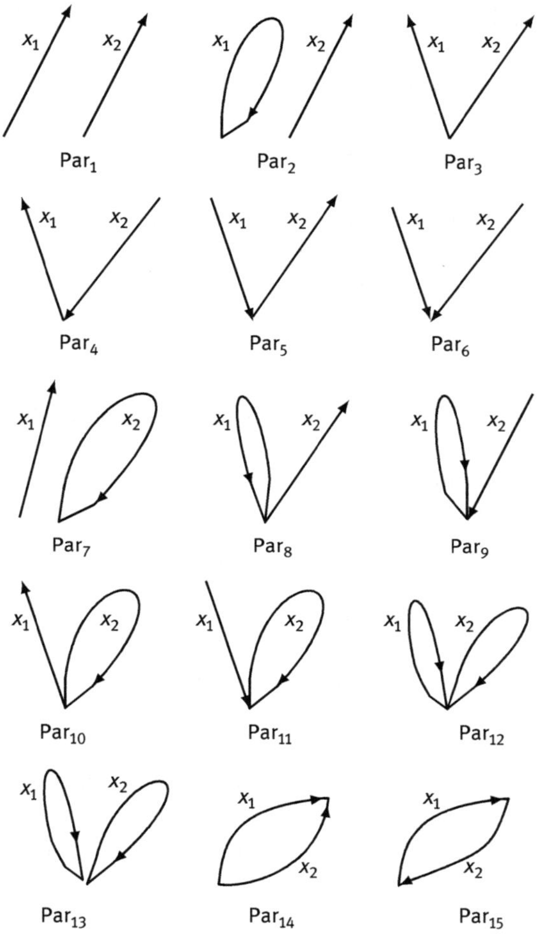

Figure 1.2: Directed pregraphs of 2 edges.

Among the 15 directed pregraphs, Par_3, Par_4, Par_5 and Par_6 are 1 pregraph; Par_8 and Par_9 are 1 pregraph; Par_{10} and Par_{11} are 1 pregraph; Par_{14} and Par_{15} are 1 pregraph; and others are 5 pregraphs. Thus, there are nine pregraphs in all (as shown in Figure 1.2).

Now, $\mathrm{Par} = \{P_1, P_2, \ldots\}$ and $\mathscr{B}$ are, respectively, seen as a mapping $z \mapsto P_i$, $z \in P_i$, $i \geq 1$ and a mapping $z \mapsto \bar{z}$, $\bar{z} \neq z$, $\{z, \bar{z}\} \subseteq B(X)$. The *composition* of two mappings α and β on a set $\mathscr{X}$ is defined to be the mapping

$$(\alpha\beta)z = \bigcup_{y \in \beta z} \alpha y, \ z \in \mathscr{X}. \tag{1.3}$$

Let $\Psi_{\{\mathrm{Par},\mathscr{B}\}}$ be the semigroup generated by $\mathrm{Par} = \mathrm{Par}(X)$ and $\mathscr{B} = B(X)$. Since the mappings $\alpha = \mathrm{Par}$ and $\mathscr{B}$ have the property that $y \in \alpha z \Leftrightarrow z \in \alpha y$, it can be checked that for any $z, y \in B(X)$, what is determined by $\exists\gamma \in \Psi_{\{\text{Par},\mathscr{B}\}}$, $z \in \gamma y$, is an equivalence. If $B(X)$ itself is an equivalent class, then the semigroup $\Psi_{\{\text{Par},\mathscr{B}\}}$ is called *transitive* on $\mathscr{X}$ as shown in (1.1). A (directed) pregraph with $\Psi_{\{\text{Par},\mathscr{B}\}}$ transitive on $\mathscr{X}$ is called a *(directed) graph*.

A (directed) pregraph $G = (V, E)$ is called *connected* if for any two vertices $u, v \in V$, there exists a sequence of edges $e_1, e_2, \dots, e_s$ for the two ends (one tail and one head) of e_i, $i = 2, 3, \dots, s-1$, are in common with those of respective e_{i-1} and e_{i+1} where u and v are, respectively, the other ends of e_1 and e_s. Such a sequence of edges is called a *trail* between u and v. A trail without edge repetition is a *walk*. A walk without vertex repetition is a *path*. A trail, walk or path with $u = v$ is, respectively, a *travel*, *tour* or *circuit*.

Theorem 1.1 A (directed) pregraph is a (directed) graph if, and only if, it is connected.

Proof Necessity. Since $\text{Par}^k = \text{Par}$, $k \geq 1$, and $\mathscr{B}^k = \mathscr{B}$, $k \geq 1$, by the transitivity, for any two elements $y, z \in \mathscr{X}$, there exists γ such that $z \in \gamma y$ and there exists an integer $n \geq 0$ such that

$$\gamma = (\mathscr{B}\text{Par})^n\mathscr{B} = \underbrace{(\mathscr{B}\text{Par})\cdots(\mathscr{B}\text{Par})}_{n}\mathscr{B}, \tag{1.4}$$

where $\mathscr{B}$Par appears for n times. Therefore, the (directed) pregraph is connected.

Sufficiency. If a (directed) pregraph is connected, i.e., for any two elements $z, y \in \mathscr{X}$, their incident vertices $u, v \in V$ have edges $e_1, e_2, \dots, e_s$, such that e_i, $i = 2, 3, \dots, s-1$, is in common with e_{i-1} and e_{i+1}. Of course, u and v are, respectively, the ends of e_1 and e_s. Thus, $y \in \gamma z$ where $\gamma = (\text{Par}\mathscr{B})^s\mathscr{B}$. This implies that the semigroup $\Psi_{\{\text{Par},\mathscr{B}\}}$ is transitive on $\mathscr{X}$. Therefore, the (directed) pregraph is a (directed) graph. □

It is seen from the theorem that (directed) graphs here are, in fact, connected (directed) graphs in most textbooks. Because disconnectedness is rarely necessary to consider, for convenience all graphs, embeddings and then maps in what follows are defined within connectedness in this book.

A *network* N is such a graph $G = (V, E)$ with a real function $w(e) \in \mathbf{R}$ for $e \in E$ on E, and hence written as $N = (G; w)$. Usually, a network N is denoted by the graph G itself if no confusion occurs.

Finite recursion principle (FRP) On a finite set A, choose $a_0 \in A$ as the initial element at the 0th step. Assume a_i is chosen at the ith, $i \geq 0$, step with a given rule. If not all elements available from a_i are not yet chosen, choose one of them as a_{i+1} at the $i+1$st step by the rule; then a chosen element will be encountered in finite steps unless all elements of A are chosen.

Finite restrict recursion principle (FRRP) On a finite set A, choose $a_0 \in A$ as the initial element at the 0th step. Assume a_i is chosen at the ith, $i \geq 0$, step with a given rule. If a_0 is not available from a_i, choose one of the elements available from a_i

as a_{i+1} at the $i + 1$st step by the rule, then a_0 will be encountered in finite steps unless all elements of A are chosen.

The two principles above are very useful in finite sets, graphs and networks, even in a wide range of combinatorial optimizations.

A $G = (V, E)$ with $V = V_1 + V_2$ of both V_1 and V_2 independent, i.e., its vertex set is partitioned into two parts with each part having no pair of adjacent vertices, is called *bipartite*.

Theorem 1.2 A graph $G = (V, E)$ is bipartite if, and only if, G has no circuit with odd number of edges.

Proof Necessity. Since G is bipartite, start from $v_0 \in V$ initially chosen and then by the rule from the vertex just chosen to one of its adjacent vertices via an edge unused and then marked by used, according to the finite recursion principle, an even circuit (from bipartite), or no circuit at all, can be found. From the arbitrariness of v_0 and the way going on, no circuit of G is with odd number of edges.

Sufficiency. Since all circuits are even, start from marking an arbitrary vertex by 0 and then by the rule from a vertex marked by $b \in B = \{0, 1\}$ to mark all its adjacent vertices by $\overline{b} = 1 - b$, according to the finite recursion principle the vertex set is partitioned into $V_0 = \{v \in V|$ marked by $0\}$ and $V_1 = \{v \in V|$ marked by $1\}$. By the rule, V_0 and V_1 are both independent and hence G is bipartite. □

From this theorem, a graph without circuit is bipartite. In fact, from the transitivity, any graph without circuit is a tree.

On a pregraph, the number of elements incident to a vertex is called the *degree* of the vertex. A pregraph of all vertices with even degree is said to be *even*. If an even pregraph is a graph, then it is called a *Eulerian graph*.

Theorem 1.3 A pregraph $G = (V, E)$ is even if, and only if, there exist circuits $C_1, C_2, \ldots, C_n$, on G such that

$$E = C_1 + C_2 + \cdots + C_n, \tag{1.5}$$

where n is a nonnegative integer.

Proof Necessity. Since all the degrees of vertices on G are even, any pregraph obtained by deleting the edges of a circuit from G is still even. From the finite recursion principle, there exist a nonnegative integer n and circuits $C_1, C_2, \ldots, C_n$, on G such that (1.5) is satisfied.

Sufficiency. Because a circuit contributes 2 to the degree of each of its incident vertices, (1.5) guarantees each of vertices on G has even degree. Hence, G is even. □

The set of circuits $\{C_i|1 \leq i \leq n\}$ of G in (1.5) is called a *circuit partition*, or written as Cir=Cir(G). Two direct conclusions of Theorem 1.3 are very useful . One is the case that G is a graph. The other is that G is a directed pregraph. Their forms and proofs are left for the reader.

Let $N = (G; w)$ be a network where $G = (V, E)$ and $w(e) = -w(e) \in Z_n = \{0, 1, \ldots, n-1\}$, i.e., mod n, $n \geq 1$, integer group. For examples, $Z_1 = \{0\}$, $Z_2 = B = \{0, 1\}$ etc. Suppose $x_v = -x_v \in Z_n$, $v \in V$, are variables. Let us discuss the system of equations

$$x_u + x_v = w(e) \ (\mathrm{mod}\ n),\ e = (u, v) \in E, \tag{1.6}$$

on Z_n.

Theorem 1.4 System of eq. (1.6) has a solution on Z_n if, and only if, there is no circuit C such that

$$\sum_{e \in C} w(e) \neq 0 \ (\mathrm{mod}\ n) \tag{1.7}$$

on N.

Proof Necessity. Assume C is a circuit satisfying (1.7) on N. Because the restricted part of (1.6) on C has no solution, the whole system of eq. (1.6) has to be no solution either. Therefore, N has no such circuit. This is a contradiction to the assumption

Sufficiency. Let $x_0 = a \in Z_n$, start from $v_0 \in V$ reached. Assume $v_i \in V$ reached and $x_i = a_i$ at step i. Choose one of $e_i = (v_i, v_{i+1}) \in E$ without used (otherwise, backward 1 step as the step i). Choose v_{i+1} reached and e_i used with $a_{i+1} = a_i + w(e_i)$ at step $i+1$. If a circuit as $\{e_0, e_1, \ldots, e_l\}$, $e_j = (v_j, v_{j+1})$, $0 \leq j \leq l$, $v_{l+1} = v_0$, occurs within a permutation of indices, then from (1.7)

$$\begin{aligned} a_{l+1} &= a_l + w(e_l) \\ &= a_{l-1} + w(e_{l-1}) + w(e_l) \\ &\vdots \\ &= a_0 + \sum_{j=0}^{l} w(e_j) \\ &= a_0. \end{aligned}$$

Because the system of equations obtained by deleting all the equations for all the edges on the circuit from (1.6) is equivalent to the original system of eq. (1.6), in virtue of the finite recursion principle a solution of (1.6) can always be extracted . □

When $n = 2$, this theorem has a variety of applications. In Liu [64], some applications can be seen. Further, its extension on a non-Abelian group can also be done while the system of equations are not yet linear but quadratic.

1.2 Surfaces

In topology, a surface is a compact two-dimensional manifold without boundary. In fact, it can be seen as what is obtained by identifying each pair of edges on a polygon of even edges pairwise.

For example, in Figure 1.3, two ends of each dot line are the same point. Figure 1.3(a) is a sphere, or the plane when the infinity is seen as a point. Figure 1.3(b) is the projective plane. From the symmetry of the sphere, a surface can also be seen as a sphere cutting off a polygon with pairwise edges identified.

The two surfaces in Figure 1.3 are formed by a polygon of two edges pairwise as a.

Surface closed curve axiom A closed curve on a surface has one of the two possibilities: one side and two sides.

A curve with two sides is called a *double side curve*; otherwise, a *single side curve*. As shown in Figure 1.3, any closed curve on a sphere is a double side curve (In fact, this is the Jordan curve axiom). However, it is different from the sphere for the projective plane, there are both a single (shown by a dotted line) and a double side curve.

How do we justify whether a closed curve on a surface is of single side, or not?

In order to answer this question, the concept of contractibility of a curve has to be clarified. If a closed curve on a surface can be continuously contracted along one side into a point, then it is said to be *contractible*, or *homotopic* to 0.

It is seen that a single side curve is never homotopic to 0 and a double side curve is not always homotopic to 0. For example, in Figure 1.4(a), i.e., the torus, each of the dotted lines is of double side but not contractible. Figure 1.4(b), i.e., the Klein bottle, all the dotted lines are of single side, and hence, none of them is contractible.

A surface with all closed curves of double side is called *orientable*; otherwise, *nonorientable*.

For example, in Figure 1.3, the sphere is orientable and the projective plane is nonorientable. In Figure 1.4, the torus is orientable and the Klein bottle is nonorientable.

The maximum number of closed curves cutting along without destroying the continuity on a surface is called the *pregenus* of the surface.

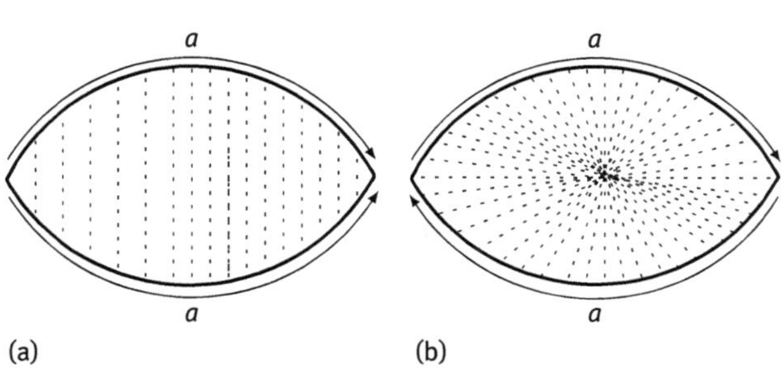

Figure 1.3: (a) Sphere and (b) Projective plane.

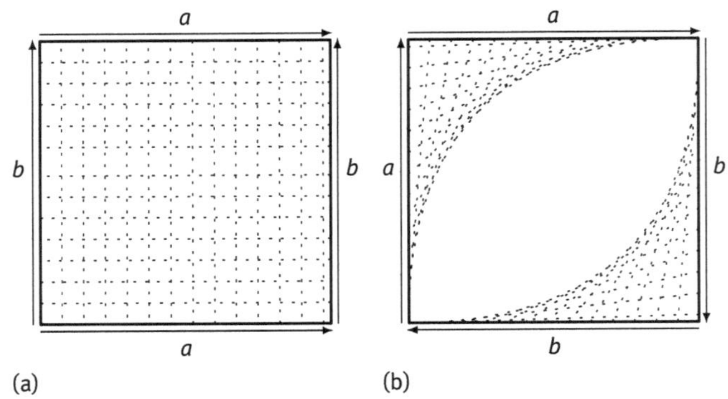

Figure 1.4: (a) Torus and (b) Klein bottle.

In view of Jordan curve axiom, there is no such closed curve on the sphere. Thus, the pregenus of sphere is 0. On the projective plane, only one such curve is available (each dotted line is such a closed curve in Figure 1.3) and hence the pregenus of projective plane is 1.

Similarly, the pregenera of torus and Klein bottle are both 2 as shown in Figure 1.4.

Theorem 1.5 The pregenus of an orientable surface is a nonnegative even number.

A formal proof cannot be done until Chapter 5. Based on this theorem, the genus of an orientable surface can be defined to be half its pregenus, called the *orientable genus*. The genus of a nonorientable surface, called *nonorientable genus*, is its pregenus itself.

The sphere is written as aa^{-1} where a^{-1} is in the opposite direction of a on the boundary of the polygon. Thus, the projective plane, torus and Klein bottle are, respectively, aa, $aba^{-1}b^{-1}$ and $aabb$. In general,

$$\begin{aligned} O_p &= \prod_{i=1}^{p} a_i b_i a_i^{-1} b_i^{-1} \\ &= a_1 b_1 a_1^{-1} b_1^{-1} a_2 b_2 a_2^{-1} b_2^{-1} \ldots a_p b_p a_p^{-1} b_p^{-1} \end{aligned} \tag{1.8}$$

and

$$Q_q = \prod_{i=1}^{q} a_i a_i = a_1 a_1 a_2 a_2 \cdots a_q a_q \tag{1.9}$$

denote, respectively, a surface of orientable genus p and a surface of nonorientable genus q. Of course, O_0, Q_1, O_1 and Q_2 are, respectively, the sphere, projective plane, torus and Klein bottle.

It is easily checked that whenever an even polygon is with a pair of its edges in the same direction, the polygon represents a nonorientable surface. Thus, all O_p, $p \geq 0$, are orientable and all Q_q, $q \geq 1$, are nonorientable.

Forms (1.8) and (1.9) are said to be *standard*.

If the *form* of a surface is defined by its orientability and its genus, then the operations 1–3 and their inverses shown in Figures 1.5, 1.6, 1.7, do not change the surface form.

In fact, what is determined under these operations is just a topological equivalence, denoted by $\sim_{\text{top}}$.

Notice that A and B are all linear order of letters and permitted to be empty as degenerate case in these operations.

The parentheses stand for cyclic order when more than one cyclic orders occur for distinguishing one from the other.

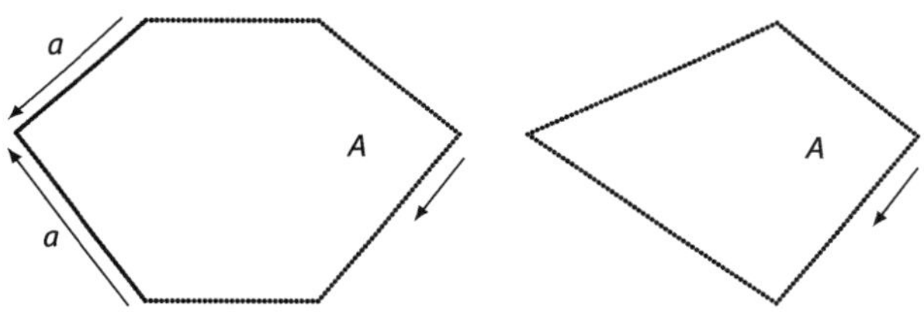

Figure 1.5: Operation 1: $Aaa^{-1} \Longleftrightarrow A$.

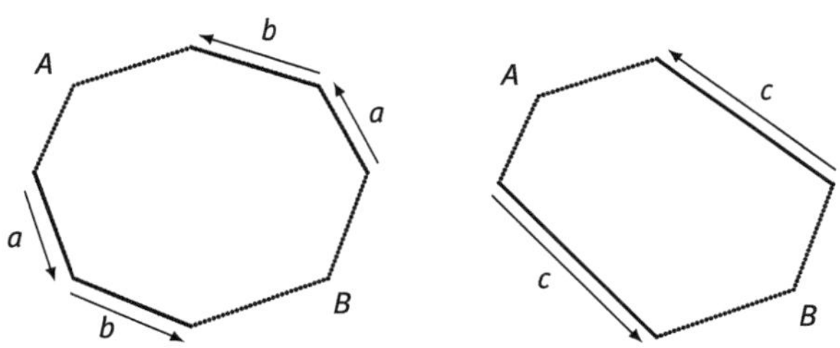

Figure 1.6: Operation 2: $AabBab \Longleftrightarrow AcBc$.

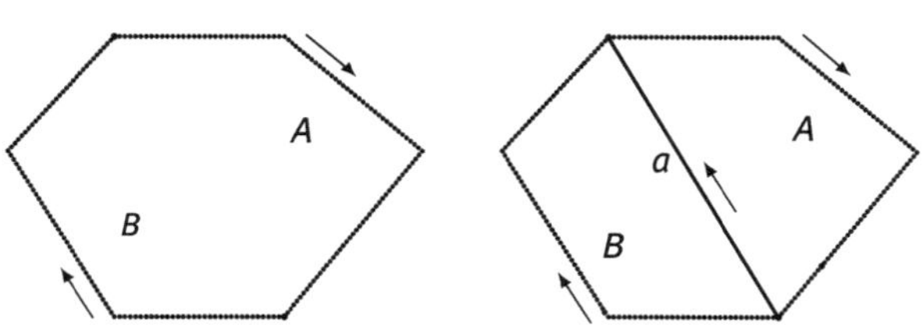

Figure 1.7: Operation 3: $AB \Longleftrightarrow (Aa)(a^{-1}B)$.

Relation 0 On a surface (A, B), if A is a surface itself then $(A, B) = ((A)x(B)x^{-1}) = ((A)(B)) = (AB)$.

Relation 1 $(AxByCx^{-1}Dy^{-1}) \sim_{\text{top}} ((ADCB)(xyx^{-1}y^{-1}))$.

Relation 2 $(AxBx) \sim_{\text{top}} ((AB^{-1})(xx))$.

Relation 3 $(Axxyzy^{-1}z^{-1}) \sim_{\text{top}} ((A)(xx)(yy)(zz))$.

In the four relations, A, B, C and D are permitted to be empty. $B^{-1} = b_s^{-1} \cdots b_3^{-1}b_2^{-1}b_1^{-1}$ is also called the inverse of $B = b_1b_2b_3 \cdots b_s$, $s \geq 1$. Parentheses are always omitted when unnecessary to distinguish cyclic or linear order.

On a surface S, the operation of cutting off a quadrangle $aba^{-1}b^{-1}$ and then identifying each pair of edges with the same letter is called a *handle*, as shown in Figure 1.8(a).

If the quadrangle $aba^{-1}b^{-1}$ is replaced by aa, then such an operation is called a *crosscap*, as shown in Figure 1.8(b).

Theorem 1.6 shows the result of doing a handle on an orientable surface.

Theorem 1.6 What is obtained by doing a handle on an orientable surface is still orientable with its genus 1 added.

Proof Suppose S is the surface obtained, then

$$\begin{aligned}
S &\sim_{\text{top}} \prod_{i=1}^{p} a_ib_ia_i^{-1}b_i^{-1}xa_{p+1}b_{p+1}a_{p+1}^{-1}b_{p+1}^{-1}x^{-1} \text{ (Relation 0)} \\
&\sim_{\text{top}} \prod_{i=1}^{p} a_ib_ia_i^{-1}b_i^{-1}xx^{-1}a_{p+1}b_{p+1}a_{p+1}^{-1}b_{p+1}^{-1} \text{ (Relation 1)} \\
&\sim_{\text{top}} \prod_{i=1}^{p} a_ib_ia_i^{-1}b_i^{-1}a_{p+1}b_{p+1}a_{p+1}^{-1}b_{p+1}^{-1} \text{ (Operation 1)} \\
&= \prod_{i=1}^{p+1} a_ib_ia_i^{-1}b_i^{-1}.
\end{aligned}$$

This finishes the proof. □

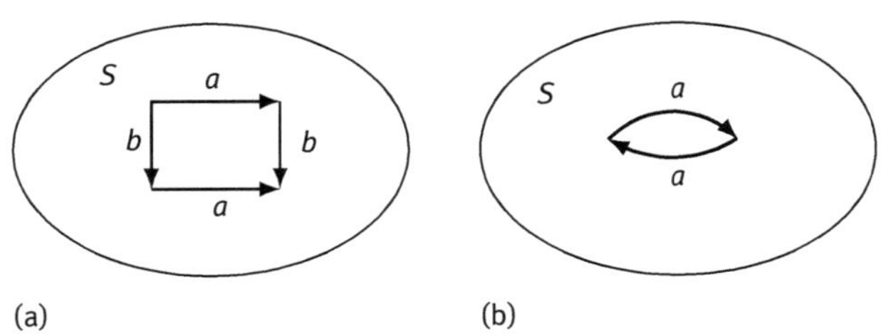

Figure 1.8: (a) Handle and (b) Crosscap.

In the above proof, x and x^{-1} are a line connecting the two boundaries to represent the surface as a polygon shown in Figure 1.8. This procedure can also be seen as the degenerate case of operation 3.

In what follows, observe the result by doing a crosscap on an orientable surface.

Theorem 1.7 On an orientable surface of genus p, $p \geq 0$, what is obtained by doing a crosscap is nonorientable with its genus $2p + 1$.

Proof Suppose N is the surface obtained, then

$$N \sim_{\text{top}} \prod_{i=1}^{p} a_i b_i a_i^{-1} b_i^{-1} x a a x^{-1} \text{ (Relation 0)}$$

$$\sim_{\text{top}} \prod_{i=1}^{p} a_i b_i a_i^{-1} b_i^{-1} x x^{-1} c_1 c_1 \text{ (Relation 2)}$$

$$\sim_{\text{top}} \prod_{i=2}^{p} a_i b_i a_i^{-1} b_i^{-1} x x^{-1} c_1 c_1 c_2 c_2 c_3 c_3 \text{ (Relation 3)}$$

$$\sim_{\text{top}} \prod_{i=1}^{2p+1} c_i c_i \text{ (Relation 3 by } p - 1 \text{ times).}$$

The proof is done. □

By doing a handle on a nonorientable surface, two more genus should be added with the same nonorientability.

Theorem 1.8 On a nonorientable surface, what is obtained by doing a handle is nonorientable with its genus 2 added.

Proof Suppose N is the obtained surface, then

$$N \sim_{\text{top}} \prod_{i=1}^{q} a_i a_i x a b a^{-1} b^{-1} x^{-1} \text{ (Relation 0)}$$

$$\sim_{\text{top}} \prod_{i=1}^{q} a_i a_i x x^{-1} a b a^{-1} b^{-1} \text{ (Relation 1)}$$

$$\sim_{\text{top}} \prod_{i=1}^{q} a_i a_i a b a^{-1} b^{-1} \text{ (Operation 1)}$$

$$\sim_{\text{top}} \prod_{i=1}^{q+2} c_i c_i \text{ (Relation 3).}$$

The proof is finished. □

By doing a crosscap on a nonorientable surface, one more genus should be produced with the same nonorientability.

Theorem 1.9 On a nonorientable surface, what is obtained by doing a crosscap is nonorientable with its genus 1 added.

Proof Suppose N is the obtained surface, then

$$\begin{aligned} N &\sim_{\text{top}} \prod_{i=1}^{q} a_i a_i x a a x^{-1} \text{ (Relation 0)} \\ &\sim_{\text{top}} \prod_{i=1}^{q} a_i a_i x x^{-1} a a \text{ (Relation 2)} \\ &\sim_{\text{top}} \prod_{i=1}^{q} a_i a_i a a \text{ (Operation 1)} \\ &\sim_{\text{top}} \prod_{i=1}^{q+1} a_i a_i \text{ (Relation 3).} \end{aligned}$$

The proof is completed. □

1.3 Embedding

Let $G = (V, E)$, $V = \{v_1, v_2, \dots, v_n\}$, be a graph. A point in the three-dimensional space is represented by a real number t as the parameter, e.g., $(x, y, z) = (t, t^2, t^3)$. Write the vertices as

$$v_i = (x_i, y_i, z_i) = (t_i, t_i^2, t_i^3)$$

such that $t_i \neq t_j$, $i \neq j$, $1 \leq i, j \leq n$, and an edge as

$$(u, v) = u + \lambda v, \ 1 \leq \lambda \leq 1,$$

i.e., the straight line segment between u and v. Because for any four vertices v_i, v_j, v_l and v_k,

$$\det \begin{pmatrix} x_i - x_j & x_i - x_l & x_i - x_k \\ y_i - y_j & y_i - y_l & y_i - y_k \\ z_i - z_j & z_i - z_l & z_i - z_k \end{pmatrix} = \det \begin{pmatrix} t_i - t_j & t_i - t_l & t_i - t_k \\ t_i^2 - t_j^2 & t_i^2 - t_l^2 & t_i^2 - t_k^2 \\ t_i^3 - t_j^3 & t_i^3 - t_l^3 & t_i^3 - t_k^3 \end{pmatrix}$$
$$= (t_i - t_j)(t_i - t_l)(t_i - t_k)(t_k - t_l)(t_k - t_j)(t_j - t_l) \neq 0,$$

i.e., the four points are not coplanar, any two edges in G has no intersection as an inner point.

A representation of a graph on a space with vertices as points and edges as curves pairwise with no intersection as an inner point is called an *embedding* of the graph in the space. If all edges are straight line segments in an embedding, then it is called a *straight line embedding*. Thus, any graph has a straight line embedding in the three-dimensional space. Similarly, a *surface embedding* of graph G is a continuous injection μG of an embedding of G on the three-dimensional space to a surface S such that each connected component of $S - \mu G$ is homotopic to 0. The connected component is called a *face* of the embedding. In early books, a surface embedding is also called a *cellular embedding*. Because only a surface embedding is concerned with in what follows, an embedding is always meant a surface embedding if not necessary to specify.

A graph (connected!) without circuit is called a *tree*. A *spanning tree* of a graph is such a subgraph that is a tree with the same order as the graph. Usually, a spanning tree of a graph is in short called a tree on the graph. For a tree on a graph, the number of edges on the tree and the number of edges not on the tree are only dependent on the order of the graph. They are, respectively, called the *rank* and the *corank* of the graph. The corank is also called the *Betti number*, or *cyclic number* by some authors.

The following procedure can be used for finding an embedding on a surface.

First, given a cyclic order of all semiedges at each vertex of G, called a *rotation*, find a tree (spanning, of course) T on G and distinguish all the edges not on T by letters. Then, replace each edge not on T by two articulate edges with the same letter.

From this procedure, G is transformed into $\tilde{G}$ without changing the rotation at each vertex except for new vertices that are all articulate. Because $\tilde{G}$ is a tree, according to the rotation, all lettered articulate edges of $\tilde{G}$ form a polygon with β pairs of edges, and hence a surface in correspondence with a choice of indices on each pair of the same letter. For convenience, $\tilde{G}$ with a choice of indices of pair in the same letter is called a *joint tree* of G.

Theorem 1.10 A graph $G = (V, E)$ can always be embedded into a surface of orientable genus at most $\lfloor \beta/2 \rfloor$, or of nonorientable genus at most β, where β is the Betti number of G.

Proof It is seen that any joint tree of G is an embedding of G on the surface determined by its associate polygon. From (1.8) for the orientable case, the surface has its genus at most $\lfloor 2\beta/4 \rfloor = \lfloor \beta/2 \rfloor$. From (1.9) for the nonorientable case, the surface has its genus at most $2\beta/2 = \beta$. □

In Figure 1.9, graph G and one of its joint trees are shown. Here, the spanning tree T is represented by edges without letter. a, b and c are edges not on T. Because the polygon is

$$\begin{aligned} abcacb &\sim_{\text{top}} c^{-1}b^{-1}cbaa \text{ (Relation 2)} \\ &\sim_{\text{top}} aabbcc \text{ (Theorem 1.7)}, \end{aligned}$$

the joint tree is, in fact, an embedding of G on a nonorientable surface of genus 3.

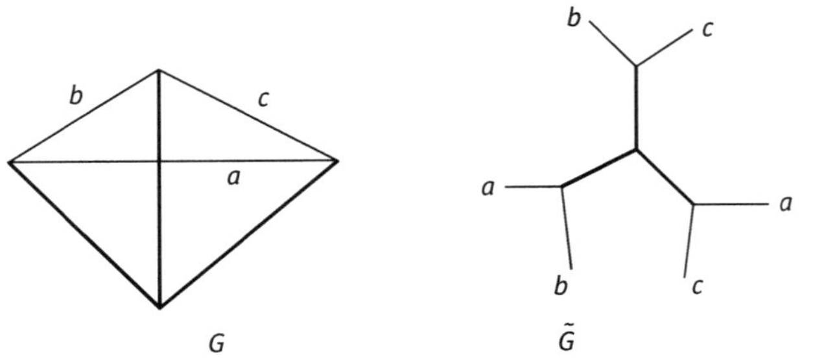

Figure 1.9: Graph and its joint tree.

Because any graph with given rotation can always be immersed in the plane in agreement with the rotation, each edge has two sides. As known, embeddings of a graph on surfaces are distinguished by the rotation of semiedges at each vertex and the choice of indices of the two semiedges on each edge of the graph whenever edges are labelled by letters. *Same indices* of the two semiedges of an edge stand for from one side of the edge to the same itself, and *Different indices* stand for from one side to the other on a face boundary of an embedding.

Theorem 1.11 A tree can only be embedded on the sphere. Any graph G except tree can be embedded on a nonorientable surface. Any graph G can always be embedded on an orientable surface. Let $n_O(G)$ be the number of distinct embeddings on orientable surfaces. Then the number of embeddings on all surfaces is

$$2^{\beta(G)} n_O(G), \ n_O(G) = \prod_{i \geq 2} ((i-1)!)^{n_i}, \tag{1.10}$$

where $\beta(G)$ is the Betti number and n_i is the number of vertices of degree i in G.

Proof On a surface of genus not 0, only a graph with at least a circuit is possible to have an embedding. Because a tree has no circuit, it can be embedded only on the sphere. Because a graph not a tree has at least one circuit, from Theorem 1.10, the second and the third statements are true. Since distinct planar embeddings of a joint tree of G with the indices of each different letter correspond to distinct embeddings of G on orientable surfaces and the number of distinct planar embeddings of joint trees is

$$n_O(G) = \prod_{i \geq 2} ((i-1)!)^{n_i}.$$

Further, since the indices of letters on the $\beta(G)$ edges has $2^{\beta(G)}$ of choices for a given orientable embedding and among them only one choice corresponds to an orientable embedding, the fourth statement is true. □

For an embedding $\mu(G)$ of G on a surface, let $\nu(\mu G)$, $\epsilon(\mu G)$ and $\phi(\mu G)$ be, respectively, its vertex number, or *order*, edge number, or *size* and face number, or *coorder*.

Theorem 1.12 For a surface S, all embeddings $\mu(G)$ of a graph G have $\mathrm{Eul}(\mu G) = \nu(\mu G) - \epsilon(\mu G) + \phi(\mu G)$ the same, dependent only on S and independent of G. Further,

$$\mathrm{Eul}(\mu G) = \begin{cases} 2-2p,\ p \geq 0, \\ \quad \text{when } S \text{ has orientable genus } p; \\ 2-q,\ q \geq 1, \\ \quad \text{when } S \text{ has nonorientable genus } q. \end{cases} \tag{1.11}$$

Proof For an embedding $\mu(G)$ on S, if it has at least two faces, then by connectedness, it has two faces with a common edge. From the finite recursion principle, by the inverse of Operation 3, an embedding $\mu(G_1)$ of G_1 on S with only one face on S is found. It is easy to check that $\mathrm{Eul}(\mu G) = \mathrm{Eul}(\mu G_1)$. Similarly, by the inverse of Operation 2, an embedding $\mu(G_0)$ of G_0 on S with only one vertex is found. It is also easy to check that $\mathrm{Eul}(\mu G) = \mathrm{Eul}(\mu G')$. Further, by Operation 1 and Relations 1–3, it is seen that $\mathrm{Eul}(\mu G_0) = \mathrm{Eul}(O_p)$, $p \geq 0$; or $\mathrm{Eul}(Q_q)$, $q \geq 1$ accordingly as S is an orientable surface in (1.8); or not in (1.9). From the arbitrariness of G, the first statement is proved.

By calculating the order, size and coorder of O_p, $p \geq 0$; or Q_q, $q \geq 1$, (1.11) is soon obtained. So, the second statement is proved. □

According to this theorem, for an embedding $\mu(G)$ of graph G, $\mathrm{Eul}(\mu G)$ is called its *Euler characteristic*, or of the surface it is on. Further, $g(\mu G)$ is the genus of the surface $\mu(G)$ it is on.

If a graph G is with the minimum length of circuits σ, then from Theorem 1.12, the genus $\gamma(G)$ of an orientable surface G can be embedded satisfying the inequality:

$$1 - \frac{\nu(G) - \epsilon(1 - \frac{2}{\sigma})}{2} \leq \gamma(G) \leq \left\lfloor \frac{\beta}{2} \right\rfloor \tag{1.12}$$

and the genus $\tilde{\gamma}(G)$ of a nonorientable surface G can be embedded on satisfying the inequality:

$$2 - \left(\nu(G) - \epsilon\left(1 - \frac{2}{\sigma}\right)\right) \leq \tilde{\gamma}(G) \leq \beta. \tag{1.13}$$

If a graph has an embedding with its genus attaining the lower (upper) bound in (1.12) and (1.13), then it is called *down(up)-embeddable*. In fact, a graph is up-embeddable on nonorientable, or orientable surfaces accordingly as it has an embedding with only one face, or at most two faces.

Theorem 1.13 All graphs but trees are up-embeddable on nonorientable surfaces.

Further, if a graph has an embedding of nonorientable genus l and an embedding of nonorientable genus k, $l < k$, then for any i, $l < i < k$, it has an embedding of nonorientable genus i.

Proof For an arbitrary embedding of a graph G on a nonorientable surface, let T be its corresponding joint tree. From the nonorientability, the associate $2\beta(G)$-gon P has at least one letter with different indices(or same power of its two occurrences!). If $P = Q_q$, $q = \beta(G)$, then the embedding is an up-embedding in its own right. Otherwise, by Relation 2, or Relation 3 if necessary, whenever $s^{-1}s$ or $stst$ occurs, it is, respectively, replaced by ss or $sts^{-1}t$. In virtue of no letter missed in the procedure, from the finite recursion principle, $P' = Q_q$, $q = \beta(G)$, is obtained. This is the first statement.

From the arbitrariness of starting embedding in the procedure of proving the first statement by only using Relation 2 instead of Relation 3 ($AststB \sim_{\text{top}} Ass^{-1}Btt$ by Relation 2), because the genus of the surface is increased by 1, from the finite recursion principle (FRP), the second statement is true. □

The second statement of this theorem is also called the *interpolation theorem*. The orientable form of interpolation theorem is first given by Duke [29]. The maximum (minimum) of the genus of surfaces (orientable or nonorientable) a graph can be embedded on is called the *maximum genus* (*minimum genus*) of the graph. Theorem 1.13 shows that graphs but trees all have their maximum genus on nonorientable surfaces and the Betti number with the interpolation theorem. The proof would be the simplest one. However, for the orientable case, it is far from simple. Many results have been obtained since 1978, see Liu [60, 61]), Liu and Liu [116], Huang and Liu [42] and Li and Liu [49] in this aspect. On the determination of minimum genus of a graph, only a few of graphs with certain symmetry are done, see Chapter 12 in Liu [64, 65].

1.4 Abstract representation

Let $G = (V, X)$, $V = \{v_1, v_2, \ldots, v_n\}$,

$$X = \{x_1, x_2, \ldots, x_m\} \subseteq V\{\times\}V = \{\{u, v\} | \forall u, v \in V\},$$

be a graph. For an embedding $\mu(G)$ of G on a surface, each edge has not only two ends as in G but also two sides. Let α be the operation from one side to the other and β be the operation from one end to the other. From the symmetry between the two ends and between the two sides,

$$\alpha^2 = \beta^2 = 1, \tag{1.14}$$

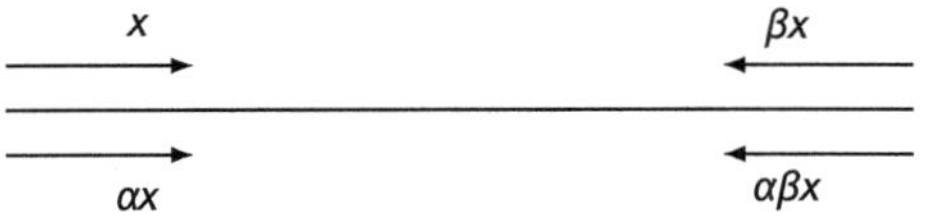

Figure 1.10: An edge sticking on K.

where 1 is the identity. By considering that the result from one side to the other and then to the other end and the result from one end to the other and then to the other side are the same, i.e.,

$$\beta\alpha = \alpha\beta. \tag{1.15}$$

Further, it can be seen that $K = \{1, \alpha, \beta, \gamma\}$, $\gamma = \alpha\beta$, is a group, called the *Klein group*, where

$$\begin{aligned}(\alpha\beta)^2 &= (\alpha\beta)(\alpha\beta) = (\alpha\beta\alpha)\beta \\ &= (\alpha\alpha\beta)\beta = (\alpha\alpha)(\beta\beta) = 1.\end{aligned} \tag{1.16}$$

Thus, an edge $x \in X$ of G in an embedding $\mu(G)$ of G becomes $Kx = \{x, \alpha x, \beta x, \gamma x\}$, as shown in Figure 1.10.

In fact, let

$$\mathscr{X} = \sum_{i=1}^{m} Kx_i, \tag{1.17}$$

where summation stands for the disjoint union, then α and β can both be seen as a permutation on $\mathscr{X}$, i.e.,

$$\alpha = \prod_{i=1}^{m}(x_i, \alpha x_i)(\beta x_i, \gamma x_i), \ \beta = \prod_{i=1}^{m}(x_i, \beta x_i)(\alpha x_i, \gamma x_i).$$

The vertex x is dealt with the rotation as

$$\{(x, \mathscr{P}x, \mathscr{P}^2x, \ldots), (\alpha x, \alpha\mathscr{P}^{-1}x, \alpha\mathscr{P}^{-2}x, \ldots)\}, \tag{1.18}$$

shown in Figure 1.11 (when its degree is 4).

It is seen that $\mathscr{P}$ is also a permutation on $\mathscr{X}$. The set of elements in each cycle of this permutation is called an *orbit* of an element in the cycle. For example, the orbit of element x under permutation $\mathscr{P}$ is denoted by $(x)_{\mathscr{P}}$. From (1.18),

$$(x)_{\mathscr{P}} \bigcap (\alpha x)_{\mathscr{P}} = \emptyset, x \in \mathscr{X}. \tag{1.19}$$

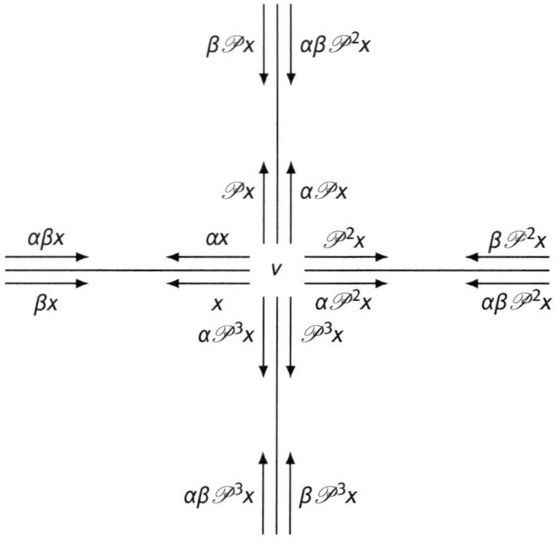

Figure 1.11: The rotation at a vertex.

The two cycles at a vertex in an embedding have a relation as

$$\begin{aligned}&(\alpha x, \mathscr{P}\alpha x, \mathscr{P}^2\alpha x, \ldots)\\ &= (\alpha x, \alpha\mathscr{P}^{-1}x, \alpha\mathscr{P}^{-2}x, \ldots)\\ &= \alpha(x, \mathscr{P}^{-1}x, \mathscr{P}^{-2}x, \ldots).\end{aligned} \tag{1.20}$$

For convenience, one of the two cycles is chosen to represent the vertex: i.e.,

$$(x, \mathscr{P}x, \mathscr{P}^2x, \ldots),$$

or

$$(\alpha x, \alpha\mathscr{P}^{-1}x, \alpha\mathscr{P}^{-2}x, \ldots).$$

Theorem 1.14 $\alpha\mathscr{P} = \mathscr{P}^{-1}\alpha$.

Proof By multiplying the two sides of (1.20) by α from the left and then comparing the second terms on the two sides,

$$\alpha\mathscr{P}\alpha = \mathscr{P}^{-1}.$$

By multiplying its two sides by α from the right, the theorem is soon obtained. □

Since α and β are both permutations on $\mathscr{X}$, $\gamma = \alpha\beta$ and $\mathscr{P}^* = \mathscr{P}\gamma$ are permutations on $\mathscr{X}$ as well. Let

$$(x, \mathscr{P}^*x, \mathscr{P}^{*2}x, \ldots)$$

be the cycle of $\mathscr{P}^*$ involving x. From the symmetry between βx and x, the cycle of $\mathscr{P}^*$ involving βx is

$$(\beta x, \mathscr{P}^*\beta x, \mathscr{P}^{*2}\beta x, \ldots),$$

which has the same number of elements as that involving x does.

Because $\mathscr{P}^*(\beta x) = \mathscr{P}\alpha\beta(\beta x) = \mathscr{P}\alpha x$ and from Theorem 1.14

$$\begin{aligned}\mathscr{P}\alpha x &= \alpha\mathscr{P}^{-1}x = \alpha\gamma(\gamma\mathscr{P}^{-1})x \\ &= \alpha\gamma\mathscr{P}^{*-1}x \\ &= \beta\mathscr{P}^{*-1}x,\end{aligned}$$

we have

$$\mathscr{P}^*(\beta x) = \beta\mathscr{P}^{*-1}x. \tag{1.21}$$

Furthermore, because $\mathscr{P}^{*2}(\beta x) = \mathscr{P}^*(\mathscr{P}^*(\beta x))$ and from (1.21)

$$\mathscr{P}^*(\mathscr{P}^*(\beta x)) = \mathscr{P}^*(\beta\mathscr{P}^{*-1}x),$$

by (1.21) for $\mathscr{P}^{*-1}x$ instead of x, we have

$$\mathscr{P}^{*2}(\beta x) = \beta(\mathscr{P}^{*-1}(\mathscr{P}^{*-1}x)) = \beta(\mathscr{P}^{*-2}x).$$

On the basis of the finite restrict recursion principle, a cycle is found. Therefore,

$$(\beta x, \mathscr{P}^*\beta x, \mathscr{P}^{*2}\beta x, \ldots) = \beta(x, \mathscr{P}^{*-1}x, \mathscr{P}^{*-2}x, \ldots). \tag{1.22}$$

This implies

$$(x)_{\mathscr{P}^*} \bigcap (\beta x)_{\mathscr{P}^*} = \emptyset, \tag{1.23}$$

for $x \in \mathscr{X}$.

Theorem 1.15 $\beta\mathscr{P}^* = \mathscr{P}^{*-1}\beta$.

Proof A direct result of (1.22). □

Based on (1.22), it is seen that the face involving x of the embedding represented by $\mathscr{P}$ is

$$\{(x, \mathscr{P}^*x, \mathscr{P}^{*2}x, \ldots), (\beta x, \mathscr{P}^{*-1}\beta x, \mathscr{P}^{*-2}\beta x, \ldots)\}. \tag{1.24}$$

Similar to vertices, based on (1.22) and (1.23), the face can be represented by one of the two cycles in (1.24).

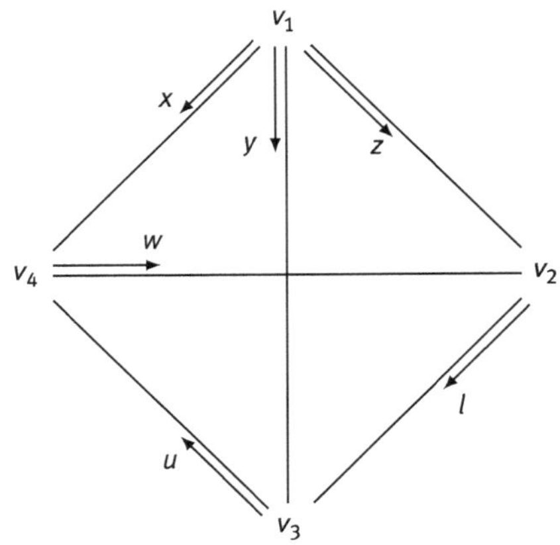

Figure 1.12: A rotation of K_4.

Example 1.3 Let $G = K_4$, i.e., the complete set of order 4. Given its rotation $\{(x, y, z), (\beta z, l, \gamma w), (\gamma l, u, \beta y), (\beta x, w, \gamma u)\}$, as shown in Figure 1.12.

Its two faces are $(x, \beta u, \beta l, \gamma z)$ and

$$(y, \alpha u, \alpha w, \alpha l, \gamma y, z, \beta w, \gamma x).$$

Thus, it is an embedding of K_4 on the torus

$$O_1 = (ABA^{-1}B^{-1})$$

as shown in Figure 1.13.

Further, another rotation of K_4 is chosen for getting another embedding of K_4.

Example 1.4 (Continuous to Example 1.3) Another embedding of K_4 is shown in Figure 1.14. Its rotation is

$$\{(x, y, z), (\beta z, l, \gamma w), (u, \gamma y, \gamma l), (\beta x, w, \gamma u)\}.$$

Its two faces are

$$(x, \beta u, \beta l, \gamma z)$$

and

$$(\alpha x, w, \beta z, \alpha y, \alpha u, \alpha w, \alpha l, \beta y).$$

This is an embedding of K_4 on the Klein bottle:

$$N_2 = (ABA^{-1}B) \sim_{top} = (AABB),$$

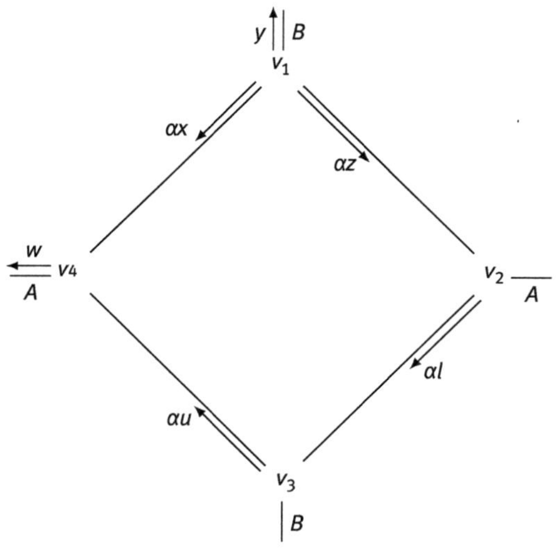

Figure 1.13: Embedding determined by rotation.

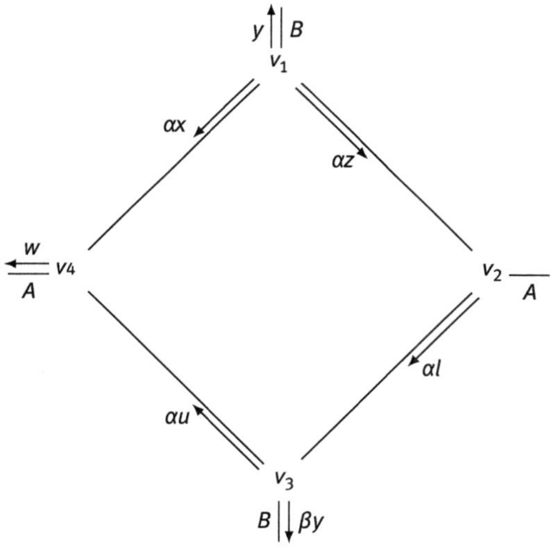

Figure 1.14: Embedding distinguished by rotation.

as shown in Figure 1.14.

Such an idea is preferable to deal with combinatorial maps via algebraic but neither geometric nor topological approaches.

1.5 Notes

The set (repetition at most twice of elements permitted) of edges appearing on a travel can be shown to have a partition of each subset forming still a subtravel except probably the travel itself. Such a partition is called a *decomposition* of a travel into subtravels. However, it is not yet known if any travel can be decomposed into tours except only the case that its induced graph has a cut-edge.

1.5.1 Prove, or disprove, the conjecture that a travel with at most twice occurrences of an edge in a graph has a decomposition into tours if, and only if, the induced subgraph of the travel is without cut-edge.

Notice 1.1 In Dong and Liu [28], only a special case is done. However, on the basis of Liu [80, 82], we have seen that the conjecture in 1.5.1 is proved.

Theorem I.1 A travel with at most twice occurrences of an edge in a graph has a decomposition into tours if, and only if, the induced subgraph of the travel is without cut-edge.

Because a circuit is restricted from a tour by no repetition of a vertex, the following conjecture would look stronger than the last one.

1.5.2 Prove, or disprove, the conjecture that a travel with at most twice occurrences of an edge in a graph has a decomposition into circuits if, and only if, the induced subgraph of the travel is without cut-edge.

However, it can be shown from Theorem 1.3 that any tour has a decomposition into circuits. The above two conjectures are, in fact, equivalent. Because a cut-edge is never on a circuit, the necessity is always true. A travel with three occurrences of an edge permitted does not have a decomposition into circuits in general. For example, on the graph determined by Par= $\{\{x(0), y(0)\}, \{x(1), y(1)\}\}$, the travel $xx^{-1}xy^{-1}$ where $x = \langle x(0), x(1)\rangle$ and $y = \langle y(0), y(1)\rangle$ has no circuit decomposition.

Furthermore, the two conjectures are apparently right when the graph is planar because each face boundary of its planar embedding is generally a tour whenever without cut-edge.

Notice 1.2 From Liu [80, 82], Notice 1.1 induces that the conjecture in 1.5.2 is true on the basis of Theorem 1.1.

Theorem I.2 A travel with at most twice occurrences of an edge in a graph has a decomposition into circuits if, and only if, the induced subgraph of the travel is without cut-edge.

1.5.3 For a given graph G and an integer p, $p \geq 0$, find the number $n_p(G)$ of embeddings of G on the orientable surface of genus p.

The aim is at the genus distribution of embeddings of G on orientable surfaces, i.e., the polynomial

$$P_O(G) = \sum_{p=0}^{\lfloor \sigma/2 \rfloor} n_p(G)x^p,$$

where σ is the Betti number of G.

For $p = 0$, $n_0(G)$ can be done based on Liu [65]. If G is planar, E1.11 in Liu [90](2010) provides the result for 2-connected case. Some other type can also be seen as in MacLane [123]. As to justify if a graph is planar, a theory can be seen from Chapters 3, 5 and 7 in Liu [64, 65]. Some other types can also be seen as in Kuratowski [46], Lefschetz [48], MacLane [123], Whitney [166], and so on.

Generally speaking, it is not easy to get the complete answer in a short period of time. However, the following approach would be available to access this problem. Choose a special type of graphs, for instance, a *wheel* (a circuit C_n all of whose vertices are adjacent to an extra vertex), a *generalized Halin graph* (a circuit with a disjoint tree except for all articulate vertices forming the vertex set of the circuit) and so forth.

Of course, the technique and theoretical results in Section 1.3 can be employed to calculate the number of distinct embeddings of a graph by hand, or by computer.

Notice 1.3 In Zhao and Liu [184, 186], the genus polynomials of tree-like graphs (vertex spanning subgraph is a tree) and circle-like graphs (vertex spanning graph is unicircuit) are, respectively, obtained by the joint tree method sketched in Section 1.3.

1.5.4 *Orientable single peak conjecture*. The coefficients of the polynomial in 1.5.3 are of *single peak*, i.e., they are from increase to decrease as p runs from 0 to $\lfloor \sigma/2 \rfloor (\geq 3)$, $n_p(G)$.

The purpose here is to prove, or disprove the conjecture not necessary to get all $n_p(G)$, $0 \leq p \leq \lfloor \sigma/2 \rfloor (\geq 3)$.

Notice 1.4 In Zhao and Liu [185], the polynomial obtained in Notice 1.3 is shown to be of single peak.

1.5.5 Determine the number of distinct embeddings, which have one, or two faces, of a graph on orientable surfaces.

Notice 1.5 The purpose of 1.5.5 is to determine the number of distinct up-embeddings of a graph, if any. Few papers are only related to the number. Almost on what type of graphs are up-embeddable is shown in Liu [60, 61], Cai et al. [8], Chai and Liu [13], Chen and Liu [16], Dong et al. [25], Dong et al. [27], Lv and Liu [118–120] and so forth.

1.5.6 For a given graph G and an integer q, $q \geq 1$, find the number $\tilde{n}_q(G)$ of distinct embeddings on nonorientable surfaces of genus q.

The aim is at the genus polynomial of embeddings of G on nonorientable surfaces:

$$P_N(G) = \sum_{q=1}^{\sigma} \tilde{n}_q(G)x^q,$$

where σ is the Betti number of G.

Some preinvestigations for G is that a wheel, or a generalized Halin graph, can first be done.

Notice 1.6 The numbers of distinct embeddings for some particular types of graphs on surfaces of small genera have been found on the basis of the technique sketched in 1.3 shown as in Yang and Liu [172–176].

1.5.7 For a graph G, justify if it is embeddable on the projective plane, and then determine $\tilde{n}_1(G)$ according to the connectivity of G.

Notice 1.7 Several methods can be used to justify if a graph can be embedded in the projective plane such as shown in Glover et al. [32], Archdeacon [2], Hao and Liu [38] based on Liu [96] and Archdeacon and Siran [3](1998), Abrams and Slilaty [1] and in the case of nonorientable genus 1 for Liu [92–95]. However, what is the simplest in usage is not yet known.

1.5.8 For a graph embeddable on the projective plane, determine how many sets of circuits such that for each, all of its circuits are essential if, and only if, one of them is essential in an embedding of G on the projective plane.

Notice 1.8 In fact, this is for providing a classification of cycles and further circuits on a given embedding of a graph by establishing a certain type of equivalence. One might be awaked from the theoretical foundation for classification on cycles and cocycles in Chapter 3 in Liu [64] and Chapter 4 in Liu [77].

1.5.9 *Nonorientable single peak conjecture*. The coefficients of the polynomial in 1.5.6 are of single peak in the interval $[o, \sigma]$, where σ is the Betti number of G.

Notice 1.9 On the structure of genus (or handle) polynomial of a graph, Stahl in Stahl [149] conjectured that the zeros of genus polynomials are real. However, this conjecture was disproved in Chen and Liu [17].

1.5.10 For a given type of graphs $\mathcal{G}$ and an integer p, find the number of distinct embeddings of graphs in $\mathcal{G}$ on the orientable surface of genus p. Further, determine the polynomial

$$P_O(\mathcal{G}) = \sum_{p=0}^{\lfloor \sigma(\mathcal{G})/2 \rfloor} n_p(\mathcal{G}) x^p,$$

where $\sigma(\mathcal{G}) = \max\{\sigma(G) | G \in \mathcal{G}\}$.

Notice 1.10 A number of types of graphs have been investigated for determining their handle polynomials by employing the technique, called joint trees, in 1.3. For instance, Liu [13, 72], Hao and Liu [37], Li and Liu [51], Shao and Liu [144], Wan and Liu [157–160], Wan et al. [155], Wan et al. [156], Zeng and Liu [177], Zeng et al. [178], Zhu and Liu [187]. This topic is motivated from Gross and Furst [33] and then Kwak and Lee [47].

1.5.11 For a given type of graphs $\mathcal{G}$ and an integer q, $q \geq 1$, find the number of embeddings of graphs in $\mathcal{G}$ on the nonorientable surface of genus q. Further, determine the crosscap polynomial

$$P_N(\mathcal{G}) = \sum_{q=1}^{\sigma(\mathcal{G})} \tilde{n}_q(\mathcal{G})x^q,$$

where $\sigma(\mathcal{G}) = \max\{\sigma(G)|G \in \mathcal{G}\}$.

Notice 1.11 From Notice 1.9, one might think of crosscap polinomials whose zeros are real, or have a zero real except the trivial zero. However, both the assertions are still not true from Chen and Liu [17]. On the crosscap polinomial of a graph, we have, in principle, seen that the crosscap polynomial of a graph is the sum of the croscap polynomials, each of which is generated by an orientable embedding, over all orientable embeddings of the graph, shown in Liu [73, pp. 202–266], or [74, pp. 279–304]. Direct results are, no doubt, necessary to ask for some types of graphs, as shown in Chen et al. [18], Wang and Liu [162], Yang and Liu [172].

1.5.12 For a set of graphs with some fixed invariants, extract sharp bounds (lower or upper) of the orientable minimum genus and sharp bounds (lower or upper) of orientable maximum genus.

Here, invariants are chosen from the *order* (vertex number), *size* (edge number), *chromatic number* (the minimum number of colors by which vertices of a graph can be colored such that adjacent vertices have distinct colors), *crossing number* (the minimum number of crossing inner points among all planar immersions of a graph), *thickness* (the minimum number of subsets among all partitions of the edge set such that each of the subsets induces a planar graph) and so forth.

Notice 1.12 Known recent results are referred to as in Chen and Liu [14], Dong and Liu [24], Huang and Liu [44], Li and Liu [49].

1.5.13 For a set of graphs and a set of invariants fixed, provide sharp bounds (lower or upper) of minimum nonorientable genus of embeddings of graphs in the set.

Notice 1.13 A graph emphasized with algebraic elements begins from my series of lectures in the Graduate School of Academia Sinica in Liu [89], and then developed particularly in a series of monographs as shown in Liu [64, 65, 67–69, 72–74, 76–79, 90].

2 Abstract Maps

- A ground set is formed by the Klein group $K = \{1, \alpha, \beta, \alpha\beta\}$ sticking on a finite set X.
- A basic permutation is such a permutation on the ground set that no element x is in the same cycle with αx.
- The conjugate axiom on a map is determined by each vertex consisting of two conjugate cycles for α, as well as by each face consisting of two conjugate cycles for β.
- The transitive axiom on a map is from the connectedness of its underlying graph.
- An included angle is determined by either a vertex with one of its incident faces or a face with one of its incident vertices.

2.1 Ground sets

Given a finite set $X = \{x_1, x_2, \ldots, x_m\}$, called the *basic set*, its elements are distinct. Two operations α and β on X are defined as for any $x \in X$, $\alpha x \neq \beta x$, $\alpha x, \beta x \notin X$ and $\alpha^2 x = \alpha(\alpha x) = x$, $\beta^2 x = \beta(\beta x) = x$. Further, define $\alpha\beta = \beta\alpha = \gamma$ such that for any $x \in X$, $\gamma x \neq \alpha x, \beta x$ and $\gamma x \notin X$.

Let $\alpha X = \{\alpha x | \forall x \in X\}$, $\beta X = \{\beta x | \forall x \in X\}$ and $\gamma X = \{\gamma x | \forall x \in X\}$, then α, β and γ determine a *bijection* between X and, respectively, αX, βX and γX. Bijection means a one to one correspondence between two sets of the same cardinality. In other words,

$$
\begin{aligned}
&X \bigcap \alpha X = X \bigcap \beta X = X \bigcap \gamma X = \emptyset;\\
&\alpha X \bigcap \beta X = \beta X \bigcap \gamma X = \gamma X \bigcap \alpha X = \emptyset;\\
&|\alpha X| = |\beta X| = |\gamma X| = |X|.
\end{aligned}
\tag{2.1}
$$

The set $X \cup \alpha X \cup \beta X \cup \gamma X$, or briefly $\mathscr{X} = \mathscr{X}(X)$, is called the *ground set*.

Now, observe set $K = \{1, \alpha, \beta, \gamma\}$. Its elements are seen as permutations on the ground set $\mathscr{X}$. Here, 1 is the identity. From $\alpha^2 = \beta^2 = 1$ and $\alpha\beta = \beta\alpha$, $\gamma^2 = (\alpha\beta)(\alpha\beta) = \alpha(\beta\beta)\alpha = \alpha^2 = 1$, and hence

$$
\begin{aligned}
\alpha &= \prod_{i=1}^{m}(x_i, \alpha x_i) \prod_{i=1}^{m}(\beta x_i, \gamma x_i);\\
\beta &= \prod_{i=1}^{m}(x_i, \beta x_i) \prod_{i=1}^{m}(\alpha x_i, \gamma x_i);\\
\gamma &= \prod_{i=1}^{m}(x_i, \gamma x_i) \prod_{i=1}^{m}(\beta x_i, \alpha x_i).
\end{aligned}
\tag{2.2}
$$

DOI 10.1515/9783110481846-002

It is easily seen that K is a group, called *Klein group* because it is isomorphic to the group of four elements discovered by Klein in geometry.

For any $x \in \mathscr{X}$, let $Kx = \{x, \alpha x, \beta x, \gamma x\}$, called a *quadricell.*

Theorem 2.1 For any basic set X, its ground set is

$$\mathscr{X} = \sum_{x \in X} Kx, \tag{2.3}$$

where the summation represents the disjoint union of sets.

Proof From (2.1), for any $x, y \in X$ and $x \neq y$,

$$Kx \bigcap Ky = \{x, \alpha x\beta x, \gamma x\} \bigcap \{y, \alpha y, \beta y, \gamma y\} = \emptyset.$$

From (2.1) again,

$$\bigcup_{x \in X} Kx = X + \alpha X + \beta X + \gamma X = \mathscr{X}.$$

Therefore, (2.3) is true. □

Furthermore, since for any $x \in X$,

$$K(\alpha x) = K(\beta x) = K(\gamma x) = Kx, \tag{2.4}$$

we have

$$\mathscr{X} = \sum_{y \in \alpha X} Ky = \sum_{z \in \beta X} Kz = \sum_{t \in \gamma X} Kt.$$

This implies that the four elements in a quadricell are with symmetry.

2.2 Basic permutations

Let $\mathrm{Per}(\mathscr{X})$, or briefly Per, be a permutation on the ground set $\mathscr{X}$. Because of bijection, according to the finite strict recursion principle, for any $x \in \mathscr{X}$, there exists a minimum positive integer $k(x)$ such that

$$\mathrm{Per}^{k(x)} x = x,$$

i.e., Per contains the *cyclic permutation* , or in short *cycle*,

$$(x)\mathrm{Per} = (x, \mathrm{Per}^2 x, \ldots, \mathrm{Per}^{k(x)-1} x).$$

Write $\{x\}$Per as the set of all elements in the cycle (x)Per. Such a set is called the *orbit* of x under permutation Per. The integer $k(x)$ is called the *order* of x under permutation Per.

If for any $x \in \mathscr{X}$,

$$\alpha x \notin \{x\}\text{Per}, \tag{2.5}$$

then the permutation Per is said to be *basic* to α.

Example 2.1 From (2.2), for permutations α and β on the ground set, α is not basic, but β is basic.

Let Par= $\{X_1, X_2, \ldots, X_s\}$ be a partition on the ground set $\mathscr{X}$, then

$$\text{Per} = \prod_{i=1}^{s}(X_i) \tag{2.6}$$

determines a permutation on $\mathscr{X}$, called *induced* from the partition Par. Here, (X_i), $1 \leq i \leq s$, stands for a cyclic order arranged on X_i. This shows that a partition $\{X_i|1 \leq i \leq s\}$ has

$$\prod_{s=1}^{s}(|X_i| - 1)! = \prod_{i\geq 3}((i-1)!)^{n_i} \tag{2.7}$$

induced permutations. In (2.7), n_i, $i \geq 3$, is the number of subsets in Par with i elements.

Theorem 2.2 Let Par= $\{X_i|1 \leq i \leq s\}$ be a partition on the ground set $\mathscr{X}(X)$. If Par has an induced permutation basic, then all of its induced permutations are basic. Further, a partition Par has its induced permutation basic if, and only if, for $x \in \mathscr{X}$, there does not exist $Y \in$ Par such that

$$\{x, \alpha x\} \text{ or } \{\beta x, \gamma x\} \subseteq Y \bigcap Kx. \tag{2.8}$$

Proof Because the basicness of a permutation is independent of the order on cycles, the first statement is proved.

Assume an induced permutation Per of a partition Par is basic. From (2.5), for any $x \in X$, in virtue of

$$Kx = \{x, \alpha x\} + \{\beta x, \gamma x\},$$

no $Y \in$ Par exists such that (2.8) is satisfied. This is the necessity of the second statement.

Conversely, because for any $x \in \mathscr{X}$, no $Y \in$ Par exists such that (2.8) is satisfied, it is only possible that x and αx are in distinct subsets of partition Par. Therefore, $\alpha x \notin \{x\}$Per. Based on (2.5), this is the sufficiency of the second statement. □

On the basis of this theorem, induced basic permutations can be easily extracted from a partition of the ground set.

Example 2.2 Let $\mathscr{X} = \{x, \alpha x, \beta x, \gamma x\} = Kx$. There are 15 partitions on $\mathscr{X}$ as

$$
\begin{aligned}
\mathrm{Par}_1 &= \{\{x\}, \{\alpha x\}, \{\beta x\}, \{\gamma x\}\};\\
\mathrm{Par}_2 &= \{\{x, \alpha x\}, \{\beta x\}, \{\gamma x\}\};\\
\mathrm{Par}_3 &= \{\{x, \beta x\}, \{\alpha x\}, \{\gamma x\}\};\\
\mathrm{Par}_4 &= \{\{x, \gamma x\}, \{\alpha x\}, \{\beta x\}\};\\
\mathrm{Par}_5 &= \{\{\beta x, \alpha x\}, \{x\}, \{\gamma x\}\};\\
\mathrm{Par}_6 &= \{\{x\}, \{\alpha x, \gamma x\}, \{\beta x\}\};\\
\mathrm{Par}_7 &= \{\{x\}, \{\alpha x\}, \{\beta x, \gamma x\}\};\\
\mathrm{Par}_8 &= \{\{x, \alpha x, \beta x\}, \{\gamma x\}\};\\
\mathrm{Par}_9 &= \{\{x, \beta x, \gamma x\}, \{\beta x\}\};\\
\mathrm{Par}_{10} &= \{\{x, \alpha x, \gamma x\}, \{\alpha x\}\};\\
\mathrm{Par}_{11} &= \{\{x\}, \{\alpha x, \beta x, \gamma x\}\};\\
\mathrm{Par}_{12} &= \{\{x, \alpha x, \beta x, \gamma x\}\};\\
\mathrm{Par}_{13} &= \{\{x, \alpha x\}, \{\beta x, \gamma x\}\};\\
\mathrm{Par}_{14} &= \{\{x, \beta x\}, \{\alpha x, \gamma x\}\};\\
\mathrm{Par}_{15} &= \{\{x, \gamma x\}, \{\alpha x, \beta x\}\}.
\end{aligned}
$$

From Theorem 2.2, induced basic permutations can only be extracted from Par_1, Par_3, Par_4, Par_5, Par_6, Par_{14} and Par_{15} among them. Since each of these partitions has no subset with at least three elements, from (2.7) it induces only one basic permutation. Hence, seven basic permutations are induced in all.

Based on Theorem 2.2, a partition that induces a basic permutation is called *basic* as well.

For a partition Par on $\mathscr{X}$, if every pair of x and αx, $x \in \mathscr{X}$, deals with the element x in Par, then this partition determines a pregraph if any. For example, in Example 2.2, there are only

$$
\begin{aligned}
\mathrm{Par}_1 = \mathrm{Par}_2 = \mathrm{Par}_7 = \mathrm{Par}_{13} &= \{\{x\}, \{\beta x\}\};\\
\mathrm{Par}_{12} = \mathrm{Par}_{14} = \mathrm{Par}_{15} &= \{\{x, \beta x\}\}
\end{aligned}
$$

form 2 premaps of size 1 and others meaningless among the 15 partitions. Further, each of the two premaps is a graph. The result is the same as in Example 1.1.

2.3 Conjugate axiom

Let Per_1 and Per_2 be two permutations on the ground set $\mathscr{X}$. If for $x \in \mathscr{X}$,

$$(\mathrm{Per}_2 x)_{\mathrm{Per}_1} = \mathrm{Per}_2(x)_{\mathrm{Per}_1^{-1}} = \mathrm{Per}_2(x)^{-1}_{\mathrm{Per}_1}, \tag{2.9}$$

then the two orbits $(x)_{\mathrm{Per}_1}$ and $(\mathrm{Per}_2 x)_{\mathrm{Per}_1}$ of Per_1 are said to be *conjugate*.

For a permutation $\mathscr{P}$ on the ground set $\mathscr{X}$, if

$$\alpha\mathscr{P} = \mathscr{P}^{-1}\alpha, \tag{2.10}$$

then $(\mathscr{P}, \alpha)$(or for the sake of brevity, $\mathscr{P}$) is called satisfying the *conjugate axiom.*

Theorem 2.3 For a basic permutation $\mathscr{P}$ on the ground set $\mathscr{X}$, the two orbits $(x)_{\mathscr{P}}$ and $(\alpha x)_{\mathscr{P}}$ for any $x \in \mathscr{X}$ are conjugate if, and only if, $(\mathscr{P}, \alpha)$ satisfies the conjugate axiom.

Proof Necessity. Because orbits $(x)_{\mathscr{P}}$ and $(\alpha x)_{\mathscr{P}}$ are conjugate, from (2.9), $(\alpha x)_{\mathscr{P}} = \alpha(x)^{-1}_{\mathscr{P}}$. Hence, $\mathscr{P}\alpha x = \alpha\mathscr{P}^{-1}x$, i.e., $\mathscr{P}\alpha = \alpha\mathscr{P}^{-1}$. This implies (2.10).

Sufficiency. Since $\mathscr{P}$ satisfies (2.10),

$$\begin{aligned}\mathscr{P}(\alpha x) &= \mathscr{P}(\alpha\mathscr{P})\mathscr{P}^{-1}x = \mathscr{P}(\mathscr{P}^{-1}\alpha)\mathscr{P}^{-1}x\\ &= (\mathscr{P}\mathscr{P}^{-1})\alpha\mathscr{P}^{-1}x = \alpha\mathscr{P}^{-1}x.\end{aligned}$$

By induction, assume that $\mathscr{P}^l(\alpha x) = \alpha\mathscr{P}^{-l}x$, $l \geq 1$, then we have

$$\begin{aligned}\mathscr{P}^{l+1}(\alpha x) &= \mathscr{P}\mathscr{P}^l(\alpha x) = \mathscr{P}\mathscr{P}^{-l}x\\ &= \mathscr{P}(\alpha\mathscr{P}\mathscr{P}^{-1})\mathscr{P}^{-l}x \text{ (then by (2.10))}\\ &= (\mathscr{P}\mathscr{P}^{-1})\alpha\mathscr{P}^{-(l+1)}x\\ &= \alpha\mathscr{P}^{-(l+1)}x.\end{aligned}$$

Hence, $(\alpha x)_{\mathscr{P}} = \alpha(x)^{-1}_{\mathscr{P}}$. From (2.9), orbits $(x)_{\mathscr{P}}$ and $(\alpha x)_{\mathscr{P}}$ are conjugate. This is the sufficiency. □

Unlike Theorem 2.2, for a partition on the ground set $\mathscr{X}$, from one of its induced permutations satisfying the conjugate axiom, it cannot be deduced to others.

A *premap*, denoted by $(\mathscr{X}_{\alpha,\beta}, \mathscr{P})$, is such a basic permutation $\mathscr{P}$ on the ground set $\mathscr{X}$ that the conjugate axiom is satisfied for $(\mathscr{P}, \alpha)$.

Example 2.3 By no means any basic partition is in companion with a basic permutation. Among the seven basic partitions as shown in Example 2.2, only the induced permutations of Par_1, Par_{14} and Par_{15} are premaps.

Because $(\mathscr{P}, \beta)$ is not necessary to satisfy the conjugate axiom on a premap $(\mathscr{X}, \mathscr{P})$, α is called the *first operation* and β the *second* . Thus, $\mathscr{X}$ should be precisely written as $\mathscr{X}_{\alpha,\beta}$ if necessary.

Based on the basicness and Theorem 2.3, any premap $\mathscr{P}$ has the form as

$$\prod_{x \in X_{\mathscr{P}}} (x)_{\mathscr{P}} (x)^*_{\mathscr{P}}, \tag{2.11}$$

where $X_{\mathscr{P}}$ is the set of distinct representatives for the conjugate pairs $\{\{x\}_{\mathscr{P}}, \{x\}^*_{\mathscr{P}}\}$ of cycles in $\mathscr{P}$. And further,

$$\mathscr{X} = \sum_{x \in X_{\mathscr{P}}} \{x\}_{\mathscr{P}} \{x\}^*_{\mathscr{P}}. \tag{2.12}$$

For convenience, one of two conjugate orbits in $\{\{x\}_{\mathscr{P}}, \{x\}^*_{\mathscr{P}}\}$ is chosen to stand for the pair itself as a *vertex* of the premap.

Example 2.4 Let $X = \{x_1, x_2\}$, then

$$\mathscr{X} = \{x_1, \alpha x_1, \beta x_1, \gamma x_1, x_2, \alpha x_2, \beta x_2, \gamma x_2\}.$$

Choose

$$\mathscr{P}_1 = (x_1, \beta x_1)(\alpha x_1, \gamma x_1)(x_2)(\alpha x_2)(\beta x_2)(\gamma x_2)$$

and

$$\mathscr{P}_2 = (x_1, \beta x_1, x_2)(\alpha x_1, \alpha x_2, \gamma x_1)(\beta x_2)(\gamma x_2).$$

The former has three vertices $(x_1, \beta x_1)$, (x_2) and (βx_2). The latter has two vertices $(x_1, \beta x_1, x_2)$ and (βx_2).

Lemma 2.1 If permutation $\mathscr{P}$ on $\mathscr{X}_{\alpha,\beta}$ is a premap, then

$$\mathscr{P}^* \beta = \beta \mathscr{P}^{*-1}, \tag{2.13}$$

where $\mathscr{P}^* = \mathscr{P}\gamma$, $\gamma = \alpha\beta$.

Proof Because $\mathscr{P}^*\beta = \mathscr{P}\alpha\beta\beta = \mathscr{P}\alpha$, from the conjugate axiom,

$$\begin{aligned}\mathscr{P}^*\beta &= \alpha\mathscr{P}^{-1}\\ &= \beta\beta\alpha\mathscr{P}^{-1}\ (\beta^2 = 1)\\ &= \beta((\mathscr{P}\alpha\beta)^{-1})\\ &= \beta\mathscr{P}^{*-1}.\end{aligned}$$

Therefore, the lemma holds. □

This lemma tells us that although β does not satisfy the conjugate axiom for permutation $\mathscr{P}$ in general, β does satisfy the conjugate axiom for permutation $\mathscr{P}^*$.

Lemma 2.2 If permutation $\mathscr{P}$ on $\mathscr{X}_{\alpha,\beta}$ is a premap, then permutation $\mathscr{P}^* = \mathscr{P}\gamma$ is basic for β.

Proof Because the four elements in a quadricell are distinct, $x \neq \beta x$.
Case 1 $\mathscr{P}^*x \neq \beta x$. Otherwise, from $\mathscr{P}^*x = \beta x$,

$$\mathscr{P}\alpha(\beta x) = \mathscr{P}^*x = \beta x.$$

A contradiction to that $\mathscr{P}$ is basic for α.
Case 2 $(\mathscr{P}^*)^2x \neq \beta x$. Otherwise, $\mathscr{P}^*x = \mathscr{P}^{*-1}\beta x$. From Lemma 2.1,

$$\mathscr{P}^*x = \beta\mathscr{P}^*x.$$

A contradiction is that $\mathscr{P}^*x$ and $\beta\mathscr{P}^*x$ are in the same quadricell.

In general, assume by induction that Case l: $(\mathscr{P}^*)^lx \neq \beta x$, $1 \leq l \leq k$, $k \geq 2$, is proved. To prove

Case $k+1$ $(\mathscr{P}^*)^{k+1}x \neq \beta x$. Otherwise, $\mathscr{P}^{*k}x = \mathscr{P}^{*-1}\beta x$. From Lemma 2.1,

$$\mathscr{P}^{*k-1}(\mathscr{P}^*x) = \beta(\mathscr{P}^*x).$$

This is a contradiction to the induction hypothesis.

In all, the lemma is proved. □

Theorem 2.4 Permutation $\mathscr{P}$ on $\mathscr{X}$ is a premap $(\mathscr{X}_{\alpha,\beta}, \mathscr{P})$ if, and only if, permutation $\mathscr{P}^* = \mathscr{P}\gamma$ is a premap $(\mathscr{X}_{\beta,\alpha}, \mathscr{P}^*)$.

Proof Necessity. Since permutation $\mathscr{P}$ is a premap $(\mathscr{X}_{\alpha,\beta}, \mathscr{P})$, $\mathscr{P}$ is basic for α and satisfies the conjugate axiom. From Lemma 2.1 and Lemma 2.2, $\mathscr{P}^*$ is basic for β and satisfies the conjugate axiom. Hence, $\mathscr{P}^*$ is a premap $(\mathscr{X}_{\beta,\alpha}, \mathscr{P}^*)$.

Sufficiency. Because $\mathscr{P}^{**} = \mathscr{P}^*\gamma = \mathscr{P}$, the sufficiency is right. □

On the basis of Theorem 2.4, the vertices of premap $\mathscr{P}^*$ are defined to be the *faces* of premap $\mathscr{P}$. The former is called the *dual* of the latter. Since $\mathscr{P}^{**} = \mathscr{P}$, the latter is also the dual of the former.

Example 2.5 For the two premaps $\mathscr{P}_1$ and $\mathscr{P}_2$ as shown in Example 2.4, we have

$$\mathscr{P}_1^* = (x_1, \alpha x_1)(\beta x_1, \gamma x_1)(x_2, \gamma x_2)(\beta x_2, \alpha x_2)$$

and

$$\mathscr{P}_2^* = (x_1, \alpha x_1, x_2, \gamma x_2)(\beta x_1, \alpha x_2, \beta x_2, \gamma x_1).$$

Because $\mathscr{P}_1^*$ has two vertices $(x_1, \alpha x_1)$ and $(x_2, \gamma x_2)$ and $\mathscr{P}_2^*$ has one vertex $(x_1, \alpha x_1, x_2, \gamma x_2)$, we see that premaps $\mathscr{P}_1$ and $\mathscr{P}_2$ have, respectively, two faces and one face.

2.4 Transitive axiom

For a set of permutations $T = \{\tau_i | 1 \le i \le k\}, k \ge 1$ on $\mathscr{X}$, let

$$\Psi_T = \{\psi | \psi = \prod_{l=1}^{s} \prod_{j=1}^{k} \tau_{\pi_l(j)}^{i_j(\pi_l)},\ i_j(\pi_l) \in Z, \pi_l \in \Pi, s \ge 1\}, \tag{2.14}$$

where Z is the set of integers and Π is the set of all permutations on $\{1, 2, \ldots, k\}$, $k \ge 1$.

Since all elements in Ψ_T are permutations on $\mathscr{X}$, they are closed for composition (or in other word, multiplication) with the associative law but without the commutative law.

Further, it is seen that a permutation is in Ψ_T if, and only if, its inverse is in Ψ. The identity is the element in Ψ_T when all $i_j(\pi) = 0$, $\pi \in \Pi$ in (2.14). Therefore, Ψ_T is a group in its own right, called the *generated group* of T.

Let Ψ be a permutation group on $\mathscr{X}$. If for any $x, y \in \mathscr{X}$, there exists an element $\psi \in \Psi$ such that $x = \psi y$, then the group Ψ is said to be *transitive*, or in other words, the group Ψ satisfies the *transitive axiom*.

Now, consider a binary relation on $\mathscr{X}$, denoted by $\sim_\Psi$, which for any $x, y \in \mathscr{X}$,

$$x \sim_\Psi y \iff \exists\, \psi \in \Psi,\ x = \psi y. \tag{2.15}$$

Because the relation $\sim_\Psi$ determined by (2.15) for a permutation group Ψ on $\mathscr{X}$ is an equivalence, $\mathscr{X}$ is classified into classes as $\mathscr{X}/\sim_\Psi$.

Theorem 2.5 A permutation group Ψ on $\mathscr{X}$ is transitive if, and only if, for the equivalence $\sim_\Psi$ determined by (2.15), $|\mathscr{X}/\sim_\Psi| = 1$.

Proof Necessity. From the transitivity, for any $x, y \in \mathscr{X}$, there exists $\psi \in \Psi$ such that $x = \psi y$. In view of (2.15), for any $x, y \in \mathscr{X}$, $x \sim_\Psi y$. Hence, for $\sim_\Psi$, $|\mathscr{X}/\sim_\Psi| = 1$.

Sufficiency. Because $|\mathscr{X}/\sim_\Psi| = 1$, for any $x, y \in \mathscr{X}$, there exists $\psi \in \Psi$ such that $x = \psi y$. Therefore, the permutation group Ψ on $\mathscr{X}$ is transitive. □

For a premap, the pregraph with the same vertices and edges as the premap is called its *under pregraph*. Conversely, the premap is a *upper premap* of its under pregraph.

Let $(\mathscr{X}_{\alpha,\beta}, \mathscr{P})$ be a premap. If permutation group Ψ_J, where $J = \{\alpha, \beta, \mathscr{P}\}$, on the ground set $\mathscr{X}_{\alpha,\beta}$ is transitive, i.e., with the *transitive axiom*, then the premap is called a *map*.

Lemma 2.3 Let $M = (\mathscr{X}_{\alpha,\beta}, \mathscr{P})$ be a premap. For any $x, y \in \mathscr{X}_{\alpha,\beta}$, there exists $\psi \in \Psi_J$, where $J = \{\alpha, \beta, \mathscr{P}\}$, such that $x = \psi y$ if, and only if, there is a path from the vertex v_x (x is in) to the vertex v_y (y is in) in the under pregraph of M.

Proof Necessity. In view of (2.14) with the conjugate axiom, write

$$\psi = \alpha^{\delta_1} \mathscr{P}^{l_1} \alpha^{\sigma_1} \beta \alpha^{\delta_2} \mathscr{P}^{l_2} \alpha^{\sigma_2} \beta \cdots \beta \alpha^{\delta_s} \mathscr{P}^{l_s} \alpha^{\sigma_s},$$

where $\delta_i, \sigma_i \in \{0, 1\}$, $1 \le i \le s$, and $l_i \in Z$, $1 \le i \le s$. Because $\alpha^{\delta_s} \mathscr{P}^{l_s} \alpha^{\sigma_s}$ and $\alpha^{\delta_1} \mathscr{P}^{l_1} \alpha^{\sigma_1}$ are, respectively, acting on vertices v_y and v_x, ψ determines a trail from v_y to v_x of $s - 1$ edges. Since there is a trail between two vertices if, and only if, there is a path between them, the necessity is done.

Sufficiency. let $\langle v_s, v_{s-1}, \ldots, v_1 \rangle$, $v_s = v_y$, $v_1 = v_x$, be a path from v_y to v_x in the under pregraph of M. Then, there exist $\delta_i, \sigma_i \in \{0, 1\}$, $1 \le i \le s$, and $l_i \in Z$, $1 \le i \le s$, such that

$$\psi = \alpha^{\delta_1} \mathscr{P}^{l_1} \alpha^{\sigma_1} \beta \cdots \beta \alpha^{\delta_s} \mathscr{P}^{l_s} \alpha^{\sigma_s}$$

and $x = \psi y$. From (2.14), the sufficiency is done. □

Theorem 2.6 A premap is a map if, and only if, its under pregraph is a graph.

Proof From the transitive axiom and Lemma 2.3, its under pregraph is a graph. This is the necessity. Conversely, from the connectedness and Lemma 2.3, the premap satisfies the transitive axiom and hence a graph. This is the sufficiency. □

Example 2.6 In Figure 1.2, the pregraph determined by Par_7 has two upper premaps:

$$\mathscr{P}_1 = (x_1)(\alpha x_1)(\gamma x_1)(\beta x_1)(x_2, \gamma x_2)(\alpha x_2, \beta x_2)$$

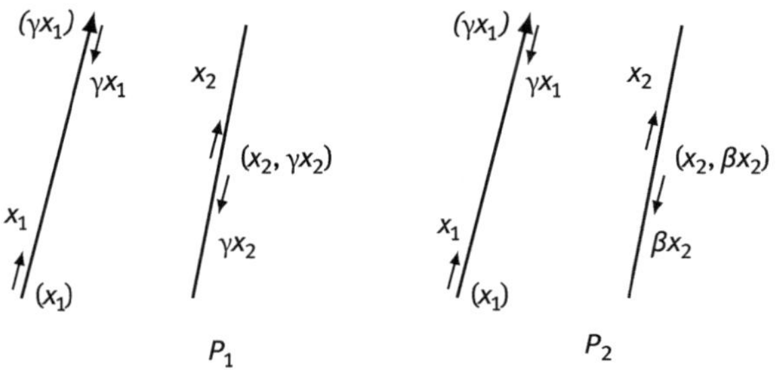

Figure 2.1: Two upper premaps.

and

$$\mathscr{P}_1 = (x_1)(\alpha x_1)(\gamma x_1)(\beta x_1)(x_2, \beta x_2)(\alpha x_2, \gamma x_2)$$

as shown in Figure 2.1.

From Theorem 2.6, none of the two upper premaps is a map in Figure 2.1.

However, the pregraph determined by Par_{12} is a graph in Figure 1.2. From Theorem 2.6, each of its upper premaps is a map as shown in the following example.

Example 2.7 In Figure 2.2, there are $2^2 3! = 24$ distinct embeddings of the graph determined by Par_{12} in Figure 1.2.

On the associate (or boundary) polygon of the joint tree, the pair of a letter is defined to be of distinct powers when x and γx appear; the same power otherwise.

In Figure 2.2, the graph determined by Par_{12} (Figure 1.2) has six orientable embeddings (a–f). Here, (a), (b), (d) and (f) are the same map on the sphere $O_0 \sim_{\mathrm{top}} (x_1 x_1^{-1})$. And, ($c$) and ($e$) are the same map on the torus $O_1 \sim_{\mathrm{top}} (x_1 x_2 x_1^{-1} x_2^{-1})$. Hence, such six distinct embeddings are, in fact, two maps.

Among the 18 nonorientable embeddings, 10 are on the projective plane and 8 are on the Klein bottle. On the projective plane, (g), (h), (j), (l), (m), (n), (p) and (r) are the same map ($N_1 \sim_{\mathrm{top}} (x_1 x_1^{-1} x_2 x_2)$). And, ($u$) and ($w$) are another map ($N_1 = (x_1 x_2 x_1 x_2)$). On the Klein bottle, (i), (k), (o) and (q) are the same map ($N_2 = (x_1 x_2 x_1^{-1} x_2)$). And, ($s$), ($t$), ($v$) and ($x$) are another map ($N_2 \sim_{\mathrm{top}} (x_1 x_1 x_2 x_2)$). Therefore, there are only 4 maps among the 18 embeddings.

2.5 Included angles

Let $M = (\mathscr{X}_{\alpha,\beta}(X), \mathscr{P})$ be a premap. Write $k = |\mathscr{X}_{\alpha,\beta}(X)/\sim_{\Psi_J}|$, where $J = \{\alpha, \beta, \mathscr{P}\}$, and

$$\mathscr{X}_{\alpha,\beta}(X) = \sum_{i=1}^{k} \mathscr{X}_{\alpha_i,\beta_i}(X_i),\ X = \sum_{i=1}^{k} X_i,$$

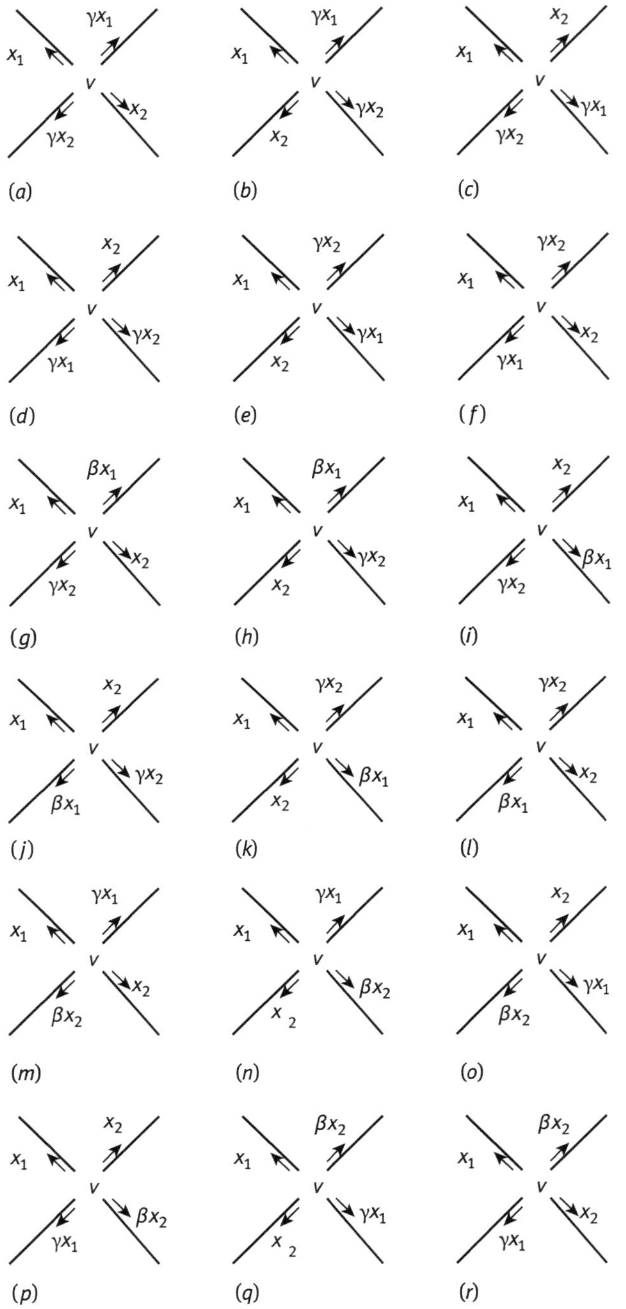

Figure 2.2: All embeddings of a graph, (a) O_0, (b) O_0, (c) O_1, (d) O_0, (e) O_1 (f) O_0, (g) N_1, (h) N_1, (i) N_2, (j) N_1, (k) N_2, (l) N_1, (m) N_1, (n) N_1, (o) N_2, (p) N_1, (q) N_2, (r) N_1, (s) N_2, (t) N_2, (u) N_1, (v) N_2, (w) N_1, (x) N_2

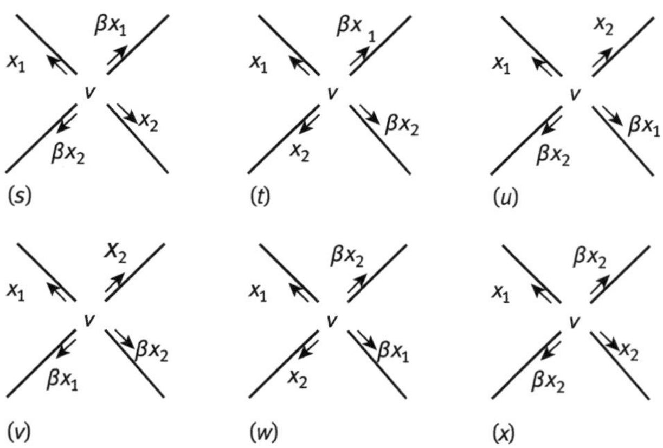

Figure 2.2: continued

where $\mathscr{X}_{\alpha_i,\beta_i}(X) \in \mathscr{X}_{\alpha,\beta}/\sim_{\Psi_J}$, α_i and β_i are, respectively, α and β restricted on $\mathscr{X}_{\alpha_i,\beta_i}(X_i)$, $i = 1, 2, \ldots, k$. Further,

$$M = \sum_{i=1}^{k} M_i,\ M_i = (\mathscr{X}_{\alpha_i,\beta_i}(X_i), \mathscr{P}_i), \tag{2.16}$$

where M_i is a map and $\mathscr{P}_i$ is $\mathscr{P}$ restricted on $\mathscr{X}_{\alpha_i,\beta_i}(X_i)$, $i = 1, 2, \ldots, k$.

This enables us only to discuss maps without loss of generality.

Lemma 2.4 Any map $(\mathscr{X}, \mathscr{P})$ has that $(\mathscr{P}\alpha)^2 = 1$.

Proof From the conjugate axiom,

$$\begin{aligned}(\mathscr{P}\alpha)^2 &= (\mathscr{P}\alpha)(\mathscr{P}\alpha) = \mathscr{P}(\alpha\mathscr{P})\alpha\\ &= (\mathscr{P}\mathscr{P}^{-1})(\alpha\alpha) = 1.\end{aligned}$$

This is the conclusion of the lemma. □

Lemma 2.5 Any map $(\mathscr{X}, \mathscr{P})$ has that $(\mathscr{P}^*\beta)^2 = 1$.

Proof Because $\mathscr{P}^*\beta = \mathscr{P}\gamma\beta = \mathscr{P}\alpha(\beta\beta) = \mathscr{P}\alpha$, from Lemma 2.4, the conclusion is obtained. □

On the basis of the above two lemmas, on a map $M = (\mathscr{X}_{\alpha,\beta}, \mathscr{P})$, any $x \in \mathscr{X}_{\alpha,\beta}$ has that

$$(x, \mathscr{P}\alpha x) = (\mathscr{P}^*\beta x, x). \tag{2.17}$$

Thus, $\langle x, \mathscr{P}\alpha x\rangle$, or $\langle \mathscr{P}^*\beta x, x\rangle$, is called an *included angle* of the map. For an edge $Kx = \{x, \alpha x, \beta x, \gamma x\}$ of M, $\{x, \alpha x\}$ and $\{\beta x, \gamma x\}$ are its two *ends*, or *semiedges*. And, $\{x, \beta x\}$ and $\{\alpha x, \gamma x\}$ are its two *sides*, or *cosemiedges*.

Theorem 2.7 For a $(\mathscr{X}(X), \mathscr{P})$, let $V = \{v|v = \{x\}_{\mathscr{P}} \cup \{\alpha x\}_{\mathscr{P}}, \forall x \in X\}$ and $F = \{f|f = \{x\}_{\mathscr{P}^*} \cup \{\beta x\}_{\mathscr{P}^*}, \forall x \in X\}$. If x_v and x_f are, respectively, in v and f as representatives, then

$$\begin{aligned}\mathscr{X} &= \sum_{v\in V}(\{x_v, \alpha\mathscr{P}^{-1}x_v\} + \{\mathscr{P}x_v, \alpha x_v\} + \cdots \\ &\quad + \{\mathscr{P}^{-1}x_v, \alpha\mathscr{P}^{-2}x_v\}) \\ &= \sum_{f\in F}(\{x_f, \beta\mathscr{P}^{*-1}x_f\} + \{\mathscr{P}^* x_f, \beta x_f\} + \cdots \\ &\quad + \{\mathscr{P}^{*-1}x_f, \beta\mathscr{P}^{*-2}x_f\}).\end{aligned} \tag{2.18}$$

Proof From $\mathscr{X} = \sum_{v\in V} v$ and the conjugate axiom,

$$\begin{aligned}v &= \{x_v, \mathscr{P}\alpha x_v\} + \{\mathscr{P}x_v, \mathscr{P}\alpha\mathscr{P}x_v\} + \cdots \\ &\quad + \{\mathscr{P}^{-1}x_v, \mathscr{P}\alpha\mathscr{P}^{-1}x_v\} \\ &= \{x_v, \alpha\mathscr{P}^{-1}x_v\} + \{\mathscr{P}x_v, \alpha x_v\} + \cdots \\ &\quad + \{\mathscr{P}^{-1}x_v, \alpha\mathscr{P}^{-2}x_v\}.\end{aligned}$$

This is the first equality.

The second equality can similarly be derived from $\mathscr{X} = \sum_{f\in F} f$ and Lemma 2.1 (the conjugate axiom for $\mathscr{P}^*$ with β). □

It is, from the theorem, seen that the numbers of included angles, semiedges and cosemiedges are, each, equal to the sum of degrees of vertices. Since every edge has exactly two semiedges, this number is two times the size of the map.

2.6 Notes

From Theorem 1.10 in Chapter 1, any graph has an embedding on a surface (orientable or nonorientable). However, if an embedding is restricted to a particular property, then the existence is still necessary to investigate. If a map has each of its faces partitionable into circuits, then it is called a *favorable map*. If a graph has an embedding which is a favorable map, then the embedding is also said to be *favorable* in Liu [71].

2.6.1 Conjecture. Any graph without cut-edge has a favorable embedding.

It is easily checked and proved that a graph with a cut-edge does not have a favorable embedding. However, no graph without cut-edge is exploded to have no favorable embedding yet. Some types of graphs have been shown to satisfy this conjecture such as K_n, $n \geq 3$; $K_{m,n}$, $m, n \geq 2$; Q_n, $n \geq 2$; and planar graphs without cut-edge.

Notice 2.1 On the basis of Liu [80–83], a favorable embedding is also a light embedding, or a weak embedding as well. Hence, the conjecture has been proved from the the weak embedding conjecture shown to be true in Liu [83]. This is shown to be complete in the stage of theoretizatoin, and hence, we have

Theorem II.1 Any graph without cut-edge has a favorable embedding.

However, the efficientization and intelligentization are still problems.

A map (or an embedding) that has no face itself with a common edge is said to be *preproper*. It can be shown that all preproper maps are favorable. However, the converse case is unnecessary to be true.

2.6.2 Conjecture. Any graph without cut-edge has a preproper embedding.

Similarly, it is also known that any graph with a cut-edge does not have a preproper embedding. And, K_n, $n \geq 3$; $K_{m,n}$, $m, n \geq 2$; Q_n, $n \geq 2$; planar graphs without cut-edge; etc. are shown to satisfy the conjecture as well.

Notice 2.2 Because a face without common edge itself induces that the face boundary can be partitioned into circuits, a preproper embedding is also a favorable embedding. From Notice 2.1, the conjecture is confirmatively done, and hence we have

Theorem II.2 Any graph without cut-edge has a preproper embedding.

However, the efficientization and intelligentization are still problems.

Furthermore, if a map has each of its faces a circuit itself, then it is called a *proper map*, or *strong map*. Likewise, *proper embedding*, or *strong embedding* . It can be shown that all proper maps are preproper. However, the converse case is unnecessary to be true.

2.6.3 Conjecture. Any graph without cut-edge has a proper embedding.

For proper embeddings as well, it is known that any graph with a cut-edge does not have a proper embedding. And, K_n, $n \geq 3$; $K_{m,n}$, $m, n \geq 2$; Q_n, $n \geq 2$; planar graphs without cut-edge; etc. are all shown to satisfy this conjecture.

Although conjectures 2.6.1–2.6.3 are stronger to stronger, because 2.6.3 is not yet shown to be true, the two formers are still meaningful.

Notice 2.3 From Liu [83], it is seen that any nonseparable graph has a strong embeding, and hence a proper embedding. It is seen that a separable graph without cut-edge does not have a proper one. Therefore, the condition of the conjecture is nonseparable instead of no cut-edge. Now, we have

Theorem II.3 Any nonseparable graph has a proper embedding.

However, the efficientization and intelligentization are still problems.

If a favorable (proper) embedding of a graph of order n has at most $n - 1$ faces, then it is said to be of *small face*.

Now, it is known that triangulations on the sphere have a small face proper embedding shown in Liu and Liu [56]. Because triangulations of order n have exactly $3n - 6$ edges and $2n - 4$ faces, all the small face embeddings are not yet on the sphere for $n \geq 4$.

2.6.4 Conjecture. Any graph of order at least six without cut-edge has a small face proper embedding.

Because it is proved that K_5 has a proper embedding only on the surfaces of orientable genus 1 (torus) and nonorientable genus at most 2 (Klein bottle) in Wei and Liu [164], they have at least five faces and hence are not of small face.

Notice 2.4 It looks no direct way yet to get the conjecture.

2.6.5 Conjecture. Any nonseparable graph of order n has a proper embedding with at most n faces.

Notice 2.5 No perspective way yet to get the conjecture.

If a map has only i-faces and $i+1$-faces, $3 \le i \le n-1$, then it is said to be *semi-regular*.

2.6.6 Conjecture. Any nonseparable graph of order n, $n \ge 7$, has a semi-regular proper embedding.

In fact, if a graph without cut-edge has a cut-vertex, then it can be decomposed into nonseparable blocks, none of which is a link itself. If this conjecture is proved, then it is also right for a graph without cut-edge. Some relationships among these conjectures and more of new developments can be seen in Liu [71].

Notice 2.6 It looks no further results for this conjecture yet.

2.6.7 For an integer $i \ge 3$, provide a necessary and sufficient condition for a graph having an i-embedding, or i^*-embedding. Particularly, when $i = 3, 4$ and 5.

First, start from $i = 3$ with a given type of graph. For instance, choose $G = K_n$, the complete graph of order n. For 3^*-embedding, on the basis of Theorem 1.12 (called *Euler formula*), a necessary condition for K_n, $n \ge 3$, having an 3^*-embedding is

$$n - \frac{n(n-1)}{2} + \phi = 2 - 2p \text{ and } 3\phi = n(n-1),$$

where ϕ and p are, respectively, the face number of an 3^*-embedding and the genus of the orientable surface the embedding is on. It is known that the condition is still sufficient.

If nonorientable surfaces are considered, then the necessary condition is

$$n - \frac{n(n-1)}{2} + \phi = 2 - q \text{ and } 3\phi = n(n-1),$$

i.e.,

$$q = \frac{(n-3)(n-4)}{6},$$

where q is the nonorientable genus of the surface and 3^*-embedding is not sufficient anymore for $q \ge 1$. When $n = 7$, K_7 would have a 3^*-embedding on the surface of nonorientable genus $q = 2$. However, it is shown that K_7 is not embeddable on the

Klein bottle, see Lemma 4.1 in Liu [70]. It has been proved that except only for this case, the necessary condition is also sufficient.

More other types of graphs, such as the complete bipartite graph $K_{m,n}$, n-cube Q_n and so forth can also be seen in Liu [70].

Notice 2.7 No further results on this conjecture occur up to now.

2.6.8 For $3 \le i, j \le 6$, recognize if a graph has an (i, j)-embedding (or (i^*, j^*)-embedding).

More generally, investigate the upper or/and the lower bounds of i and j such that for a given type of graphs we have an (i^*, j^*)-embedding.

Notice 2.8 It seems no further results on this conjecture appear up to now.

2.6.9 Given two integers i, j not less than 7, justify if a graph has an (i, j)-embedding (or (i^*, j^*)-embedding).

Notice 2.9 It seems no further results for this conjecture appear as yet.

If a proper map has any pair of its faces with at most one edge in common, then it is called a *polygonal map*.

2.6.10 Conjecture. Any 3-connected graph has an embedding which is a polygonal map.

From (a) and (b) in Figure 2.3, this conjecture is not valid for nonseparable graphs. The two graphs are nonseparable. The graph in (a) has a multi-edge, but that in (b) does not. It can be checked that none of them has a polygonal embedding.

Notice 2.10 No further result on this conjecture occurs up to now.

If an embedding of a graph has its genus (orientable or nonorientable) minimum among all the embeddings of the graph, then it is called a *minimum* (*orientable or nonorientable*) *genus embedding*. Based on the Euler formula, a minimum genus embedding has its face number maximum. So, a minimum (orientable or nonorientable) genus embedding is also called a *maximum* (*orientable or nonorientable*) *face number embedding*. Maximum face number implies that the average length of faces is smaller, and hence the possibility of faces being circuits is greater. This once caused to

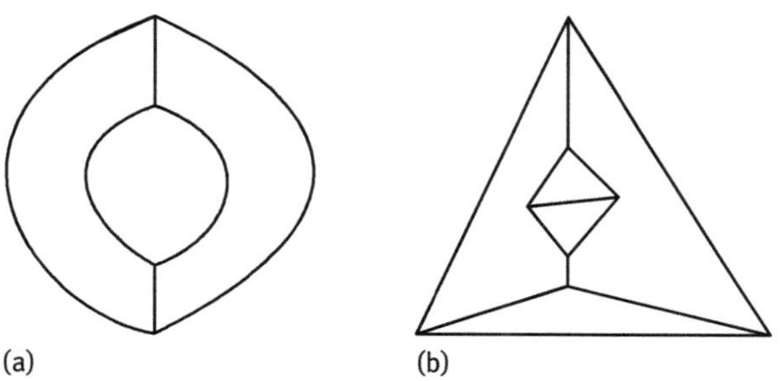

(a) (b)

Figure 2.3: Two graphs without polygonal embedding.

guess that minimum genus embeddings were all proper. However, a nonproper minimum genus embedding of a specific graph can be constructed by making one face as greater as possible with all other faces as less as possible. In fact, for torus and projective plane, all maximum face number embeddings are shown to be proper. For surfaces of big genus, a specific type of graphs were provided for all of their maximum face number embeddings nonproper in Zha [179].

2.6.11 Conjecture. Any nonseparable regular graph has a maximum face number embedding that is a proper map.

Notice 2.11 No further result on this conjecture occurs up to now.

A further suggestion is to find an embedding with nearly equal lengths of faces. The difference between the maximum length and the minimum length of faces in a map is call the *equilibrium* of the map. An embedding of a graph with its equilibrium minimum is called an *equilibrious embedding* of the graph.

2.6.12 Conjecture. Any 3-connected graph has an equilibrious embedding that is proper.

Notice 2.12 No further result on this conjecture occurs up to now.

An approach to access the conjecture is still for some types of graphs, e.g., planar graphs, Halin graphs, Hamiltonian graphs and further graphs embeddable to a surface with given genus.

3 Duality

- The dual of a map $(\mathscr{X}_{\alpha,\beta}, \mathscr{P})$ is the map $(\mathscr{X}_{\beta,\alpha}, \mathscr{P}\alpha\beta)$ and vice versa.
- The deletion of an edge in a map is the contraction of the corresponding edge in the dual map and vice versa.
- The addition of an edge to, the inverse of deleting an edge in, a map is splitting off its corresponding edge on, the inverse of contracting an edge in, the dual map and vice versa.
- The deletion of an edge with its inverse, the addition, and the dual of deletion, the contraction of an edge with its inverse, splitting off an edge are restricted on the same surface to form basic transformations.

3.1 Dual maps

On the basis of Section 2.2, for a basic permutation $\mathscr{P}$ on the ground set $\mathscr{X}_{\alpha,\beta}$, $(\mathscr{X}_{\alpha,\beta}, \mathscr{P})$ is a premap if, and only if, $(\mathscr{X}_{\beta,\alpha}, \mathscr{P}^*)$ is a premap where $\mathscr{P}^* = \mathscr{P}\gamma$, $\gamma = \alpha\beta$ (Theorem 2.4. The latter is called the *dual* of the former. Since

$$\mathscr{P}^{**} = \mathscr{P}^*\beta\alpha = (\mathscr{P}\alpha\beta)\beta\alpha = \mathscr{P}(\alpha\beta\beta\alpha) = \mathscr{P},$$

the former is also the dual of the latter.

Because the transitivity of two elements in the ground set $\mathscr{X}_{\alpha,\beta}$ on a premap $(\mathscr{X}_{\alpha,\beta}, \mathscr{P})$ under the group Ψ_J, where $J = \{\mathscr{P}, \alpha, \beta\}$, determine an equivalence, denoted by $\sim_{\Psi_J}$, the restriction of $\mathscr{P}$ on a class

$$\mathscr{X}_{\alpha,\beta}/\sim_{\Psi_J}$$

is called a *transitive block*.

Theorem 3.1 Premap $M_2 = (\mathscr{X}, \mathrm{Per}_2)$ is the dual of premap $M_1 = (\mathscr{X}, \mathrm{Per}_1)$ if, and only if,

$$\mathscr{X}/\sim_{\Psi_{J_2}} = \mathscr{X}/\sim_{\Psi_{J_1}}, \tag{3.1}$$

where

$$J_2 = \{\mathrm{Per}_2, \alpha, \beta\},\ J_1 = \{\mathrm{Per}_1, \alpha, \beta\},$$

and

$$\mathrm{Per}_2 = \mathrm{Per}_1\gamma = \mathrm{Per}_1{}^*,\ \gamma = \alpha\beta.$$

DOI 10.1515/9783110481846-003

Proof Necessity. Since M_2 is the dual of M_1, $\mathrm{Per}_2 = \mathrm{Per}_1\alpha\beta$. From $\mathrm{Per}_2 = \mathrm{Per}_1\alpha\beta \in \Psi_{J_1}$, $\Psi_{J_2} = \Psi_{J_1}$. Hence, (3.1) holds. This is the necessity.

Sufficiency. Since M_1 is a premap and $\mathrm{Per}_2 = \mathrm{Per}_1\alpha\beta$, M_2 is also a premap, and then the dual of M_1 by considering Theorem 2.4. This is the sufficiency. □

From this theorem, the duality between M_1 and M_2 induces a one-to-one correspondence between their transitive blocks in dual pair. Because each transitive block is a map, it leads what the *dual map* of a map is. The representation of a premap by its transitive blocks is called its *transitive decomposition*.

Example 3.1 Map

$$\tilde{L}_1 = (\{x, \alpha x, \beta x, \gamma x\}, (x, \beta x)),\ \ \gamma = \alpha\beta,$$

and its dual

$$\tilde{L}_1^* = (\{x, \beta x, \alpha x, \gamma x\}, (x, \alpha x)).$$

Or, in the form as

$$\tilde{L}_1 = (e, v) \text{ and } \tilde{L}_1^* = (e^*, f),$$

where $v = (x, \beta x)$, $f = (x, \alpha x)$,

$$e = \{x, \alpha x, \beta x, \gamma x\},\ \ \gamma = \alpha\beta,$$

and

$$e^* = \{x, \beta x, \alpha x, \gamma x\},$$

as shown in Figure 3.1.

In the following figure, two As and two Bs are, respectively, identified on the surface.

This figure shows what a dual pair of maps looks like. It is a generalization of a dual pair of maps on the plane.

A map with its under pregraph a selfloop is called a *loop map*. It is seen that both $\tilde{L}_1$ and its dual $\tilde{L}_1^*$ in Figure 3.1 are loop maps.

In a premap $M = (\mathscr{X}, \mathscr{P})$, if a vertex $v = \{(x)_{\mathscr{P}}, (\alpha x)_{\mathscr{P}}\}$ is transformed into two vertices

$$\begin{aligned} v_1 &= \{(x,\ \mathscr{P}x,\ \dots,\ \mathscr{P}^j x),\ (\alpha x,\ \alpha\mathscr{P}^j x,\ \dots, \alpha\mathscr{P}x)\} \\ &= \{(x)_{\mathscr{P}'}, (\alpha x)_{\mathscr{P}'}\} \end{aligned}$$

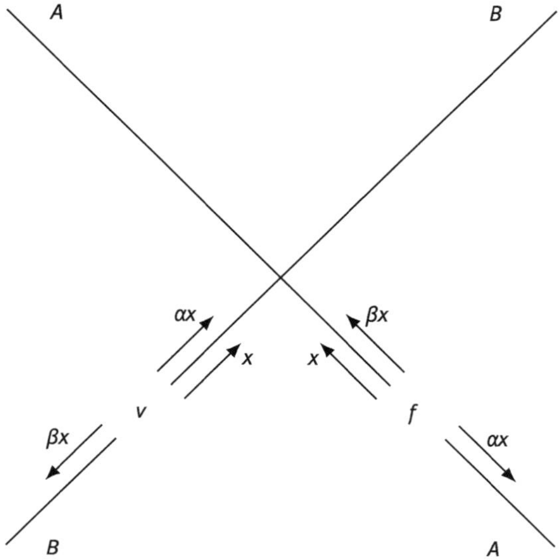

Figure 3.1: Map and its dual.

and

$$\begin{aligned} v_2 &= \{(\mathscr{P}^{j+1}x, \mathscr{P}^{j+2}x, \ldots, x\mathscr{P}^{-1}x), (\alpha\mathscr{P}^{j+1}x, \alpha\mathscr{P}^{-1}x, \ldots, \alpha\mathscr{P}^{j+2}x)\} \\ &= \{(\mathscr{P}^{j+1}x)_{\mathscr{P}'}, (\alpha\mathscr{P}^{j+1}x)_{\mathscr{P}'}\},\ j \geq 0, \end{aligned}$$

with other vertices unchanged for permutation $\mathscr{P}$ becoming permutation $\mathscr{P}'$. It is seen that permutation $\mathscr{P}'$ is basic and with the conjugate axiom as well. Hence, $M' = (\mathscr{X}, \mathscr{P}')$ is also a premap. Such an operation is called *cutting* a vertex. If elements at v_1 are not transitive with elements at v_2 in M', then elements at v_1 and elements at v_2 are said to be *cuttable* in M. The vertex v is called a *cutting vertex* in M; otherwise, *noncuttable*.

If there are two elements cuttable in a map, then the map is said to be cuttable in its own right.

In virtue of Theorem 3.1, cuttability and noncuttability are concerned with only maps without loss of the generality of premaps.

Lemma 3.1 A map M is cuttable if, and only if, its dual M^* is cuttable.

Proof Necessity. From $M = (\mathscr{X}, \mathscr{P})$ cuttable, vertex

$$v = (x,\ \mathscr{P}x,\ \ldots,\ \mathscr{P}^{-1}x)$$

is assumed to cut into two vertices as

$$v_1 = (x, \ \mathcal{P}x, \ \dots, \mathcal{P}^j x)$$

and

$$v_2 = (\mathcal{P}^{j+1}x, \ \mathcal{P}^{j+2}x, \dots, \mathcal{P}^{-1}x)$$

for obtaining premap $M' = (\mathcal{X}, \mathcal{P}') = M_1 + M_2$, where v_i is on $M_i = (\mathcal{X}_i, \mathcal{P}_i)$, $\mathcal{X} = \mathcal{X}_1 + \mathcal{X}_2$, $\mathcal{P}_i$ is the restriction of $\mathcal{P}'$ on $\mathcal{X}_i$, where $i = 1, 2$. It can be checked that M_1 and M_2 are both maps. Thus, on $M^* = (\mathcal{X}, \mathcal{P}^*)$, vertex

$$v^* = (x, A, \gamma\mathcal{P}^j x, \ \mathcal{P}^{j+1}x, B, \gamma\mathcal{P}^{-1}x)$$

can be cut into two vertices

$$v_1^* = (x, A, \gamma\mathcal{P}^j x)$$

and

$$v_2^* = (\mathcal{P}^{j+1}x, B, \gamma\mathcal{P}^{-1}x),$$

where A and B are, respectively, linear orders of elements in $\mathcal{X}_1$ and $\mathcal{X}_2$. This attains the premap $M^{*\prime} = (\mathcal{X}, \mathcal{P}^{*\prime}) = M_1^* + M_2^*$, where $\mathcal{X} = \mathcal{X}_1 + \mathcal{X}_2$, $\mathcal{P}^*{}_i$ is the restriction of $\mathcal{P}^{*\prime}$ on $\mathcal{X}_i$, where $i = 1, 2$. The necessity is obtained.

Sufficiency. From the duality, it is deduced from the necessity. □

If two elements in the ground set of a map M are not transitive in M' obtained by cutting a vertex on M, they are said to be *cuttable*; otherwise, *noncuttable*. It can be checked that the noncuttability determines an equivalence, denoted by $\sim$ nc on the ground set of M. The restriction of M on each

$$\mathcal{X}_{\alpha,\beta}/\sim \mathrm{nc}$$

is called a *noncuttable block*.

If all noncuttable blocks and all cutting vertices of a map deal with vertices such that two vertices are adjacent if, and only if, one is a noncuttable block and the other is a cutting vertex incident to the block, then the graph obtained in this way is called a *cutting graph* of the map. It is easily shown that the cutting graph of a map is always a tree.

For a face $f = (x)_{\mathcal{P}\gamma}$ of a premap $M = (\mathcal{X}, \mathcal{P})$, if there is an integer $l \geq 0$ for transforming f into

$$f_1 = (x, \dots, (\mathcal{P}\gamma)^l x)$$

and

$$f_2 = ((\mathscr{P}\gamma)^{l+1}x, \ldots, (\mathscr{P}\gamma)^{-l}x)$$

such that x and $(\mathscr{P}\gamma)^{l+1}x$ are not transitive at all, then f is called a *cutting face* of M. From the procedure in the proof of Lemma 3.1, for a cutting vertex of a premap, there is a corresponding cutting face in the dual of the premap.

Theorem 3.2 Two maps M and N are mutually dual if, and only if, their cutting graphs are the same and the corresponding noncuttable blocks are mutually dual such that a cutting vertex of one corresponds to a cutting face of the other.

Proof Necessity. Because maps M and N are mutually dual, from the procedure in the proof of Lemma 3.1, there is a one-to-one correspondence between their noncuttable blocks such that two corresponding blocks are mutually dual. There is also a one-to-one correspondence between their cutting vertices such that the cyclic orders of their blocks at two corresponding cutting vertices are in correspondence. Therefore, their cutting graphs are the same. This is the necessity.

Sufficiency. Because maps M and N have the same cutting graph, a tree of course, in virtue of the correspondence between cutting vertices and cutting faces, the sufficiency is deduced from Lemma 3.1. □

Note 3.1 Two trees are said to be the same in the theorem when trees as maps (planar of course) are the same but not the isomorphism of trees as graphs (the latter can be deduced from the former but is not necessarily true from the latter to the former).

On a premap $M = (\mathscr{X}, \mathscr{P})$, if an edge Kx is incident with two faces, i.e.,

$$\gamma x \notin \{x\}_{\mathscr{P}\gamma} \cup \{\beta x\}_{\mathscr{P}\gamma},$$

then it is said to be *single*; otherwise, i.e.,

$$\gamma x \in \{x\}_{\mathscr{P}\gamma} \cup \{\beta x\}_{\mathscr{P}\gamma}$$

(with only one face), *double*. An edge with distinct ends is called a *link*; otherwise, a *loop*. Clearly, an edge can be *single link*, *single loop*, *double link* and *double loop*. Further, a double link is called a *harmonic link*, or *singular link* accordingly as $\gamma x \in (x)_{\mathscr{P}\gamma}$ or not. Similarly, a single loop is called a *harmonic loop*, or *singular loop* accordingly as $\gamma x \in (x)_{\mathscr{P}\gamma}$ or not.

Theorem 3.3 For an edge $e_x = \{x, \alpha x, \beta x, \gamma x\}$ of premap $M = (\mathscr{X}_{\alpha,\beta}, \mathscr{P})$ and its corresponding edge $e_x^* = \{x, \beta x, \alpha x, \gamma x\}$ of the dual $M^* = (\mathscr{X}_{\beta,\alpha}, \mathscr{P}^*)$, $\mathscr{P}^* = \mathscr{P}\gamma$,

(i) e_x is a single link if, and only if, e^* is a single link;
(ii) e_x is a harmonic link if, and only if, e^* is a harmonic loop;

(iii) e_x is a singular link if, and only if, e^* is a singular loop;
(iv) e_x is a double loop if, and only if, e^* is a double loop.

Proof Necessity.
(i) Because e_x is a link, $(x)_{\mathscr{P}}$ and $(\gamma x)_{\mathscr{P}}$ belong to distinct vertices. And because e_x is a single edge, $(x)_{\mathscr{P}\gamma}$ and $(\gamma x)_{\mathscr{P}\gamma}$ belong to distinct faces. By the duality, e_x^* is a single link as well.
(ii) Because e_x is a double link, in spite of $(x)_{\mathscr{P}}$ and $(\gamma x)_{\mathscr{P}}$ belonging to distinct vertices, γx, or $\alpha x \in (x)_{\mathscr{P}^*}$. And because of harmonic link, the only opportunity is $\gamma x \in (x)_{\mathscr{P}^*}$. From the duality, e_x^* is a harmonic loop.
(iii) Because e_x is a singular link, in spite of $(x)_{\mathscr{P}}$ and $(\gamma x)_{\mathscr{P}}$ belonging to distinct vertices, $\alpha x \in (x)_{\mathscr{P}^*}$. In virtue of α as the second operation of M^*, e_x^* is a singular loop.
(iv) Because e_x is a double loop, $\beta x \in (x)_{\mathscr{P}}$ and $\alpha x \in (x)_{\mathscr{P}^*}$. From the symmetry between α and β, M and M^*, e_x^* is a double loop as well.

Sufficiency. From the symmetry in duality, i.e., $M = (M^*)^*$, it is obtained from the necessity. □

On the basis of this theorem, the classification and the dual relationship among edges are shown in Figure 3.2.

From Figure 3.2, harmonic links will be classified into segmentation edges and terminal links and harmonic loops into shearing loops and terminal loops in 3.2 have additional two dual pairs of edges: segmentation edges and shearing loops, terminal links and terminal loops.

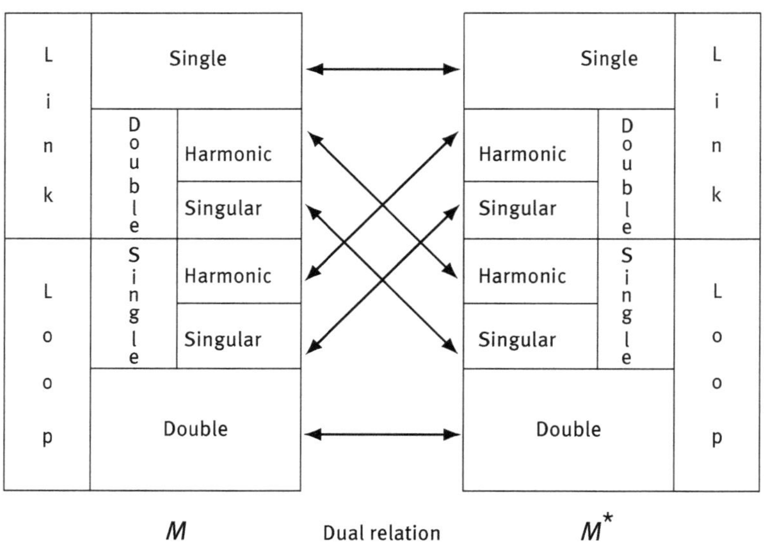

Figure 3.2: Duality between edges.

3.2 Deletion of an edge

Let $M = (\mathscr{X}_{\alpha,\beta}(X), \mathscr{P})$ be a premap and $e_x = Kx = \{x, \alpha x, \beta x, \gamma x\}$, $x \in X$, an edge.

What is obtained by *deleting* the edge e_x from M is denoted by

$$M - e_x = (\mathscr{X}_{\alpha,\beta}(X) - Kx, \mathscr{P}_{-x}), \tag{3.2}$$

where $\mathscr{P}_{-x}$ is the permutation restricted from $\mathscr{P}$ on $\mathscr{X}_{\alpha,\beta}(X) - Kx$.

Lemma 3.2 Permutation $\mathscr{P}_{-x}$ is determined in the following way as when e_x is not a selfloop,

$$\mathscr{P}_{-x}y = \begin{cases} \mathscr{P}x \text{ (and } \alpha\mathscr{P}^{-1}x), & \text{if } y = \mathscr{P}^{-1}x \text{ (and } \alpha\mathscr{P}x); \\ \mathscr{P}\gamma x \text{ (}\alpha\mathscr{P}^{-1}\gamma x), & \text{if } y = \mathscr{P}^{-1}\gamma x \text{ (and } \alpha\mathscr{P}\gamma x); \\ \mathscr{P}y, & \text{otherwise,} \end{cases} \tag{3.3}$$

and when e_x is a selfloop with $\gamma x \in (x)_{\mathscr{P}}$,

$$\mathscr{P}_{-x}y = \begin{cases} \mathscr{P}x \text{ (and } \alpha\mathscr{P}^{-1}x), & \text{if } y = \mathscr{P}^{-1}x \text{ (and } \alpha\mathscr{P}x); \\ \mathscr{P}\gamma x \text{ (and } \alpha\mathscr{P}^{-1}\gamma x), & \text{if } y = \mathscr{P}^{-1}x \text{ (and } \alpha\mathscr{P}\gamma x); \\ \mathscr{P}y, & \text{otherwise,} \end{cases} \tag{3.4}$$

otherwise, i.e., $\gamma x \notin (x)_{\mathscr{P}}$, γx is replaced by βx in (3.4).

Proof When e_x is not a selfloop. Because only vertices $(x)_{\mathscr{P}}$ and $(\gamma x)_{\mathscr{P}}$ are, respectively, changed in $M - e_x$ from M as

$$(\mathscr{P}^{-1}x)_{\mathscr{P}_{-x}} = (\mathscr{P}^{-1}x, \mathscr{P}^2 x, \ldots, \mathscr{P}^{-2}x)$$

and

$$(\mathscr{P}^{-1}\gamma x)_{\mathscr{P}_{-x}} = (\mathscr{P}^{-1}\gamma x, \mathscr{P}\gamma x, \ldots, \mathscr{P}^{-2}\gamma x)$$

(Figure 3.3(a)⇒(b)). This implies (3.3).

When e_x is a selfloop with $\gamma x \in (x)_{\mathscr{P}}$. Because only vertex $(x)_{\mathscr{P}}$ is changed in $M - e_x$ from M as

$$(\mathscr{P}^{-1}x)_{\mathscr{P}_{-x}} = (\mathscr{P}^{-1}x, \mathscr{P}x, \ldots, \mathscr{P}^{-1}\gamma x, \mathscr{P}\gamma x, \ldots, \mathscr{P}^{-2}x)$$

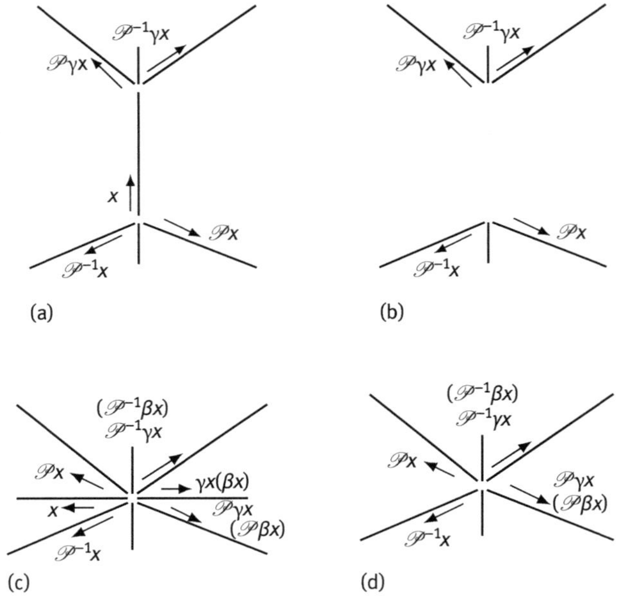

Figure 3.3: Deletion of an edge.

(Figure 3.3(c)→(d)), or

$$(\mathscr{P}^{-1}x)_{\mathscr{P}_{-x}} = (\mathscr{P}^{-1}x, \mathscr{P}x, \dots, \mathscr{P}^{-1}\beta x, \mathscr{P}\beta x, \dots, \mathscr{P}^{-2}x)$$

(Figure 3.3(c)→(d) in parentheses) accordingly as $\gamma x \in (x)_{\mathscr{P}}$ or not.

The former is (3.4). The latter is what is obtained from (3.4) with γx replaced by βx. □

In Figure 3.3 (a) and (c) are parts of the original map and (b) and (d), the results by deleting the edge Kx.

Further, Figures 3.4–3.10 are all like this without specification.

Lemma 3.3 For a premap $M = (\mathscr{X}, \mathscr{P})$, $M - e_x = (\mathscr{X} - Kx, \mathscr{P}_{-x})$ is also a premap. And, the number of transitive blocks in $M - e_x$ is not less than that in M.

Proof Because $\mathscr{P}$ is basic for α, from Lemma 3.2 $\mathscr{P}_{-x}$ is also basic for α. Because $\mathscr{P}$ satisfies the conjugate axiom for α, from Lemma 3.2 and Theorem 2.3, $\mathscr{P}_{-x}$ also satisfies the conjugate axiom for α. The first statement is done.

Because any nontransitive pair of elements in M is never transitive in $M - e_x$, the second statement is done. □

If e_x is an edge of a premap M such that $M - e_x$ has more transitive blocks than M has, then e_x is called a *segmentation edge*. If an edge has its one end formed by only one semiedge of the edge itself, then it is called a *terminal link*. From the symmetry of elements in a quadricell, (x) can be assumed as the 1-vertex incident with a terminal link without loss of generality. Since $(\mathscr{P}\gamma)\gamma x = \mathscr{P}x = x$, $\gamma x \in (x)_{\mathscr{P}\gamma}$. Hence, a terminal link is always a harmonic link. However, a harmonic link is unnecessary to be a terminal link. This point can be seen in the following theorem.

Theorem 3.4 For a map $M = (\mathscr{X}, \mathscr{P})$, $M - e_x = (\mathscr{X} - Kx, \mathscr{P}_{-x})$ is a map if, and only if, e_x is not a harmonic link of M except for terminal link.

Proof When $\langle x, \gamma x\rangle \subseteq (x)_{\mathscr{P}\gamma}$, i.e., e_x is a terminal link, no isolated vertex in any premap, from Lemma 3.3, $M - e_x = (\mathscr{X} - Kx, \mathscr{P}_{-x})$ is a map. In what follows, this case is not considered again.

Necessity. Suppose $M - e_x$ is a map, but e_x is a harmonic link of M. Because $\mathscr{P}\gamma^{-1}x$ and $\mathscr{P}\gamma^{-1}\gamma x \neq x(\langle x, \gamma x\rangle \not\subseteq (x)_{\mathscr{P}\gamma})$ for group $\Psi_{J'}$, where $J' = \{\mathscr{P}_{-x}, \alpha, \beta\}$, are not transitive on the set $\mathscr{X} - Kx$, $M - e_x$ is not a map. This is a contradiction to the assumption.

Sufficiency. Because M is a map, $\mathscr{P}_{-x}$ is basic. From Theorem 2.3, $\mathscr{P}_{-x}$ satisfies the conjugate axiom. Then, based on Figure 3.2, two cases should be discussed for the transitivity.

(i) When e_x is a single edge (including single loops!) or singular link of M. Because e_x is not a cut-edge of its under graph $G(M)$, $G(M - e_x)$ is connected. From Theorem 2.6, $\mathscr{P}_{-x}$ for group $\Psi_{J'}$ is transitive on the set $\mathscr{X} - Kx$. Thus, $M - e_x$ is a map.

(ii) When e_x is a double loop of M. Because

$$((\mathscr{P}\gamma)^{-1}x)_{\mathscr{P}_{-x}} = ((\mathscr{P}\gamma)^{-1}x, \mathscr{P}\alpha x, \ldots, \beta\mathscr{P}\beta x, \mathscr{P}\gamma x, \ldots, (\mathscr{P}\gamma)^{-2})\alpha x),$$

we have $\mathscr{P}x = \beta(\mathscr{P}\gamma)x$ with $(\mathscr{P}\gamma)^{-1}\alpha x$ and $\mathscr{P}\beta x = \beta(\beta\mathscr{P}\beta x)$ are transitive in $M - e_x$. Hence, $M - e_x$ is a map as well. □

From the proof of the theorem, a much fundamental conclusion is soon deduced.

Corollary 3.1 In a map M, an edge e_x is a segmentation edge if, and only if, e_x is a harmonic link except for terminal links. And, e_x is a harmonic link if, and only if, it is a cut-edge of graph $G(M)$.

Proof To prove the first statement.

Necessity. Because e_x is a segmentation edge, $G(M - e_x)$ is not connected. e_x is only a link. On the basis of Figure 3.2, e_x is also a double edge. And, because e_x is not singular, e_x is only harmonic. Clearly, a terminal link is not a segmentation edge in its own right.

Sufficiency. Because e_x is not a terminal link, $\mathscr{P}x$ is distinct form x and γx is distinct from $\mathscr{P}\gamma x$. And because e_x is a harmonic link, $\mathscr{P}x$ and $\mathscr{P}\gamma x$ are not transitive in $M - e_x$. Thus, e_x is a segmentation edge.

To prove the second statement. Because it can be shown that e_x is a terminal link of M if, and only if, e_x is an articulate edge of graph $G(M)$, a cut-edge as well. This statement is deduced from the first statement. □

Let $M = (\mathscr{X}_{\alpha,\beta}(X), \mathscr{P})$ be a pregraph and $e_x = Kx = \{x, \alpha x, \beta x, \gamma x\}$, $x \in X$, be an edge. The *contraction* of e_x from M, denoted by

$$M \bullet e_x = (\mathscr{X}_{\alpha,\beta}(X) - Kx, \mathscr{P}_{\bullet x}),$$

is defined to be the dual of $M^* - e_x^*$, where $e_x^* = \{x, \beta x, \alpha x, \gamma x\}$, the corresponding edge of e_x in the dual M^* of M. In other words, $\mathscr{P}_{\bullet x} = \mathscr{P}^*_{-x}\gamma$.

Lemma 3.4 $\mathscr{P}_{\bullet x}$ is determined by the following (i–iii):

(i) When e_x is a link. For $y \in \mathscr{X}_{\alpha,\beta}(X) - Kx$,

$$\mathscr{P}_{\bullet x}y = \begin{cases} \mathscr{P}\gamma x \text{ (and } \alpha\mathscr{P}^{-1}x), & \text{if } y = \mathscr{P}^{-1}x \text{ (and } \alpha\mathscr{P}\gamma x); \\ \mathscr{P}x \text{ (and } \alpha\mathscr{P}^{-1}\gamma x), & \text{if } y = \mathscr{P}^{-1}\gamma x \text{ (and } \alpha\mathscr{P}x); \\ \mathscr{P}y, & \text{otherwise,} \end{cases} \tag{3.5}$$

as shown in Figure 3.4(a)⟹(b).

(ii) When e_x is a harmonic loop. For $y \in \mathscr{X}_{\alpha,\beta}(X) - Kx$,

$$\mathscr{P}_{\bullet x}y = \begin{cases} \mathscr{P}\gamma x \text{ (and } \alpha\mathscr{P}^{-1}x), & \text{if } y = \mathscr{P}^{-1}x \text{ (and } \alpha\mathscr{P}\gamma x); \\ \mathscr{P}x \text{ (and } \alpha\mathscr{P}^{-1}\gamma x), & \text{if } y = \mathscr{P}^{-1}\gamma x \text{ (and } \alpha\mathscr{P}x); \\ \mathscr{P}y, & \text{otherwise,} \end{cases} \tag{3.6}$$

as shown in Figure 3.4(c)⟹(d).

(iii) When e_x is a singular, or double loop. For $y \in \mathscr{X}_{\alpha,\beta}(X) - Kx$,

$$\mathscr{P}_{\bullet x}y = \begin{cases} \alpha\mathscr{P}^{-1}\beta x \text{ (and } \alpha\mathscr{P}^{-1}x), & \text{if } y = \mathscr{P}^{-1}x \text{ (and } \mathscr{P}^{-1}\beta x); \\ \mathscr{P}\beta x \text{ (and } \mathscr{P}x), & \text{if } y = \alpha\mathscr{P}x \text{ (and } \alpha\mathscr{P}\beta x); \\ \mathscr{P}y, & \text{otherwise,} \end{cases} \tag{3.7}$$

as shown in Figure 3.4(e)⟹(f).

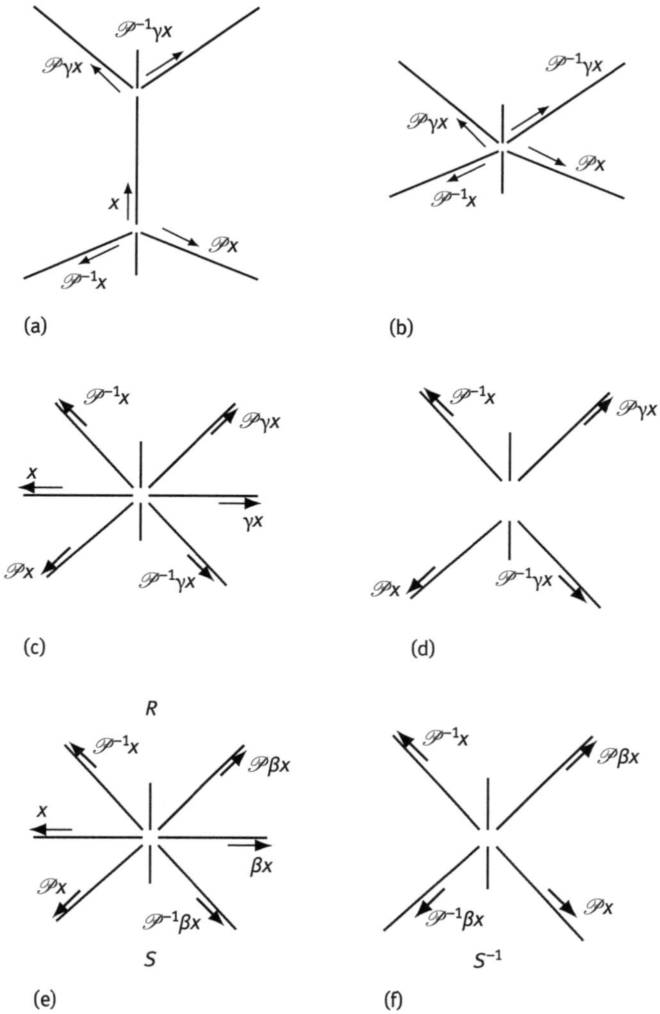

Figure 3.4: Contraction of an edge.

Proof

(i) When e_x is a link. In the dual M^* of M, from the duality,

$$(x)_{\mathscr{P}^*} = (x, \mathscr{P}\gamma x, (\mathscr{P}\gamma)^2 x, \dots, (\mathscr{P}\gamma)^{-1} x)$$

and

$$(\gamma x)_{\mathscr{P}^*} = (\gamma x, \mathscr{P}x, (\mathscr{P}\gamma)^2 \gamma x, \dots, (\mathscr{P}\gamma)^{-1} \gamma x),$$

or

$$(x)_{\mathscr{P}^*} = (x, \mathscr{P}\gamma x, \dots, (\mathscr{P}\gamma)^{-1} \gamma x, \gamma x, \mathscr{P}x, \dots, (\mathscr{P}\gamma)^{-1} x),$$

and hence $\mathscr{P}^*_{-x}$ is only different from $\mathscr{P}^* = \mathscr{P}\gamma$ at vertices

$$(\mathscr{P}\gamma x)_{\mathscr{P}^*_{-x}} = (\mathscr{P}\gamma x, (\mathscr{P}\gamma)^2 x, \dots, (\mathscr{P}\gamma)^{-1} x)$$

and

$$(\mathscr{P}x)_{\mathscr{P}^*_{-x}} = (\mathscr{P}x, (\mathscr{P}\gamma)^2 \gamma x, \dots, (\mathscr{P}\gamma)^{-1} \gamma x)$$

or at vertex

$$(\mathscr{P}x)_{\mathscr{P}^*_{-x}} = (\mathscr{P}\gamma x, \dots, (\mathscr{P}\gamma)^{-1}\gamma x, \mathscr{P}x, \dots, (\mathscr{P}\gamma)^{-1} x)$$

with their conjugations accordingly as e_x is single or double. By considering $\mathscr{P}_{\bullet x} = \mathscr{P}^*_{-x}\gamma$,

$$\begin{aligned}\mathscr{P}_{\bullet x}(y) &= \mathscr{P}_{\bullet x}(\mathscr{P}^{-1}\gamma x) = \mathscr{P}^*_{-x}\gamma(\mathscr{P}^{-1}\gamma x)\\ &= \mathscr{P}^*_{-x}(\mathscr{P}\gamma^{-1}\gamma x) = \mathscr{P}x\end{aligned}$$

for $y = \mathscr{P}^{-1}\gamma x$ and

$$\begin{aligned}\mathscr{P}_{\bullet x}(y) &= \mathscr{P}_{\bullet x}(\mathscr{P}^{-1}x) = \mathscr{P}^*_{-x}\gamma(\mathscr{P}^{-1}x)\\ &= \mathscr{P}^*_{-x}(\mathscr{P}\gamma)^{-1}x = \mathscr{P}\gamma x\end{aligned}$$

for $y = \mathscr{P}^{-1}x$. From the conjugate axiom, the cases for $y = \alpha\mathscr{P}^{-1}x$ and $y = \alpha(\mathscr{P}\gamma x)$ in the parentheses of (3.5) are also obtained. Then, for other y,

$$\mathscr{P}_{\bullet x}(y) = \mathscr{P}^*_{-x}\gamma y = (\mathscr{P}\gamma)\gamma y = \mathscr{P}y$$

in both the cases. Therefore, (3.5) is true.

(ii) When e_x is a harmonic loop. In a similar way to (3.5) for e_x single, (3.6) is also obtained.

(iii) When e_x is a singular, or double loop. In the dual M^* of M,

$$(x)_{\mathscr{P}^*} = (x, \mathscr{P}\gamma x, (\mathscr{P}\gamma)^2 x, \dots, (\mathscr{P}\gamma)^{-1} x)$$

and

$$(\alpha x)_{\mathscr{P}^*} = (\alpha x, \mathscr{P}\gamma\alpha x, (\mathscr{P}\gamma)^2\alpha x, \dots, (\mathscr{P}\gamma)^{-1}\alpha x),$$

or

$$(x)_{\mathscr{P}^*} = (x, \mathscr{P}\gamma x, \dots, (\mathscr{P}\gamma)^{-1}\alpha x, \alpha x, \mathscr{P}\gamma\alpha x, \dots, (\mathscr{P}\gamma)^{-1} x),$$

and hence

$$(x)_{\mathscr{P}^*_{-x}} = (\mathscr{P}\gamma x, (\mathscr{P}\gamma)^2 x, \ldots, (\mathscr{P}\gamma)^{-1} x)$$

and

$$(\alpha x)_{\mathscr{P}^*_{-x}} = (\mathscr{P}\gamma\alpha x, (\mathscr{P}\gamma)^2 \alpha x, \ldots, (\mathscr{P}\gamma)^{-1} \alpha x)$$

or

$$(x)_{\mathscr{P}^*_{-x}} = (\mathscr{P}\gamma x, \ldots, (\mathscr{P}\gamma)^{-1}\alpha x, \mathscr{P}\gamma\alpha x, \ldots, (\mathscr{P}\gamma)^{-1} x)$$

with their conjugations as per e_x is singular or double. By considering $\mathscr{P}_{\bullet x} = \mathscr{P}^*_{-x}\gamma$,

$$\begin{aligned}\mathscr{P}_{\bullet x}(y) &= \mathscr{P}_{\bullet x}(\alpha\mathscr{P}x) = \mathscr{P}^*_{-x}\gamma(\alpha\mathscr{P}x)\\ &= \mathscr{P}^*_{-x}\gamma(\mathscr{P}^{-1}\alpha x) = \mathscr{P}^*_{-x}((\mathscr{P}\gamma)^{-1}\alpha x)\\ &= \mathscr{P}\gamma\alpha x = \mathscr{P}\beta x\end{aligned}$$

for $y = \alpha\mathscr{P}x$ and

$$\begin{aligned}\mathscr{P}_{\bullet x}(y) &= \mathscr{P}_{\bullet x}(\mathscr{P}^{-1}x) = \mathscr{P}^*_{-x}\gamma(\mathscr{P}^{-1}x)\\ &= \mathscr{P}^*_{-x}((\mathscr{P}\gamma)^{-1}x) = \mathscr{P}\gamma x\\ &= \mathscr{P}\alpha\beta x = \alpha\mathscr{P}^{-1}\beta x\end{aligned}$$

for $y = \mathscr{P}^{-1}x$. From the conjugate axiom, the cases for $y = \mathscr{P}\beta x$ and $\alpha\mathscr{P}\beta x$ are also obtained. Then, for all other y,

$$\mathscr{P}_{\bullet x}(y) = \mathscr{P}^*_{-x}\gamma y = (\mathscr{P}\gamma)\gamma y = \mathscr{P}y.$$

This is (3.7). □

From Lemma 3.4, it is seen that in the constriction of edge e_x on a premap only if e_x is not a selfloop, two vertices $(x)_{\mathscr{P}}$ and $(\gamma x)_{\mathscr{P}}$ are composed of one vertex

$$(\mathscr{P}^{-1}x)_{\mathscr{P}_{\bullet x}} = (\mathscr{P}^{-1}x, \mathscr{P}\gamma x, \ldots, \mathscr{P}^{-1}\gamma x, \mathscr{P}x, \ldots, \mathscr{P}^{-2}x)$$

(Figure 3.4(a)⇒(b)); if e_x is a harmonic loop, vertex $(x)_{\mathscr{P}}$ is divided into two vertices

$$(\mathscr{P}^{-1}x)_{\mathscr{P}_{\bullet x}} = (\mathscr{P}^{-1}x, \mathscr{P}\gamma x, \ldots, \mathscr{P}^{-2}x)$$

and

$$(\mathscr{P}^{-1}\gamma x)_{\mathscr{P}_{\bullet x}} = (\mathscr{P}^{-1}\gamma x, \mathscr{P}x, \ldots, \mathscr{P}^{-2}\gamma x)$$

(Figure 3.4(c)⇒(d)); and if e_x is a singular, or double loop, vertex $(x)_{\mathscr{P}}$ becomes vertex

$$(\mathscr{P}x)_{\mathscr{P}_{\bullet x}} = (\mathscr{P}x, \ldots, \mathscr{P}^{-1}\beta x, \alpha\mathscr{P}^{-1}x, \ldots, \alpha\mathscr{P}\beta x)$$

(Figure 3.4(e)⇒(f)).

Lemma 3.5 For a premap M, $M \bullet e_x$ is always a premap. And, the number of transitive blocks in $M \bullet e_x$ is not less than that in M.

Proof From Lemma 3.3 and the duality, the first statement is true. Because any non-transitive pair of elements in M is never transitive in $M \bullet e_x$ from Lemma 3.4, the second statement is true. □

If a harmonic loop e_x has $(x)_{\mathscr{P}\gamma} = (x)$, or $(\gamma x)_{\mathscr{P}\gamma} = (\gamma x)$, then it is called a *terminal loop*. If the two elements of a co-semiedge appear in a vertex in succession, then the edge is called a *twist loop*.

Lemma 3.6 For an edge e_x of a map M, e_x is a terminal loop if, and only if, e_x^* is an terminal link in M^*. And e_x is a twist loop if, and only if, e_x^* is a twist loop.

Proof A direct result deduced from the duality. □

Theorem 3.5 For an edge e_x of a map $M = (\mathscr{X}_{\alpha,\beta}(X), \mathscr{P})$, $M \bullet e_x$ is a map if, and only if, e_x is not a harmonic loop but terminal loop.

Proof Because for a terminal loop e_x, $M \bullet e_x$ is always a map. In what follows, this case is excluded.

Necessity. Suppose $M \bullet e_x$ is a map but e_x is a harmonic loop. Since e_x^* is a harmonic link in M^*(Figure 3.2), from Theorem 3.4 and Lemma 3.1, $\mathscr{P}^{-1}x$ and $\mathscr{P}x$, respectively, belong to two distinct transitive blocks of M. From Lemma 3.4(ii), $M \bullet e_x$ has two transitive blocks. This contradicts the fact that $M \bullet e_x$ is a map.

Sufficiency. Since e_x is not a harmonic loop, only two cases should be considered: either e_x is not a loop or e_x is a singular loop. For the former, in spite of a single or double edge, from Lemma 3.4(i), $M \bullet e_x$ is a map. For the latter, from Lemma 3.4(iii), $M \bullet e_x = M - e_x$ is also a map. □

If a loop e_x has that $\mathscr{P}^{-1}x$ and $\mathscr{P}x$ are in distinct noncuttable blocks, then it is called a *shearing loop*. From Theorem 3.5, all shearing loops are harmonic. However, the converse case is unnecessarily true.

Corollary 3.2 In a map M, an edge e_x is a shearing loop if, and only if, e_x^* is a harmonic, but not terminal loop in M^*.

Proof A direct result of Theorem 3.5. □

Theorem 3.6 The dual of premap $M - e_x$ is the premap $M^* \bullet e_x^*$, where M^* is the dual of M and e_x^* in M^* is the corresponding edge of e_x in M.

Proof (1) Because $M^* \bullet e_x^*$ is the dual of $(M^*)^* - e^{**}_x = M - e_x$, by the symmetry of the duality the theorem holds. □

However, if the contraction of e_x on M is defined by (3.5–7), then the theorem can also be proved.

Proof (2) Based on Figure 3.2, four cases should be discussed.

(i) In M, e_x is a single link, and hence e_x^* is a single link in M^*. Since

$$(\mathscr{P}\gamma x)_{\mathscr{P}_{-x\gamma}} = (\mathscr{P}\gamma x)_{\mathscr{P}^*_{\bullet x}},\ (\mathscr{P}x)_{\mathscr{P}_{-x}} = (\mathscr{P}x)_{\mathscr{P}^*_{\bullet x\gamma}}$$

and $(\mathscr{P}\gamma x)_{\mathscr{P}_{-x}} = (\mathscr{P}\gamma x)_{\mathscr{P}^*_{\bullet x\gamma}}$,

$$(M - e_x)^* = M^* \bullet e_x^*.$$

(ii) In M, e_x is a harmonic link, and hence e_x^* is a harmonic loop in M^* (dually, in M, e_x is a harmonic loop, and hence e_x^* is a harmonic link in M^*). Now, $(x)_{\mathscr{P}\gamma} = (x)_{\mathscr{P}^*}$. Accordingly as e_x is a terminal link or not, a transitive block of M becomes one or two transitive blocks in $M - e_x$. Meanwhile, Accordingly as e_x^* is a terminal loop or not, a transitive block of M^* becomes one or two transitive blocks of $M^* \bullet e_x^*$. By considering the changes in vertices and faces, $(M - e_x)^* = M^* \bullet e_x^*$ is found.

(iii) In M, e_x is a singular link, and hence e_x^* is a singular loop in M^* (dually, in M, e_x is a singular loop, and hence e_x^* is a singular link in M^*). Since

$$(\mathscr{P}^{-1}x)_{\mathscr{P}_{-x}} = (\mathscr{P}^{-1}x)_{\mathscr{P}^*_{\bullet x\gamma}}$$

and

$$(\mathscr{P}^{-1}\gamma x)_{\mathscr{P}_{-x}} = (\mathscr{P}^{-1}\gamma x)_{\mathscr{P}^*_{\bullet x\gamma}},$$

in view of $(\mathscr{P}\gamma x)_{\mathscr{P}_{-x\gamma}} = (\mathscr{P}\gamma x)_{\mathscr{P}^*_{\bullet x}}$, we have $(M - e_x)^* = M^* \bullet e_x^*$.

(iv) In M, e_x is a double loop, and hence e_x^* is a double loop in M^*. Since

$$(\mathscr{P}x)_{\mathscr{P}_{-x}} = (\mathscr{P}x)_{\mathscr{P}^*_{\bullet x\gamma}}$$

and

$$(\mathscr{P}x)_{\mathscr{P}_{-x\gamma}} = (\mathscr{P}x)_{\mathscr{P}^*_{\bullet x}},$$

we have $(M - e_x)^* = M^* \bullet e_x^*$. □

Corollary 3.3 In a map, an edge is a harmonic link if, and only if, the corresponding edge in its dual is a harmonic loop. And an edge is a segmentation edge if, and only if, the corresponding edge in its dual is a shearing loop.

Proof A direct result of Theorem 3.6. □

Example 3.2 Map $M = (Kx + Ky + Kz, \mathscr{P})$, where

$$\mathscr{P} = (x, \beta y, \gamma z)(y, z, \gamma x)$$

and its dual $M^* = (K^*x + K^*y + K^*z, \mathscr{P}^*)$, where

$$\mathscr{P}^* = \mathscr{P}\gamma = (x, y, \alpha x, \alpha z, \gamma y, z),$$

are, respectively, shown in Figure 3.5(a) and (b). Here, $K = \{1, \alpha, \beta, \gamma\}$ and $K^* = \{1, \beta, \alpha, \gamma\}$ are used to distinguish α and β.

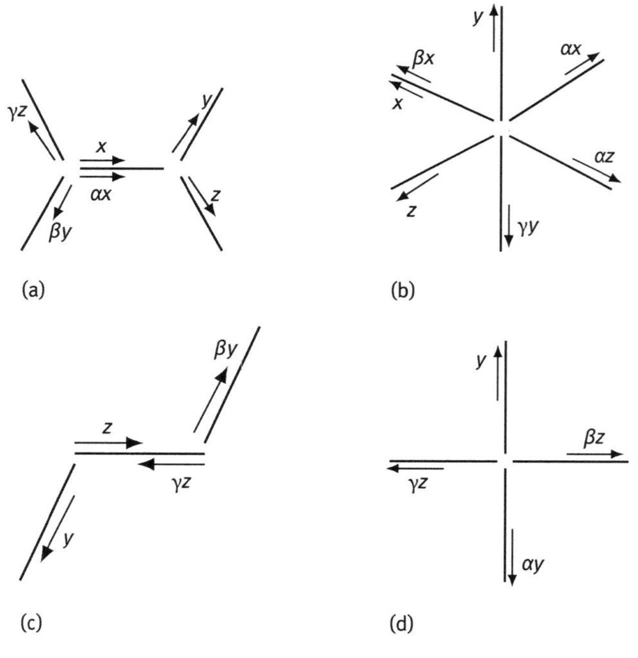

Figure 3.5: Duality between deletion and contraction. (a) M, (b) M^*, (c) $M - e_x$ and (d) $(M - e_x)^*$.

Map $M - e_x = (Ky + Kz, \mathscr{P}_{-x})$, where $\mathscr{P}_{-x} = (\beta y, \gamma z)(y, z)$ and its dual $(M - e_x)^* = (K^*y + K^*z, (\mathscr{P}_{-x})^*)$, where $(\mathscr{P}_{-x})^* = \mathscr{P}_{-x}\gamma = (y, \beta z, \alpha y, \gamma z)$, are, respectively, shown in Figure 3.5(c) and (d). It is easily seen that $(M - e_x)^* = M^* \bullet e_x^*$.

3.3 Addition of an edge

Let $M = (\mathscr{X}_{\alpha,\beta}, \mathscr{P})$ be a premap, $e_x = Kx = \{x, \alpha x, \beta x, \gamma x\}$, and $x \notin \mathscr{X}_{\alpha,\beta}$. Write as

$$M + e_x = (\mathscr{X}_{\alpha,\beta} + Kx, \mathscr{P}_{+x}),$$

where $\mathscr{P}_{+x}$ is determined from $\mathscr{P}$ in the following manner. For any $y \in \mathscr{X}_{\alpha,\beta}$ and two given angles $\langle l, \mathscr{P}\alpha l\rangle$ and $\langle t, \mathscr{P}\alpha t\rangle$, if l and t are not at the same vertex, or at the same vertex and e_x as a harmonic loop (assume $t \in (l)_{\mathscr{P}}$ without loss of generality), then

$$\mathscr{P}_{+x}y = \begin{cases} t \text{ (and } \alpha x), & \text{if } y = x \text{ (and } \alpha t); \\ \mathscr{P}\alpha t \text{ (and } x), & \text{if } y = \alpha x \text{ (and } \alpha\mathscr{P}\alpha t); \\ l \text{ (and } \beta x), & \text{if } y = \gamma x \text{ (and } \alpha l); \\ \mathscr{P}\alpha l \text{ (and } \gamma x), & \text{if } y = \beta x \text{ (and } \alpha\mathscr{P}\alpha l); \\ \mathscr{P}y, & \text{otherwise,} \end{cases} \tag{3.8}$$

shown in Figure 3.6(a), otherwise, i.e., e_x is a double, or singular loop (assume $t \in (l)_{\mathscr{P}}$ without loss of generality),

$$\mathscr{P}_{+x}y = \begin{cases} t \text{ (and } \alpha x), & \text{if } y = x \text{ (and } \alpha t); \\ \mathscr{P}\alpha t \text{ (and } x), & \text{if } y = \alpha x \text{ (and } \alpha\mathscr{P}\alpha t); \\ l \text{ (and } \beta x), & \text{if } y = \gamma x \text{ (and } \alpha l); \\ \mathscr{P}\alpha l \text{ (and } \beta x), & \text{if } y = \gamma x \text{ (and } \alpha\mathscr{P}\alpha l); \\ \mathscr{P}y, & \text{otherwise,} \end{cases} \tag{3.9}$$

as shown in Figure 3.6(b).

Such a transformation from M into $M + e_x$ is called *appending an edge* e_x.

Lemma 3.7 For a premap $M = (\mathscr{X}, \mathscr{P})$, $M + e_x = (\mathscr{X} + Kx, \mathscr{P}_{+x})$ is also a premap. And the number of transitive blocks in $M + e_x$ is not greater than that in M.

Proof From (3.8) and (3.9), $\mathscr{P}_{+x}$ is basic. In virtue of Theorem 2.3, it suffices to show that the orbits of $\mathscr{P}_{+x}$ are partitioned into conjugate pairs for the conjugate axiom. In fact, if l and t are at distinct vertices, then $\mathscr{P}_{+x}$ is obtained from $\mathscr{P}$ in replacing two vertices $(t)_{\mathscr{P}}$ and $(l)_{\mathscr{P}}$ by respective

$$(t)_{\mathscr{P}_{+x}} = (t, \mathscr{P}t, \dots, \alpha\mathscr{P}\alpha t, x)$$

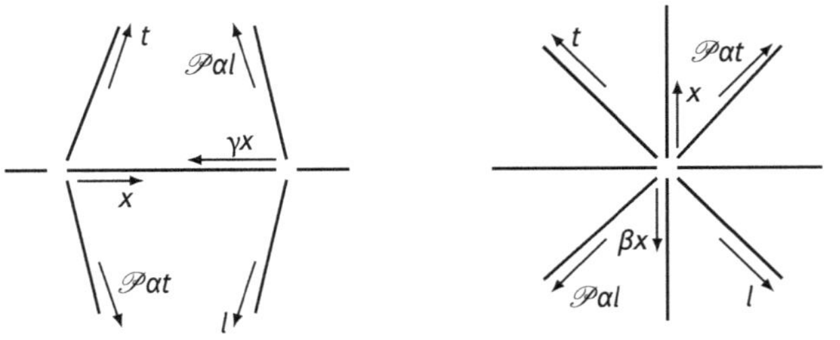

(a) (b)

Figure 3.6: Appending an edge. (a) For (3.8) and (b) For (3.9).

and

$$(l)_{\mathscr{P}_{+x}} = (l, \mathscr{P}l, \ldots, \alpha\mathscr{P}\alpha l, \gamma x),$$

or βx is substituted for γx. If l and t are at the same vertex, then $\mathscr{P}_{+x}$ is obtained from $\mathscr{P}$ in replacing the vertex $(l)_{\mathscr{P}}$ by

$$(l)_{\mathscr{P}_{+x}} = (l, \mathscr{P}l, \ldots, \alpha\mathscr{P}\alpha t, x, t, \mathscr{P}t, \ldots, \alpha\mathscr{P}\alpha l, \gamma x)$$

or

$$(l)_{\mathscr{P}_{+x}} = (l, \mathscr{P}l, \ldots, \alpha\mathscr{P}\alpha t, x, t, \mathscr{P}t, \ldots, \alpha\mathscr{P}\alpha l, \beta x)$$

accordingly as e_x is a harmonic loop or not. This shows that the orbits of $\mathscr{P}_{+x}$ are partitioned into conjugate pairs for α. □

Note 3.2 On the degenerate case $t = l$, if e_x is a harmonic loop, then

$$(l)_{\mathscr{P}_{+x}} = (l, \mathscr{P}l, \ldots, \alpha\mathscr{P}\alpha l, \gamma x, x);$$

otherwise, i.e., e_x is a twist loop,

$$(l)_{\mathscr{P}_{+x}} = (l, \mathscr{P}l, \ldots, \alpha\mathscr{P}\alpha l, \beta x, x).$$

Theorem 3.7 For a premap, not a map, $M = (\mathscr{X}_{\alpha,\beta}, \mathscr{P})$, the number of transitive blocks of $M + e_x$ is less than that of M if, and only if, e_x is a segmentation edge. If M is a map, then $M + e_x$ is also a map.

Proof Since the number of components of graph $G(M + e_x)$ is less than that of $G(M)$ if, and only if, e_x is a cut-edge that is not articulate, the first statement is deduced from Corollary 3.1.

Because the transitivity between two elements in the ground set of M under appending an edge is unchanged, the second statement is valid. □

Note 3.3 Let $M' = (\mathscr{X} + Kx, \mathscr{P}') = M + e_x$ for $M = (\mathscr{X}, \mathscr{P})$. Because $\mathscr{P}'_{-x} = \mathscr{P}$, M is obtained by the deletion of the edge e_x from M', i.e., $M = M' - e_x$. Therefore, the operation of appending an edge on a premap is the inverse of the corresponding edge deletion.

Now, another operation for increasing by an edge on a premap is considered. This is splitting an edge seen as the inverse of edge contraction.

Let $M = (\mathscr{X}_{\alpha,\beta}, \mathscr{P})$ be a premap. Suppose $\langle l, \mathscr{P}\alpha l\rangle$ and $\langle t, \mathscr{P}\alpha t\rangle$ are two angles. For $x \notin \mathscr{X}_{\alpha,\beta}$, let $M \circ e_x = (\mathscr{X}_{\alpha,\beta} + Kx, \mathscr{P}_{\circ x})$, where $\mathscr{P}_{\circ x}$ is determined by $\mathscr{P}$ in the following manner. The transformation from M into $M \circ e_x$ is called *splitting* an edge e_x and e_x, the *splitting edge* of M.

Lemma 3.8 Let $l \in \{t\}_{\mathscr{P}} \cup \{\alpha t\}_{\mathscr{P}}$. If $l \notin (t)_{\mathscr{P}\gamma} \cup (\beta t)_{\mathscr{P}\gamma}$, then

$$\mathscr{P}_{\circ x} y = \begin{cases} \mathscr{P}\alpha t \text{ (or } \alpha x), & \text{if } y = x \text{ (or } \alpha\mathscr{P}\alpha t); \\ l \text{ (or } x), & \text{if } y = \alpha x \text{ (or } \alpha l), \\ \mathscr{P}\alpha l \text{ (or } \beta x), & \text{if } y = \gamma x \text{ (or } \alpha\mathscr{P}\alpha l); \\ t \text{ (or } \gamma x), & \text{if } y = \beta x \text{ (or } \alpha t); \\ \mathscr{P}y, & \text{otherwise.} \end{cases} \tag{3.10}$$

The edge e_x is a single link as shown in Figure 3.7; otherwise, i.e., $l \in (t)_{\mathscr{P}\gamma} \cup (\beta t)_{\mathscr{P}\gamma}$, then

$$\mathscr{P}_{\circ x} y = \begin{cases} \mathscr{P}\alpha t \text{ (or } \alpha x), & \text{if } y = x \text{ (or } \alpha\mathscr{P}\alpha t); \\ l \text{ (or } x), & \text{if } y = \alpha x \text{ (or } \alpha l); \\ t \text{ (or } \beta x), & \text{if } y = \beta x \text{ (or } \alpha t); \\ \mathscr{P}\alpha l \text{ (or } \beta x), & \text{if } y = \gamma x \text{ (or } \alpha\mathscr{P}\alpha l); \\ \mathscr{P}y, & \text{otherwise,} \end{cases} \tag{3.11}$$

or γx is replaced by βx to attain e_x as a singular or harmonic link shown in Figure 3.8.

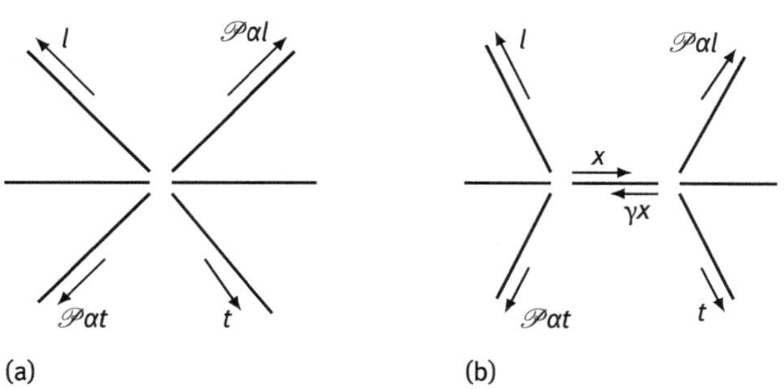

Figure 3.7: $l \notin (t)_{\mathscr{P}\gamma} \cup (\beta t)_{\mathscr{P}\gamma}$. (a) M and (b) $M \circ e_x$.

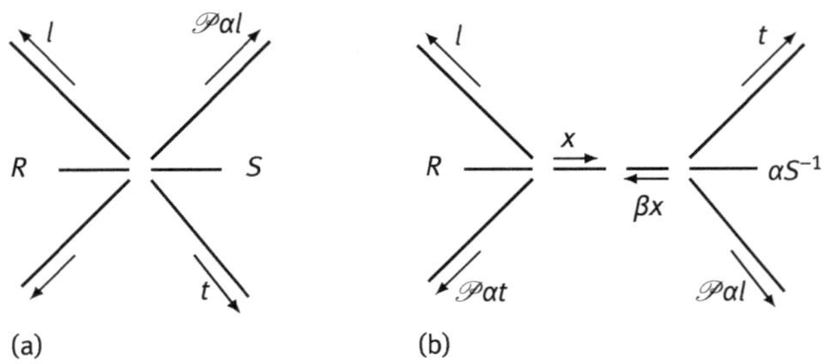

Figure 3.8: $l \in (t)_{\mathscr{P}\gamma} \cup (\beta t)_{\mathscr{P}\gamma}$. (a) M and (b) $M \circ e_x$.

Proof Since $l \in \{t\}_{\mathscr{P}} \cup \{\alpha t\}_{\mathscr{P}}$, l and t are at the same vertex. Thus, e_x is a link. If $l \notin (t)_{\mathscr{P}\gamma} \cup (\beta t)_{\mathscr{P}\gamma}$, i.e., l is in a face different from that t is in, or in other words, e_x is single, then by the reason as $(\mathscr{P}_{\circ x})_{\bullet x}$ is different from only the vertex

$$(\mathscr{P}_{\circ x}x)_{(\mathscr{P}_{\circ x})_{\bullet x}} = (\mathscr{P}\alpha t, \dots, \alpha l, \mathscr{P}\alpha l, \dots, \alpha t),$$

where $\mathscr{P}_{\circ x}x = \mathscr{P}\alpha t$ as shown in Figure 3.6, from Lemma 3.4(i), $(\mathscr{P}_{\circ x})_{\bullet x} = \mathscr{P}$. Otherwise, accordingly as e_x is singular or harmonic, $(\mathscr{P}_{\circ x})_{\bullet x}$ is different from only the vertex

$$(\mathscr{P}_{\circ x}x)_{(\mathscr{P}_{\circ x})_{\bullet x}} = (\mathscr{P}\alpha t, \dots, \alpha l, \mathscr{P}\alpha l, \dots, \alpha t)$$

or

$$(\mathscr{P}\alpha t, \dots, \alpha l, t, \dots, \alpha\mathscr{P}\alpha l)$$

as shown in Figure 3.8. From Lemma 3.4(i) again, $(\mathscr{P}_{\circ x})_{\bullet x} = \mathscr{P}$. □

Lemma 3.9 Let $l \notin \{t\}_{\mathscr{P}} \cup \{\alpha t\}_{\mathscr{P}}$. If t and l are not transitive on M, then

$$\mathscr{P}_{\circ x}y = \begin{cases} \mathscr{P}\alpha l \text{ (or } \alpha x), & \text{if } y = x \text{ (or } \alpha\mathscr{P}\alpha l); \\ t \text{ (or } x), & \text{if } y = \alpha x \text{ (or } \alpha t); \\ \mathscr{P}\alpha t \text{ (or } \beta x), & \text{if } y = \gamma x \text{ (or } \alpha\mathscr{P}\alpha t); \\ l \text{ (or } \gamma x), & \text{if } y = \beta x \text{ (or } \alpha l); \\ \mathscr{P}y, & \text{otherwise,} \end{cases} \tag{3.12}$$

as shown in Figure 3.9, where e_x is a harmonic loop.

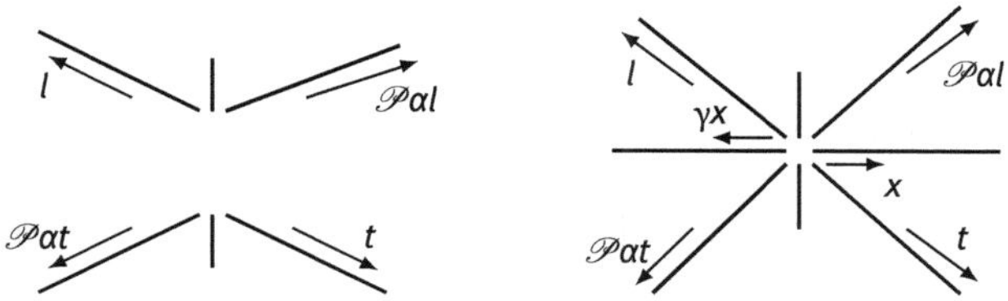

Figure 3.9: t are l nontransitive. (a) M and (b) $M \circ e_x$.

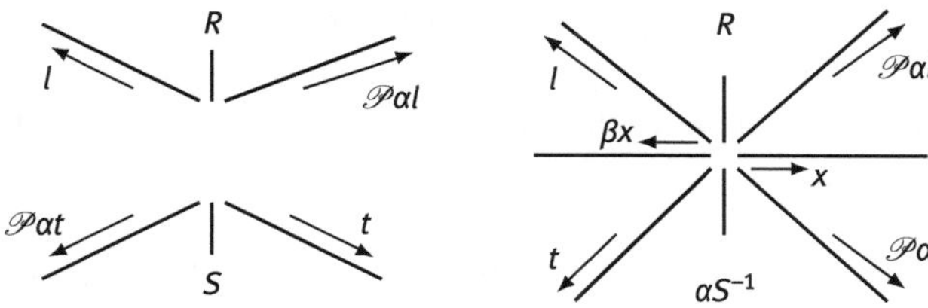

Figure 3.10: t are l nontransitive. (a) M and (b) $M \circ e_x$.

If t and l are transitive on M, then

$$\mathscr{P}_{\circ x}y = \begin{cases} \mathscr{P}\alpha l \text{ (or } \alpha x), & \text{if } y = x \text{ (or } \alpha\mathscr{P}\alpha l); \\ \mathscr{P}\alpha t \text{ (or } x), & \text{if } y = \alpha x \text{ (or } \alpha\mathscr{P}\alpha t); \\ l \text{ (or } \beta x), & \text{if } y = \gamma x \text{ (or } \alpha l); \\ t \text{ (or } \gamma x), & \text{if } y = \beta x \text{ (or } \alpha t); \\ \mathscr{P}y, & \text{otherwise,} \end{cases} \tag{3.13}$$

as shown in Figure 3.10, where e_x is a singular or harmonic loop accordingly as it is incident with two faces or one face.

Proof Since $l \in \{t\}_{\mathscr{P}} \cup \{\alpha t\}_{\mathscr{P}}$, l and t are at the same vertex. Thus, e_x is a link. If $l \notin (t)_{\mathscr{P}\gamma} \cup (\beta t)_{\mathscr{P}\gamma}$, i.e., l is in a face different from that t is in, or in other words, e_x is single, then by the reason as $(\mathscr{P}_{\circ x})_{\bullet x}$ is different from only the vertex

$$(\mathscr{P}_{\circ x}x)_{(\mathscr{P}_{\circ x})_{\bullet x}} = (\mathscr{P}\alpha t, \ldots, \alpha l, \mathscr{P}\alpha l, \ldots, \alpha t),$$

where $\mathscr{P}_{\circ x}x = \mathscr{P}\alpha t$, as shown in Figure 3.7, from Lemma 3.4(i), $(\mathscr{P}_{\circ x})_{\bullet x} = \mathscr{P}$. Otherwise, accordingly as e_x is singular or harmonic, $(\mathscr{P}_{\circ x})_{\bullet x}$ is different from only the vertex

$$(\mathscr{P}_{\circ x}x)_{(\mathscr{P}_{\circ x})_{\bullet x}} = (\mathscr{P}\alpha t, \ldots, \alpha l, \mathscr{P}\alpha l, \ldots, \alpha t)$$

or

$$(\mathscr{P}\alpha t, \ldots, \alpha l, t, \ldots, \alpha\mathscr{P}\alpha l)$$

as shown in Figure 3.8. From Lemma 3.4(i) again, $(\mathscr{P}_{\circ x})_{\bullet x} = \mathscr{P}$. □

Lemma 3.10 If $M = (\mathscr{X}, \mathscr{P})$ is a premap, then for any $x \in \mathscr{X}$, $M \circ e_x = (\mathscr{X} + Kx, \mathscr{P}_{\circ x})$ is also a premap. And $M \circ e_x$ has the number of its transitive blocks not greater than M has.

Proof From Lemmas 3.8–3.9, permutation $\mathscr{P}_{\circ x}$ is basic and partitioned into conjugate pairs for α. Then, by Theorem 2.3, $M \circ x$ is also a premap. This is the first statement.

Because splitting an edge does not change the transitivity of any pair of elements in $\mathscr{X}$, the second statement holds. □

Lemma 3.11 Edge e_x is a harmonic loop on $M \circ e_x$ if, and only if, $\mathscr{P}_{\circ x}x$ and $\mathscr{P}_{\circ x}\gamma x$ are not transitive on M.

Proof Necessity. Because the splitting edge e_x is a harmonic loop on M, it is in the case (3.12). As shown in Figure 3.9, $\mathscr{P}_{\circ x}\beta x (= l)$ and $\mathscr{P}_{\circ x}\alpha x (= t)$ are not transitive on M. Hence, by the symmetry among elements in Kx, $\mathscr{P}_{\circ x}x$ and $\mathscr{P}_{\circ x}\gamma x$ are not transitive on M either.

Sufficiency. Because $\mathscr{P}_{\circ x}\beta x (= l)$ and $\mathscr{P}_{\circ x}\alpha x (= t)$ are not transitive on M, only $l \notin \{t\}_{\mathscr{P}} \cup \{\alpha t\}_{\mathscr{P}}$ is possible. This is the case for (3.12). Thus, e_x is a harmonic loop. □

Theorem 3.8 For a premap $M = (\mathscr{X}_{\alpha,\beta}, \mathscr{P})$ not a map, the number of transitive blocks in $M \circ e_x$ is less than that in M if, and only if, e_x is a harmonic loop. If M is a map, then $M \circ e_x$ is also a map.

Proof From Lemma 3.11, the number of transitive blocks in $M \circ e_x$ is less than that in M if, and only if, e_x is a segmentation edge in $M + e_x$. This is the first statement. Because splitting an edge in a map does not change the transitivity, the second statement is obtained. □

Lemma 3.12 Edge $e_x = \{x, \alpha x, \beta x, \gamma x\}$ is appended in premap M if, and only if, edge $e_x^* = \{x, \beta x, \alpha x, \gamma x\}$ is split to in premap M^*.

Proof Necessity. (1) If e_x is a single edge of $M + e_x$, i.e., $\gamma x \notin (x)_{\mathscr{P}_{+x\gamma}} \cup (\beta x)_{\mathscr{P}_{+x\gamma}}$, then $(x)_{\mathscr{P}^*_{+x}}$ and $(\gamma x)_{\mathscr{P}^*_{+x}}$ are two vertices on $(M + e_x)^*$. Hence, from (3.10) and (3.11), e_x^* is the splitting edge from the vertex

$$(\mathscr{P}^*_{+x}x)_{\mathscr{P}^*} = (\mathscr{P}^*_{+x}x, \ldots, (\mathscr{P}^*_{+x})^{-1}x, \mathscr{P}^*_{+x}\gamma x, \ldots, (\mathscr{P}^*_{+x})^{-1}\gamma x)$$

on M^*.

(2) Otherwise e_x is a double edge on $M + e_x$. Thus, $\gamma x \in (x)_{\mathscr{P}_{+x}\gamma} \cup (\beta x)_{\mathscr{P}_{+x}\gamma}$. Two cases should be considered.

(i) If $\gamma x \in (x)_{\mathscr{P}_{+x}\gamma}$, then e_x is a segmentation edge on $M+e_x$. From Corollary 3.1, $\mathscr{P}^*_{+x}x$ and $\mathscr{P}^*_{+x}\gamma x$ are not transitive on M. Furthermore, from (3.12), e^*_x is a splitting edge (a harmonic loop) on M^*.

(ii) If $\gamma x \in (\beta x)_{\mathscr{P}_{+x}\gamma}$, then from (3.13), e^*_x is a splitting edge on M^*.

Sufficiency. (1) If e^*_x is not a loop on $M^* \circ e^*_x$, then from (3.10–11), M has a face

$$(\mathscr{P}^*_{\circ x}x)_{\mathscr{P}\gamma} = (\mathscr{P}^*_{\circ x}x, \ldots, (\mathscr{P}^*_{\circ x})^{-1}x, \mathscr{P}^*_{\circ x}\gamma x, \ldots, (\mathscr{P}^*_{\circ x})^{-1}\gamma x).$$

From (3.6), e_x is an appending edge between

$$\text{angles } \langle \alpha\mathscr{P}^{*\,-1}_{\circ x}x, \mathscr{P}^*_{\circ x}\gamma x\rangle \text{ and } \langle \alpha\mathscr{P}^{*\,-1}_{\circ x}\gamma x, \mathscr{P}^*_{\circ x}x\rangle$$

in a face of M, as shown in (3.8) and (3.9).

(2) Otherwise e^*_x is a loop. From (3.12) and (3.13), there are two faces on M with an angle each. Such two angles determine the appending edge e_x on M. □

Theorem 3.9 The dual of a premap $M + e_x$ is the premap $M^* \circ e^*_x$, where M^* is the dual of M and e^*_x is the dual edge in $M^* \circ e^*_x$ corresponding to e_x in $M + e_x$.

Proof A directed result of Lemma 3.4. □

From what has been discussed above, both the following diagrams

$$\begin{array}{ccc} (\mathscr{X}_{\alpha,\beta}, \mathscr{P}) & \xrightarrow{-e_x} & (\mathscr{X}_{\alpha,\beta} - Kx, \mathscr{P}_{-e_x}) \\ {\scriptstyle *}\updownarrow & & {\scriptstyle *}\updownarrow \\ (\mathscr{X}^*, \mathscr{P}^*) & \xleftarrow{\circ e^*_x} & (\mathscr{X}^* - K^*x, \mathscr{P}^*_{\bullet e^*_x}) \end{array} \tag{3.14}$$

and

$$\begin{array}{ccc} (\mathscr{X}_{\alpha,\beta}, \mathscr{P}) & \xleftarrow[+e_x]{} & (\mathscr{X}_{\alpha,\beta} - Kx, \mathscr{P}_{-e_x}) \\ {\scriptstyle *}\updownarrow & & {\scriptstyle *}\updownarrow \\ (\mathscr{X}^*, \mathscr{P}^*) & \xrightarrow[\bullet e^*_x]{} & (\mathscr{X}^* - K^*x, \mathscr{P}^*_{\bullet e^*_x}) \end{array} \tag{3.15}$$

are commutative.

Example 3.3 Map $M = (Kx + Ky, \mathscr{P})$, $\mathscr{P} = (x, y)(\gamma y, \gamma x)$ and its dual $M^* = (K^*x + K^*y, \mathscr{P}^*)$,

$$\mathscr{P}^* = \mathscr{P}\gamma = (x, y, \alpha x, \gamma y)(y, \gamma x),$$

are, respectively, shown in Figure 3.11(a) and (b). Notice that α and β have distinguished roles in $K = \{1, \alpha, \beta, \gamma\}$ and $K^* = \{1, \beta, \alpha, \gamma\}$.

Map $M + e_z = (Kx + Ky + Kz, \mathscr{P}_{+z})$,

$$\mathscr{P}_{+z} = (x, y, z)(\gamma x, \beta z, \gamma y)$$

and its dual $(M + e_z)^* = (K^*x + K^*y + K^*z, (\mathscr{P}_{+z})^*)$,

$$(\mathscr{P}_{+z})^* = \mathscr{P}_{+z}\gamma = (x, \beta z, \gamma y, \gamma z)(\gamma x, \beta y),$$

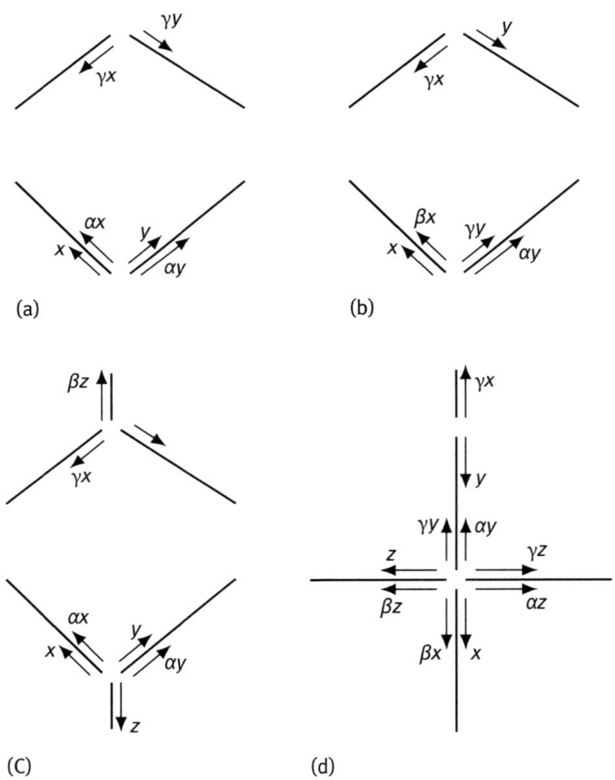

Figure 3.11: Duality between appending and splitting an edge. (a) M, (b) M^*, (c) $M + e_z$ and (d) $(M + e_z)^*$.

are, respectively, shown in Figure 3.11(c) and (d). According to (3.13), e_z^* is the splitting edge at the pair of angles $\langle x, \alpha y\rangle$ and $\langle \gamma y, \beta x\rangle$ in M^*. Therefore,

$$M^* \circ e_z^* = (K^*x + K^*y + K^*z, (z, \beta x, \alpha z, \gamma y)(y, \gamma x)).$$

If βy is seen as y, then $M^* \circ e_z^* = (M + e_z)^*$.

Based on Theorem 3.9, in a premap $M = (\mathscr{X}, \mathscr{P})$, splitting an edge e_x attains $M \circ e_x$, which is just $M^* + e_x^*$ obtained by appending the edge e_x^* in its dual $M^* = (\mathscr{X}^*, \mathscr{P}^*)$.

3.4 Basic transformation

In a premap $M = (\mathscr{X}_{\alpha,\beta}, \mathscr{P})$, the deletion of a single edge e_x is called *basic deleting* an edge and its result is denoted by $M -_{\mathrm{b}} e_x$. The contraction of a link e_x is called *basic contracting* an edge, and its result is denoted by $M \bullet_{\mathrm{b}} e_x$. The two operations are together called *basic subtracting* an edge. Similarly, appending a single edge is called *basic appending* an edge, and splitting a link is called *basic splitting* an edge. Such two operations are together called *basic adding* an edge. Apparently, $M +_{\mathrm{b}} e_x$ and $M \circ_{\mathrm{b}} e_x$ are the results of basic adding an edge e_x on M in their own right. Basic subtracting and basic adding an edge are together called *basic transformation*. From what we have known above, a premap becomes another premap under basic transformation.

Theorem 3.10 Suppose M' is a premap obtained by basic transformation from premap M, then M' is a map if, and only if, M is a map.

Proof Because a single edge is never a harmonic link, from Theorem 3.4, the theorem holds for basic deleting an edge. Because a link is never a harmonic loop, from Theorem 3.5, the theorem holds for basic contracting an edge. Then, from Theorem 3.7–3.8, the theorem holds for basic adding an edge. □

Furthermore, for basic transformation, the following conclusion can also be drawn.

Theorem 3.11 Let $M = (\mathscr{X}, \mathscr{P})$ be a map and $M^* = (\mathscr{X}^*, \mathscr{P}^*)$, its dual. Then, for any single edge e_x in M, $(M -_{\mathrm{b}} e_x)^* = M^* \bullet_{\mathrm{b}} e_x^*$ and for a single edge e_x not in M, $(M +_{\mathrm{b}} e_x)^* = M^* \circ_{\mathrm{b}} e_x^*$. Conversely, for any link e_x in M, $M \bullet_{\mathrm{b}} e_x = M^* -_{\mathrm{b}} e_x^*$ and for a link e_x not in M, $M \circ \mathrm{b}e_x = M^* +_{\mathrm{b}} e_x^*$.

Proof Based on the duality between edges as shown in Figure. 3.2, the statements are meaningful. From Theorem 3.6 and Theorem 3.9, the first statement is true. In virtue of the duality, the second statement is true. □

From this theorem, the following two diagrams are seen to be commutative:

$$\begin{array}{ccc} (\mathscr{X}_{\alpha,\beta}, \mathscr{P}) & \xrightarrow{\quad -_{\mathrm{b}}e_x \quad} & (\mathscr{X}_{\alpha,\beta} - Kx, \mathscr{P}_{-_{\mathrm{b}}e_x}) \\ {\scriptstyle *}\updownarrow & & {\scriptstyle *}\updownarrow \\ (\mathscr{X}^*, \mathscr{P}^*) & \xleftarrow{\quad \circ_{\mathrm{b}}e_x^* \quad} & (\mathscr{X}^* - K^*x, \mathscr{P}^*_{\bullet_{\mathrm{b}}e_x^*}) \end{array} \tag{3.16}$$

and

$$\begin{array}{ccc} (\mathscr{X}_{\alpha,\beta}, \mathscr{P}) & \xleftarrow[\quad +_{\mathrm{b}}e_x \quad]{} & (\mathscr{X}_{\alpha,\beta} - Kx, \mathscr{P}_{-_{\mathrm{b}}e_x}) \\ {\scriptstyle *}\updownarrow & & {\scriptstyle *}\updownarrow \\ (\mathscr{X}^*, \mathscr{P}^*) & \xrightarrow[\quad \bullet_{\mathrm{b}}\, e_x^* \quad]{} & (\mathscr{X}^* - K^*x, \mathscr{P}^*_{\bullet_{\mathrm{b}}e_x^*}) \end{array} \tag{3.17}$$

On the basis of basic transformation, an equivalence can be established for classifying maps in agreement with the classification of surfaces.

3.5 Notes

For a map, if the basic deletion of an edge cannot be done anymore, then the map is said to be *basic deleting edge irreducible*. Similarly, if the basic contraction of an edge cannot be done on a map anymore, then the map is said to be *basic contracting irreducible*.

3.5.1 Given the size, determine the number of self-dual maps as an integral function of the size, or provide a way to list all the self-dual maps of the same size and deduce a relation among the numbers of different sizes.

Notice 3.1 No research result is known for the number yet. It looks the only way known by employing joint trees, as shown topologically in Liu [72, pp. 14–17], combinatorially in Liu [73, pp. 195–201] and algebraically in Liu [77, pp. 25–39] and Liu [79, pp. 20–42].

3.5.2 Given the size, determine the number of maps which are all basic deleting irreducible as an integral function of the size, or provide a way to list all such maps with the same size and deduce a relation among the numbers of different sizes.

Notice 3.2 No research result is known for the number yet.

3.5.3 Given the size, determine the number of maps which are all basic contracting irreducible as an integral function of the size, or provide a way to list all such maps with the same size and deduce a relation among the numbers of different sizes.

Notice 3.3 No research result is known for the number yet.

3.5.4 For any given graph, determine the number of maps which are all basic deleting irreducible with the same under graph, or provide a way to list all such maps with the same size and deduce a relation among the numbers of different sizes.

Notice 3.4 No research result is known for the number yet.

3.5.5 For any given map, determine the number of all basic deleting irreducible maps obtained from the map by basic deletion, or provide a way to list all such maps with the same size and deduce a relation among the numbers of different sizes.

Notice 3.5 No research result is known for the number yet.

3.5.6 For a given graph, determine the number of maps which are all basic contracting irreducible with the same under graph, or provide a way to list such maps and deduce a relation among the numbers of different sizes.

Notice 3.6 No research result is known for the number yet.

3.5.7 For a given map, determine the number of all basic contracting irreducible maps obtained from the map by basic contraction, or provide a way to list such maps and deduce a relation among the numbers of different sizes.

Notice 3.7 No research result is known for the number yet.

If a map is basic both deleting and contracting irreducible, then it is said to be *basic subtracting irreducible*.

3.5.8 Given the size, determine the number of basic subtracting irreducible maps as an integral function of the size, or provide a way to list all such maps with the same size and deduce a relation among the numbers of different sizes.

Notice 3.8 No research result is known for the number yet.

3.5.9 For a given graph, determine the number of maps which are all basic subtracting irreducible with the same under graph, or provide a way to list such maps and deduce a relation among the numbers of different sizes.

Notice 3.9 No research result is known for the number yet.

3.5.10 Find a relation between triangulations and quadrangulations.

Notice 3.10 No valid relationship is known for their numbers yet.

If a map has each of its faces pentagon, then it is called a *quinquangulation*. Similarly, the meaning of a *hexagonalization*.

3.5.11 Justify whether or not a triangulation has a spanning quinquangulation or hexagonalization. If do, determine its number.

Notice 3.11 No relevant result is known for the existence yet.

3.5.12 *Even assigned conjecture*: A bipartite graph without cut-edge has an upper map even assigned.

Notice 3.12 No research result is known for the existence yet. Referred to as in Liu [91, pp. 60, 100 and 102].

4 Orientability

- The orientability is determined by the orientation for each edge with two sides; otherwise, nonorientability.
- The basic equivalence is defined via basic transformations to show that the orientability is an invariant in an equivalent class. This equivalence is, in fact, the elementary equivalence on surfaces.
- The Euler characteristic is also shown to be an invariant in an equivalent class.
- Two examples show that none of the orientability (nonorientability as well) and the Euler characteristic can determine the equivalent class.

4.1 Orientation

Let $M = (\mathscr{X}_{\alpha,\beta}, \mathscr{P})$ be a map (from Theorem 2.6, without loss of generality for a premap), and Ψ_I, $I = \{\gamma, \mathscr{P}\}$, $\gamma = \alpha\beta = \beta\alpha$, be the group generated by the set of permutations I. Now, it is known that the number of orbits of $\mathscr{P}$ on $\mathscr{X}_{\alpha,\beta}$ is double the number of vertices on M and the number of orbits of $\mathscr{P}\gamma$ on $\mathscr{X}_{\alpha,\beta}$ is double the number of faces on M. Because $\mathscr{P}, \mathscr{P}\gamma \in \Psi_I$, the number of orbits of the group Ψ_I on $\mathscr{X}_{\alpha,\beta}$ is not greater than either of the both.

Lemma 4.1 The number of orbits of the group Ψ_I on $\mathscr{X}_{\alpha,\beta}$ is not more than two.

Proof Because $\mathscr{P}\gamma \in \Psi_I$, for any $x \in \mathscr{X}_{\alpha,\beta}$, $\{x\}_{\mathscr{P}\gamma} \subseteq \{x\}_{\Psi_I}$. Here, $\{x\}_{\mathscr{P}\gamma}$ and $\{x\}_{\Psi_I}$ are the orbits of, respectively, the permutation $\mathscr{P}\gamma$ and the group Ψ_I on $\mathscr{X}_{\alpha,\beta}$. For any chosen element $x \in \mathscr{X}_{\alpha,\beta}$, from $\mathscr{P} \in \Psi$, for any $y \in \{x\}_{\mathscr{P}\gamma}$, $\{y\}_{\mathscr{P}\gamma} \subseteq \{x\}_{\Psi_I}$, and from $\gamma \in \Psi$,

$$\{\gamma y\}_{\mathscr{P}\gamma} \subseteq \{\gamma y\}_{\Psi_I} \subseteq \{x\}_{\Psi_I}.$$

In view of Theorem 2.6, at least half of elements at each vertex belong to $\{x\}_{\Psi_I}$. Therefore, $\{x\}_{\Psi_I}$ contains at least half of elements in $\mathscr{X}_{\alpha,\beta}$.

Similarly, $\{\alpha x\}_{\Psi_I}$ contains at least half of elements in $\mathscr{X}_{\alpha,\beta}$.

In consequence, based on the basicness of $\mathscr{P}$ for α, Ψ_I has at most two orbits on $\mathscr{X}_{\alpha,\beta}$. □

According to this lemma, a map $M = (\mathscr{X}_{\alpha,\beta}, \mathscr{P})$ has only two possibilities: group Ψ_I is with two, or one, orbits on $\mathscr{X}_{\alpha,\beta}$. The former is called *orientable*, and the later *nonorientable*.

From the proof of the lemma, an efficient algorithm can be established for determining all the orbits of group Ψ_I on the ground set.

Actually, in an orientable map, because Ψ_I has two orbits for α, the ground set is partitioned into two parts of equal size. It can be seen from Lemma 4.1 that each

DOI 10.1515/9783110481846-004

quadricell (i.e., edge) is distinguished by two elements in each of the two orbits. And the two elements of an edge in the same orbit have to be with different ends of the edge. Thus, each of the two orbits determines the under graph of the map.

Example 4.1 Consider map $M = (\mathscr{X}, \mathscr{P})$, where

$$\mathscr{X} = Kx + Ky + Kz + Ku + Kv + Kw$$

and

$$\mathscr{P} = (x, y, z)(\gamma z, u, v)(\gamma v, \gamma y, w)(\gamma w, \gamma u, \gamma x)$$

as shown in Figure 4.1(a). Its two faces are

$$(x, \gamma w, \gamma v, \gamma z)$$

and

$$(\gamma x, y, w, \gamma u, v, \gamma y, z, u).$$

In fact, for this map, group Ψ_I has two orbits: one is

$$\{x, \gamma w, \gamma v, \gamma z, \gamma x, y, w, \gamma u, v, \gamma y, z, u\}$$

and the other is what is obtained from it by multiplying α to each of its elements. Thus, M is orientable. Figure 4.1(b) shows that M is an embedding of the complete graph of order 4 on the torus $(yuy^{-1}u^{-1})$.

Corollary 4.1 If Ψ_I, $I = \{\mathscr{P}, \alpha\beta\}$, has two orbits on $\mathscr{X}_{\alpha,\beta}$, then they are conjugate for both α and β.

Proof It is known from Lemma 4.1 that the two obits have the same number of elements, i.e., half of $\mathscr{X}_{\alpha,\beta}$. Because $y \in \{x\}_{\Psi_I}$ if, and only if, $\alpha y \in \{\alpha x\}_{\Psi_I}$ and for any Kx,

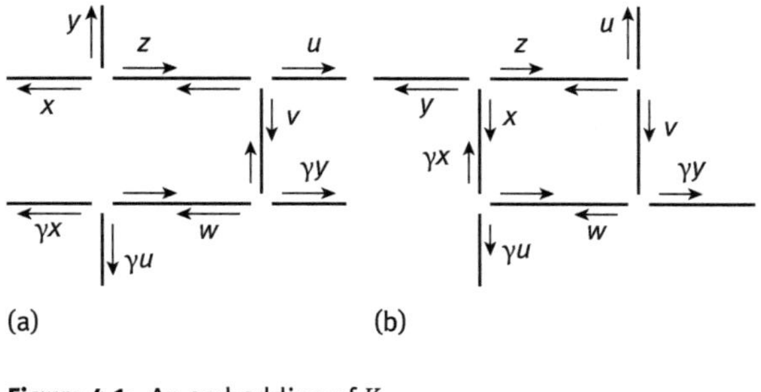

Figure 4.1: An embedding of K_4.

αx and βx are always in the same orbit of Ψ_I, this implies that $\{\alpha x\}_{\Psi_I} = \{\beta x\}_{\Psi_I}$ different from $\{x\}_{\Psi_I}$ and hence the conclusion. □

Example 4.2 Consider map $N = (\mathscr{X}, \mathscr{Q})$, where

$$\mathscr{X} = Kx + Ky + Kz + Ku + Kv + Kw$$

and

$$\mathscr{Q} = (x, y, z)(\gamma z, u, v)(\gamma v, \beta y, w)(\gamma w, \gamma u, \gamma x)$$

as shown in Figure 4.2(a).

That is obtained from the map M in Figure 4.1(a) in the replacement of cycle $(\gamma v, \gamma y, w)$ by cycle $(\gamma v, \beta y, w)$. Here, N has also two faces

$$(x, \gamma w, \gamma v, \gamma z)$$

and

$$(\gamma x, y, \beta v, \alpha u, \beta w, \gamma y, z, u).$$

Because $\beta y \in \{y\}_{\mathscr{Q}\gamma} \subseteq \{y\}_{\Psi_{\{\gamma,\mathscr{Q}\}}}$, from the Corollary 4.1, group $\Psi_{\{\gamma,\mathscr{Q}\}}$ has only one orbit, i.e.,

$$\{x\}_{\Psi_{\{\gamma,\mathscr{Q}\}}} = \mathscr{X}.$$

Therefore, N is nonorientable. It is seen from Figure 4.2(b) that N is an embedding of the complete graph of order 4 on the surface

$$(yuyu^{-1}) \sim_{\text{top}} (yyuu),$$

i.e., Klein bottle.

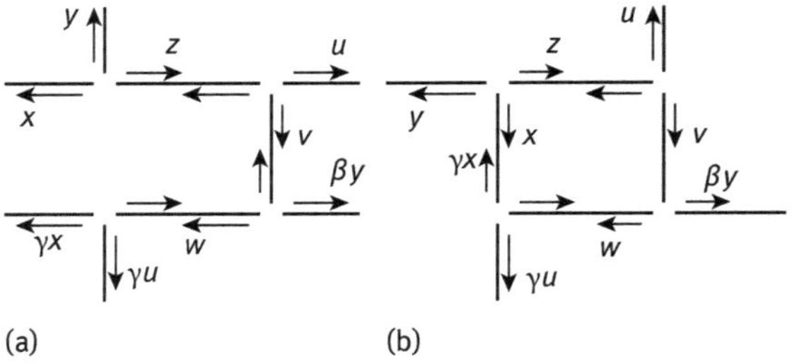

Figure 4.2: An embedding of K_4 on the Klein bottle.

Theorem 4.1 A map $M = (\mathscr{X}, \mathscr{P})$ is nonorientable if, and only if, there exists an element $x \in \mathscr{X}$ such that $\beta x \in \{x\}_{\Psi_I}$, or $\alpha x \in \{x\}_{\Psi_I}$, where $I = \{\gamma, \mathscr{P}\}$.

Proof Necessity. Suppose $\alpha x \notin \{x\}_{\Psi_I}$, then Ψ_I has at least two orbits. However, from Lemma 4.1, it has exactly two orbits. Thus, M is never nonorientable. The necessity holds.

Sufficiency. Because $\beta x \in \{x\}_{\Psi_I}$, from Corollary 4.1, it is only possible to have $\{x\}_{\Psi_I} = \mathscr{X}$, i.e., Ψ_I has only one orbit. Hence, M is nonorientable. This is the sufficiency. □

This theorem enables us to justify the nonorientability and hence the orientability of a map much simple. If there exists a face $(x)_{\mathscr{P}\gamma}$, denoted by $\mathscr{S}_x$, such that $\alpha x \in \mathscr{S}_x$, or there exists a vertex $(x)_{\mathscr{P}}$, denoted by $\mathscr{S}_x$, such that $\beta x \in \mathscr{S}_x$ on M, then M is nonorientable (as shown in Example 4.2). Otherwise, from $y \in \mathscr{S}_x$ via acting $\mathscr{P}$, or γ, for getting $z \notin \mathscr{S}_x$, $\mathscr{S}_x$ is extended into

$$\mathscr{S}_x \cup \{z\}_{\mathscr{P}\gamma} \cup \{z\}_{\mathscr{P}},$$

which is seen as a new $\mathscr{S}$ to see if $y, \alpha y \in \mathscr{S}$, or $y, \beta y \in \mathscr{S}$. If it does, then M is nonorientable; otherwise, do the extension until $|\mathscr{S}| = |\mathscr{X}|/2$, or $\mathscr{S} = \mathscr{X}$.

Theorem 4.2 A map $M = (\mathscr{X}, \mathscr{P})$ is orientable if, and only if, its dual $M^* = (\mathscr{X}^*, \mathscr{P}^*)$ is orientable.

Proof Because $\mathscr{P}^* = \mathscr{P}\gamma \in \Psi_{\{\gamma,\mathscr{P}\}}(\gamma = \alpha\beta = \beta\alpha)$, $\Psi_{\{\gamma,\mathscr{P}\}} = \Psi_{\{\gamma,\mathscr{P}^*\}}$. So, for any $x \in \mathscr{X} = \mathscr{X}^*$, $\{x\}_{\Psi_{\{\gamma,\mathscr{P}\}}} = \{x\}_{\Psi_{\{\gamma,\mathscr{P}^*\}}}$. This is the conclusion of the theorem. □

4.2 Basic equivalence

First, observe the effect for the orientability of a map via basic transformation.

For a map $M = (\mathscr{X}, \mathscr{P})$ and its edge e_x, let $M -_{\text{b}} e_x$ and $M \bullet_{\text{b}} e_x$ be, respectively, obtained by basic deleting and basic contracting the edge e_x on M. From Theorem 3.10, both $M -_{\text{b}} e_x$ and $M \bullet_{\text{b}} e_x$ are maps.

Lemma 4.2 If M' is the map obtained by basic subtracting an edge from M, then M' is orientable if, and only if, M is orientable.

Proof First, to prove the theorem for $M' = M -_{\text{b}} e_x$.

Necessity. From $M' = M - e_x = (\mathscr{X}', \mathscr{P}')$ orientable, group $\Psi' = \Psi_{\{\gamma,\mathscr{P}'\}}$ has two orbits on $\mathscr{X}' = \mathscr{X} - Kx$, i.e., $\{\mathscr{P}x\}_{\Psi'}$ and $\{\mathscr{P}\alpha x\}_{\Psi'}$. Because e_x is single, $\mathscr{P}\gamma x \in \{\mathscr{P}x\}_{\Psi'}$ and $\mathscr{P}\beta x \in \{\mathscr{P}\alpha x\}_{\Psi'}$. So, group $\Psi = \Psi_{\{\gamma,\mathscr{P}\}}$ has two orbits

$$\{x\}_{\Psi} = \{\mathscr{P}x\}_{\Psi'} \cup \{x, \gamma x\}$$

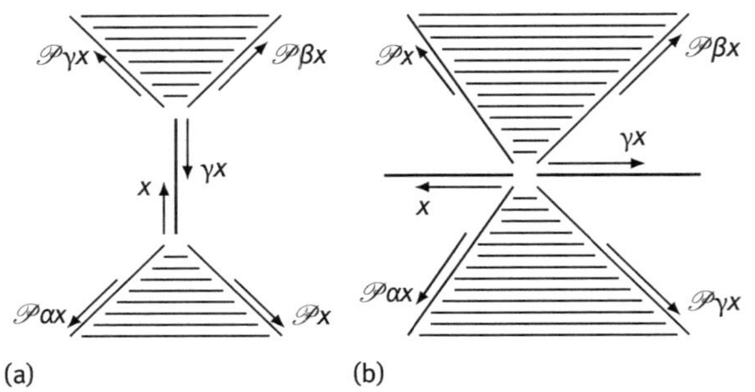

Figure 4.3: Basic deleting an edge. (a) Single link e_x and (b) Single loop e_x.

and

$$\{\alpha x\}_\Psi = \{\mathscr{P}\alpha x\}_{\Psi'} \cup \{\alpha x, \beta x\}$$

on $\mathscr{X}$, i.e., M is orientable.

Sufficiency. Because e_x is a single link (Figure 4.3(a)), or single loop (Figure 4.3(b)), in virtue of that group Ψ has two orbits $\{x\}_\Psi$ and $\{\alpha x\}_\Psi$ on $\mathscr{X}$, group Ψ' has two orbits

$$\{\mathscr{P}x\}_{\Psi'} = \{x\}_\Psi - \{x, \gamma x\}$$

and

$$\{\mathscr{P}\alpha x\}_{\Psi'} = \{\alpha x\}_\Psi - \{\alpha x, \beta x\}$$

on $\mathscr{X}'$, i.e., M' is orientable.

Then, to prove the theorem for $M' = M \bullet_{\text{b}} e_x$. On the basis of Theorem 3.11, the result is directly deduced from that for $M' = M -_{\text{b}} e_x$. □

Whenever the two new angles occur in the deletion of an edge with four angles lost, the edge appending as the inverse of deletion is done between the two angles. And then the same case comes for basic deleting and basic appending an edge. In this sense, Lemma 4.3 in what follows is seen as a direct result of Lemma 4.2. However, it is still proved in an independent way.

For basic appending an edge, since the edge is only permitted to be a single link or a single loop, this operation is, in fact, done by putting the edge in the same face.

Let map $M = (\mathscr{X}, \mathscr{P})$ have a face

$$(y)_{\mathscr{P}\gamma} = (y_0, y_1, \ldots, y_s),$$

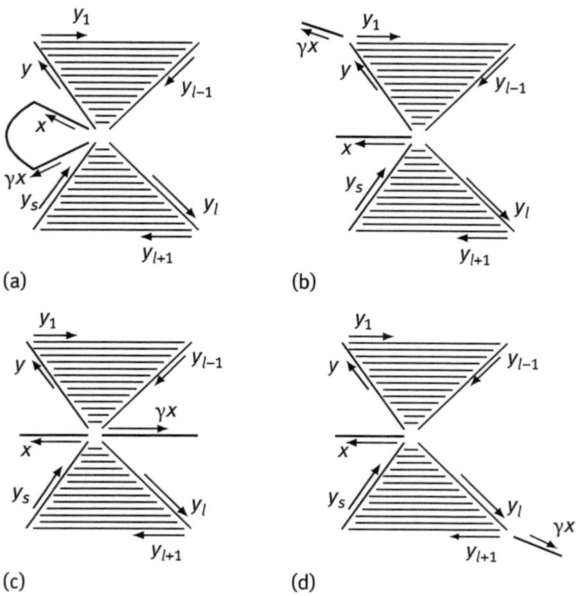

Figure 4.4: Basic appending an edge. (a) $M +_0 e_x$, (b) $M +_1 e_x$, (c) $M +_l e_x$, (d) $M +_{l+1} e_x$.

where $y_0 = y$, $y_1 = (\mathscr{P}\gamma)y, \ldots, y_s = (\mathscr{P}\gamma)^{-1}y$. Denote

$$M +_i e_x = M +_{\mathrm{b}} e_x$$

when appending the edge e_x in between angles $\langle y, \mathscr{P}\alpha y\rangle$ and $\langle y_i, \mathscr{P}\alpha y_i\rangle$, $0 \le i \le s$. From (3.10), $M +_i e_x = M +_{\mathrm{b}} e_x$, $0 \le i \le s$, are all maps (Figure 4.4).

Lemma 4.3 Maps $M +_i e_x = M +_{\mathrm{b}} e_x$, $0 \le i \le s$, are orientable if, and only if, M is orientable.

Proof Necessity. Since $M' = M +_i e_x = (\mathscr{X}', \mathscr{P}')$, $0 \le i \le s$, are all orientable, group $\Psi' = \Psi_{\{\gamma, \mathscr{P}'\}}$ has two orbits $\{x\}_{\Psi'}$ and $\{\alpha x\}_{\Psi'}$ on $\mathscr{X}' = \mathscr{X} + Kx$. Because e_x is a single link (Figure 4.4(a) and 4.4(c)), or single loop (Figure 4.4(b) and 4.4(d)),

$$\mathscr{P}'x \in \{x\}_{\Psi'} \text{ and } \mathscr{P}'\alpha x \in \{\alpha x\}_{\Psi'}.$$

Hence, group $\Psi = \Psi_{\{\gamma, \mathscr{P}\}}$ has two orbits

$$\{\mathscr{P}'x\}_{\Psi} = \{x\}_{\Psi'} - \{x, \gamma x\}$$

and

$$\{\mathscr{P}'\alpha x\}_{\Psi} = \{\alpha x\}_{\Psi'} - \{\alpha x, \beta x\}$$

on $\mathscr{X}$. This implies that M is orientable.

Sufficiency. Since e_x is a single link (Figure 4.4(a) and 4.4(c)), or single loop (Figure 4.4(b) and 4.4(d)), the two orbits

$$\{x\}_{\Psi'} = \{y\}_{\Psi} + \{x, \gamma x\}$$

and

$$\{\alpha x\}_{\Psi'} = \{\alpha y\}_{\Psi} + \{\alpha x, \beta x\}$$

of group Ψ' on $\mathscr{X}'$ are deduced from the two orbits $\{y\}_{\Psi}$ and $\{\alpha y\}_{\Psi}$ of group Ψ on $\mathscr{X}$. Therefore, M' is orientable. □

As for basic splitting an edge, whenever the two new angles occur in the contraction of an edge with four angles lost, the edge splitting seen as the inverse of contraction is done between the two angles.

Next, consider how to list all possibilities for basic splitting from a given angle.

For a map $M = (\mathscr{X}, \mathscr{P})$, let

$$(y)_{\mathscr{P}} = (y_0, y_1, \ldots, y_{l-1}, y_l, \ldots, y_s),$$

$y_0 = y$, $s \geq 0$, be a vertex. Denote by $M \circ_i e_x$ the result obtained from M by basic splitting an edge between angles $\langle y, \mathscr{P}\alpha y\rangle$ and $\langle y_i, \alpha y_{i-1}\rangle$, where $y = y_0$ and $\mathscr{P}\alpha y = \alpha y_s$. From Theorem 3.10, $M \circ_i e_x = M +_{\mathrm{b}} e_x$, $0 \leq i \leq s$, are all maps (Figure 4.5).

Lemma 4.4 For a map $M = (\mathscr{X}, \mathscr{P})$ and $x \notin \mathscr{X}$, map $M \circ_i e_x = M +_{\mathrm{b}} e_x$, $0 \leq i \leq s$, is orientable if, and only if, M is orientable.

Proof Necessity. Because $M' = M \circ_i e_x = (\mathscr{X}', \mathscr{P}')$, $0 \leq i \leq s$, is orientable, group $\Psi' = \Psi_{\{\gamma, \mathscr{P}'\}}$ has two obits $\{x\}_{\Psi'}$ and $\{\alpha x\}_{\Psi'}$ on $\mathscr{X}' = \mathscr{X} + Kx$. Since e_x is a single link (Figure 4.5(b–d)), or a double link (Figure 4.5(a) and 4.5(c)),

$$\mathscr{P}'x \in \{x\}_{\Psi'} \text{ and } \mathscr{P}'\alpha x \in \{\alpha x\}_{\Psi'}.$$

Therefore, group $\Psi = \Psi_{\{\gamma, \mathscr{P}\}}$ has two orbits

$$\{\mathscr{P}'x\}_{\Psi} = \{x\}_{\Psi'} - \{x, \gamma x\} \text{ and } \{\mathscr{P}'\alpha x\}_{\Psi} = \{\alpha x\}_{\Psi'} - \{\alpha x, \beta x\}$$

on $\mathscr{X}$. This implies that M is orientable.

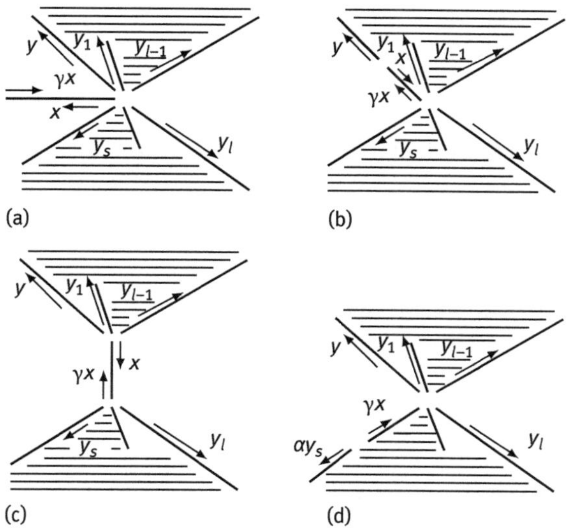

Figure 4.5: Basic splitting an edge. (a) $M \circ_0 e_x$, (b) $M \circ_1 e_x$, (c) $M \circ_l e_x$, (d) $M \circ_s e_x$.

Sufficiency. Because e_x is a single link (Figure 4.5(b–d)), or a double link (Figure 4.5(a) and (c)), the two orbits $\{x\}_{\Psi'} = \{y\}_{\Psi} + \{x, \gamma x\}$ and $\{\alpha x\}_{\Psi'} = \{\alpha y\}_{\Psi} + \{\alpha x, \beta x\}$ of group Ψ' on $\mathscr{X}'$ are deduced from the two orbits $\{y\}_{\Psi}$ and $\{\alpha y\}_{\Psi}$ of group Ψ on $\mathscr{X}$. Therefore, M' is orientable. □

Corollary 4.2 If M' is the map obtained by basic adding an edge from map M, then M' is orientable if, and only if, M is orientable.

Proof A direct result of Lemmas 4.3 and 4.4. □

The operation of basic appending an edge between two successive angles of a face in a map is also called *increasing duplition* (Figure 4.4(b) and (d)), and its inverse operation, *decreasing duplition*. And dually, the operation of basic splitting an edge is also called *increasing subdivision* (Figure 4.5(b) and (d)), and its inverse, *decreasing subdivision*.

Corollary 4.3 A premap M' obtained by increasing duplition, increasing subdivision, decreasing duplition or decreasing subdivision from a map M is still a map with the same orientability of M.

Proof The results for decreasing duplition and decreasing subdivision are derived from Lemma 4.2. Those for increasing duplition and increasing subdivision are derived from Corollary 4.2. □

If map M_1 can be obtained from map M_2 via a series of basic adding and/or basic subtracting an edge, then they are called mutually *basic equivalence*, denoted by $M_1 \sim_{\mathrm{bc}} M_2$.

Theorem 4.3 If maps $M_1 \sim_{\mathrm{bc}} M_2$, then M_1 is orientable if, and only if, M_2 is orientable.

Proof A direct result of Lemma 4.2 and Corollary 4.2. □

Since $\sim_{\mathrm{bc}}$ is an equivalent relation, maps are partitioned into *classes of basic equivalence*, in short *equivalent class*. Theorem 4.3 shows that the orientability of maps is an invariant in the same equivalent class.

4.3 Euler characteristic

For a map $M = (\mathscr{X}, \mathscr{P})$, let $\nu = \nu(M)$, $\epsilon = \epsilon(M)$ and $\phi = \phi(M)$ be, respectively, the order (vertex number), size (edge number) and *co-order* (face number) of M, then

$$\chi(M) = \nu - \epsilon + \phi \tag{4.1}$$

is called the *Euler characteristic* of M.

Theorem 4.4 Let M^* be the dual of a map M, then

$$\chi(M^*) = \chi(M). \tag{4.2}$$

Proof Because $\nu(M^*) = \phi(M)$, $\epsilon(M^*) = \epsilon(M)$ and $\phi(M^*) = \nu(M)$, (4.2) is obtained from (4.1). □

Lemma 4.5 For a map $M = (\mathscr{X}, \mathscr{P})$ and an edge e_x, $x \in \mathscr{X}$, let $M - e_x$ and $M \bullet e_x$ be, respectively, obtained from M by deleting and contracting the edge e_x, then

$$\chi(M) = \begin{cases} \chi(M - e_x), & \text{if } e_x \text{ is single;} \\ \chi(M \bullet e_x), & \text{if } e_x \text{ is a link.} \end{cases} \tag{4.3}$$

Proof From Theorems 3.11 and 4.4, it is only necessary to consider either $M - e_x$ or $M \bullet e_x$. Here, the former is chosen. To prove $\chi(M - e_x) = \chi(M)$ for e_x single.

Because e_x is single, $\nu(M - e_x) = \nu(M)$, $\epsilon(M - e_x) = \epsilon(M) - 1$ and $\phi(M - e_x) = \phi(M) - 1$. From (4.1),

$$\begin{aligned} \chi(M - e_x) &= \nu(M) - (\epsilon(M) - 1) + (\phi(M) - 1) \\ &= \nu(M) - \epsilon(M) + \phi(M) \\ &= \chi(M). \end{aligned}$$

This is just what we wanted to get. □

Corollary 4.4 For any map M, $\chi(M) \leq 2$.

Proof By induction on the co-order $\phi(M)$. If M has only one face, i.e., $\phi(M) = 1$, then

$$\chi(M) = \nu(M) - \epsilon(M) + 1.$$

In view of the connectedness,

$$\epsilon(M) \geq \nu(M) - 1.$$

In consequence,

$$\chi(M) \leq \nu(M) - (\nu(M) - 1) + 1 = 2.$$

Thus, the conclusion is true for $\phi(M) = 1$.

In general, $\phi(M) \geq 2$. Because of the transitivity on a map, there exists a single edge e_x on M. From Lemma 4.5, $M' = M - e_x$ has $\chi(M') = \chi(M)$. Since $\phi(M') = \phi(M) - 1$, by the induction hypothesis $\chi(M') \leq 2$. That is, $\chi(M) \leq 2$, the conclusion. □

For an indifferent reception, the order, size and co-order of a map can be much greater as the map is much enlarged. The conclusion would be unimaginable. In fact, since the deletion of a single edge does not change the connectivity with the Euler characteristic unchanged and the size of a connected graph never less than its order −1, this conclusion becomes reasonable.

Corollary 4.5 For basic subtracting an edge e_x on a map M, $\chi(M -_{\mathrm{b}} e_x) = \chi(M)$ and $\chi(M \bullet_{\mathrm{b}} e_x) = \chi(M)$.

Proof A direct result of Lemma 4.5. □

Lemma 4.6 For a map $M = (\mathscr{X}, \mathscr{P})$ and an edge e_x, where $x \notin \mathscr{X}$, let $M + e_x$ and $M \circ e_x$ be obtained from M via, respectively, appending and splitting the edge e_x, then

$$\chi(M) = \begin{cases} \chi(M + e_x), & \text{if } e_x \text{ is single;} \\ \chi(M \circ e_x), & \text{if } e_x \text{ is a link.} \end{cases} \tag{4.4}$$

Proof From Theorems 3.11 and 4.4, only necessary to consider for one of $M + e_x$ and $M \circ e_x$. The former is chosen. To prove $\chi(M + e_x) = \chi(M)$.

Because e_x is single, so $\nu(M + e_x) = \nu(M)$, $\epsilon(M + e_x) = \epsilon(M) + 1$ and $\phi(M + e_x) = \phi(M) + 1$. From (4.1),

$$\begin{aligned}\chi(M + e_x) &= \nu(M) - (\epsilon(M) + 1) + (\phi(M) + 1)\\ &= \nu(M) - \epsilon(M) + \phi(M)\\ &= \chi(M).\end{aligned}$$

Therefore, the lemma is true. □

Corollary 4.6 For basic adding an edge e_x on a map M, $\chi(M +_{\mathrm{b}} e_x) = \chi(M)$ and $\chi(M \circ_{\mathrm{b}} e_x) = \chi(M)$.

Proof A direct result of Lemma 4.6. □

For a map $M = (\mathscr{X}, \mathscr{P})$ and an edge e_x, where $x \in \mathscr{X}$, let $M_{[\circ x]}$ and $M_{[+x]}$ be obtained from M by, respectively, increasing subdivision and increasing duplition for edge e_x, and $M_{[\bullet x]}$ and $M_{[-x]}$ by, respectively, decreasing subdivision and decreasing duplition for edge e_x. From Corollary 4.3, they are all maps.

Corollary 4.7 For increasing subdivision and increasing duplition,

$$\chi(M_{[\circ x]}) = \chi(M);\ \chi(M_{[+x]}) = \chi(M) \tag{4.5}$$

and for decreasing subdivision and decreasing duplition,

$$\chi(M_{[\bullet x]}) = \chi(M);\ \chi(M_{[-x]}) = \chi(M). \tag{4.6}$$

Proof Because increasing subdivision and increasing duplition are a special type of basic adding an edge, from Corollary 4.5, (4.5) holds. Because decreasing subdivision and decreasing duplition are a special type of basic subtracting an edge, from Corollary 4.6, (4.6) holds. The corollary is obtained. □

The following theorem shows that the Euler characteristic is an invariant in the basic equivalent classes of maps.

Theorem 4.5 If maps $M_1 \sim_{\mathrm{bc}} M_2$, then

$$\chi(M_1) = \chi(M_2). \tag{4.7}$$

Proof Because the basic transformation consists of basic subtracting and basic adding an edge, from Corollary 4.5 and Corollary 4.6, (4.7) is obtained. □

4.4 Pattern examples

Pattern 4.1 Consider the map $M = (\mathscr{X}, \mathscr{P})$, where $\mathscr{X} = Kx + Ky + Kz + Ku + Kw + Kl$ and

$$\mathscr{P} = (x, y, z)(\alpha l, \gamma z, \beta w)(\beta l, \gamma y, \alpha u)(w, \gamma u, \beta x),$$

as shown in Figure 1.13.

By deleting the single edge e_x on M, let $M_1 = (\mathscr{X}_1, \mathscr{P}_1) = M -_{\mathrm{b}} e_x$, then $\mathscr{X}_1 = Ky + Kz + Ku + Kw + Kl$ and

$$\mathscr{P}_1 = (y, z)(\alpha l, \gamma z, \beta w)(\beta l, \gamma y, \alpha u)(w, \gamma u).$$

By contracting the double link e_z on M_1, let $M_2 = (\mathscr{X}_2, \mathscr{P}_2) = M_1 \bullet_{\mathrm{b}} e_z$, then $\mathscr{X}_2 = Ky + Ku + Kw + Kl$ and

$$\mathscr{P}_2 = (y, \beta w, \alpha l)(\beta l, \gamma y, \alpha u)(w, \gamma u).$$

By contracting the double link e_l on M_2, let $M_3 = (\mathscr{X}_3, \mathscr{P}_3) = M_2 \bullet_{\mathrm{b}} e_l$, then $\mathscr{X}_3 = Ky + Ku + Kw$ and $\mathscr{P}_3 = (y, \beta w, \gamma y, \alpha u)(w, \gamma u)$.

By contracting the double link e_u on M_3, let $M_4 = (\mathscr{X}_4, \mathscr{P}_4) = M_3 \bullet_{\mathrm{b}} e_u$, then $\mathscr{X}_4 = Ky + Kw$ and $\mathscr{P}_4 = (y, \beta w, \gamma y, \alpha w)$.

Now, M_4 has only one vertex and only one face and hence any basic transformation for subtracting an edge cannot be done. It is a map on the torus (Figure 4.6).

Pattern 4.2 Again, consider the map $N = (\mathscr{X}, \mathscr{Q})$, where $\mathscr{X} = Kx+Ky+Kz+Ku+Kw+Kl$ and

$$\mathscr{Q} = (x, y, z)(\alpha l, \gamma z, \beta w)(\beta l, \beta y, \alpha u)(w, \gamma u, \beta x),$$

as shown in Figure 1.14.

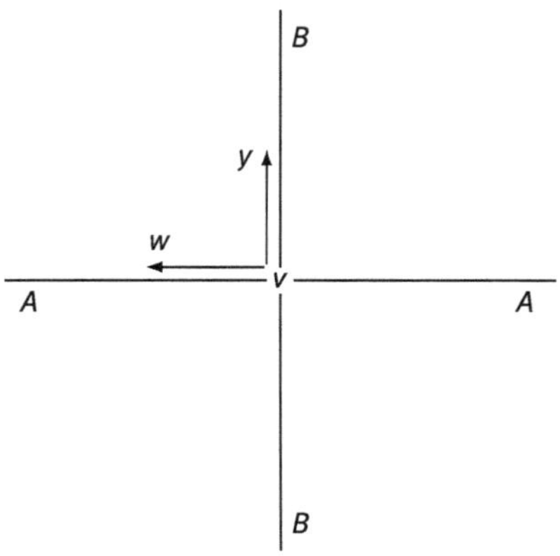

Figure 4.6: A map basic equivalent to M.

By deleting the single edge e_x on M, let $N_1 = (\mathscr{X}_1, \mathscr{Q}_1) = N -_{\rm b} e_x$, then $\mathscr{X}_1 = Ky + Kz + Ku + Kw + Kl$ and

$$\mathscr{Q}_1 = (y, z)(\alpha l, \gamma z, \beta w)(\beta l, \beta y, \alpha u)(w, \gamma u).$$

By contracting the double link e_z on N_1, let $N_2 = (\mathscr{X}_2, \mathscr{Q}_2) = N_1 \bullet_{\rm b} e_z$, then $\mathscr{X}_2 = Ky + Ku + Kw + Kl$ and

$$\mathscr{Q}_2 = (y, \beta w, \alpha l)(\beta l, \beta y, \alpha u)(w, \gamma u).$$

By contracting the double link e_l on N_2, let $N_3 = (\mathscr{X}_3, \mathscr{Q}_3) = N_2 \bullet_{\rm b} e_l$, then $\mathscr{X}_3 = Ky + Ku + Kw$ and

$$\mathscr{Q}_3 = (y, \beta w, \beta y, \alpha u)(w, \gamma u).$$

Finally, by contracting the double link e_u on N_3, let $N_4 = (\mathscr{X}_4, \mathscr{Q}_4) = N_3 \bullet_{\rm b} e_u$, then $\mathscr{X}_4 = Ky + Kw$ and $\mathscr{Q}_4 = (y, \beta w, \beta y, \alpha w)$.

Now, the basic transformation cannot be done anymore on N_4. N_4 is a map on the Klein bottle, as shown in Figure 4.7.

From the two patterns, it is seen that $M_4 \not\sim_{\rm bc} N_4$, and hence $M \not\sim_{\rm bc} N$. Although their Euler characteristic are the same, i.e.,

$$\chi(M) = \chi(M_4) = 1 - 2 + 1 = \chi(N_4) = \chi(N),$$

their orientability are different.

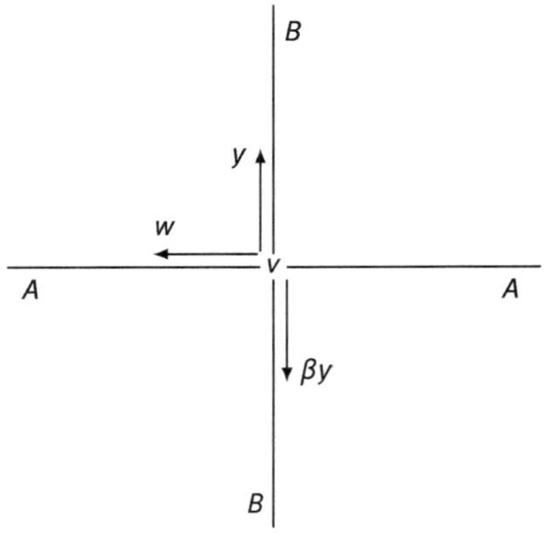

Figure 4.7: A map basic equivalent to N.

4.5 Notes

4.5.1 Characterize a map whose under graph has an upper map with Euler characteristic 1.

Notice 4.1 Only one case, i.e., the projective plane, should be considered. However, it is different from characterizing a graph that is embeddable in the projective plane. Here, we only emphasize the inner constructions of a map itself.

4.5.2 Characterize that the under graph of a map has an upper map with Euler characteristic 0.

Notice 4.2 Two cases have to be considered. One is the torus (the orientable surface of genus 1) and the other, Klein bottle (nonorientable surface of genus 2). However, it is different from characterizing a graph that is embeddable in the torus and the Klein bottle as shown in Notice 4.1.

4.5.3 For any orientable map, characterize that the under graph of the map has an upper orientable map with Euler characteristic 0.

Notice 4.3 The difference from Notice 4.2 above is in that only an orientable map is considered here.

4.5.4 For a vertex regular map and a given integer $g \leq 1$, characterize that the under graph of the map has an upper map with Euler characteristic g.

Notice 4.4 A generalization of 4.5.1.

4.5.5 For a vertex regular orientable map and a given integer $g \leq 0$, characterize that the under graph of the map has an upper map with its Euler characteristic $2g$.

Notice 4.5 A generalization of 4.5.2.

A graph which has a spanning circuit is called a *Hamiltonian graph*. Such a spanning circuit is called a *Hamiltonian circuit* of the graph. If a map has its under graph Hamiltonian, then it is called a *Hamiltonian map*.

4.5.6 For a Hamiltonian map and a given integer $g \leq 1$, characterize that the under graph of the map has an upper nonorientable map with Euler characteristic g.

Notice 4.6 A restriction of 4.5.4 for Hamiltonian maps.

4.5.7 For a Hamiltonian map and a given integer $g \leq 0$, characterize that the under graph of the map has an upper orientable map with Euler characteristic $2g$.

Notice 4.7 A restriction of 4.5.6 for Hamiltonian maps.

For a vertex 3-map (or cubic map), if it has only i-face and j-face, $i \neq j$, $i, j \geq 3$, then it is called an *$(i, j)_f$-map*, or *$(i, j)_f$-gulation*.

4.5.8 For a given integer $g \leq 1$, determine the number of $(3, 4)_f$-map of order n ($n \geq 1$) with Euler characteristic g.

Notice 4.8 Referred to as in Liu [67, pp. 117–141], Dong and Liu [21], Cai and Liu [10, 11], Ren and Liu [132, 133], Pan and Liu [130] and Liu [86, 87], etc.

4.5.9 For a given integer $g \leq 1$, determine the number of $(4, 5)_f$-map of order n ($n \geq 1$) with Euler characteristic g.

Notice 4.9 On the basis of Notice 4.8, referred to as in Liu [78, pp. 97–119, 85, 97; pp. 171–209].

4.5.10 For a given integer $g \leq 1$, determine the number of $(5, 6)_f$-map of order n ($n \geq 1$) with Euler characteristic g.

Notice 4.10 On the basis of Notice 4.9, referred to as in Liu [97, pp. 339–436].

4.5.11 Given a graph G of order n ($n \geq 4$), determine the condition for G to have an upper $(n-1, n)_f$-map.

Notice 4.11 On the basis of 4.5.8 and 4.5.9, try to find a way for doing it.
On a $(n-1, n)_f$-map of order n ($n \geq 4$), let ϕ_1 be the number of $(n-1)$-faces. If its Euler characteristic is $g \leq 1$, then n and ϕ_1 should satisfy the following condition:

$$(n-1)|(n(n-g)+\phi_1), \qquad (4.8)$$

i.e., $n-1$ is a factor of $n(n-g)+\phi_1$.

4.5.12 Given an integer $g \leq 1$, for any positive numbers n and ϕ_1 satisfying (4.8), determine if there exists a $(n-1, n)_f$-map with Euler characteristic g.

Notice 4.12 No theoretical investigation has been carried out yet.

5 Orientable Maps

- Any irreducible orientable map under basic subtracting edges is said to be a butterfly. However, an equivalent class may have more than one butterflies.
- The simplified butterflies are for the standard orientable maps to show that each equivalent class has at most one simplified butterfly.
- Reduced rules are for transforming a map (not unnecessarily orientable) into another butterfly, if orientable, in the same equivalent class. A basic rule is extracted for deriving all other rules.
- Only principles for orientable maps are capable of transforming any map to a simplified butterfly in the same equivalent class. Hence, each equivalent class has at least one simplified butterfly.
- Orientable genus instead of the Euler characteristic is an invariant in an equivalent class to show that orientable genus itself determines the equivalent class.

5.1 Butterflies

On the basis of Chapter 4, this chapter discusses orientable maps with a standard form in each of basic equivalent classes. If an orientable map has only one vertex and only one face, then it is called a *butterfly*.

Lemma 5.1 In each of basic equivalent classes, there exists a map with only one vertex.

Proof For a map $M = (\mathscr{X}_{\alpha,\beta}, \mathscr{P})$, if M has at least two vertices, from the transitive axiom, there exists an $x \in \mathscr{X}_{\alpha,\beta}$ such that $(x)_{\mathscr{P}}$ and $(\gamma x)_{\mathscr{P}}$, $\gamma = \alpha\beta$, determine two distinct vertices. Because e_x is a link, by basic contracting e_x, $M' = M \bullet_{\mathrm{b}} e_x \sim_{\mathrm{bc}} M$. Then, M' has one vertex less than M does. In view of Theorem 3.10, M' is also a map. If M' does not have only one vertex, the procedure is permitted to go on with M' instead of M. By the finite recursion principle, a map M' with only one vertex can be found such that $M' \sim_{\mathrm{bc}} M$. This is the lemma. □

A map with only one vertex is also called a *single vertex map*, or in brief, a *petal bundle*.

Lemma 5.2 In a basic equivalent class of maps, there exists a map with only one face.

Proof For a map $M = (\mathscr{X}_{\alpha,\beta}, \mathscr{P})$, if M has at least two faces, from the transitive axiom, there exists an $x \in \mathscr{X}_{\alpha,\beta}$ such that $(x)_{\mathscr{P}\gamma}$ and $(\gamma x)_{\mathscr{P}\gamma}$, $\gamma = \alpha\beta$, determine two distinct faces. Because e_x is single, by basic deleting e_x, $M' = M -_{\mathrm{b}} e_x \sim_{\mathrm{bc}} M$. Now, M' has

DOI 10.1515/9783110481846-005

one face less than M has. From Theorem 3.10, M' is still a map. Thus, if M' is not with only one face, this procedure is allowed to go on with M' instead of M. By the finite recursion principle, a map M' with only one face can be finally found such that $M' \sim_{\rm bc} M$. The lemma is proved. □

In fact, on the basis of Theorem 3.6, Lemmas 5.1 and 5.2 are mutually dual. Furthermore, what should be noticed is the independence of the orientability for the two lemmas.

Theorem 5.1 For any orientable map M, there exists a butterfly H such that $H \sim_{\rm bc} M$.

Proof If M has at least two vertices, from Lemma 5.1, there exists a single vertex map $L \sim_{\rm bc} M$. In virtue of Theorem 4.3, L is still orientable. If L has at least two faces, from Lemma 5.2, there exists a single face map $H \sim_{\rm bc} L$. In virtue of Theorem 4.3, H is still orientable. Since H has, finally, both one vertex and one face, H is a butterfly. Therefore, $H \sim_{\rm bc} L \sim_{\rm bc} M$. This is the theorem. □

This theorem enables us only to discuss butterflies for the basic equivalence classes of maps without loss of generality.

5.2 Simplified butterflies

Let $O_k = (\mathscr{X}_k, \mathscr{J}_k)$, $k \geq 0$, where

$$\mathscr{X}_k = \begin{cases} \emptyset, & k = 0; \\ \sum_{i=1}^{k} (Kx_i + Ky_i), & k \geq 1 \end{cases} \tag{5.1}$$

and

$$\mathscr{J}_k = \begin{cases} (\emptyset), & k = 0; \\ \left(\prod_{i=1}^{k} \langle x_i, y_i, \gamma x_i, \gamma y_i \rangle \right), & k \geq 1. \end{cases} \tag{5.2}$$

It is easy to check that all O_k, where $k \geq 0$, are maps. And, they are called *O-standard maps*. When $k = 1$ and 2, O_1 and O_2 are, respectively, given in Figure 5.1(a) and (b).

Note 5.1 When $k = 0$, $O_0 = (\emptyset, \emptyset)$ is seen as the degenerate case of a map with no edge. For example, what is obtained by basic deleting an edge on $\hat{L}_0 = (Kx, (x, \gamma x))$ is just O_0. Usually, it is seen as the map with only one vertex without edge, and is called the *trivial map*.

Lemma 5.3 For any $k \geq 0$, O-standard map O_k is orientable.

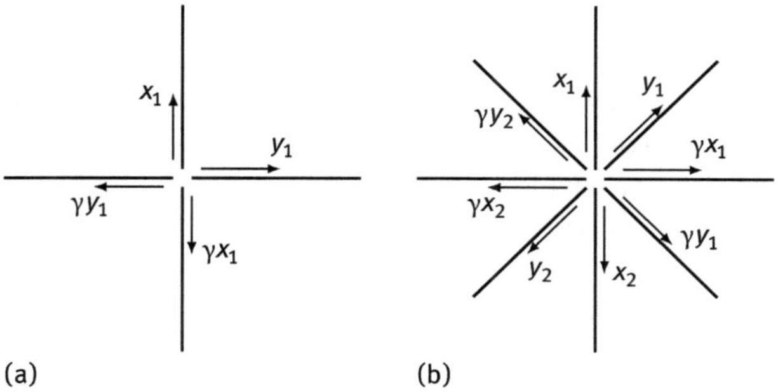

Figure 5.1: Two O-standard maps. (a) O_1 and (b) O_2.

Proof When $k = 0$, from $O_0 = (Kx, (x, \gamma x)) -_{\mathrm{b}} e_x$,

$$O_0 \sim_{\mathrm{bc}} (Kx, (x, \gamma x)).$$

Because $\{x\}_{\Psi_{\{(x,\gamma x),\gamma\}}} = \{x, \gamma x\}$ and $\{\alpha x\}_{\Psi_{\{(x,\gamma x),\gamma\}}} = \{\alpha x, \beta x\}$ are two orbits, $(Kx, (x, \gamma x))$ is orientable. In view of Theorem 4.3, O_0 is orientable.

For $k \geq 1$, assume, by induction, that $O_{k-1} = (\mathscr{X}_{k-1}, \mathscr{J}_{k-1})$ is orientable. From (5.1) and (5.2), group $\Psi_{\{\mathscr{J}_{k-1},\gamma\}}$ has two orbits as

$$\{x_1\}_{\Psi_{\{\mathscr{J}_{k-1},\gamma\}}} = \{x_i, \gamma x_i | 1 \leq i \leq k-1\}$$

and

$$\{\alpha x_1\}_{\Psi_{\{\mathscr{J}_{k-1},\gamma\}}} = \{\alpha x_i, \beta x_i | 1 \leq i \leq k-1\}.$$

For $O_k = (\mathscr{X}_k, \mathscr{J}_k)$, from (5.2), $\mathscr{J}_k = (\langle \mathscr{J}_{k-1} \rangle, x_k, y_k, \gamma x_k, \gamma y_k)$. Group $\Psi_{\{\mathscr{J}_k,\gamma\}}$ has only two orbits as

$$\begin{aligned}\{x_1\}_{\Psi_{\{\mathscr{J}_k,\gamma\}}} &= \{x_1\}_{\Psi_{\{\mathscr{J}_{k-1},\gamma\}}} \cup \{x_k, y_k, \gamma x_k, \gamma y_k\} \\ &= \{x_i, \gamma x_i | 1 \leq i \leq k-1\} \cup \{x_k, y_k, \gamma x_k, \gamma y_k\} \\ &= \{x_i, \gamma x_i | 1 \leq i \leq k\}\end{aligned}$$

and

$$\begin{aligned}\{\alpha x_1\}_{\Psi_{\{\mathscr{J}_k,\gamma\}}} &= \{\alpha x_1\}_{\Psi_{\{\mathscr{J}_{k-1},\gamma\}}} \cup \{\alpha x_k, \alpha y_k, \beta x_k, \beta y_k\} \\ &= \{\alpha x_i, \beta x_i | 1 \leq i \leq k-1\} \cup \{\alpha x_k, \alpha y_k, \beta x_k, \beta y_k\} \\ &= \{\alpha x_i, \beta x_i | 1 \leq i \leq k\}.\end{aligned}$$

Therefore, O_k, $k \geq 1$, are all orientable. □

Lemma 5.4 For any $k \geq 0$, O-standard map O_k has only one face.

Proof When $k = 0$, since $O_0 = (Kx, (x, \gamma x)) -_{\mathrm{b}} e_x$, O_0 should have one face that is composed of the two faces (x) and (αx) of $(Kx, (x, \gamma x))$. Therefore, O_0 has only one face (seen as a degenerate case because of no edge).

For any $O_k = (\mathscr{X}_k, \mathscr{J}_k)$, $k \geq 1$, from (5.1) and (5.2),

$$(x_1)_{\mathscr{J}_k\gamma} = (x_1, \gamma y_1, \gamma x_1, y_1, \ldots, x_k, \gamma y_k, \gamma x_k, y_k)$$

is a face of O_k. However, since

$$|\{x_1\}_{\mathscr{J}_k\gamma}| = \frac{1}{2}|\mathscr{X}_k|,$$

O_k has only this face. □

From (5.2), each O-standard map has only one vertex (O_0 is the degenerate case of no incident edge). Based on the above two lemmas, any O-standard map is a butterfly. Because of the simplicity in form for them, they are called *simplified butterflies*. Since for any $k \geq 0$, simplified butterfly O_k has $2k$ edges, one vertex and one face, its Euler characteristic is

$$\chi(O_k) = 2 - 2k. \tag{5.3}$$

Lemma 5.5 For any two simplified butterflies O_i and O_j, $i, j \geq 0$, $O_i \sim_{\mathrm{bc}} O_j$ if, and only if, $i = j$.

Proof Because the sufficiency, i.e., the former $O_i \sim_{\mathrm{el}} O_j$ is derived from the latter $i = j$, is natural, only necessary to prove the necessity.

By contradiction. Suppose $i \neq j$, but $O_i \sim_{\mathrm{bc}} O_j$. Because of the basic equivalence, from Theorem 4.5,

$$\chi(O_i) = \chi(O_j).$$

However, from (5.3) and the condition $i \neq j$,

$$\chi(O_i) = 2 - 2i \neq 2 - 2j = \chi(O_j).$$

This is a contradiction. □

Theorem 5.2 In each of the basic equivalent classes of orientable maps, there is at most one map that is a simplified butterfly.

Proof By contradiction. Suppose simplified butterflies O_i and O_j, $i \neq j$, $i, j \geq 0$, are in the same class. However, this is a contradiction to Lemma 5.5. □

In the following two sections, it will be seen that in each basic equivalent class of orientable maps, there is at least one map that is a simplified butterfly.

On the basis of Theorem 4.5, two butterflies of the same size have the same Euler characteristic. Are they all simplified butterflies? However, the answer is negative!

Example 5.1 Observe map $M = (\mathscr{X}, \mathscr{J})$, where

$$\mathscr{X} = Kx_1 + Ky_1 + Kx_2 + Ky_2$$

and

$$\mathscr{J} = (x_1, y_1, x_2, y_2, \gamma x_1, \gamma y_1, \gamma x_2, \gamma y_2).$$

Because the face

$$(x_1)_{\mathscr{J}\gamma} = (x_1, \gamma y_1, x_2, \gamma y_2, \gamma x_1, y_1, \gamma x_2, y_2)$$

has eight elements, half the elements of ground set, M has only one face. Hence, M is a butterfly, but not a simplified butterfly. Actually, the simplified butterfly with the same Euler characteristic of M is O_2.

5.3 Reduced rules

Although butterflies are necessary to find a representative for each basic equivalent class of orientable maps, single vertex maps are restricted in this section for such a classification based on Lemma 5.1.

For convenience, the basic equivalence between two maps are not distinguished from that between their basic permutations. In other words, $(\mathscr{X}_1, \mathscr{P}_1) \sim_{\mathrm{bc}} (\mathscr{X}_2, \mathscr{P}_2)$ stands for $\mathscr{P}_1 \sim_{\mathrm{bc}} \mathscr{P}_2$.

Lemma 5.6 For a single vertex map $M = (\mathscr{X}, \mathscr{J})$, if

$$\mathscr{J} = (R, x, \gamma x, S),$$

where R and S are two linear orders on $\mathscr{X}$, then

$$\mathscr{J} \sim_{\mathrm{bc}} (R, S), \tag{5.4}$$

as shown in Figure 5.2(a) and (b).

Proof Because $\mathscr{J} = (R, x, \gamma x, S)$,

$$(\mathscr{J}\gamma)\gamma x = \mathscr{J}x = \gamma x,$$

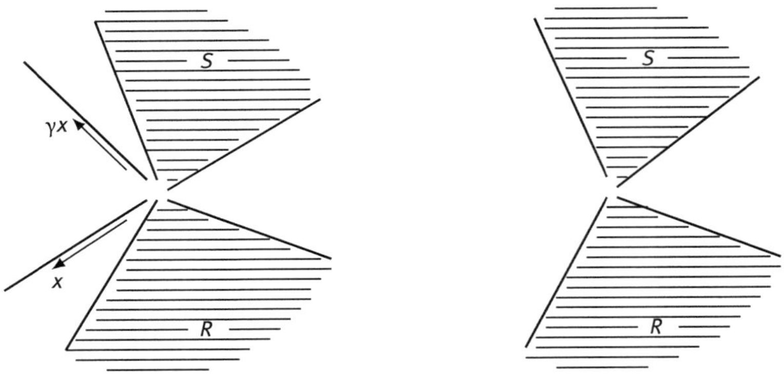

(a) (b)

Figure 5.2: Reduced rule (5.4).

i.e., $(\gamma x)_{\mathscr{J}\gamma} = (\gamma x)$ is a face. Because e_x is a single edge, by basic deleting e_x on M,

$$M' = M -_{\mathrm{b}} e_x = (\mathscr{X} - Kx, \mathscr{J}'),$$

where $\mathscr{J}' = (R, S)$.

From $M \sim_{\mathrm{bc}} M'$, $\mathscr{J} \sim_{\mathrm{bc}} \mathscr{J}' = (R, S)$. □

This lemma enables us to transform a single vertex map into another single vertex map by one face decreased in a basic equivalent class.

Lemma 5.7 For $(\mathscr{X}_{\alpha,\beta}, \mathscr{J})$, if $\mathscr{J} = (A, x, B, y, C, \gamma x, D, \gamma y, E)$, where A, B, C, D and E are all linear orders on $\mathscr{X}$, then

$$\mathscr{J} \sim_{\mathrm{bc}} (A, D, C, B, E, x, y, \gamma x, \gamma y), \tag{5.5}$$

as shown in Figure 5.3(a) and (b).

Proof Four steps are considered for each step as a claim. □

Claim 1 $\mathscr{J} \sim_{\mathrm{bc}} (E, A, x, z, D, C, \gamma x, \gamma z, B)$.

Proof For the angle pair $(\alpha x, \mathscr{J} x)$ and $(\beta x, \mathscr{J} \gamma x)$ of

$$\mathscr{J} = (A, x, B, y, C, \gamma x, D, \gamma y, E),$$

by basic splitting e_z (a link), get

$$\mathscr{J} \sim_{\mathrm{bc}} \mathscr{J}_1 = (D, \gamma y, E, A, x, z)(\gamma z, B, y, C, \gamma x).$$

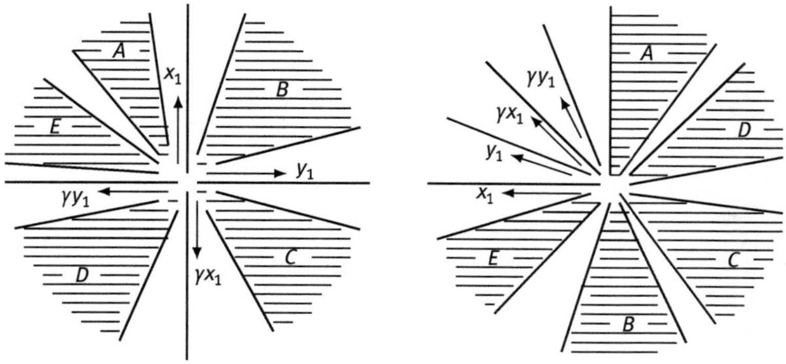

(a) (b)

Figure 5.3: Reduced rule (5.5).

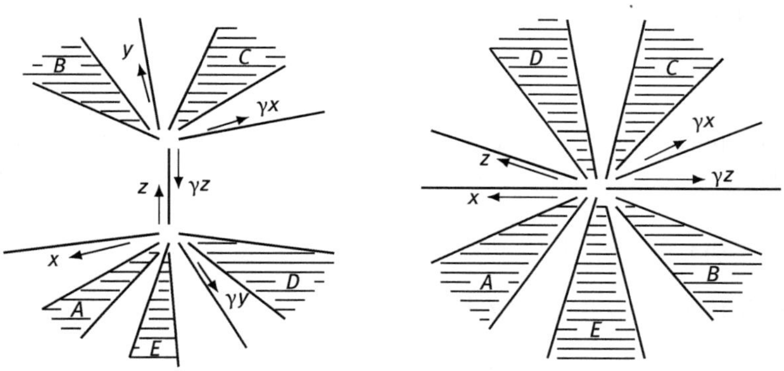

(a) (b)

Figure 5.4: For Claim 1.

Then, since e_y is a link, by basic contracting e_y on $\mathcal{J}_1$, get

$$\mathcal{J}_1 \sim_{\text{bc}} \mathcal{J}_2 = (E, A, x, z, D, C, \gamma x, \gamma z, B).$$

This is the conclusion of Claim 1. Here, $\mathcal{J}_1$ and $\mathcal{J}_2$ are, respectively, shown in Figure 5.4(a) and (b).

□

Claim 2 $\mathcal{J}_2 \sim_{\text{bc}} (y, A, x, B, E, \gamma y, D, C, \gamma x)$.

Proof For the pair of angle $\langle \alpha z, \mathcal{J}_2 z \rangle$ and angle between E and A on $\mathcal{J}_2$ = $(E, A, x, z, D, C, \gamma x, \gamma z, B)$, by basic splitting e_y (a link), get

$$\mathcal{J}_2 \sim_{\text{bc}} \mathcal{J}_3 = (A, x, z, y)(\gamma y, D, C, \gamma x, \gamma z, B, E).$$

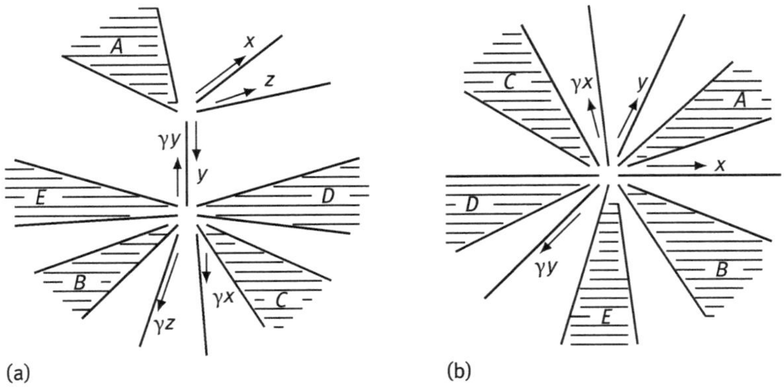

Figure 5.5: For Claim 2.

Then, since e_z is a link, by basic contracting e_z on $\mathcal{J}_3$, get

$$\mathcal{J}_3 \sim_{\text{bc}} \mathcal{J}_4 = (y, A, x, B, E, \gamma y, D, C, \gamma x).$$

This is the conclusion of Claim 2. Here, $\mathcal{J}_3$ and $\mathcal{J}_4$ are, respectively, shown in Figure 5.5(a) and (b). □

Claim 3 $\mathcal{J}_4 \sim_{\text{bc}} (B, E, z, A, y, \gamma z, \gamma y, D, C)$.

Proof For the angle pair $(\alpha y, \mathcal{J}_4 y)$ and $(\alpha \mathcal{J}_4^{-1} \gamma y, \gamma y)$ on

$$\mathcal{J}_4 = (y, A, x, B, E, \gamma y, D, C, \gamma x),$$

by basic splitting e_z (a link), get

$$\mathcal{J}_4 \sim_{\text{bc}} \mathcal{J}_5 = (A, x, B, E, z)(\gamma z, \gamma y, D, C, \gamma x, y).$$

Then, since e_x is a link, by basic contracting e_x on $\mathcal{J}_3$, get

$$\mathcal{J}_5 \sim_{\text{bc}} \mathcal{J}_6 = (B, E, z, A, y, \gamma z, \gamma y, D, C).$$

This is the conclusion of Claim 3. Here, $\mathcal{J}_5$ and $\mathcal{J}_6$ are, respectively, shown in Figure 5.6(a) and (b). □

Claim 4 $\mathcal{J}_6 \sim_{\text{bc}} (A, D, C, B, E, z, x, \gamma z, \gamma x)$.

Proof For the angle pair $(\alpha z, \mathcal{J}_6 z)$ and $(\alpha y, \gamma z)$ of

$$\mathcal{J}_6 = (B, E, z, A, y, \gamma z, \gamma y, D, C),$$

by basic splitting e_x (a link), get

$$\mathscr{J}_6 \sim_{\mathrm{bc}} \mathscr{J}_7 = (A, y, x)(\gamma x, \gamma z, \gamma y, D, C, B, E, z).$$

Then, since e_y is a link, by basic contracting e_y on $\mathscr{J}_7$, get

$$\mathscr{J}_7 \sim_{\mathrm{bc}} \mathscr{J}_8 = (A, D, C, B, E, z, x, \gamma z, \gamma x).$$

This is the conclusion of Claim 4. Here, $\mathscr{J}_7$ and $\mathscr{J}_8$ are, respectively, shown in Figure 5.7(a) and (b).

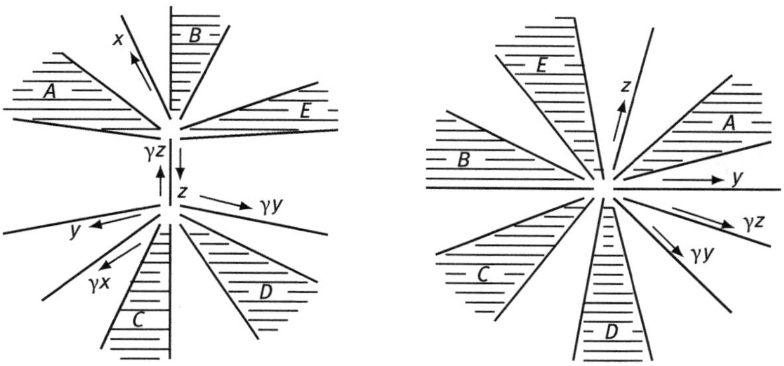

(a) (b)

Figure 5.6: For Claim 3.

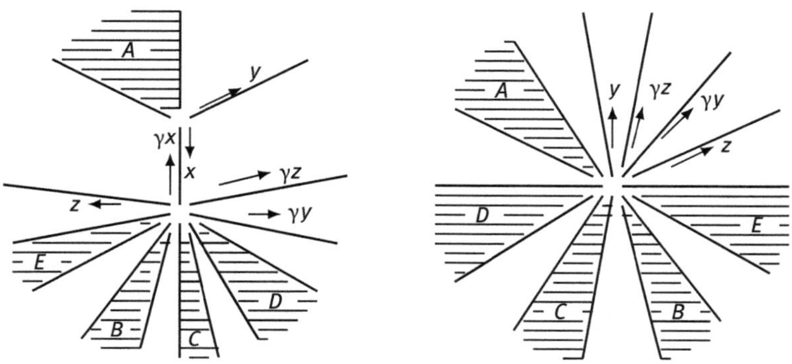

(a) (b)

Figure 5.7: For Claim 4.

On the basis of the four claims above,

$$\mathcal{J} \sim_{\text{bc}} \mathcal{J}_2 \sim_{\text{bc}} \mathcal{J}_4 \sim_{\text{bc}} \mathcal{J}_6 \sim_{\text{bc}} \mathcal{J}_8$$
$$= (A, D, C, B, E, x, y, \gamma x, \gamma y).$$

This is (5.5). □

Note that Lemmas 5.6 and 5.7 are valid for both orientable and nonorientable maps. They are called *reduced rules* for maps. More precisely, they are explained as in the following.

Reduced rule 1 A map with its basic permutation $\mathcal{J}$ is basic equivalent to what is obtained by leaving off such successive elements $\langle x, \gamma x\rangle$, $x \in \mathcal{X}$, on $\mathcal{J}$.

Reduced rule 2 A map with its basic permutation $\mathcal{J}$ in the form $(A, x.B, y, C, \gamma x, D, \gamma y, E)$ is basic equivalent to what is obtained by interchanging the linear order B between x and y and the linear order D between γx and γy, and then leaving off $x, y, \gamma x$ and γy and putting $\langle x, y, \gamma x, \gamma y\rangle$ behind E on $\mathcal{J}$.

5.4 Orientable principles

This section is centralized on discussing the basic equivalent classes of orientable maps. The main purpose is to extract that there is at least one simplified butterfly in each class. From the first section of this chapter, it is known that each class is considered for only butterflies without loss of generality.

Theorem 5.3 For a butterfly $M = (\mathcal{X}_{\alpha,\beta}, \mathcal{J})$, $\gamma = \alpha\beta$, if no $x, y \in \mathcal{X}$ exist such that $\mathcal{J} = (A, x, B, y, C, \gamma x, D, \gamma y, E)$, where A, B, C, D and E are all linear orders on $\mathcal{X}_{\alpha,\beta}$, then

$$M \sim_{\text{bc}} O_0, \tag{5.6}$$

i.e., the trivial map (the degenerate simplified butterfly without edge).

Proof For convenience, in a cyclic permutation $\mathcal{J}$ on $\mathcal{X}_{\alpha,\beta}$, if two elements $x, y \in \mathcal{X}_{\alpha,\beta}$ are in the form $\mathcal{J} = (A, x, B, y, C, \gamma x, D, \gamma y, E)$, then they are said to be *interlaced*; otherwise, *parallel.* □

Claim: If any two elements are parallel on $\mathcal{J}$, then there is an element $x \in \mathcal{X}_{\alpha,\beta}$ such that $\langle x, \gamma x\rangle \subseteq \mathcal{J}$, i.e., $\langle x, \gamma x\rangle$ is a segment of $\mathcal{J}$ itself.

Proof By contradiction. If no such element exists on $\mathcal{X}_{\alpha,be}$, then for any $x_1 \in \mathcal{X}$, there is a nonempty linear order B_1 on $\mathcal{X}$ such that $\mathcal{J} = (A_1, x, B_1, \gamma x, C_1)$, where A_1 and C_1 are some linear orders on $\mathcal{X}_{\alpha,\beta}$. Because $B_1 \neq \emptyset$, for any $x_2 \in B_1$, on the basis of orientability and x_2 and x_1 parallel, the only possibility is $\gamma x_2 \in B_1$. From the known condition, there is also a linear order $B_2 \neq \emptyset$ on $\mathscr{X}_{\alpha,\beta}$ such that $B_1 = \langle A_2, x_2, B_2, \gamma x_2, C_2\rangle$, where A_2 and C_2 are segments on B_1, i.e., some linear orders on $\mathscr{X}$. Such a procedure can only go on to the infinity. This is a contradiction to the finiteness of $\mathscr{X}_{\alpha,\beta}$. Hence, the claim is true.

If $\mathscr{J} \neq \emptyset$, then from the claim, there exists an element x in $\mathscr{J}$ such that $\mathscr{J} = (A, x, \gamma x, B)$. However, because $(\gamma x)_{\mathscr{J}\gamma} = (\gamma x)$ is a face in its own right, $\mathscr{J}$ has to be with at least two faces. This is a contradiction to that $M = (\mathscr{X}, \mathscr{J})$ is a butterfly. Hence, the only possibility is $\mathscr{J} = \emptyset$, i.e., (5.6) holds. □

Actually, this theorem, including the claim in its proof, is valid for any orientable single vertex map. Therefore, it can be seen that the reduced rule (Lemma 5.6) and the following corollary is valid for any map (orientable or nonorientable) as well.

Corollary 5.1 Let $S = \langle A, x, \gamma x, B\rangle$ be a segment on a vertex of a map M. And let M' be obtained from M by substituting $\langle A, B\rangle$ for S and afterward deleting Kx from the ground set. Then, M' is a map and $M' \sim_{\mathrm{bc}} M$.

Proof Because it is easy to check that $M' = M -_{\mathrm{b}} M$, from Theorem 3.10, M' is a map. This is the first statement. In view of basic deletion of an edge as a basic transformation, $M' \sim_{\mathrm{bc}} M$. □

Corollary 5.2 Let S be a segment at a vertex of a map M. If for each element x in S, γx is also in S and any two elements in S are not interlaced, then there exists an element y in S such that $S = \langle A, y, \gamma y, B\rangle$.

Proof In the same way of proving Theorem 5.3, the conclusion is soon obtained. □

Theorem 5.4 In a butterfly $M = (\mathscr{X}_{\alpha,\beta}, \mathscr{J})$, if there are $x, y \in \mathscr{X}_{\alpha,\beta}$ such that $\mathscr{J} = (A, x, B, y, C, \gamma x, D, \gamma y, E)$, then there is an integer $k \geq 1$ such that

$$M \sim_{\mathrm{bc}} O_k, \tag{5.7}$$

i.e., the simplified butterfly with $2k$ edges.

Proof Based on reduced rule 2 (Lemma 5.7),

$$\mathscr{J} \sim_{\mathrm{bc}} (A, D, C, B, E, x, y, \gamma x, \gamma y).$$

Let $H = \langle A, D, C, B, E\rangle$. From Corollary 5.1, assume H is not in the form as S without loss of generality. From Corollary 5.2 and Theorem 5.3, H has two possibilities: $H = \emptyset$, or there exist two elements x_1 and y_1 interlaced in H.

If the former, then $\mathscr{J} \sim_{\text{bc}} (x, y, \gamma x, \gamma y)$, i.e., $M \sim_{\text{el}} O_1$. Otherwise, i.e., the latter, then $\mathscr{J} = (A_1, x_1, B_1, y_1, C_1, \gamma x_1, D_1, \gamma y_1, E_1)$. It should be noted that $E_1 = \langle F_1, x, y, \gamma x, \gamma y \rangle$. In this case, from Lemma 5.7,

$$\begin{aligned}\mathscr{J} &\sim_{\text{bc}} (A_1, D_1, C_1, B_1, E_1, x_1, y_1, \gamma x_1, \gamma y_1)\\ &= (A_1, D_1, C_1, B_1, F_1, x, y, \gamma x, \gamma y, x_1, y_1, \gamma x_1, \gamma y_1).\end{aligned}$$

Let $H_1 = \langle A_1, D_1, C_1, B_1, F_1 \rangle$, then for H_1 instead of H, go on with the procedure. According to the principle of finite recursion, there exists an integer $k \geq 1$ such that (5.7) holds. □

This theorem shows that each basic equivalent class of orientable maps has at least one map that is a simplified butterfly.

By considering Theorem 5.2, each basic equivalent class of orientable maps has an integer $k \geq 0$ such that the simplified butterfly of size k is in the class.

5.5 Orientable genus

Although Euler characteristic of a map is an invariant for basic transformation, a basic equivalent class of maps cannot be determined by itself. This is shown from the map M in Example 4.1 of Section 4.4 and the map N in Example 4.2 of Section 4.4. They both have the same Euler characteristic. However, they are not in the same basic equivalent class of maps because M is orientable and N is nonorientable.

Now, a further invariant should be considered for a class of orientable maps under the basic equivalence.

Theorem 5.5 For any orientable map $M = (\mathscr{X}, \mathscr{P})$, there is an integer $k \geq 0$ such that its Euler characteristic is

$$\chi(M) = 2 - 2k. \tag{5.8}$$

Proof From Theorems 5.1 and 4.5, only necessary to discuss butterflies. From Theorems 5.2 and 5.4, M has an integer $k \geq 0$ such that $M \sim_{\text{bc}} O_k$. Therefore, from (5.3),

$$\chi(M) = 2 - 2k.$$

This is (5.8). □

Corollary 5.3 The Euler characteristic of an orientable map $M = (\mathscr{X}, \mathscr{P})$ is always an even number, i.e.,

$$\chi(M) = 0 \pmod 2. \tag{5.9}$$

Proof A direct result of Theorem 5.5. □

Since Euler characteristic is an invariant of a basic equivalent class, the integer k in (5.8) is an invariant as well. From Theorem 5.5, k determines a basic equivalent class for orientable maps. Since each orientable map in the basic equivalent class determined by k can be seen as an embedding of its under graph on the orientable surface of genus k, k is also called the *genus*, or more precisely, *orientable genus* of the map. Of course, only an orientable map has the orientable genus.

From what has been discussed above, it is seen that although Euler characteristic cannot determine the basic equivalent class for all maps, the Euler characteristic can certainly determine the basic equivalent class for orientable maps.

5.6 Notes

If the travel formed by a face in a map can be partitioned into *tours* (travel without edge repetition), then the face is said to be *pan-tour*. A map with all of its faces pan-tour is called a *pan-tour map*. Because any tour can be partitioned into circuits, a pan-tour map is, in fact, a favorable map as mentioned in 2.6. A pre-proper embedding corresponds to what is called a *tour map* because each face forms a tour in its under graph.

5.6.1 Characterize and recognize that a graph has an upper map that is an orientable pan-tour map.

Notice 5.1 It looks suffice to observe bridgeless, or in other words without cut-edge (or articulate edge) from Theorem II.1 in Notice 2.1 or Theorem II.2 in Notice 2.2. However, it has not yet been proved.

5.6.2 Characterize and recognize that a graph has an upper map that is an orientable tour map.

Notice 5.2 It looks suffice to observe bridgeless, or in other words without cut-edge (or articulate edge) from Theorem II.1 in Notice 2.1 or Theorem II.2 in Notice 2.2. However, it has not yet been proved.

5.6.3 *Orientable pan-tour conjecture.* Prove, or disprove, that any nonseparable graph has an upper map that is an orientable pan-tour map.

Notice 5.3 No result is obtained. Referred to as in Liu [82, 83].

5.6.4 *Orientable tour conjecture.* Prove, or disprove, that any nonseparable graph has an upper map that is an orientable tour map.

Notice 5.4 No result is obtained. Referred to as in Liu [82, 83].

5.6.5 Characterize and recognize that a graph has an upper map that is an orientable pre-proper map.

Notice 5.5 The statement in 5.6.5 is stronger than what have been done in Liu [82, 83].

5.6.6 *Orientable proper map conjecture.* Prove, or disprove, that any nonseparable graph has an upper map that is an orientable proper map.

Notice 5.6 The statement in 5.6.6 is stronger than what have been done in Liu [82, 83]. However, the two statements are equivalent.

The *orientable minimum pan-tour genus* , usually called *orientable pan-tour genus*, of a graph is the minimum among all orientable genera of its upper pan-tour maps. Similarly, the *orientable pan-tour maximum genus* of a graph is the maximum among all orientable genera of its upper pan-tour maps.

5.6.7 Determine the orientable maximum pan-tour genus of a graph.

Notice 5.7 Relevant results are referred to as in Chen and Liu [14], Dong and Liu [23, 24], Dong et al. [26], Huang and Liu [42, 44], Li and Liu [49], Wang and Liu [161].

5.6.8 Determine the orientable maximum tour genus of a graph.

Notice 5.8 Relevant results are referred to as in Chen and Liu [14], Dong and Liu [23, 24], Dong et al. [26], Huang and Liu [42, 44], Li and Liu [49], Wang and Liu [161].

5.6.9 Determine the orientable pan-tour genus of a graph.

Notice 5.9 Relevant results are referred to as in Wei and Liu [164], Shao and Liu [141, 143].

5.6.10 Determine the orientable tour genus of a graph.

Notice 5.10 Relevant results are referred to as in WeiEL and Liu [164], ShaoZL and Liu [141, 143].

Although many progresses have been made on determining the orientable maximum genus of a graph, the study on determining orientable (maximum) pan-tour genus, or orientable (maximum) tour genus of a graph does not lead to any notable result yet. This suggests to investigate their bounds (upper or lower) for some class of graphs.

5.6.11 Characterize the class of graphs in which each graph has its orientable maximum pan-tour genus equal to its orientable maximum genus. Find the least upper bound of the absolute difference between the orientable maximum pan-tour genus and the orientable maximum genus for a class of graphs with the two genera not equal.

Notice 5.11 No relevant result is known yet.

5.6.12 Characterize the class of graphs in which each graph has its orientable maximum tour genus equal to its orientable maximum genus. Find the least upper bound of the absolute difference between the orientable maximum pan-tour genus and the orientable maximum genus for a class of graphs with the two genera not equal.

Notice 5.12 No relevant result is known yet.

5.6.13 Characterize the class of graphs in which each graph has its orientable pan-tour genus equal to its orientable genus. Find the least upper bound of the absolute difference between the orientable pan-tour genus and the orientable genus for a class of graphs with the two genera not equal.

Notice 5.13 No relevant result is known yet.

5.6.14 Characterize the class of graphs in which each graph has its orientable tour genus equal to its orientable genus. Find the least upper bound of the absolute difference between the orientable tour genus and the orientable genus for a class of graphs with the two genera not equal.

Notice 5.14 No relevant result is known yet.

6 Nonorientable Maps

- Any irreducible nonorientable map under basic subtraction of an edge is defined to be a barfly. However, an equivalent class may have more than one barflies.
- The simplified barflies are for standard nonorientable maps to show that each equivalent class has at most one simplified barfly.
- Nonorientable rules are for transforming a barfly into another barfly in the same equivalent class. A basic rule is extracted for deriving from one to all others.
- Principles only for nonorientable maps are clarified to transform any nonorientable map to a simplified barfly in the same equivalent class. Hence, each equivalent class has at least one simplified barfly.
- Nonorientable genus instead of the Euler characteristic is an invariant in an equivalent class to show that nonorientable genus itself determines the equivalent class.

6.1 Barflies

This chapter concentrates on discussing the basic equivalent classes of nonorientable maps by extracting a representative for each class. On the basis of Lemmas 5.1 and 5.2, only maps with a single vertex and a single face are considered for this purpose without loss of generality. A nonorientable map with both a single vertex and a single face is called a *barfly*. The barfly with only one edge is the map comprising a single twist loop, i.e., $N^{(1)} = (Kx, (x, \beta x))$.

Example 6.1 Two barflies of size two. Let $N_1^{(2)} = (Kx + Ky, \mathscr{I}_1)$ and $N_2^{(2)} = (Kx + Ky, \mathscr{I}_2)$ (shown, respectively, in Figure 6.1(a) and (b)), where

$$\mathscr{I}_1 = (, x, y, \gamma x, \beta y), \ \mathscr{I}_2 = (x, \beta x, y, \beta y).$$

Because

$$(x)_{\mathscr{I}_1\gamma} = (x, \beta y, \alpha x, \alpha y)$$

and

$$(x)_{\mathscr{I}_2\gamma} = (x, \alpha x, y, \alpha y),$$

each of $N_1^{(2)}$ and $N_2^{(2)}$ has exactly one face. And, since

$$(x)_{\Psi_{\{\mathscr{I}_1,\gamma\}}} = (x)_{\Psi_{\{\mathscr{I}_2,\gamma\}}} = Kx + Ky,$$

they are both nonorientable.

DOI 10.1515/9783110481846-006

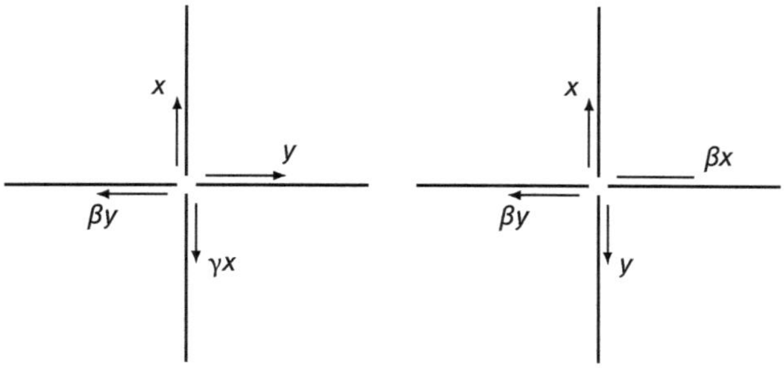

Figure 6.1: Barflies with two edges. (a) $N_1^{(2)}$ and (b) $N_2^{(2)}$.

As mentioned in the previous section, for convenience, the scope of maps considered here for the specific purpose should be enlarged to all nonorientable one vertex maps from barflies.

Lemma 6.1 For a single vertex map $M = (\mathscr{X}_{\alpha,.be}, \mathscr{P})$, M is nonorientable if, and only if, there exists an $x \in \mathscr{X}_{\alpha,\beta}$ such that $\beta x \in \{x\}_{\mathscr{P}}$.

Proof Necessity. By contradiction. Assume that for any $x \in \mathscr{X}_{\alpha,\beta}$, there always has $\gamma x \in \{x\}_{\mathscr{P}}$, where $\gamma = \alpha\beta$, then $\alpha x \notin \{x\}_{\Psi_{\{\mathscr{P},\gamma\}}}$. From 4.1, M is not nonorientable.

Sufficiency. Since $x \in \mathscr{X}_{\alpha,\beta}$ and $\beta x \in \{x\}_{\mathscr{P}}$, from Corollary 4.1 and only one vertex, $\Psi_{\{\mathscr{P},\gamma\}}$ has only one orbit on $\mathscr{X}$. Hence, M is nonorientable. □

This lemma can easily be employed for checking the nonorientability of a one vertex map.

Example 6.2 Six barflies of three edges. Let $N_i^{(3)} = (Kx+Ky+Kz, \mathscr{I}_i)$, where $i = 1, 2, \ldots, 6$ (shown, respectively, in Figure 6.2(a–f) and

$$\mathscr{I}_1 = (x, \beta x, y, \beta y, z, \beta z),$$

$$\mathscr{I}_2 = (x, y, z, \beta y, \beta x, \beta z),$$

$$\mathscr{I}_3 = (x, y, z, \beta z, \gamma x, \beta y),$$

$$\mathscr{I}_4 = (x, \beta x, y, z, \beta z, \beta y),$$

$$\mathscr{I}_5 = (x, y, \beta x, z, \beta y, \beta z),$$

$$\mathscr{I}_6 = (x, y, z, \beta x, \gamma y, \gamma z).$$

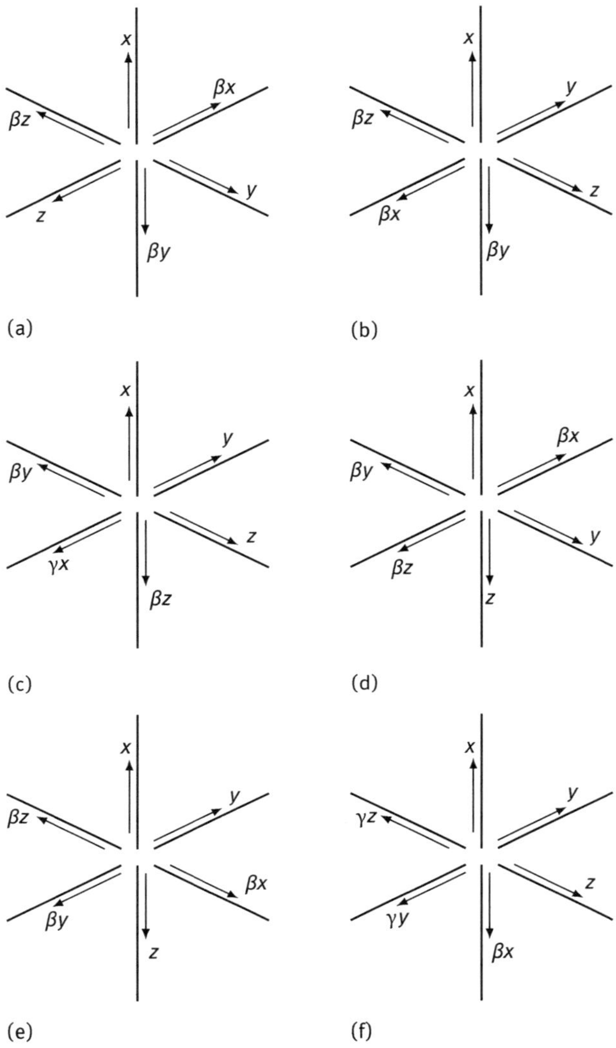

Figure 6.2: Barflies of three edges. (a) $N_1^{(3)}$, (b) $N_2^{(3)}$, (c) $N_3^{(3)}$, (d) $N_4^{(3)}$, (e) $N_5^{(3)}$, and (f) $N_6^{(3)}$.

Because $\beta x \in \{x\}_{\mathscr{I}_i} \subseteq \{x\}_{\Psi_{\{\mathscr{I}_i,\gamma\}}}$, $\gamma = \alpha\beta$, where $i = 1, 2, \ldots, 6$, from Lemma 6.1, they are all nonorientable. Since

$$(x)_{\mathscr{I}_{1}\gamma} = (x, \alpha x, y, \alpha y, z, \alpha z),$$

$$(x)_{\mathscr{I}_{2}\gamma} = (x, \gamma y, z, \gamma x, y, \alpha z),$$

$$(x)_{\mathscr{I}_3\gamma} = (x, \beta y, \alpha x, \gamma z, \beta z, \alpha y),$$
$$(x)_{\mathscr{I}_4\gamma} = (x, \alpha x, y, \gamma z, \beta z, \alpha y),$$
$$(x)_{\mathscr{I}_5\gamma} = (x, \alpha y, \beta z, , \gamma x, y, \alpha z),$$
$$(x)_{\mathscr{I}_6\gamma} = (x, \alpha z, \beta y, \alpha x, \gamma y, z),$$

the maps $(Kx + Ky + Kz, \mathscr{I}_i)$, where $i = 1, 2, \dots, 6$, are all with only one face. Therefore, they are all barflies.

Theorem 6.1 For any nonorientable map M, there exists a barfly N such that

$$M \sim_{\mathrm{bc}} N. \tag{6.1}$$

Proof From Lemma 5.1, by basic transformation M can be transformed into a single vertex map. From Theorem 4.3, in view of the nonorientability of M, the single vertex map is also nonorientable. Then, from Lemma 5.2, by basic transformation the single vertex can be transformed into a single face map. From Theorem 4.3, this map is also nonorientable and hence a barfly N satisfies (6.1). □

This theorem enables us to restrict ourselves only to transform a barfly into another barfly under the basic equivalence.

6.2 Simplified barflies

Let $Q_l = (\mathscr{X}_l, \mathscr{I}_l)$, $l \geq 1$, where

$$\mathscr{X}_l = \sum_{i=1}^{l} Kx_i, \tag{6.2}$$

and

$$\mathscr{I}_l = \left(\prod_{i=1}^{l} \langle x_i, \beta x_i \rangle \right). \tag{6.3}$$

They are called *N-standard map*. When $k = 1, 2, 3$ and 4, $Q_1 = N_1^{(1)}$, $Q_2 = N_2^{(2)}$, $Q_3 = N_3^{(3)}$ and $Q_4 = N_4^{(4)}$ are, respectively, shown in Figure 6.3(a–d).

Lemma 6.2 For any $l \geq 1$, N-standard maps Q_l are all nonorientable.

Proof Because all Q_l, $l \geq 1$, are single vertex map and $\beta x_1 \in \{x_1\}_{\mathscr{I}_l}$, $l \geq 1$, from Lemma 6.1, they are all nonorientable. □

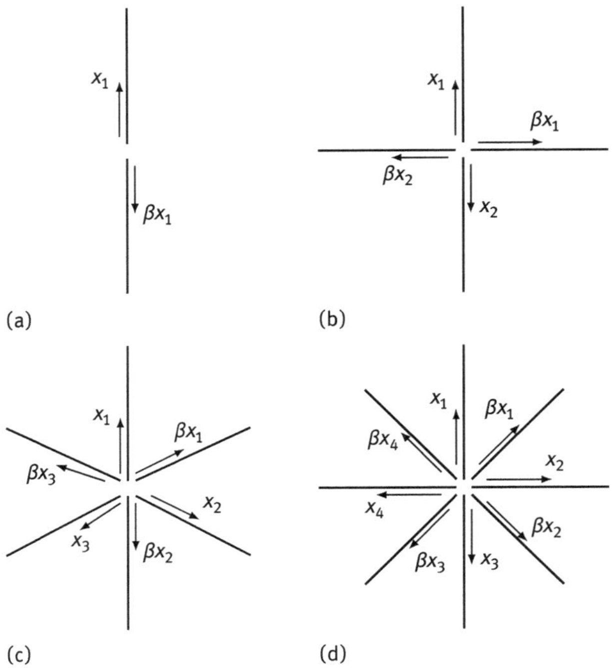

Figure 6.3: Simplified barfly. (a) N_1, (b) N_2, (c) N_3, and (d) N_4.

Lemma 6.3 For any $l \geq 1$, N-standard maps Q_l are all with only one face.

Proof Because $Q_1 = N^{(1)}$, $Q_2 = N_2^{(2)}$ and $Q_3 = N_1^{(3)}$, from the two examples above, they are all with only one face. Their faces are $(x_1)_{\mathscr{I}_1\gamma} = (x_1, \alpha x_1)$, $(x_1)_{\mathscr{I}_2\gamma} = (\langle x_1\rangle_{\mathscr{I}_1\gamma}, x_2, \alpha x_2) = (x_1, \alpha x_1, x_2, \alpha x_2)$ and $(x_1)_{\mathscr{I}_3\gamma} = (\langle x_1\rangle_{\mathscr{I}_2\gamma}, x_3, \alpha x_3) = (x_1, \alpha x_1, x_2, \alpha x_2, x_3, \alpha x_3)$.

Assume, by induction, that

$$(x_1)_{\mathscr{I}_{l-1}\gamma} = (x_1, \alpha x_1, x_2, \alpha x_2, \ldots, x_{l-1}, \alpha x_{l-1}),$$

for $l \geq 4$. Since $\mathscr{I}_l = (\langle x_1\rangle_{\mathscr{I}_{l-1}}, x_l, \beta x_l)$,

$$(x_1)_{\mathscr{I}_l\gamma} = (\langle x_1\rangle_{\mathscr{I}_{l-1}\gamma}, (\mathscr{I}_l\gamma)\alpha x_{l-1}, \ldots).$$

And since $(\mathscr{I}_l\gamma)\alpha x_{l-1} = \mathscr{I}_l\beta x_{l-1} = x_l$, $(\mathscr{I}_l\gamma)x_l = \mathscr{I}_l\alpha(\beta x_l) = \alpha x_l$ and $(\mathscr{I}_l\gamma)\alpha x_l = \mathscr{I}_l\beta x_l = x_1$,

$$\begin{aligned}(x_1)_{\mathscr{I}_l\gamma} &= (\langle x_1\rangle_{\mathscr{I}_{l-1}\gamma}, x_l, \alpha x_l)\\ &= (x_1, \alpha x_1, \ldots, x_{l-1}, \alpha x_{l-1}, x_l, \alpha x_l).\end{aligned}$$

Therefore, all N-standard maps are with only one face. □

Because each N-standard map has only one face, from the two lemmas above, it is known that any N-standard map is a barfly. And because each of such barflies has a simpler form, it is called a *simplified barfly*.

Since for any $l \geq 1$, the simplified barfly Q_l is with one edge, one vertex and one face, its Euler characteristic is

$$\chi(Q_l) = 2 - l. \tag{6.4}$$

Theorem 6.2 For any basic equivalent class of nonorientable maps, there exists at most one map that is a simplified barfly.

Proof By contradiction. Assume that there are two simplified barflies Q_i and Q_j, where $i \neq j$, and $i, j \geq 1$, in a basic equivalent class of nonorientable maps. From Theorems 4.5 and (6.4),

$$\chi(Q_i) = 2 - i = 2 - j = \chi(Q_j).$$

This implies $i = j$, a contradiction to the assumption. □

In the following sections, it will be shown that there exists at least one map that is a simplified barfly in each basic equivalent class of nonorientable maps.

6.3 Nonorientable rules

As mentioned above, this section is for establishing two basic rules of transforming a nonorientable single vertex map into another nonorientable single vertex map within basic equivalence.

Lemma 6.4 For a nonorientable single vertex map $M = (\mathscr{X}_{\alpha,\beta}, \mathscr{J})$, if $\mathscr{J} = (A, x, B, \beta x, C)$, where A, B and C are segments of linear order in the cycle $\mathscr{J}$ on $\mathscr{X}_{\alpha,\beta}$, then

$$\mathscr{J} \sim_{\text{bc}} (A, \alpha B^{-1}, C, x, \beta x). \tag{6.5}$$

Note 6.1 For a segment $B = \langle \mathscr{J}x, \mathscr{J}^2x, \ldots, \mathscr{J}^s x\rangle$, $\beta x = \mathscr{J}^{s+1}x$, of linear order in the cycle of $\mathscr{J}$ on $\mathscr{J}_{\alpha,\beta}$, from Theorem 2.3,

$$\begin{aligned} &\langle \alpha\mathscr{J}^s x, \mathscr{J}(\alpha\mathscr{J}^s x), \ldots, \mathscr{J}^{s-1}(\alpha\mathscr{J}^s x)\rangle \\ &= \langle \alpha\mathscr{J}^s x, \alpha\mathscr{J}^{s-1}x, \ldots, \alpha\mathscr{J}x\rangle \\ &= \alpha B^{-1}. \end{aligned} \tag{6.6}$$

Proof Two steps expressed by claims are considered for transforming a nonorientable single vertex map into another nonorientable single vertex map under the basic equivalence. □

Claim 1 $(A, x, B, \beta x, C) \sim \mathrm{bc}(A, \alpha B^{-1}, \beta y, \alpha C^{-1}, y)$.

Proof By basic splitting an edge e_y between the two angles $\langle \alpha x, \mathscr{I} x\rangle$ and $\langle C, A\rangle$ (i.e., the angle between C and A) on $\mathscr{I} = (A, x, B, \beta x, C)$,

$$\begin{aligned}\mathscr{I} &\sim_{\mathrm{bc}} (A, x, y)(\gamma y, B, \beta x, C)\\ &= (A, x, y)(\beta y, \alpha C^{-1}, \gamma x, \alpha B^{-1})\\ &= (y, A, x)(\gamma x, \alpha B^{-1}, \beta y, \alpha C^{-1}),\end{aligned}$$

as shown in Figure 6.4(a) and (b).

Because e_x is a link in $\mathscr{I}_1 = (y, A, x)(\gamma x, \alpha B^{-1}, \beta y, \alpha C^{-1})$, by basic contracting e_x,

$$\begin{aligned}\mathscr{I}_1 &\sim_{\mathrm{bc}} (y, A, \alpha B^{-1}, \beta y, \alpha C^{-1})\\ &= (A, \alpha B^{-1}, \beta y, \alpha C^{-1}, y),\end{aligned}$$

as shown in Figure 6.4(c). □

Claim 2 $\mathscr{I}_2 \sim_{\mathrm{bc}} (A, \alpha B^{-1}, C, \beta x, x)$, where

$$\mathscr{I}_2 = (A, \alpha B^{-1}, \beta y, \alpha C^{-1}, y).$$

Proof By basic splitting e_x between the two angles $(\gamma y, \mathscr{I}_2 \beta y)$ and $(\alpha y, \mathscr{I}_2 y)$ on $(A, \alpha B^{-1}, \beta y, \alpha C^{-1}, y)$,

$$\begin{aligned}\mathscr{I}_2 &\sim_{\mathrm{bc}} (A, \alpha B^{-1}, \beta y, x)(\gamma x, \alpha C^{-1}, y)\\ &= (A, \alpha B^{-1}, \beta y, x)(\alpha y, C, \beta x)\\ &= (x, A, \alpha B^{-1}, \beta y)(\alpha y, C, \beta x),\end{aligned}$$

as shown in Figure 6.4(d) and (e).

Because e_y is a link in $\mathscr{I}_3 = (x, A, \alpha B^{-1}, \beta y)(\alpha y, C, \beta x)$, by basic contracting e_y,

$$\begin{aligned}\mathscr{I}_3 &\sim_{\mathrm{bc}} (x, A, \alpha B^{-1}, C, \beta x)\\ &= (A, \alpha B^{-1}, C, \beta x, x),\end{aligned}$$

as shown in Figure 6.4(f).

From Claim 1 and Claim 2,

$$\begin{aligned}\mathscr{I} &\sim_{\mathrm{bc}} \mathscr{I}_1 \sim_{\mathrm{bc}} \mathscr{I}_2 \sim_{\mathrm{bc}} \mathscr{I}_3\\ &\sim_{\mathrm{bc}} (A, \alpha B^{-1}, C, \beta x, x).\end{aligned}$$

This is (6.5). □

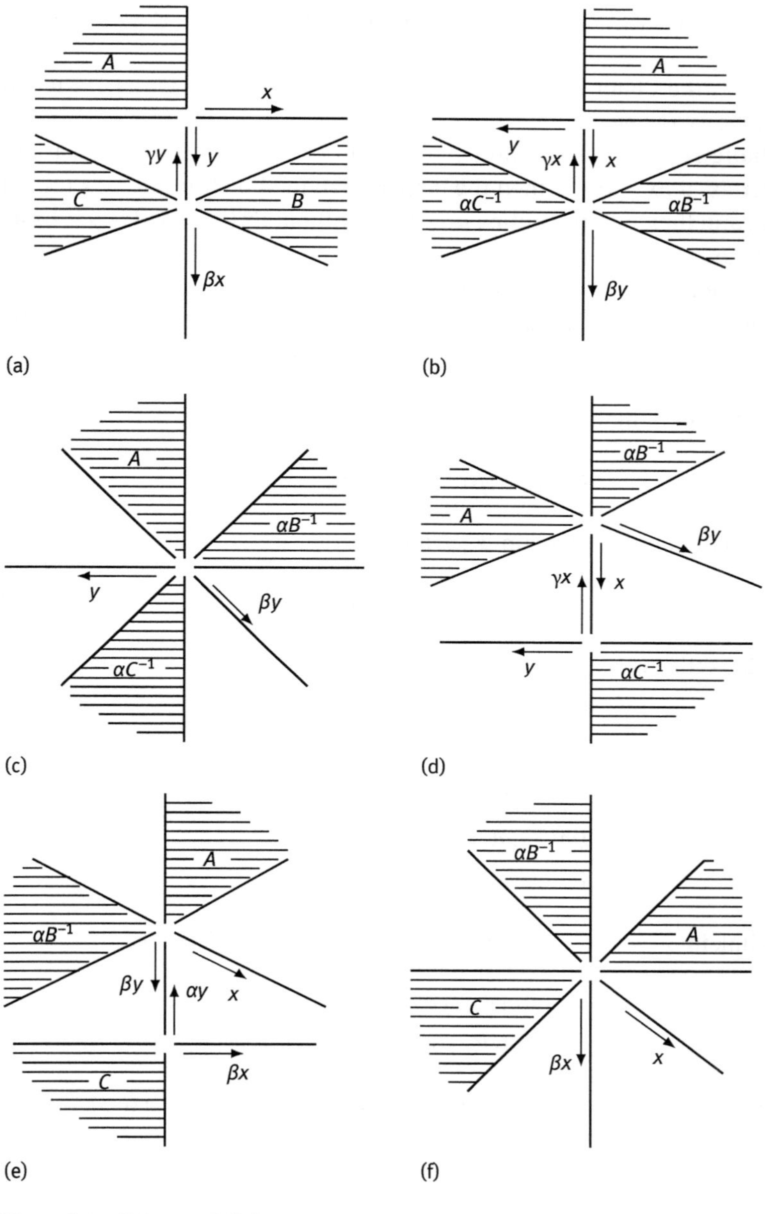

Figure 6.4: Claim 1 and Claim 2.

On the basis of the procedure in the proof of the lemma, the two claims show the following rules as basic equivalent transformations.

Nonorientable rule 1 On a nonorientable map $M = (\mathscr{X}_{\alpha,\beta}, \mathscr{P})$ unnecessary to have a single vertex, if $\beta x \in (x)_{\mathscr{P}}$, then the map M' obtained by translating x and βx in a direction via, respectively, segments C and D, and then by substituting αC^{-1} and

αD^{-1} for, respectively, C (without βx) and D (without x) on $(x)_{\mathscr{P}}$ is basic equivalent to M, i.e., $M' \sim_{\rm bc} M$.

This is, in fact, the Claim 1 above. However, from the proof of Claim 2 a much simpler rule can be extracted.

Nonorientable rule 2 On a nonorientable map $M = (\mathscr{X}, \mathscr{P})$ unnecessary to have a single vertex, if $\beta x \in (x)_{\mathscr{P}}$, then the map M' obtained by translating x (or βx) via a segment C, and then by substituting αC^{-1} for C without βx (or x) on $(x)_{\mathscr{P}}$ is basic equivalent to M, i.e., $M' \sim_{\rm bc} M$.

It is seen that nonorientable rule 1 can be done by employing nonorientable rule 2 twice. Therefore, nonorientable rule 2 is fundamental. From this point of view, the proof of Lemma 6.4 can be done only by nonorientable rule 2.

Lemma 6.5 For a nonorientable single vertex map $(\mathscr{X}_{\alpha,\beta}, \mathscr{J})$, $\gamma = \alpha\beta$, if

$$\mathscr{J} = (A, x, \beta x, y, z, \gamma y, \gamma z),$$

where A is a segment of linear order on $\mathscr{X}_{\alpha,\beta}$, then

$$\mathscr{J} \sim (A, x, \beta x, y, \beta y, z, \beta z). \tag{6.7}$$

Proof By basic splitting e_t between the two angles $(\alpha x, \beta x)$ and $(\alpha z, \gamma y)$ on $(A, x, \beta x, y, z, \gamma y, \gamma z)$,

$$\begin{aligned} \mathscr{J} &= (A, x, \beta x, y, z, \gamma y, \gamma z) \\ &\sim_{\rm bc} (\gamma y, \gamma z, A, x, t)(\gamma t, \beta x, y, z) \\ &= (\gamma y, \gamma z, A, x, t)(\gamma x, \beta t, \alpha z, \alpha y) \\ &= (t, \gamma y, \gamma z, A, x)(\gamma x, \beta t, \alpha z, \alpha y). \end{aligned}$$

Because e_x is a link in $\mathscr{J}_1 = (t, \gamma y, \gamma z, A, x)(\gamma x, \beta t, \alpha z, \alpha y)$, by contracting e_x,

$$\begin{aligned} \mathscr{J}_1 &\sim_{\rm bc} (t, \gamma y, \gamma z, A, \beta t, \alpha z, \alpha y) \\ &= (A, \beta t, \alpha z, \alpha y, t, \gamma y, \gamma z). \end{aligned}$$

By substituting $\langle A, \beta t\rangle$, $\langle \alpha y, t, \gamma y\rangle$ and $\langle \emptyset\rangle$ for , respectively, A, B and C in (6.5),

$$\mathscr{J}_1 \sim_{\rm bc} (A, \beta t, \beta y, \alpha t, y, z, \beta z).$$

Furthermore, by substituting $\langle A, \beta t\rangle$, $\langle \alpha t\rangle$ and $\langle z, \beta z\rangle$ for, respectively, A, B and C in (6.5),

$$\begin{aligned} \mathscr{J}_1 &\sim_{\rm bc} (A, t, \beta t, y, \beta y, z, \beta z) \\ &= (A, x, \beta x, y, \beta y, z, \beta z). \end{aligned}$$

This is (6.7). □

This lemma shows that in a map $M = (\mathscr{X}_{\alpha,\beta}, \mathscr{P})$, $\gamma = \alpha\beta$, if

$$(x)_{\mathscr{P}} = (x, \beta x, y, z, \gamma y, \gamma z, A),$$

then the map obtained by substituting $\langle y, \beta y, z, \beta z\rangle$ for $\langle y, z, \gamma y, \gamma z\rangle$ on $(x)_{\mathscr{P}}$ is basic equivalent to M, i.e., $M' \sim_{\mathrm{bc}} M$. This is usually called *nonorientable rule* 3.

Actually, nonorientable rule 3 can be also deduced from nonorientable rule 2. Although nonorientable rule 2 is fundamental, nonorientable rules 1 and 3 are more convenient for recursion.

6.4 Nonorientable principles

In this section, barflies are only considered for this classification because it has been known that there is no loss of generality for general nonorientable maps.

Lemma 6.6 In a barfly $N = (\mathscr{X}_{\alpha,\beta}, \mathscr{J})$, there exists an element $x \in \mathscr{X}_{\alpha,\beta}$ such that

$$\mathscr{J} = (A, x, B, \beta x, C), \tag{6.8}$$

where A, B and C are segments of $\mathscr{J}$ on $\mathscr{X}_{\alpha,\beta}$.

Proof By contradiction. Since A, B and C are permitted to be empty, if no $x \in \mathscr{X}$ such that $\mathscr{J}$ satisfies (6.8), then from only one vertex, for any $x \in \mathscr{X}$, it is only possible that $\gamma x \in (x)_{\mathscr{J}}$ and $\beta x \notin (x)_{\mathscr{J}}$. Therefore, $(x)_{\Psi_{\{\mathscr{J},\gamma\}}}$ and $(\beta x)_{\Psi_{\{\mathscr{J},\gamma\}}}$ are the two orbits of $\Psi_{\{\mathscr{J},\gamma\}}$ on $\mathscr{X}_{\alpha,\beta}$. Thus, M is orientable. This is a contradiction to the nonorientability of a barfly. □

Theorem 6.3 For any barfly $N = (\mathscr{X}_{\alpha,\beta}, \mathscr{J})$, there exists an integer $l \geq 1$ such that

$$\mathscr{J} \sim_{\mathrm{bc}} Q_l. \tag{6.9}$$

Proof From Lemmas 6.6 and 6.4, it can assumed that

$$\mathscr{J} \sim_{\mathrm{bc}} (A, \prod_{j=1}^{i} \langle x_j, \beta x_j\rangle),$$

where i is as great as possible in this form. Naturally, $i \geq 1$.

From the maximality of i and only one vertex, $x \in A$ if, and only if, $\gamma x \in A$.

Two cases have to be discussed.

Case 1 If no element in A is interlaced, then from Corollary 5.2 and Corollary 5.1, (6.9) holds. Here, $l = i$.

Case 2 Otherwise, by Lemma 5.7 (the reduced rule). It can be assumed that

$$\mathcal{I} \sim_{\mathrm{bc}} (B, \prod_{j=1}^{i} \langle x_j, \beta x_j \rangle \prod_{j=1}^{t} \langle y_j, z_j, \gamma y_j, \gamma z_j \rangle),$$

where t is as great as possible in this form. Naturally, $t \geq 1$. From the maximality of t, no element in B is interlaced. By Corollary 5.2 and Corollary 5.1,

$$\mathcal{I} \sim_{\mathrm{bc}} (\prod_{j=1}^{i} \langle x_j, \beta x_j \rangle \prod_{j=1}^{t} \langle y_j, z_j, \gamma y_j, \gamma z_j \rangle).$$

By nonorientable rule 3,

$$\mathcal{I} \sim \mathrm{bc}(\prod_{j=1}^{2t+i} \langle x_j, \beta x_j \rangle) = \mathcal{I}_l.$$

From (6.2) and (6.3), this is (6.9), where $l = 2t + i$. □

On the basis of Theorems 6.1 and 6.3, it is know that there is at least one simplified barfly in each of the basic equivalent classes for nonorientable maps.

6.5 Nonorientable genus

Now, let us go back to general nonorientable maps for the invariants of determining the basic equivalent classes for nonorientable maps.

Theorem 6.4 For any nonorientable map $N = (\mathcal{X}, \mathcal{P})$ in a basic equivalent class, there is an integer $l \geq 1$ such that the Euler characteristic

$$\chi(N) = 2 - l. \tag{6.10}$$

Proof From Theorem 6.3, there is a simplified barfly in a basic equivalent class of barflies. From Theorem 6.1, in each basic equivalent class of nonorientable maps, there is an integer $l \geq 1$ such that Q_l is in this class. On the other hand, from Theorem 6.2, only Q_l is in this class. Therefore, from (6.4) and Theorem 4.5, (6.10) is obtained. □

This integer $l = 2 - \chi(N) \geq 1$ called the *nonorientable genus* of the class N is in, or of N.

Now, it is seen from this chapter and Chapters 4 and 5 that if the orientability of a map is defined to be 1 when the map is orientable, and −1 when the map is nonorientable, then the *relative genus* of a map is the product of its orientability and its *absolute genus* (orientable genus, if the map is orientable; nonorientable genus, if the map is nonorientable). Thus, a basic equivalent class of maps (orientable and nonorientable) is determined by only its relative genus.

6.6 Notes

Similarly to Chapter 5, among all nonorientable embeddings of a graph, the one with minimum (maximum) of absolute genus is called a *minimum (maximum) genus embedding*.

The genus of a minimum (maximum) genus embedding on nonorientable surfaces for a graph is called the *minimum (maximum) nonorientable genus* of the graph.

The minimum nonorientable genus of a graph is also called the *nonorientable genus* of the graph. If the minimum genus embedding is a nonorientable pan-tour (favorable) map, the the genus is called the *nonorientable pan-tour (favorable) genus.*

And the likes, *nonorientable pan-tour maximum genus*, *nonorientable tour genus* (or *nonorientable preproper genus*), *nonorientable tour maximum genus*, etc.

6.6.1 Justify and recognize if a graph has a nonorientable embedding that is a pan-tour map.

Notice 6.1 From Liu [80–83], the graph is bridgeless or a single loop.

6.6.2 Justify and recognize if a graph has a nonorientable embedding that is a tour map.

Notice 6.2 From Liu [80–83], the graph is bridgeless or a single loop.

6.6.3 Determine the least upper bound and the greatest lower bound of the nonorientable pan-tour genus (or genera) for a graph (or a set of graphs).

Notice 6.3 For some relevant results, refer to Chen and Liu [15], Dong and Liu [24], Huang and Liu [42–44], Li and Liu [50].

6.6.4 Determine the least upper bound and the greatest lower bound of the nonorientable tour genus (or genera) for a graph (or a set of graphs).

Notice 6.4 For some relevant results, see Chen and Liu [15], Dong and Liu [24], Huang and Liu [42–44], Li and Liu [50].

6.6.5 Determine the least upper bound and the greatest lower bound of the nonorientable proper genus (or genera) for a graph (or a set of graphs).

Notice 6.5 For some relevant results, see Chen and Liu [15], Dong and Liu [24], Huang and Liu [42–44], Li and Liu [50].

Because there is no possibility of getting a result as simple as that shown in (6.12) of Liu [90] for determining the nonorientable pan-tour maximum genus, nonorientable tour maximum genus and nonorientable proper maximum genus of a graph in general, only some types of graphs are available to be considered for such a kind of result.

6.6.6 Determine the least upper bound and the greatest lower bound of the nonorientable pan-tour maximum genus (or genera) for a graph (or a set of graphs).

Notice 6.6 For some relevant results, refer to in Dong and Liu [24], Huang and Liu [42–44], Li and Liu [50], Lv and Liu [121, 122].

6.6.7 Determine the least upper bound and the greatest lower bound of the nonorientable tour maximum genus (or genera) for a graph (or a set of graphs).

Notice 6.7 For some relevant results, refer to Dong and Liu [24], Huang and Liu [42–44], Li and Liu [50], Lv and Liu [121, 122].

6.6.8 Determine the least upper bound and the greatest lower bound of the nonorientable proper maximum genus (or genera) for a graph (or a set of graphs).

Notice 6.8 For some relevant results, refer to Dong and Liu [24], Huang and Liu [42–44], Li and Liu [50], Lv and Liu [121, 122].

6.6.9 *Nonorientable pan-tour conjecture* (prove, or disprove). Any nonseparable graph has a nonorientable embedding, which is a pan-tour map.

Notice 6.9 The conjecture should have an additional condition that the graph is not a loop itself.

6.6.10 *Nonorientable tour map conjecture* (prove, or disprove). Any nonseparable graph has a nonorientable embedding, which is a tour map.

Notice 6.10 The conjecture should have an additional condition that the graph is not a loop itself.

6.6.11 *Nonorientable proper map conjecture* (prove, or disprove). Any nonseparable graph has a nonorientable embedding, which is a proper map.

Notice 6.11 The conjecture should have an additional condition that the graph is not a loop itself.

6.6.12 *Nonorientable small face proper map conjecture* (prove, or disprove). A nonseparable graph of order n has a nonorientable embedding, which is a proper map with $n - 1$ faces.

Notice 6.12 Apparently, the graph is not allowed to be the 1-*circuit*, i.e., the loop itself.

7 Isomorphisms of Maps

- An isomorphism is defined for the classification of maps. A map is dealt with an isomorphic class of embeddings of the undergraph of the map.
- Two maps are isomorphic if, and only if, their dual maps are isomorphic with the same isomorphism.
- Two types of efficient algorithms are designed for identifying if two maps are isomorphic.
- Primal trail codes, or dual trail codes, are used for justifying the isomorphism of two maps.
- Two pattern examples show how to recognize and justify if two maps are isomorphic.

7.1 Commutativity

In view of topology, the basic equivalent classes of maps are, in fact, a type of topological equivalent classes of two-dimensional compact manifolds without boundary, or in brief surfaces.

Two embeddings of a graph explained in Chapter 1 are distinct if they are treated as one-dimensional complexes to be nonequivalent under a topological equivalence.

If a map is dealt with an embedding of a graph on a surface, then two distinct maps are, of course, distinct embeddings of their undergraph. However, the converse case is not necessarily true.

This chapter introduces a type of combinatorial equivalence that is still seen as a type of topological equivalence but different from that for embeddings of a graph.

In general, the equivalence between two maps can be deduced from that between two embeddings of their undergraph. However, the coverse case is not necessarily true.

For two maps $M_1 = (\mathscr{X}_{\alpha,\beta}(X_1), \mathscr{P}_1)$ and $M_2 = (\mathscr{X}_{\alpha,\beta}(X_2), \mathscr{P}_2)$, if there exists a one-to-one correspondence (i.e., bijection)

$$\tau : \mathscr{X}_{\alpha,\beta}(X_1) \longrightarrow \mathscr{X}_{\alpha,\beta}(X_2)$$

between $\mathscr{X}_{\alpha,\beta}(X_1)$ and $\mathscr{X}_{\alpha,\beta}(X_2)$ such that for any $x \in \mathscr{X}_{\alpha,\beta}(X_1)$,

$$\tau(\alpha x) = \alpha\tau(x),\ \tau(\beta x) = \beta\tau(x),\ \tau(\mathscr{P}_1 x) = \mathscr{P}_2\tau(x), \tag{7.1}$$

then τ is called an *isomorphism* from M_1 to M_2.

Lemma 7.1 If τ is an isomorphism from M_1 to M_2, then its inverse τ^{-1} exists, and τ^{-1} is an isomorphism from M_2 to M_1.

DOI 10.1515/9783110481846-007

Proof Since τ is a bijection, τ^{-1} exists. And, τ^{-1} is also a one-to-one correspondence from M_2 to M_1. For any $y \in \mathscr{X}_{\alpha,\beta}(X_2)$, let $x = \tau^{-1}y \in \mathscr{X}_{\alpha,\beta}(X_1)$. Because $y = \tau x$ and τ is an isomorphism from M_1 to M_2, from (7.1),

$$\tau(\alpha x) = \alpha y, \ \tau(\beta x) = \beta y, \ \tau(\mathscr{P}_1 x) = \mathscr{P}_2 y.$$

Further, because τ^{-1} exists, so

$$\tau^{-1}(\alpha y) = \alpha x = \alpha(\tau^{-1}y),$$
$$\tau^{-1}(\beta y) = \beta x = \beta(\tau^{-1}y),$$
$$\tau^{-1}(\mathscr{P}_2 y) = \mathscr{P}_1 x = \mathscr{P}_1(\tau^{-1}y).$$

This implies τ^{-1} is an isomorphism from M_2 to M_1. □

Based on this lemma, τ, or τ^{-1} can be called an isomorphism between M_1 and M_2.

Example 7.1 Let $M_1 = (\mathscr{X}_1, \mathscr{P}_1)$, where

$$\mathscr{X}_1 = Kx_1 + Ky_1 + Kz_1 + Ku_1$$

and

$$\mathscr{P}_1 = (x_1, y_1, z_1)(u_1, \gamma z_1)(\beta u_1, \gamma x_1)(\gamma y_1),$$

and $M_2 = (\mathscr{X}_2, \mathscr{P}_2)$, where

$$\mathscr{X}_2 = Kx_2 + Ky_2 + Kz_1 + Ku_2$$

and

$$\mathscr{P}_2 = (y_2, z_2, x_2)(\gamma u_2, \beta x_2)(\alpha u_2, \beta z_2)(\beta y_2)$$

as shown in Figure 7.1.

First, let $\tau(x_1) = x_2$. From the first two relations in (7.1) and the property of Klein group, if τ is an isomorphism between M_1 and M_2, then

$$\tau(\alpha x_1) = \alpha(\tau x_1) = \alpha x_2,$$
$$\tau(\beta x_1) = \beta(\tau x_1) = \beta x_2,$$
$$\tau(\gamma x_1) = \gamma(\tau x_1) = \gamma x_2,$$

i.e., $\tau(Kx_1) = Kx_2$.

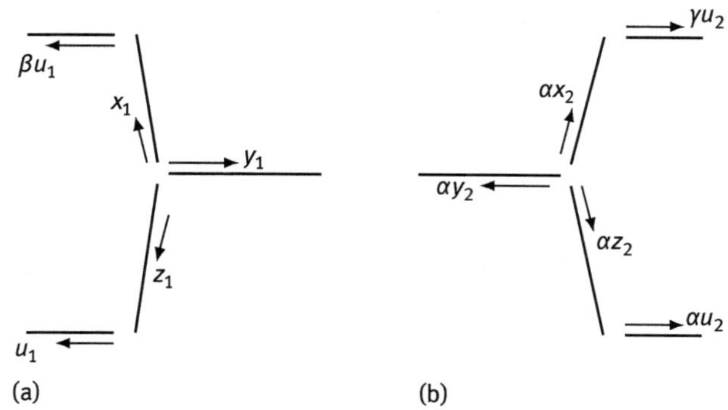

Figure 7.1: Two isomorphic maps. (a) M_1 and (b) M_2.

Then, from the third relation of (7.1),

$$\tau(y_1) = \tau(\mathscr{P}_1 x_1) = \mathscr{P}_2 \tau(x_1) = \mathscr{P}_2 x_2 = y_2.$$

Thus, $\tau(Ky_1) = Ky_2$. Similarly, from

$$\tau(z_1) = \tau(\mathscr{P}_1 y_1) = \mathscr{P}_2 \tau(y_1) = \mathscr{P}_2 y_2 = z_2,$$

$\tau(Kz_1) = Kz_2$, and from

$$\tau(u_1) = \tau(\mathscr{P}_1 \gamma z_1) = \mathscr{P}_2 \tau(\gamma z_1) = \mathscr{P}_2 \gamma z_2 = u_2,$$

$\tau(Ku_1) = Ku_2$.

Finally, check that if the one-to-one correspondence τ from $\mathscr{X}_1$ to $\mathscr{X}_2$ satisfies $\tau \mathscr{P}_1 = \mathscr{P}_2$. In fact, from the conjugate axiom, it is only necessary to have

$$\begin{aligned} \tau \mathscr{P}_1 &= (\tau x_1, \tau y_1, \tau z_1)(\tau u_1, \tau \gamma z_1)(\tau \beta u_1, \tau \gamma x_1)(\tau \gamma y_1) \\ &= (x_2, y_2, z_2)(u_2, \gamma z_2)(\beta u_2, \gamma x_2)(\gamma y_2) \\ &= (y_2, z_2, x_2)(\gamma u_2, \beta x_2)(\alpha u_2, \beta z_2)(\beta y_2) \\ &= \mathscr{P}_2. \end{aligned}$$

Therefore, τ is an isomorphism between M_1 and M_1.

Note 7.1 If the two maps M_1 and M_2 in Example 7.1 are, respectively, seen as embeddings of their undergraphs G_1 and G_2, then they are distinct. If Kx is represented by $x = (x^{+1}, x^{-1})$, where $x^{+1} = \{x, \alpha x\}$ and $x^{-1} = \{\beta x, \gamma x\}$, then the vertices of μG_1 have their rotation as

$$(x_1^{+1}, y_1^{+1}, z_1^{+1}), (z_1^{+-1}, u_1^{+1}), (x_1^{-1}, u_1^{+-1}), (y_1^{-1}),$$

and hence μG_1 is on the projective plane (u_1, u_1). And the vertices of μG_2 have their rotation as

$$(x_2^{+1}, z_2^{+1}, y_2^{+1}), (z_1^{+-1}, u_1^{+1}), (x_1^{-1}, u_1^{+-1}), (y_1^{-1}),$$

and hence μG_2 is on the projective plane (u_1, u_1) as well.

However, the induced one-to-one correspondence $\tau|_\mu(\tau|_\mu(s_1) = s_2, s = x, y, z, u)$ from $\mu(G_1)$ to $\mu(G_2)$ has

$$\tau|_\mu(x_1^{+1}, y_1^{+1}, z_1^{+1}) = (x_2^{+1}, y_2^{+1}, z_2^{+1}) \neq (x_2^{+1}, z_2^{+1}, y_2^{+1}).$$

This implies that μG_1 and μG_2 are distinct.

Theorem 7.1 Let $M_1 = (\mathscr{X}_{\alpha,\beta}(X_1), \mathscr{P}_1)$ and $M_2 = (\mathscr{X}_{\alpha,\beta}(X_2), \mathscr{P}_2)$ be two maps. For a bijection $\tau : \mathscr{X}_{\alpha,\beta}(X_1) \longrightarrow \mathscr{X}_{\alpha,\beta}(X_2)$, τ is an isomorphism if, and only if, the diagram

$$\begin{array}{ccc} \mathscr{X}_{\alpha,\beta}(X_1) & \xrightarrow{\quad\tau\quad} & \mathscr{X}_{\alpha,\beta}(X_2) \\ \eta_1 \big\downarrow & & \big\downarrow \eta_2 \\ \mathscr{X}_{\alpha,\beta}(X_1) & \xrightarrow{\quad\tau\quad} & \mathscr{X}_{\alpha,\beta}(X_2) \end{array} \tag{7.2}$$

for $\eta_1 = \eta_2 = \alpha$, $\eta_1 = \eta_2 = \beta$, and for $\eta_1 = \mathscr{P}_1$ and $\eta_2 = \mathscr{P}_2$, is all *commutative*, i.e., all paths with the same initial object and the same terminal object have the same effect.

Proof Necessity. From the first relation in (7.1), for any $x \in \mathscr{X}_{\alpha,\beta}(X_1)$, $\tau(\alpha x) = \alpha(\tau x)$. That is to say, the result of composing the mappings on the direct path

$$\mathscr{X}_{\alpha,\beta}(X_1) \xrightarrow{\alpha} \mathscr{X}_{\alpha,\beta}(X_1) \xrightarrow{\tau} \mathscr{X}_{\alpha,\beta}(X_2)$$

is the same as the result of composing the mappings on the direct path

$$\mathscr{X}_{\alpha,\beta}(X_1) \xrightarrow{\tau} \mathscr{X}_{\alpha,\beta}(X_2) \xrightarrow{\alpha} \mathscr{X}_{\alpha,\beta}(X_2).$$

Therefore, (7.2) is commutative for $\eta_1 = \eta_2 = \alpha$.

Similarly, from the second and the third relations in (7.1), the commutativity for $\eta_1 = \eta_2 = \beta$, and for $\eta_1 = \mathscr{P}_1$ and $\eta_2 = \mathscr{P}_2$ are obtained.

Sufficiency. On the basis of (7.2), the three relations in (7.1) can be induced from the commutativity for $\eta_1 = \eta_2 = \alpha$, for $\eta_1 = \eta_2 = \beta$ and for $\eta_1 = \mathscr{P}_1$ and $\eta_2 = \mathscr{P}_2$. This is the sufficiency. □

7.2 Isomorphism theorem

Because an isomorphism between two maps determines an equivalent relation, what has to be considered for the equivalence is the equivalent classes, called *isomorphic classes* of maps. Two maps are said to be different if they are in different isomorphic classes. In order to clarify the isomorphic classes of maps, invariants should be investigated.

In this and the following sections, a sequence of elements with its length half the cardinality of the ground set is discussed.

In fact, this implies that the isomorphic class can be determined by a polynomial of degree as a linear function of half the cardinality of the ground set for both orientable and nonorientable maps.

Lemma 7.2 If two maps M_1 and M_2 are isomorphic, then M_1 is orientable if, and only if, M_2 is orientable.

Proof Let $M_i = (\mathscr{X}_i, \mathscr{P}_i)$, $i = 1, 2$. Assume τ is an isomorphism from M_1 to M_2. From (7.2), $\tau\alpha = \alpha\tau$, $\tau\beta = \beta\tau$ and $\tau\mathscr{P}_1 = \mathscr{P}_2\tau$, i.e., $\tau\alpha\tau^{-1} = \alpha$, $\tau\beta\tau^{-1} = \beta$ and $\tau\mathscr{P}_1\tau^{-1} = \mathscr{P}_2$.

Necessity. Since M_1 is orientable, from Theorem 4.1, permutation group $\Psi_1 = \Psi_{\{\mathscr{P}_1,\gamma\}}$ has two orbits $(x_1)_{\Psi_1}$ and $(\alpha x_1)_{\Psi_1}$, $x_1 \in \mathscr{X}_1$ on $\mathscr{X}_1$. And since $\tau\alpha\tau^{-1} = \alpha$ and $\tau\beta\tau^{-1} = \beta$,

$$\begin{aligned}\tau\gamma\tau^{-1} &= \tau(\alpha\beta)\tau^{-1} = \tau(\alpha\tau^{-1}\tau\beta)\tau^{-1}\\ &= (\tau\alpha\tau^{-1})(\tau\beta\tau^{-1}) = \alpha\beta\\ &= \gamma.\end{aligned}$$

By considering $\tau\mathscr{P}_1\tau^{-1} = \mathscr{P}_2$, for any $\psi_1 \in \Psi_1$,

$$\tau\psi_1\tau^{-1} = \psi_2 \in \Psi_2.$$

Therefore, Ψ_2 also has two orbits on $\mathscr{X}_2$, namely, $(x_2)_{\Psi_2}$ and $(\alpha x_2)_{\Psi_2}$, where $x_2 = \tau x_1 \in \mathscr{X}_2$. This implies that M_2 is orientable as well.

Sufficiency. Because of the symmetry of τ between M_1 and M_2, the sufficiency is deduced from the necessity. □

For a map $M = (\mathscr{X}, \mathscr{P})$, where $\nu(M)$, $\epsilon(M)$ and $\phi(M)$ stand for, respectively, the *order* (vertex number), the *size* (edge number) and the *co-order* (face number) of M.

Lemma 7.3 If two maps M_1 and M_2 are isomorphic, then

$$\nu(M_1) = \nu(M_2), \quad \epsilon(M_1) = \epsilon(M_2), \quad \phi(M_1) = \phi(M_2). \tag{7.3}$$

Proof Let $M_i = (\mathscr{X}_i, \mathscr{P}_i)$, $i = 1, 2$. Assume τ is an isomorphism from M_1 to M_2. From the commutativity for $\eta_1 = \mathscr{P}_1$ and $\eta_2 = \mathscr{P}_2$ in (7.2), $\tau\mathscr{P}_1\tau^{-1} = \mathscr{P}_2$. Then, for any integer $n \geq 1$ by induction,

$$\begin{aligned}\tau(\mathscr{P}_1)^n\tau^{-1} &= \tau((\mathscr{P}_1)^{n-1}\mathscr{P}_1)\tau^{-1}\\ &= \tau((\mathscr{P}_1)^{n-1}\tau^{-1}\tau\mathscr{P}_1)\tau^{-1}\\ &= (\tau(\mathscr{P}_1)^{n-1}\tau^{-1})(\tau\mathscr{P}_1)\tau^{-1})\\ &= (\tau(\mathscr{P}_1)^{n-1}\tau^{-1})(\tau\mathscr{P}_1)\tau^{-1})\\ &= (\mathscr{P}_2)^{n-1}\mathscr{P}_2\\ &= \mathscr{P}_2^n.\end{aligned}$$

Therefore, for any $x_1 \in \mathscr{X}_1$, $\tau x_1 = x_2$,

$$\tau(x_1)_{\mathscr{P}_1} = (\tau x_1)_{\tau\mathscr{P}_1\tau^{-1}} = (x_2)_{\mathscr{P}_2}.$$

Because a one-to-one correspondence on vertices between M_1 and M_2 is induced from this, $\nu(M) = \nu(M)$.

Similarly, from $\tau\gamma\tau^{-1} = \gamma$ and $\tau(\mathscr{P}_1)^n\tau^{-1} = (\mathscr{P}_2)^n$,

$$\tau(\mathscr{P}_1\gamma)\tau^{-1} = \mathscr{P}_2\gamma.$$

Further, for any integer $n \geq 1$, $\tau(\mathscr{P}_1\gamma)^n\tau^{-1} = (\mathscr{P}_2\gamma)^n$. This provides

$$\tau(x_1)_{\mathscr{P}_1\gamma} = (\tau x_1)_{\tau(\mathscr{P}_1\gamma)\tau^{-1}} = (x_2)_{\mathscr{P}_2\gamma}$$

as a one-to-one correspondence on faces between M_1 and M_2. Therefore, $\phi(M_1) = \phi(M_2)$.

Finally, from $\tau\alpha\tau^{-1} = \alpha$ and $\tau\beta\tau^{-1} = \beta$ and hence $\tau\gamma\tau^{-1} = \gamma$, for any $x_1 \in \mathscr{X}_1$, $x_2 = \tau x_1$ implies $\tau K x_1 = K x_2$.

This provides a one-to-one correspondence on edges between M_1 and M_2. therefore, $\epsilon(M_1) = \epsilon(M_2)$. □

For a map $M = (\mathscr{X}, \mathscr{P})$, the Euler characteristic given by (4.1) is $\chi(M) = \nu(M) - \epsilon(M) + \phi(M)$, where $\nu(M)$, $\epsilon(M)$ and $\phi(M)$ are, respectively, the order, the size and the co-order of M.

Corollary 7.1 If two maps M_1 and M_2 are isomorphic, then

$$\chi(M_1) = \chi(M_2). \tag{7.4}$$

Proof A direct result of Lemma 7.3. □

For a map $M = (\mathscr{X}_{\alpha,\beta}, \mathscr{P})$, let $M^* = (\mathscr{X}^*_{\alpha,\beta}, \mathscr{P}^*)$ be the dual of M. It is, from Chapter 3, known that $M^* = (\mathscr{X}_{\beta,\alpha}, \mathscr{P}\gamma)$.

Theorem 7.2 Maps M_1 and M_2 are isomorphic if, and only if, their duals M_1^* and M_2^* are isomorphic.

Proof Let $M_i = (\mathscr{X}^{(i)}_{\alpha,\beta}, \mathscr{P}_i)$, $i = 1, 2$, then $M_i^* = (\mathscr{X}^{(i)^*}_{\alpha,\beta}, \mathscr{P}_i^*)$, $i = 1, 2$, where $\mathscr{X}^{(i)^*}_{\alpha,\beta} = \mathscr{X}^{(i)}_{\beta,\alpha}$ and $\mathscr{P}_i^* = \mathscr{P}_i\gamma$, $i = 1, 2$.

Necessity. Suppose τ is an isomorphic between M_1 and M_2, then from Theorem 7.1,

$$\tau\alpha\tau^{-1} = \alpha,\ \tau\beta\tau^{-1} = \beta,\ \tau\mathscr{P}_1\tau^{-1} = \mathscr{P}_2.$$

On the basis of this, for any $x_1 \in \mathscr{X}^{(1)^*}_{\alpha,\beta} = \mathscr{X}^{(1)}_{\beta,\alpha}$ and $x_2 = \tau x_1 \in \mathscr{X}^{(2)^*}_{\alpha,\beta} = \mathscr{X}^{(2)}_{\beta,\alpha}$,

$$\begin{aligned}\tau K^* x_1 &= \tau\{x_1, \beta x_1, \alpha x_1, \gamma x_1\} = \{\tau x_1, \tau\beta x_1, \tau\alpha x_1, \tau\gamma x_1\}\\ &= \{x_2, \beta x_2, \alpha x_2, \gamma x_2\} = K^* x_2,\end{aligned}$$

and

$$\begin{aligned}\tau\mathscr{P}_1^*\tau^{-1} &= \tau(\mathscr{P}_1\gamma)\tau^{-1} = \tau(\mathscr{P}_1\tau^{-1}\tau\gamma)\tau^{-1}\\ &= (\tau\mathscr{P}_1\tau^{-1})(\tau\gamma\tau^{-1}) = \mathscr{P}_2\gamma\\ &= \mathscr{P}_2^*.\end{aligned}$$

This implies that the diagram

$$\begin{array}{ccc}\mathscr{X}^{(1)^*}_{\alpha,\beta} & \xrightarrow{\ \tau\ } & \mathscr{X}^{(2)^*}_{\alpha,\beta}\\ \Big\downarrow{\scriptstyle\eta_1} & & \Big\downarrow{\scriptstyle\eta_2}\\ \mathscr{X}^{(1)^*}_{\alpha,\beta} & \xrightarrow{\ \tau\ } & \mathscr{X}^{(2)^*}_{\alpha,\beta}\end{array} \tag{7.5}$$

is all commutative for $\eta_1 = \eta_2 = \beta$, for $\eta_1 = \eta_2 = \alpha$, and for $\eta_1 = \mathscr{P}_1^*$ and $\eta_2 = \mathscr{P}_2^*$. Therefore, from Theorem 7.1, τ is an isomorphism between M_1^* and M_2^* in its own right.

Sufficiency. From the symmetry of duality, the sufficiency is deduced from the necessity. □

Let $M_i = (\mathscr{X}^{(i)}_{\alpha,\beta}, \mathscr{P}_i)$, and $M_i^* = (\mathscr{X}^{(i)^*}_{\alpha,\beta}, \mathscr{P}_i^*)$, where $\mathscr{X}^{(i)^*}_{\alpha,\beta} = \mathscr{X}^{(i)}_{\beta,\alpha}$ and $\mathscr{P}_i^* = \mathscr{P}_i\gamma$, $i = 1, 2$.

Corollary 7.2 A bijection $\tau : \mathscr{X}^{(1)}_{\alpha,\beta} \longrightarrow \mathscr{X}^{(2)}_{\alpha,\beta}$ is an isomorphism between maps M_1 and M_2 if, and only if, τ is an isomorphism between maps M_1^* and M_2^*.

Proof A direct result in the proof of Theorem 7.2. □

7.3 Recognition

Although some invariants are provided, they are still far from determining an isomorphism between two maps in the last section.

In fact, it will be shown in this section that an isomorphism between two maps can be determined by the number of invariants dependent on their size, i.e., a sequence of invariants in a number as a function of their size.

In order to do this, algorithms are established for justifying and identifying if two maps are isomorphic. In other words, an isomorphism can be found between two maps if any; or no isomorphism exits at all otherwise.

Generally speaking, since the ground set of a map is finite, i.e., its cardinality is 4ϵ, where ϵ is the size of the map, in a theoretical point of view, there exists a permutation that corresponds to an isomorphism among all the $(4\epsilon)!$ permutations if any, or no isomorphism at all between two maps otherwise. However, this is an impractical way even on a modern computer.

Our purpose is to establish an algorithm directly with the amount of computation as small as possible without counting all the permutations.

Here, two types of algorithms are presented. One is called *vertex-algorithm* based on (7.2). Another is called *face-algorithm* based on (7.5).

Their clue is as follows. For two maps $M_1 = (\mathscr{X}_1, \mathscr{P}_1)$ and $M_2 = (\mathscr{X}_2, \mathscr{P}_2)$, from Lemma 7.3, it is only necessary to consider $|\mathscr{X}_1| = |\mathscr{X}_2|$ because the cardinality is an invariant under an isomorphism.

First, choose $x_1 \in \mathscr{X}_1$ and $y_1 \in \mathscr{X}_2$ (a trick should be noticed here!).

Then, start, respectively, from x_1 and y_1 on M_1 and M_2 by a certain rule (algorithms are distinguished by rules). Arrange the orbits $\{x_1\}_{\Psi_{\{\mathscr{P}_1,\gamma\}}}$ and $\{y_1\}_{\Psi_{\{\mathscr{P}_2,\gamma\}}}$ as cycles. If

$$\tau(x_1)_{\Psi_{\{\mathscr{P}_1,\gamma\}}} \neq (y_1)_{\Psi_{\{\mathscr{P}_2,\gamma\}}} \tag{7.6}$$

is induced from $y_1 = \tau(x_1)$, then stop. Otherwise, choose another y_1 (a trick!). Go on the procedure for M_2 until every possible y_1 has be chosen.

Finally, if stops at the latter, then it is shown that M_1 and M_2 are not isomorphic, and denoted by $M_1 \neq M_2$; otherwise, an isomorphism between M_1 and M_2 is done from (7.6), denoted by $M_1 = M_2$.

Algorithm 7.1 Based on vertices, determine if two maps are isomorphic.

Given two maps $P = (\mathscr{X}, \mathscr{P})$ and $Q = (\mathscr{Y}, \mathscr{Q})$, their order, size and co-order are all equal (otherwise, not isomorphic!). For convenience, for any $x \in \mathscr{X}$, let $|x| = |\{x\}_{\mathscr{P}}|$, i.e., the valency of vertex $(x)_{\mathscr{P}}$.

Initiation Given $x \in \mathscr{X}$, choose $y \in \mathscr{Y}$. Let $\tau(x) = y$ and $\tau Kx = Ky$. Label both x and y by 1. Naturally, $Kx = Ky = K1 = \{1, \alpha 1, \beta 1, \gamma 1\}$ (Here, the number 1 deals with a symbol!). Label $(x)_{\mathscr{P}}$ by 0, then $x = 1$ is the first element coming to vertex 0. By (v, t_v), denote that t_v is the first element coming to vertex v.

Let S be a sequence of symbols storing numbers and symbols and l, the maximum of labels on all the edges with a label. Here, $S = \emptyset$, $l = 1$ and the minimum of labels among all labelled but not passed vertices $n = 0$. If vertex $(\gamma 1)_{\mathscr{P}} = (1)_{\mathscr{P}}$, the maximum vertex label $m = 0$; otherwise, label vertex $(\gamma 1)_{\mathscr{P}}$ by $1m = 1$.

Proceeding When all vertices are labelled as used, then go to Halt (1).

For n, let s_P and s_Q be, respectively, the number of edges without label on $(\gamma t_n)_{\mathscr{P}}$ and $(\gamma t_n)_{\mathscr{Q}}$.

If $s_P \neq s_Q$, when no y can be chosen, then go to Halt (2); otherwise, choose another y and then go to Initiation.

In the direction starting from γt_n, label those edges by $l+1, \ldots, l+s$, $s = s_P = s_Q \geq 0$ in order. Thus, two linear orders of elements with numbers labeled

$$\langle \gamma t_n, \mathscr{P}\gamma t_n, \ldots, \mathscr{P}^{-1}\gamma t_n \rangle$$

and

$$\langle \gamma t_n, \mathscr{Q}\gamma t_n, \ldots, \mathscr{Q}^{-1}\gamma t_n \rangle$$

are obtained.

If the two are not equal, when no y is available to choose, then go to Halt (2); otherwise, choose another y and then go to Initiation.

Put this linear order into S as last part and then substitute the extended sequence for S. In the meantime, label $K(l+1), K(l+2), \ldots, K(l+s)$ on P and Q. Substitute $l+s$ for l. Mark vertex n as used. Substitute $n+1$ for n. Let r be the number of vertices without label in

$$(\gamma(l+1))_{\mathscr{P}}, \ldots, (\gamma(l+s))_{\mathscr{P}},$$

and label them as $m+1, \ldots, m+r$ in order. Substitute $m+r$ for m. Go on the Proceeding.

Halt (1) Output S. (2) P and Q are not isomorphic.

About Algorithm 7.1, from the way of choosing y, each element in the ground set is passed through at most once. So there exists a constant c such that the amount of computation is at most $c|\mathscr{X}|$. Since the worst case is that y chooses all over the ground set $\mathscr{Y}$, the total amount of computation is at most $c|\mathscr{X}|^2$. Because $|\mathscr{X}| = 4\epsilon$ where ϵ is the size of the map, this amount is with its order as the size squared, i.e., $O(\epsilon^2)$.

As described above, if checking all possibilities of $|\mathscr{Y}|!$, by Stirling formula,

$$\begin{aligned} |\mathscr{Y}|! &\sim \sqrt{2\pi} e^{-|\mathscr{Y}|} |\mathscr{Y}|^{|\mathscr{Y}|-\frac{1}{2}} \\ &\gg O(e^{|\mathscr{Y}|}) \gg O(\epsilon^{\epsilon}) \\ &\gg O(\epsilon^2) \end{aligned}$$

when $|\mathscr{Y}| = |\mathscr{X}| = 4\epsilon$ is large enough. Thus, this algorithm is much efficient.

Algorithm 7.2 Based on faces, determine if two maps are isomorphic.

Given two maps $P = (\mathscr{X}_{\alpha,\beta}, \mathscr{P})$ and $Q = (\mathscr{Y}_{\alpha,\beta}, \mathscr{Q})$, their order, size and co-order are all equal (otherwise, not isomorphic!). For convenience, let $\mathscr{X} = \mathscr{X}_{\alpha,\beta}$, $\mathscr{Y} = \mathscr{Y}_{\alpha,\beta}$ and for any $x \in \mathscr{X}$, let $|x| = |\{x\}_{\mathscr{P}\gamma}|$, i.e., the valency of face $(x)_{\mathscr{P}\gamma}$ where $\gamma = \alpha\beta$.

Initiation Given $x \in \mathscr{X}$, choose $y \in \mathscr{Y}$. Let $\tau(x) = y$ and $\tau Kx = Ky$. Label both x and y by 1. Naturally, $Kx = Ky = K1 = \{1, \alpha 1, \beta 1, \gamma 1\}$ (Here, the number 1 deals with a symbol!). Label $(x)_{\mathscr{P}|ga}$ by 0, then $x = 1$ is the first element coming to face 0. By (f, t_f) denote that t_f is the first element coming to face f.

Let T be a sequence of symbols storing numbers and symbols and l, the maximum of labels over all the edges with a label. Here, $T = \emptyset$, $l = 1$ and the minimum of labels among all labeled but not passed faces $n = 0$. If face $(\gamma 1)_{\mathscr{P}\gamma} = (1)_{\mathscr{P}\gamma}$, the maximum face label $m = 0$; otherwise, label face $(\gamma 1)_{\mathscr{P}\gamma}$ by $1m = 1$.

Proceeding When all faces are labeled as used, then go to Halt (1).

For n, let s_P and s_Q be, respectively, the number of edges without label on $(\gamma t_n)_{\mathscr{P}\gamma}$ and $(\gamma t_n)_{\mathscr{Q}\gamma}$.

If $s_P \neq s_Q$, when no y can be chosen, then go to Halt (2); otherwise, choose another y and then go to Initiation.

In the direction starting from γt_n, label those edges by $l+1, \ldots, l+s$, $s = s_P = s_Q \geq 0$ in order. Thus, two linear orders of elements with numbers labeled

$$\langle \gamma t_n, \mathscr{P}\gamma\gamma t_n, \ldots, \mathscr{P}\gamma^{-1}\gamma t_n \rangle$$

and

$$\langle \gamma t_n, \mathscr{Q}\gamma\gamma t_n, \ldots, \mathscr{Q}\gamma^{-1}\gamma t_n \rangle$$

are obtained.

If the two are not equal, when no y is available to choose, then go to Halt (2); otherwise, choose another y and then go to Initiation.

Put this linear order into S as last part and then substitute the extended sequence for S. In the meantime, label $K(l+1)$, $K(l+1)$, ..., $K(l+s)$ on P and Q. Substitute $l+s$ for l. Mark n as used. Substitute $n+1$ for n. Let r be the number of vertices without label in

$$(\gamma(l+1))_{\mathscr{P}}, \ldots, (\gamma(l+s))_{\mathscr{P}},$$

and label them as $m+1, \ldots, m+r$ in order. Substitute $m+r$ for m. Go on the Proceeding.

Put this linear order into T as last part and then substitute the extended sequence for T. In the meantime, label $K(l+1)$, $K(l+2)$, ..., $K(l+s)$ on P and Q. Substitute $l+s$ for l. Mark face n as used. Substitute $n+1$ for n. Let r be the number of faces without label in

$$(\gamma(l+1))_{\mathscr{P}}, \ldots, (\gamma(l+s))_{\mathscr{P}},$$

and label them as $m + 1,\ldots,m + r$ in order. Substitute $m + r$ for m. Go on the Proceeding.

Halt (1) Output T. (2) P and Q are not isomorphic.

About Algorithm 7.2, it can be seen as the dual of Algorithm 7.1. The amount of its computation is also estimated as $O(\epsilon^2)$.

Note 7.2 These two algorithms suggest that whenever a cyclic order of edges at each vertex is given, an efficient algorithm for justifying and identifying if two graphs are isomorphic within the cyclic order at each vertex can be established. By saying an algorithm *efficient*, it is meant that there exists a constant c such that the amount of its computation is about $O(\epsilon^c)$, ϵ is the size of the graphs.

Whenever none of a cyclic order at each vertex, no efficient algorithm for an isomorphism of two graphs has been found yet. However, a new approach is, from what has been discussed here, provided for further investigation of an isomorphism between two graphs.

7.4 Justification

In this section, it is shown that the two algorithms described in the previous section can be used for justifying and identifying whether or not two maps are isomorphic.

Lemma 7.4 Let S and T are, respectively, the outputs of Algorithm 7.1 and Algorithm 7.2 at Halt (1), then

(i) Elements in S and T are all in the same orbit of group $\Psi_{\{\mathscr{P},\gamma\}}$ on $\mathscr{X}$.
(ii) S forms an orbit of group $\Psi_{\{\mathscr{P},\gamma\}}$ on $\mathscr{X}$ if, and only if, T forms an orbit of group $\Psi_{\{\mathscr{P},\gamma\}}$ on $\mathscr{X}$.
(iii) S forms an orbit of group $\Psi_{\{\mathscr{P},\gamma\}}$ on $\mathscr{X}$ if, and only if, for any $x \in S$, $\gamma x \in S$.

Proof

(i) From the proceedings of the two algorithms, it is seen that an element passes only through γ and $\mathscr{P}$ (Algorithm 7.1), or γ and $\mathscr{P}\gamma$ (Algorithm 7.2) for getting an element in S, or T. Because γ, $\mathscr{P}$, $\mathscr{P}\gamma \in \Psi_{\{\mathscr{P},\gamma\}}$ and $\gamma^2 = 1$, elements in S and T are all in the same orbit of group $\Psi_{\{\mathscr{P},\gamma\}}$ on $\mathscr{X}$.
(ii) Necessity. Because S forms an orbit of group $\Psi_{\{\mathscr{P},\gamma\}}$ on $\mathscr{X}$ and, from Algorithm 7.1, S contains half the elements of $\mathscr{X}$, by Lemma 4.1, group $\Psi_{\{\mathscr{P},\gamma\}}$ has two orbits on $\mathscr{X}$. This implies in the orientable case. Thus, from (i), T form an orbit of group $\Psi_{\{\mathscr{P},\gamma\}}$ on $\mathscr{X}$ as well.

Sufficiency. On the basis of duality, it is deduced from the necessity.

(iii) Necessity. Since S forms an orbit of group $\Psi_{\{\mathscr{P},\gamma\}}$ on $\mathscr{X}$ and S contains only half the elements of $\mathscr{X}$, by Lemma 4.1, group $\Psi_{\{\mathscr{P},\gamma\}}$ has two orbits on $\mathscr{X}$. From the orientability, for any $x \in S$, $\gamma x \in S$.

Sufficiency. Since for any $x \in S$, $\gamma x \in S$, and S only contains half the elements of $\mathscr{X}$, by Corollary 4.1, it is only possible that S itself form an orbit of group $\Psi_{\{\mathscr{P},\gamma\}}$ on $\mathscr{X}$. □

For nonorientable maps, such two algorithms have their outputs S and T also containing half the elements of $\mathscr{X}$ but not forming an orbit of group $\Psi_{\{\mathscr{P},\gamma\}}$.

Lemma 7.5 Let S and T be, respectively, the outputs of Algorithm 7.1 and Algorithm 7.2 at Halt (1). And , let G_S and G_T be, respectively, the graphs induced by elements in S and T, then $G_S = G_T = G(P)$.

Proof From Lemma 7.4(i), by the procedures of the two algorithms, because the intersection of each of S and T with any quadricell consists of two elements incident to the two ends of the edge, S and T are also incident to all edges with two ends of each edge in map P.

Therefore, $G_S = G_T = G(P)$. □

Theorem 7.3 The output S of Algorithm 7.1 at Halt (1) induces an isomorphism between maps P and Q. Halt (2) shows that maps P and Q are not isomorphic.

Proof Let τ be a mapping from $\mathscr{X}$ to $\mathscr{Y}$ such that the image and the co-image are with the same label. From the transitivity of a map, τ is a bijection. Because $\tau Kx = K\tau x$, $x \in \mathscr{X}$, so $\tau\alpha\tau^{-1} = \alpha$ and $\tau\beta\tau^{-1} = \beta$. And in the Proceeding, for labeling a vertex $(x)_{\mathscr{P}}$, $\tau(x)_{\mathscr{P}} = (\tau x)_{\mathscr{Q}}$. From Lemma 7.5, this implies that $\tau\mathscr{P}\tau^{-1} = \mathscr{Q}$. Based on Theorem 7.1, τ is an isomorphism between P and Q. This is the first statement.

By contradiction to prove the second statement. Assume that there is an isomorphism τ between P and Q. If $\tau(x) = y$, then by Algorithm 7.1 the procedure should terminate at Halt (1). However, a termination at Halt (2) shows that for any $x \in \mathscr{X}$, there is no element in $\mathscr{Y}$ corresponding to x in an isomorphism between maps P and Q, and hence it is impossible to terminate at Halt (1). This is a contradiction.

Therefore, the theorem is true. □

Although the theorem below has its proof with a similar reasoning, in order to understand the precise differences, the proof is still in a detailed explanation.

Theorem 7.4 The output T of Algorithm 7.2 at Halt (1) induces an isomorphism between maps P and Q. Halt (2) shows that maps P and Q are not isomorphic.

Proof Let τ be a mapping from $\mathscr{X}$ to $\mathscr{Y}$ such that the image and the co-image are with the same label. From the transitivity of a map, τ is a bijection. Because $\tau Kx = K\tau x$,

$x \in \mathscr{X}$, so $\tau\alpha\tau^{-1} = \alpha$ and $\tau\beta\tau^{-1} = \beta$. And in the Proceeding, for labeling a face $(x)_{\mathscr{P}\gamma}$, $\tau(x)_{\mathscr{P}\gamma} = (\tau x)_{\mathscr{Q}\gamma}$. From Lemma 7.5, this implies that $\tau\mathscr{P}\gamma\tau^{-1} = \mathscr{Q}\gamma$. Based on Theorem 7.2, τ is an isomorphism between P and Q. This is the first statement.

By contradiction to prove the second statement. Assume that there is an isomorphism τ between P and Q. If $\tau(x) = y$, then by Algorithm 7.2 the procedure should terminate at Halt (1). However, a termination at Halt (2) shows that for any $x \in \mathscr{X}$, there is no element in $\mathscr{Y}$ corresponding to x in an isomorphism between maps P and Q, and hence it is impossible to terminate at Halt (1). This is a contradiction.

Therefore, the theorem is true. □

If missing what is related to y in Algorithm 7.1 and Algorithm 7.2, then for any map $M = (\mathscr{X}, \mathscr{P})$, the procedures will always terminate at Halt (1). Thus, their outputs S and T are, respectively, called a *primal trail code* and a *dual trail code* of M. When an element x and a map P should be indicated, they are denoted by respective $S_x(P)$ and $T_x(P)$.

Theorem 7.5 Let $P = (\mathscr{X}, \mathscr{P})$ and $Q = (\mathscr{Y}, \mathscr{Q})$ be two given maps. Then, they are isomorphic if, and only if, for any $x \in \mathscr{X}$ chosen, there exists an element $y \in \mathscr{Y}$ such that $S_x(P) = S_y(Q)$, or $T_x(P) = T_y(Q)$.

Proof Necessity. Suppose τ is an isomorphism between maps $P = (\mathscr{X}, \mathscr{P})$ and $Q = (\mathscr{Y}, \mathscr{Q})$. For the given element $x \in \mathscr{X}$, let $y = \tau(x)$. From Theorem 7.3, or Theorem 7.4, $S_x(P) = S_y(Q)$, or $T_x(P) = T_y(Q)$.

Sufficiency. From Theorem 7.3, or Theorem 7.4, it is known that by Algorithm 7.1 or Algorithm 7.2, their outputs induce an isomorphism between $P = (\mathscr{X}, \mathscr{P})$ and $Q = (\mathscr{Y}, \mathscr{Q})$. □

Note 7.3 In justifying whether or not two maps are isomorphic, the initial element x can be chosen arbitrarily in one of the two maps to see if there is an element y in the other such that $S_x(P) = S_y(Q)$, or $T_x(P) = T_y(Q)$. This enables us to do for some convenience.

In addition, based on Theorem 7.5, all isomorphisms between two maps can be found if any.

7.5 Pattern examples

Here, two pattern examples are provided for further understanding the procedures of the two algorithms described in the previous section.

Pattern 7.1 Justify whether or not two maps $M_1 = (\mathscr{X}_1, \mathscr{P}_1)$ and $M_2 = (\mathscr{X}_2, \mathscr{P}_2)$ are isomorphic where

$$\mathscr{X}_1 = Kx_1 + Ky_1, \quad \mathscr{P}_1 = (x_1, y_1, \beta y_1)(\gamma x_1)$$

and

$$\mathcal{X}_2 = Kx_2 + Ky_2, \ \mathcal{P}_2 = (y_2, x_2, \beta y_2)(\gamma x_2).$$

First, for M_1, choose $x = x_1$. By Algorithm 7.1, find $S_x(M_1)$. Let

$$\mathcal{P}_1 = (x_1, y_1, \beta y_1)(\gamma x_1) = uv.$$

Initiation

$$x_1 = 1, \ Kx_1 = \{1, \alpha 1, \beta 1, \gamma 1\}, \ u = 0, v = 1,$$
$$S = \emptyset, \ l = 0, \ m = 1.$$

Proceeding

Step 1 $\mathcal{P}_1 = (1, y_1, \beta y_1)(\gamma 1)$.

$$y_1 = 2, \ Ky_1 = \{2, \alpha 2, \beta 2, \gamma 2\}, \ u = 0, \ v = 1,$$
$$S = \langle 1, 2, \beta 2\rangle, \ l = 2, \ n = 1, \ m = 1.$$

Step 2 $\mathcal{P}_1 = (1, 2, \beta 2)(\gamma 1)$.

$$u = 0, \ v = 1,$$
$$S = \langle 1, 2, \beta 2, \gamma 1\rangle, \ l = 2, \ n = 1, \ m = 1.$$

Halt (1) Output: $S_x(M_1) = S = \langle 1, 2, \beta 2, \gamma 1\rangle$.

Then, for M_2, because a link should correspond to a link and a vertex should correspond to a vertex with the same valency, y has only two possibilities for choice: x_2 and αx_2. Choose $y = x_2$. By Algorithm 7.1, find $S_y(M_2)$. Let

$$\mathcal{P}_2 = (y_2, x_2, \beta y_2)(\gamma x_2) = uv.$$

Initiation

$$x_2 = 1, \ Kx_2 = \{1, \alpha 1, \beta 1, \gamma 1\}, \ u = 0, v = 1,$$
$$S = \emptyset, \ l = 0, \ m = 1.$$

Proceeding

Step 1 $\mathcal{P}_2 = (y_1, 1, \beta y_2)(\gamma 1)$.

$$\beta y_2 = 2, \ K\beta y_2 = \{2, \alpha 2, \beta 2, \gamma 2\}, \ u = 0, \ v = 1,$$
$$S = \langle 1, 2, \beta 2\rangle, \ l = 2, \ n = 1, \ m = 1.$$

Step 2 $\mathscr{P}_2 = (2, 1, \beta 2)(\gamma 1)$.

$$u = 0,\ v = 1,$$
$$S = \langle 1, 2, \beta 2, \gamma 1\rangle,\ l = 2,\ n = 1,\ m = 1.$$

Halt (1) Output: $S_y(M_2) = S = \langle 1, 2, \beta 2, \gamma 1\rangle$.

Since $S_x(M_1) = S_y(M_2)$ and $y = x_2$, an isomorphism from M_1 to M_2 is found as τ_1:

$$\tau_1 Kx_1 = Kx_2,\ \tau_1 Ky_1 = Ky_2.$$

Then, choose $y = \alpha x_2$. By Algorithm 7.1, find $S_y(M_2)$. Let

$$\mathscr{P}_2 = (\alpha x_2, \alpha y_2, \gamma y_2)(\beta x_2) = uv.$$

Initiation

$$\alpha x_2 = 1,\ K\alpha x_2 = \{1, \alpha 1, \beta 1, \gamma 1\},\ u = 0, v = 1,$$
$$S = \emptyset,\ l = 0,\ m = 1.$$

Proceeding

Step 1 $\mathscr{P}_2 = (1, \alpha y_2, \gamma y_2)(\gamma 1)$.

$$\alpha y_2 = 2,\ K\alpha y_2 = \{2, \alpha 2, \beta 2, \gamma 2\},\ u = 0,\ v = 1,$$
$$S = \langle 1, 2, \beta 2\rangle,\ l = 2,\ n = 1,\ m = 1.$$

Step 2 $\mathscr{P}_2 = (1, 2, \beta 2)(\gamma 1)$.

$$u = 0,\ v = 1,$$
$$S = \langle 1, 2, \beta 2, \gamma 1\rangle,\ l = 2,\ n = 1,\ m = 1.$$

Halt (1) Output: $S_y(M_2) = S = \langle 1, 2, \beta 2, \gamma 1\rangle$.

Since $S_x(M_1) = S_y(M_2)$ and $y = \alpha x_2$, an isomorphism from M_1 to M_2 is found as τ_2:

$$\tau_2 Kx_1 = K\alpha x_2,\ \tau_2 Ky_1 = K\alpha y_2.$$

In consequence, there are in all two isomorphisms between M_1 and M_2 above. Since $2 \in S_x(M_1)$ but $\gamma 2 \notin S_x(M_1)$, by Lemma 7.4(iii), M_1 and M_2 are nonorientable.

Pattern 7.2 Justify whether or not $M_1 = (\mathscr{X}_1, \mathscr{P}_1)$ and $M_2 = (\mathscr{X}_2, \mathscr{P}_2)$ are isomorphic where

$$\mathscr{X}_1 = Kx_1 + Ky_1,\ \mathscr{P}_1 = (x_1, y_1, \gamma y_1)(\gamma x_1)$$

and

$$\mathscr{X}_2 = Kx_2 + Ky_2, \ \mathscr{P}_2 = (y_2, x_2, \gamma y_2)(\gamma x_2).$$

First, for M_1, choose $x = x_1$. By Algorithm 7.2, find $T_x(M_1)$. Let

$$\mathscr{P}_1\gamma = (x_1, \gamma x_1, y_1)(\gamma y_1) = fg.$$

Initiation

$$x_1 = 1, \ Kx_1 = \{1, \alpha 1, \beta 1, \gamma 1\}, \ f = 0, g = 1,$$
$$T = \emptyset, \ l = 0, \ m = 0.$$

Proceeding

Step 1 $\mathscr{P}_1\gamma = (1, \gamma 1, y_1)(\gamma y_1)$.

$$y_1 = 2, \ Ky_1 = \{2, \alpha 2, \beta 2, \gamma 2\}, \ f = 0, \ g = 1,$$
$$T = \langle 1, \gamma 1, 2\rangle, \ l = 2, \ n = 1, \ m = 1.$$

Step 2 $\mathscr{P}_1\gamma = (1, \gamma 1, 2)(\gamma 2)$.

$$f = 0, \ g = 1,$$
$$T = \langle 1, \gamma 1, 2, \gamma 2\rangle, \ l = 2, \ n = 1, \ m = 1.$$

Halt (1) Output: $T_x(M_1) = T = \langle 1, \gamma 1, 2, \gamma 2\rangle$.

Then, for M_2, because a link should correspond to a link and a vertex to a vertex with the same valency, y has only two possibilities for choosing: x_2 and αx_2. Choose $y = x_2$. By Algorithm 7.2, find $T_y(M_2)$. Let

$$\mathscr{P}_2\gamma = (x_2, \gamma x_2, \gamma y_2)(y_2) = fg.$$

Initiation

$$x_2 = 1, \ Kx_2 = \{1, \alpha 1, \beta 1, \gamma 1\}, \ f = 0, g = 1,$$
$$T = \emptyset, \ l = 0, \ m = 0.$$

Proceeding

Step 1 $\mathscr{P}_2\gamma = (1, \gamma 1, \gamma y_2)(y_2)$.

$$\gamma y_2 = 2, \ K\gamma y_2 = \{2, \alpha 2, \beta 2, \gamma 2\}, \ f = 0, \ g = 1,$$
$$T = \langle 1, \gamma 1, 2\rangle, \ l = 2, \ n = 1, \ m = 1.$$

Step 2 $\mathscr{P}_2\gamma = (1, \gamma 1, 2)(\gamma 2)$.

$$f = 0,\ g = 1,$$
$$T = \langle 1, \gamma 1, 2, \gamma 2\rangle,\ l = 2,\ n = 1,\ m = 1.$$

Halt (1) Output: $T_y(M_2) = T = \langle 1, \gamma 1, 2, \gamma 2\rangle$.

Since $T_x(M_1) = T_y(M_2)$ and $y = x_2$, an isomorphism from M_1 to M_2 is found as τ_1:

$$\tau_1 Kx_1 = Kx_2,\ \tau_1 Ky_1 = K\gamma y_2.$$

Then, choose $y = \alpha x_2$. By Algorithm 7.2, find $T_y(M_2)$. Let

$$\mathscr{P}_2\gamma = (\alpha x_2, \beta x_2, \alpha y_2)(\beta y_2) = fg.$$

Initiation

$$\alpha x_2 = 1,\ K\alpha x_2 = \{1, \alpha 1, \beta 1, \gamma 1\},\ f = 0, g = 1,$$
$$T = \emptyset,\ l = 0,\ m = 0.$$

Proceeding

Step 1 $\mathscr{P}_2\gamma = (1, \gamma 1, \alpha y_2)(\beta y_2)$.

$$\alpha y_2 = 2,\ K\gamma y_2 = \{2, \alpha 2, \beta 2, \gamma 2\},\ f = 0,\ g = 1,$$
$$T = \langle 1, \gamma 1, 2\rangle,\ l = 2,\ n = 1,\ m = 1.$$

Step 2 $\mathscr{P}_2\gamma = (1, \gamma 1, 2)(\gamma 2)$.

$$f = 0,\ g = 1,$$
$$T = \langle 1, \gamma 1, 2, \gamma 2\rangle,\ l = 2,\ n = 1,\ m = 1.$$

Halt (1) Output: $T_y(M_2) = T = \langle 1, \gamma 1, 2, \gamma 2\rangle$.

Since $T_x(M_1) = T_y(M_2)$ and $y = \alpha x_2$, an isomorphism from M_1 to M_2 is found as τ_2:

$$\tau_2 Kx_1 = K\alpha x_2,\ \tau_2 Ky_1 = K\alpha y_2.$$

In consequence, there are in all two isomorphisms between M_1 and M_2. By Lemma 7.4(iii), M_1 and M_2 are orientable.

7.6 Notes

7.6.1 Discuss whether or not there exists a number of invariants, independent of the order, size and co-oder of a map considered, within isomorphism of maps for justifying and identifying an isomorphism between two maps.

Notice 7.1 Although Sections 7.3 and 7.4 imply a polynomial invariant for justifying and identifying an isomorphism between two maps, this problem is still necessary to be solved in the intelligentization stage for usages as easily as possible.

7.6.2 For a given graph G and an integer g, determine the number of distinct embeddings of G on the surface of relative genus g, and the number of non-isomorphic maps among them.

Notice 7.2 It has been shown that joint tree model of embedding is powerfull to determine the number as in Liu [72, 73, 90, 91]. The problem aims at an explicit expresion of the numbers. Partial results for certain types of G and g smaller can be seen as in Cai et al. [9], Cai and Liu [10, 11], Li and Liu [52, 54], Liu and Liu [58, 59], Pan and Liu [130], Ren and Liu [132–134], Wan et al. [155], Xu and Liu [167, 168], Yang and Liu [174–176] and Zhang and Liu [180].

7.6.3 For a given type of graphs $\mathscr{G}$ and an integer g, find the number of distinct embeddings of graphs in $\mathscr{G}$ on the surface of relative genus g, and the number of non-isomorphic maps among them.

Notice 7.3 Joint tree model of embedding might be a powerfull tool for determining the number as in Liu [72, 73, 90, 91]. The problem aims at an explicit expresion of the numbers. Partial results for certain types of G and g smaller can be seen as in Cai et al. [9], Cai and Liu [10, 11], Li and Liu [52, 54], Liu and Liu [58, 59], Pan and Liu [130], Ren and Liu [132–134], Wan et al. [155], Xu and Liu [167, 168], Yang and Liu [174–176] and Zhang and Liu [180].

7.6.4 Determine the number of non-isomorphic triangulations of size $m \geq 3$.

Notice 7.4 According to the principle in Liu [72, pp. 221–233], only necessary to determine the *auto-polynomial* (the coefficient of ith term is the number of all maps with order i of automorphism group) for rooted triangulations over all surfaces. Relevant research might be seen as in Liu [98], Cai and Liu [10, 11], Dong and Liu [21], Li and Liu [52, 54] and Liu [67, pp. 81–115, 78, pp. 67–95].

7.6.5 Determine the number of non-isomorphic quadrangulations of size $m \geq 4$.

Notice 7.5 According to the principle in Liu [72, pp. 221–233], only necessary to determine the auto-polynomial of rooted quadrangulations of size $m \geq 4$. Relevant references might be seen as in Liu [67, pp. 81–115, 78, pp. 97–119], Pan et al. [131], Ren and Liu [132, 133] and Zhang et al. [182].

7.6.6 For an integral vector $(n_2, n_4, \ldots, n_{2i}, \ldots)$, find the number of non-isomorphic Euler planar maps each of which has n_{2i} vertices of valency $2i$, $i \geq 1$.

Notice 7.6 Relevant references as in Liu [99–102], particularly [97, pp. 341–367, pp. 409–417] Cai and Liu [12].

Because it can be shown that two graphs G_1 and G_2 are isomorphic if, and only if, for a surface they can be embedded into, there exist embeddings $\mu_1(G_1)$ and $\mu_2(G_2)$ isomorphic. This enables us to investigate the isomorphism between the two graphs. The aim is at an efficient algorithm if any.

7.6.7 Suppose map M_1 is an embedding of G_1 on an orientable surface of genus g, justify whether or not there is an embedding M_2 of graph G_2 such that M_2 and M_1 are isomorphic.

Notice 7.7 On the basis of Liu [91] with Sections 7.2—7.4, observe the orientable classifications of rooted isomorphism for G_1 and G_2.

7.6.8 Suppose map M_1 is an embedding of G_1 on a non-orientable surface of genus g, justify whether or not there is an embedding M_2 of graph G_2 such that M_2 and M_1 are isomorphic.

Notice 7.8 On the basis of Liu [91] with Sections 7.2—7.4, observe the non-orientable classifications of rooted isomorphism for G_1 and G_2.

7.6.9 According to Liu [60], any graph with at least one circuit has a non-orientable embedding with only one face. Justify whether or not two graphs G_1 and G_2 have two respective single face embeddings that are isomorphic.

Notice 7.9 The invariants on single face embeddings within isomorphism has to be emphasized.

7.6.10 Justify whether or not a graph has two distinct single face embeddings that are isomorphic maps.

Notice 7.10 It looks that the order of the automorphism group of a single face embedding should be determined first.

A graph is called *up-embeddable* if it has an orientable embedding of genus that is the integral part of half the Betti number of graph. Because of the result in Liu [60], it is unnecessary to consider the up-embeddability for non-orientable case.

7.6.11 Determine the up-embeddability and maximum orientable genus of a graph via its joint sequences.

Notice 7.11 Based on joint tree model of embedding on surfaces, as shown in Liu [91], joint sequences are all associate polyhegons.

7.6.12 For a given graph G and an integer g, justify whether or not the graph G has an embedding of relative genus g.

Notice 7.12 According to such as Liu [92](Theorem 1 for $g \geq 0$; Theorem 2 for $g < 0$), this problem has been theoretically solved in the systematization stage, and hence in the other two stages, efficientization and intelligentization, still remains to open.

8 Asymmetrization

- An automorphism of a map is an isomorphism from the map to itself. All automorphisms of a map form a group called its automorphism group. A map is, in fact, the trivialization of its automorphism group.
- A number of sharp upper bounds of automorphism group orders for a variety of maps are provided.
- The automorphism groups of simplified butterflies and those of simplified barflies are determined.
- The realization of asymetrization for a map is from rooting an element of the ground set.

8.1 Automorphisms

An isomorphism of a map to itself is called an *automorphism*. Let τ be an automorphism of map $M = (\mathscr{X}, \mathscr{P})$. If for $x \in \mathscr{X}$, $\tau(x) = y$ and $x \neq y$, then two elements x and y play the same role on M, or say, they are *symmetric*. Hence, an automorphism of a map reflects the symmetry among elements in the ground set of a map.

Lemma 8.1 Suppose τ_1 and τ_2 are two automorphisms of map M, then their composition $\tau_1\tau_2$ is also an automorphism of map M.

Proof Because τ_1 is an automorphism of $M = (\mathscr{X}_{\alpha,\beta}, \mathscr{P})$, from (7.2)

$$\tau_1\alpha\tau_1^{-1} = \alpha,\ \tau_1\beta\tau_1^{-1} = \beta,\ \tau_1\mathscr{P}\tau_1^{-1} = \mathscr{P}.$$

Similarly, for τ_2,

$$\tau_2\alpha\tau_2^{-1} = \alpha,\ \tau_2\beta\tau_2^{-1} = \beta,\ \tau_2\mathscr{P}\tau_2^{-1} = \mathscr{P}.$$

Therefore, for $\tau_1\tau_2$,

$$\begin{aligned}(\tau_1\tau_2)\alpha(\tau_1\tau_2)^{-1} &= (\tau_1\tau_2)\alpha(\tau_2^{-1}\tau_1^{-1})\\ &= \tau_1(\tau_2\alpha\tau_2^{-1})\tau_1^{-1}\\ &= \tau_1\alpha\tau_1 = \alpha,\end{aligned}$$

DOI 10.1515/9783110481846-008

$$\begin{aligned}(\tau_1\tau_2)\beta(\tau_1\tau_2)^{-1} &= (\tau_1\tau_2)\beta(\tau_2^{-1}\tau_1^{-1})\\ &= \tau_1(\tau_2\beta\tau_2^{-1})\tau_1^{-1}\\ &= \tau_1\beta\tau_1 = \beta,\end{aligned}$$

and

$$\begin{aligned}(\tau_1\tau_2)\mathscr{P}(\tau_1\tau_2)^{-1} &= (\tau_1\tau_2)\mathscr{P}(\tau_2^{-1}\tau_1^{-1})\\ &= \tau_1(\tau_2\mathscr{P}\tau_2^{-1})\tau_1^{-1}\\ &= \tau_1\mathscr{P}\tau_1 = \mathscr{P}.\end{aligned}$$

This implies that for $\tau_1\tau_2$, (7.2) is commutative. From Theorem 7.1, $\tau_1\tau_2$ is an automorphism of M as well. □

On the basis of the property on permutation composition, automorphisms satisfy the associate law for composition.

Because an automorphism τ is a bijection, it has a unique inverse denoted by τ^{-1}. Because

$$\tau^{-1}\alpha\tau = \tau^{-1}(\tau\alpha\tau^{-1})\tau = (\tau^{-1}\tau)\alpha(\tau^{-1}\tau) = \alpha,$$

and similarly,

$$\tau^{-1}\beta\tau = \beta,\ \tau^{-1}\mathscr{P}\tau = \mathscr{P},$$

from Theorem 7.1, τ^{-1} is also an automorphism.

If an element $x \in \mathscr{X}$ has $\tau(x) = x$ for a mapping (particularly, an automorphism) τ, then x is called a *fixed point* of τ. If every element is a fixed point of τ, then τ is called an *identity*. Easy to see that an identity on $\mathscr{X}$ is, of course, an automorphism of M, usually said to be *trivial*. By the property of a permutation, an identity is the unity of automorphisms, always denoted by 1.

In summary, the set of all automorphisms of a map M forms a group, called the *automorphism group* of M, denoted by Aut(M). Its *order* is the cardinality of itself, denoted by aut(M) = |Aut(M)|, i.e., the number of elements in Aut(M) because of the finiteness.

Theorem 8.1 Let τ be an automorphism of map $M = (\mathscr{X}, \mathscr{P})$. If τ has a fixed point, the $\tau = 1$, i.e., the identity.

Proof Suppose x is the fixed point, i.e., $\tau(x) = x$. Because τ is an isomorphism, From (7.1),

$$\tau(\alpha x) = \alpha\tau(x) = \alpha x,$$
$$\tau(\beta x) = \beta\tau(x) = \beta x,$$
$$\tau(\mathscr{P}x) = \mathscr{P}(\tau(x)) = \mathscr{P}x,$$

i.e., αx, βx and $\mathscr{P}x$ are all fixed points.

Then, for any $\psi \in \Psi_{\{\alpha,\beta,\mathscr{P}\}}$,

$$\tau(\psi(x)) = \psi(\tau(x)) = \psi(x).$$

Therefore, from transitive axiom, every element on $\mathscr{X}$ is a fixed point of τ. This means that τ is the identity. □

In virtue of this theorem, the automorphism induced from $\tau(x) = y$ can be represented by $\tau = (x \to y)$.

Example 8.1 Let us go back to the automorphisms of the maps described in Pattern 7.1 and Pattern 7.2.

If M_1 and M_2 in Pattern 7.1 are taken to be

$$M = (Kx + Ky, (x, y, \beta y)(\gamma x)) = M_1,$$

then it is seen that only one nontrivial automorphism $\tau = (x \to \alpha x)$ exists. Thus, its automorphism group is

$$\mathrm{Aut}(M) = \{1, (x \to \alpha x)\},$$

i.e., a group of order 2.

Then, maps M_1 and M_2 in Pattern 7.2 are taken to be

$$M = (Kx + Ky, (x, y, \gamma y)(\gamma x)) = M_2.$$

It has also only one nontrivial automorphism $\tau = (x \to \alpha x)$. So, its automorphism group is

$$\mathrm{Aut}(M) = \{1, (x \to \alpha x)\},$$

also a group of order 2.

However, maps M_1 and M_2 here are not isomorphic. In fact, it is seen that M_1 is nonorientable with relative genus −1 and M_2 is orientable of relative genus 1.

8.2 Upper bounds of group order

Because the automorphism group of a combinatorial structure with finite elements is a finite permutation group in its own right, its order must be bounded by an finite number. And because there are $n!$ permutations on a combinatorial structure of n elements, the order of its automorphism group is bounded by $n!$.

However, $n!$ is an exponential function of n according to the Stirling approximate formula; it is too large for determining the automorphism group in general.

Now, is there an constant c such that the order of automorphism group is bounded by n^c, or denoted by $O(n^c)$? If there is, then such a result would be much hopeful for the determination of the group efficiently.

As a matter of fact, if the order of automorphism group is $O(n^c)$, c is independent of n for a structure with n elements, then an efficient algorithm can be designed for justifying and recognizing if two of them are isomorphic in a theoretical point of view.

Lemma 8.2 For any map $M = (\mathscr{X}, \mathscr{P})$, the order of its automorphism group is

$$\mathrm{aut}(M) \leq |\mathscr{X}| = 4\epsilon(M), \tag{8.1}$$

where $\epsilon(M) = \frac{1}{4}|\mathscr{X}|$ is the size of M.

Proof From Theorem 8.1, M has at most $|\mathscr{X}| = 4\epsilon(M)$ automorphism, i.e., (8.1). □

The bound presented by this lemma is *sharp*, i.e., it cannot be reduced any more. For an example, the *link map* $L = (Kx, (x)(\gamma x))$. The order of its automorphism group is $4 = |Kx| = \epsilon(L)$.

Lemma 8.3 For an integer $i \geq 1$, let $v_i(M)$ be the number of i-vertices (vertex incident with i semi-edges) in map $M = (\mathscr{X}, \mathscr{P})$, then

$$\mathrm{aut}(M) \mid 2iv_i(M), \tag{8.2}$$

i.e., $\mathrm{aut}(M)$ is a factor of $2iv_i(M)$.

Proof Let $\tau \in \mathrm{Aut}(M)$ be an automorphism of M. For $x \in \mathscr{X}$, $(x)_{\mathscr{P}}$ is an i-vertex, assume $\tau(x) = y$. From the third relation of (7.1), $(y)_{\mathscr{P}}$ is also an i-vertex. Then, the elements of $\mathscr{X}_i = \{x | \forall x \in \mathscr{X}, |\{x\}_{\mathscr{P}}| = i\}$ can be classified by the equivalent relation

$$x \sim \mathrm{Aut} y \Longleftrightarrow \exists \tau \in \mathrm{Aut}(M) x = \tau y$$

induced from the group $\mathrm{Aut}(M)$.

From Theorem 8.1, Aut(G) has a bijection with every equivalent class. This implies that each class has aut(M) elements. Therefore,

$$\mathrm{aut}(M) \mid |\mathscr{X}_i|.$$

Because $|\mathscr{X}_i| = 2iv_i(M)$, (8.2) is soon obtained. □

This lemma enables us to improve, even apparently improve the bound presented by Lemma 8.1 for a map not vertex-regular (each vertex has the same valency).

Lemma 8.4 For an integer $j \geq 1$, let $\phi_j(M)$ be the number of j-faces of map $M = (\mathscr{X}, \mathscr{P})$, then

$$\mathrm{aut}(M) \mid 2j\phi_j(M), \tag{8.3}$$

i.e., aut(M) is a factor of $2j\phi_j(M)$.

Proof Let $\tau \in \mathrm{Aut}(M)$ be an automorphism of M. For $x \in \mathscr{X}$, $(x)_{\mathscr{P}\gamma}$ is a j-face, assume $\tau(x) = y$. From the first two relations of (7.1), $\tau(\gamma x) = \gamma y$. Then from this and the third relation, $\tau((\mathscr{P}\gamma)x) = (\mathscr{P}\gamma)y$. Thus, $(y)_{\mathscr{P}\gamma}$ is also a j-face. And, the elements of $\mathscr{X}_j = \{x | \forall x \in \mathscr{X}, |\{x\}_{\mathscr{P}\gamma}| = j\}$ can be classified by the equivalence

$$x \sim \mathrm{Aut} y \Longleftrightarrow \exists \tau \in \mathrm{Aut}(M) x = \tau y$$

induced from the group Aut(M).

Further, from Theorem 8.1, Aut(G) has a bijection with every equivalent class. This leads to the fact that each class has aut(M) elements. Therefore,

$$\mathrm{aut}(M) \mid |\mathscr{X}_j|.$$

Because $|\mathscr{X}_j| = 2j\phi_j(M)$, (8.3) is soon obtained. □

This lemma also enables us to improve, even apparently improve the bound presented by Lemma 8.1 for a map not *face-regular* (each face has the same valency).

Theorem 8.2 Let $v_i(M)$ and $\phi_j(M)$ be, respectively, the numbers of i-vertices and j-faces in map $M = (\mathscr{X}, \mathscr{P})$, $i, j \geq 1$, then

$$\mathrm{aut}(M) \Big| (2iv_i,\ 2j\phi_j \mid \forall i,\ i \geq 1,\ \forall j,\ j \geq 1), \tag{8.4}$$

where $(2iv_i,\ 2j\phi_j \mid \forall i,\ i \geq 1,\ \forall j,\ j \geq 1)$ represents the greatest common divisor of all the numbers in the parentheses.

Proof From Lemma 8.3, for any integer $i \geq 1$,

$$\mathrm{aut}(M) \mid 2iv_i(M).$$

From Lemma 8.4, for any integer $j \geq 1$,

$$\mathrm{aut}(M) \mid 2i\phi_j(M).$$

By combining the two relations above, (8.4) is soon found. □

Based on this theorem, the following corollary is naturally deduced.

Corollary 8.1 Let $v_i(M)$ and $\phi_j(M)$ be, respectively, the numbers of i-vertices and j-faces in map $M = (\mathscr{X}, \mathscr{P})$, $i, j \geq 1$, then

$$\mathrm{aut}(M) \leq (2iv_i,\ 2j\phi_j \mid \forall i,\ i \geq 1,\ \forall j,\ j \geq 1). \tag{8.5}$$

Proof A direct result of (8.4). □

Corollary 8.2 For map $M = (\mathscr{X}, \mathscr{P})$, $\epsilon(M)$ is its size, then

$$\mathrm{aut}(M) \mid 4\epsilon(M). \tag{8.6}$$

Proof Because

$$\begin{aligned} 4\epsilon(M) &= 2\sum_{i\geq 1} iv_i(M) = 2\sum_{j\geq 1} j\phi_j(M) \\ &= \sum_{i\geq 1} 2iv_i(M) = \sum_{j\geq 1} 2j\phi_j(M), \end{aligned}$$

we have

$$(2iv_i,\ 2j\phi_j \mid \forall i,\ i \geq 1,\ \forall j,\ j \geq 1) \,\Big|\, 4\epsilon(M).$$

Hence, from Theorem 8.2, (8.6) is soon derived. □

8.3 Determination of the group

In this section, the automorphism groups of standard maps, i.e., simplified butterflies and simplified barflies, are discussed.

First, observe the orientable case. For

$$O_1 = (\mathscr{X}_1, \mathscr{J}_1) = (Kx_1 + Ky_1, (x_1, y_1, \gamma x_1, \gamma y_1)),$$

by Algorithm 7.1,

$$\begin{aligned}
S_{x_1}(O_1) &= 1, y_1, \gamma 1, \gamma y_1 = 1, 2, \gamma 1, \gamma 2;\\
S_{\alpha x_1}(O_1) &= 1, \beta y_1, \gamma 1, \alpha y_1 = 1, 2, \gamma 1, \gamma 2;\\
S_{\beta x_1}(O_1) &= 1, \alpha y_1, \gamma 1, \beta y_1 = 1, 2, \gamma 1, \gamma 2;\\
S_{\gamma x_1}(O_1) &= 1, \gamma y_1, \gamma 1, y_1 = 1, 2, \gamma 1, \gamma 2;\\
S_{y_1}(O_1) &= 1, \gamma x_1, \gamma 1, x_1 = 1, 2, \gamma 1, \gamma 2;\\
S_{\alpha y_1}(O_1) &= 1, \alpha x_1, \gamma 1, \beta x_1 = 1, 2, \gamma 1, \gamma 2;\\
S_{\beta y_1}(O_1) &= 1, \beta x_1, \gamma 1, \alpha x_1 = 1, 2, \gamma 1, \gamma 2;\\
S_{\gamma y_1}(O_1) &= 1, x_1, \gamma 1, \gamma x_1 = 1, 2, \gamma 1, \gamma 2;
\end{aligned}$$

i.e.,

$$\begin{aligned}
S_{x_1}(O_1) &= S_{\alpha x_1}(O_1) = S_{\beta x_1}(O_1) = S_{\gamma x_1}(O_1) = S_{y_1}(O_1)\\
&= S_{\alpha y_1}(O_1) = S_{\beta y_1}(O_1) = S_{\gamma y_1}(O_1)\\
&= 1, 2, \gamma 1, \gamma 2.
\end{aligned}$$

Thus, O_1 has its automorphism group of order 8, i.e.,

$$\text{aut}(O_1) = 4 \times (2 \times 1) = 8.$$

A map with a non-trivial automorphism group is said to be *symmetrical*. If there is a map with its automorphism group of order 4 times its size, then it is said to be *completely symmetrical*. It can be seen that O_1 is completely symmetrical. However, none of O_k, $k \geq 2$, is completely symmetrical although they are all symmetrical.

Theorem 8.3 For simplified butterflies (orientable standard maps) $O_k = (\mathscr{X}_k, \mathscr{J}_k)$, $k \geq 1$, where

$$\mathscr{X}_k = \sum_{i=1}^{k} (Kx_i + Ky_i)$$

and

$$\mathscr{J}_k = \left(\prod_{i=1}^{k} \langle x_i, y_i, \gamma x_i, \gamma y_i \rangle \right),$$

we have

$$\text{aut}(O_k) = \begin{cases} 2k, & \text{if } k \geq 2;\\ 8, & \text{if } k = 1. \end{cases} \tag{8.7}$$

Proof From the symmetry between $\langle x_i, y_i, \gamma x_i, \gamma y_i\rangle$, $i \geq 2$, and $\langle x_1, y_1, \gamma x_1, \gamma y_1\rangle$ in $\mathscr{J}_k$, $k \geq 2$, it is only necessary to calculate $S_{x_1}(O_k)$, $S_{\gamma x_1}(O_k)$, $S_{y_1}(O_k)$, $S_{\gamma y_1}(O_k)$, $S_{\alpha x_1}(O_k)$, $S_{\beta x_1}(O_k)$, $S_{\alpha y_1}(O_k)$ and $S_{\beta y_1}(O_k)$ by Algorithm 7.1.

From Algorithm 7.1,

$$\begin{aligned}
S_{x_1}(O_k) &= 1, y_1, \gamma 1, \gamma y_1, \prod_{i=1}^{k}\langle x_i, y_i, \gamma x_i, \gamma y_i\rangle \\
&= 1, 2, \gamma 1, \gamma 2, \prod_{i=2}^{k}\langle (2i-1), 2i, \gamma(2i-1), \gamma 2i\rangle \\
&= \prod_{i=1}^{k}\langle (2i-1), 2i, \gamma(2i-1), \gamma 2i\rangle, \\
S_{\alpha x_1}(O_k) &= 1, \alpha\left(\prod_{i=1}^{k}\langle x_i, y_i, \gamma x_i, \gamma y_i\rangle\right)^{-1}, \beta y_1, \gamma 1, \alpha y_1 \\
&\neq S_{x_1}(O_k), \\
S_{\gamma x_1}(O_k) &= 1, \gamma y_1, \prod_{i=1}^{k}\langle x_i, y_i, \gamma x_i, \gamma y_i\rangle, \gamma 1, y_1 \\
&\neq S_{x_1}(O_k), \\
S_{\beta x_1}(O_k) &= 1, \alpha y_1, \gamma 1, \alpha\left(\prod_{i=1}^{k}\langle x_i, y_i, \gamma x_i, \gamma y_i\rangle\right)^{-1}, \beta y_1 \\
&\neq S_{x_1}(O_k), \\
S_{y_1}(O_k) &= 1, \gamma x_1, \gamma 1, \prod_{i=1}^{k}\langle x_i, y_i, \gamma x_i, \gamma y_i\rangle, x_1 \\
&\neq S_{x_1}(O_k), \\
S_{\alpha y_1}(O_k) &= 1, \alpha x_1, \alpha\left(\prod_{i=1}^{k}\langle x_i, y_i, \gamma x_i, \gamma y_i\rangle\right)^{-1}, \gamma 1, \beta x_1 \\
&\neq S_{x_1}(O_k), \\
S_{\gamma y_1}(O_k) &= 1, \prod_{i=1}^{k}\langle x_i, y_i, \gamma x_i, \gamma y_i\rangle, x_1, \gamma 1, \gamma x_1 \\
&\neq S_{x_1}(O_k), \\
S_{\beta y_1}(O_k) &= 1, \beta x_1, \alpha 1, \alpha x_1, \alpha\left(\prod_{i=1}^{k}\langle x_i, y_i, \gamma x_i, \gamma y_i\rangle\right)^{-1} \\
&= 1, 2.\gamma 1, \gamma 2, \prod_{i=2}^{k}\langle (2i-1), 2i, \gamma(2i-1), \gamma 2i\rangle \\
&= S_{x_1}(O_k).
\end{aligned}$$

Because two automorphisms are from $S_{\beta y_1}(O_k) = S_{x_1}(O_k)$, O_k has altogether $2 \times k = 2k$ automorphisms. Hence, when $k \geq 2$,

$$\mathrm{aut}(O_k) = 2k.$$

When $k = 1$, $\mathrm{aut}(O_1) = 8$ is known. □

Then, observe the nonorientable case. For

$$Q_1 = (\mathscr{X}_1, \mathscr{I}_1) = (Kx_1, (x_1, \beta x_1)),$$

by Algorithm 7.1,

$$S_{x_1}(Q_1) = 1, \beta_1;\ S_{\alpha x_1}(Q_1) = 1, \beta 1,$$
$$S_{\beta x_1}(Q_1) = 1, \beta 1;\ S_{\gamma x_1}(Q_1) = 1, \beta 1,$$

i.e., $\mathrm{aut}(Q_1) = 4$.

Theorem 8.4 For simplified barflies $Q_l = (\mathscr{X}_l, \mathscr{I}_l)$, $l \geq 1$, where

$$\mathscr{X}_l = \sum_{i=1}^{l} Kx_i$$

and

$$\mathscr{I}_l = \prod_{i=1}^{l} \langle x_i, \beta x_i \rangle,$$

we have

$$\mathrm{aut}(Q_l) = \begin{cases} 2l, & l \geq 2; \\ 4, & l = 1. \end{cases} \tag{8.8}$$

Proof From the symmetry of $\langle x_i, \beta x_i \rangle$, $i \geq 2$, and $\langle x_1, \beta x_1 \rangle$ in $\mathscr{I}_l$, $l \geq 2$, it is only necessary to calculate

$$S_{x_1}(Q_l), S_{\alpha x_1}(Q_l), S_{\beta x_1}(Q_l) \text{ and } S_{\gamma x_1}(Q_l)$$

by employing Algorithm 7.1.

From Algorithm 7.1,

$$\begin{aligned}
S_{x_1}(Q_l) &= 1, \beta 1, \prod_{i=2}^{l} \langle x_i, \beta x_i \rangle \\
&= 1, \beta 1, 2, \beta 2, \prod_{i=3}^{l} \langle x_i, \beta x_i \rangle \\
&= \prod_{i=1}^{l} \langle i, \beta i \rangle, \\
S_{\alpha x_1}(Q_l) &= 1, \alpha \left(\prod_{i=2}^{l} \langle x_i, \beta x_i \rangle \right)^{-1}, \beta 1 \\
&\neq S_{x_1}(Q_l), \\
S_{\beta x_1}(Q_l) &= 1, \prod_{i=2}^{l} \langle x_i, \beta x_i \rangle, \beta 1 \\
&\neq S_{x_1}(Q_l), \\
S_{\gamma x_1}(Q_l) &= 1, \beta 1, \alpha \left(\prod_{i=2}^{l} \langle x_i, \beta x_i \rangle \right)^{-1} \\
&= 1, \beta 1, l, \beta l, \prod_{i=l-1}^{2} \langle \gamma x_i, \alpha x_i \rangle \\
&= \prod_{i=1}^{l} \langle i, \beta i \rangle \\
&= S_{x_1}(Q_l).
\end{aligned}$$

Because two automorphisms are from $S_{\gamma x_1}(Q_l) = S_{x_1}(Q_l)$, Q_l has altogether $2 \times l = 2l$ automorphisms. Hence, when $l \geq 2$,

$$\text{aut}(Q_l) = 2l.$$

When $l = 1$, $\text{aut}(Q_1) = 4$ is known. □

Similarly, the two theorems can also be proved by employing Algorithm 7.2.

8.4 Rootings

For a given map $M = (\mathscr{X}, \mathscr{P})$, if a subset $R \subseteq \mathscr{X}$ is chosen such that an automorphism of M with R fixed, i.e., an element of R, does only correspond to an element of R, then M is called a *set rooted* map. The subset R is called the *rooted set* of M, and an element of R is called a *rooted element*.

Theorem 8.5 For a set rooted map $M^{\mathrm{R}} = (\mathscr{X}, \mathscr{P})$, R is the rooted set,

$$\mathrm{aut}(M^{\mathrm{R}}) \mid |R|. \tag{8.9}$$

Proof Assume that all elements in R are partitioned into equivalent classes under the group $\mathrm{Aut}(M^{\mathrm{R}})$. From Theorem 8.1, each class has $\mathrm{aut}(M^{\mathrm{R}})$ elements. Therefore, (8.9) is satisfied. □

Corollary 8.3 For a set rooted map $M^{\mathrm{R}} = (\mathscr{X}, \mathscr{P})$, R is the rooted set,

$$\mathrm{aut}(M^{\mathrm{R}}) \leq |R|. \tag{8.10}$$

Proof A direct result of (8.9). □

For a given map $M = (\mathscr{X}, \mathscr{P})$, if a vertex v_x, $x \in \mathscr{X}$, is chosen such that an automorphism of M has to be with v_x fixed, i.e., an element incident with v_x has to correspond to an element incident with v_x, then M is called a *vertex rooted* map. The vertex v_x is called the *rooted vertex* of M, and an element incident with v_x, *rooted element*.

Corollary 8.4 For a vertex rooted map $M^{\mathrm{vr}} = (\mathscr{X}, \mathscr{P})$, v_x is the rooted vertex,

$$\mathrm{aut}(M^{\mathrm{vr}}) \mid 2|\{x\}_{\mathscr{P}}|. \tag{8.11}$$

Proof This is (8.9) when $R = \{x\}_{\mathscr{P}} \cup \{\alpha x\}_{\mathscr{P}}$. □

For a given map $M = (\mathscr{X}, \mathscr{P})$, if face f_x, $x \in \mathscr{X}$, is chosen such that an automorphism of M has f_x fixed, i.e., an element incident with f_x should be corresponding to an element incident with f_x, then M is said to be a *face rooted* map. The face f_x is called the *rooted face* of M. An element in rooted face is called an *rooted element*.

Corollary 8.5 For a face rooted map $M^{\mathrm{fr}} = (\mathscr{X}, \mathscr{P})$ with rooted face f_x,

$$\mathrm{aut}(M^{\mathrm{fr}}) \mid 2|\{x\}_{\mathscr{P}\gamma}|. \tag{8.12}$$

Proof This is (8.9) when $R = \{x\}_{\mathscr{P}\gamma} \cup \{\alpha x\}_{\mathscr{P}\gamma}$. □

For given map $M = (\mathscr{X}, \mathscr{P})$, if edge e_x, $x \in \mathscr{X}$, is chosen such that an automorphism of M is with e_x fixed, i.e., an element in e_x is always corresponding to an element in e_x, then M is called an *edge rooted* map. Edge e_x is the *rooted edge* of M. An element in the rooted edge is also called a *rooted element*.

Corollary 8.6 For an edge rooted map $M^{\mathrm{er}} = (\mathscr{X}, \mathscr{P})$ with the rooted edge e_x,

$$\mathrm{aut}(M^{\mathrm{er}}) \mid |Kx|. \tag{8.13}$$

Proof The case of (8.9) when $R = Kx$. □

For a given map $M = (\mathscr{X}, \mathscr{P})$, an element $x \in \mathscr{X}$ is chosen such that an automorphism of M is with x as a fixed point, then M is called a *rooted map*. The element x is the *root* of M. The vertex, the edge and the face incident to the root are, respectively, called the *root vertex*, the *root edge* and the *root face*.

Corollary 8.7 For a rooted map $M^{\mathrm{r}} = (\mathscr{X}, \mathscr{P})$ with its root x,

$$\mathrm{aut}(M^{\mathrm{r}}) = 1. \tag{8.14}$$

Proof The case of (8.9) when $R = \{x\}$. □

This tells us that a rooted map does not have the symmetry at all. The way mentioned above shows such a general clue for transforming a problem with symmetry to a problem without symmetry and then doing the reversion.

Example 8.2 Map $M_1 = (Kx + Ky, (x)(\gamma x, y, \gamma y))$ has four distinct ways for choosing the root. Because M_1 has the following four primal trail codes

$$S_x = \underline{1}_0, \underline{\gamma 1, 2, \gamma 2}_1 = S_{\alpha x}, S_{\gamma x} = \underline{1, 2, \gamma 2}_0, \underline{\gamma 1}_1 = S_{\beta x}$$

$$S_y = \underline{1, \gamma 1, 2}_0, \underline{\gamma 2}_1 = S_{\beta y}, S_{\gamma y} = \underline{1, 2, \gamma 1}_0 \underline{\gamma 2}_1 = S_{\alpha y},$$

the four ways of rooting are shown in Figure 8.1(a–d), where the root is marked at its tail.

Example 8.3 Map $M_2 = (Kx + Ky, (x)(\gamma x, y, \beta y))$ has four distinct ways for choosing the root. Because M_2 has the following four primal trail codes

$$S_x = \underline{1}_0, \underline{\gamma 1, 2, \beta 2}_1 = S_{\alpha x}, S_{\gamma x} = \underline{1, 2, \beta 2}_0, \underline{\gamma 1}_1 = S_{\beta x}$$

$$S_y = \underline{1, \beta 1, 2}_0, \underline{\gamma 2}_1 = S_{\gamma y}, S_{\beta y} = \underline{1, 2, \beta 1}_0 \underline{\gamma 2}_1 = S_{\alpha y},$$

the four ways of rooting are shown in Figure 8.2(a–d), where the root is marked at its tail.

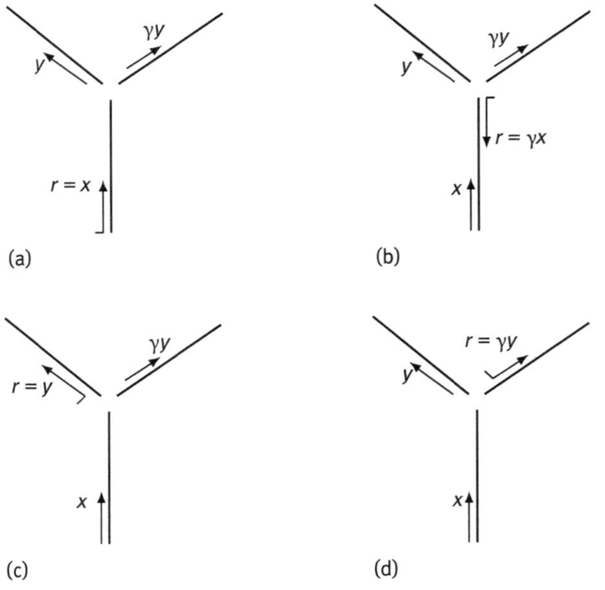

Figure 8.1: Rootings in Example 1.

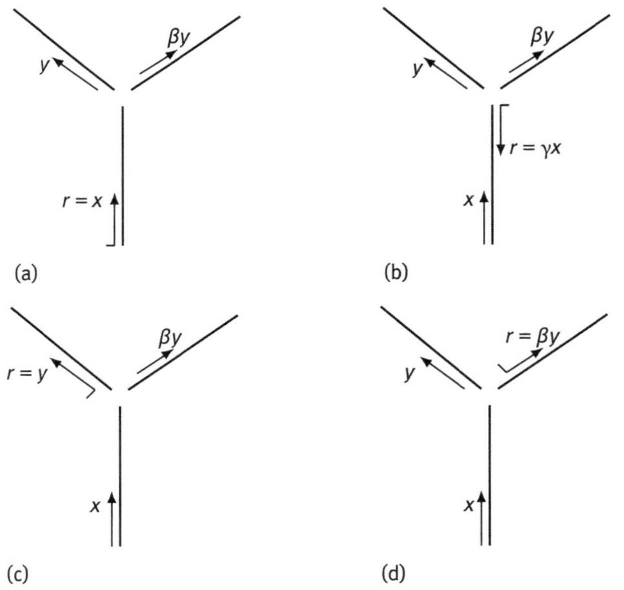

Figure 8.2: Rootings in Example 2.

8.5 Notes

8.5.1 Given three integers $m \geq 1$, g and $s \geq 1$, determine the number of primal matching maps, on page 208 of Liu [90], for relative genus g with size m and the order s of their automorphism groups.

Notice 8.1 In principle, it can be done by employing the joint trees shown in Liu [72]. However, its aim is to seek an expression as summation free, finite sum of all terms positive, or recursive form of positive terms. Basic references as Liu [73, 74], or [90].

8.5.2 Given three integers $m \geq 1$, g and $s \geq 1$, determine the number of dual matching maps, on page 208 of Liu [90], for relative genus g with size m and the order s of their automorphism groups.

Notice 8.2 In principle, it can be done by employing the joint trees shown in Liu [72]. However, its aim is to seek an expression as summation free, finite sum of all terms positive or recursive form of positive terms. Basic references as Liu [73, 74], or [90].

8.5.3 Given three integers $m \geq 1$, g and $s \geq 1$, determine the number of bi-matching maps, on page 208 of Liu [90], for relative genus g with size m and the order s of their automorphism groups.

Notice 8.3 In principle, it can be done by employing the joint trees shown in Liu [72]. However, its aim is to seek an expression as summation free, finite sum of all terms positive or recursive form of positive terms. Basic references as Liu [73, 74], or [90].

The first three problems should be considered for starting from $g = 0, 1$ and then -1. Particularly, the three problems for self-dual maps should be first studied before the general cases.

8.5.4 Find the cubic maps of size $m \geq 7$ with a given relative genus and the maximum order of their automorphism groups

Notice 8.4 Relevant results can be seen in Liu [67, pp. 117–141, 68; pp. 144–172], Cai and Liu [10, 11].

8.5.5 Find the maps of size $m \geq 1$ with a given relative genus and the order 1 of their automorphism groups.

Notice 8.5 Relevant results can be seen in Liu [67, pp. 265–272, 68; pp. 310–319] and [97, pp. 435–436].

8.5.6 Prove, or disprove, the conjecture that for a given relative genus, almost all maps have their automorphism groups of order 1.

Notice 8.6 Relevant results can be seen in Yan and Liu [171], LiuYP [67, pp. 359–365, 68; pp. 418–425], and Liu [97, pp. 435–436].

8.5.7 Given three integers $m \geq 1$, g and $s \geq 1$, determine the full cavity maps, on page 209 of Liu [90], for size m with relative genus g and the order of their automorphism groups s.

Notice 8.7 On the basis of joint trees, all such maps can be easily found. However, the aim is for seeking an explicit expression with parameters m, g and s, preferably summation free or recursion available to realization on computers.

If a map has a set of edges inducing a Hamiltonian circuit on its under graph, then it is called a *primal* H-*map* . If a map has a set of edges inducing a Hamiltonian circuit on the under graph of its dual, then it is called a *dual* H-*map* . If a map is both a primal H-map and a dual H-map, then it is called a *double* H-*map*.

8.5.8 Given three integers $m \geq 1$, g and $s \geq 1$, determine the primal H-maps of size m with relative genus g and their automorphism group of order s.

Notice 8.8 On the basis of joint trees, all such maps can be easily found. However, the aim is for seeking an explicit expression with parameters m, g and s, preferably summation free or recursion available to realization on computers.

8.5.9 Given three integers $m \geq 1$, g and $s \geq 1$, determine the dual H-maps of size m with relative genus g and their automorphism group of order s.

Notice 8.9 On the basis of joint trees, all such maps can be easily found. However, the aim is for seeking an explicit expression with parameters m, g and s, preferably summation free or recursion available to realization on computers.

8.5.10 Given three integers $m \geq 1$, g and $s \geq 1$, determine the double H-maps of size m with relative genus g and their automorphism group of order s.

Notice 8.10 On the basis of joint trees, all such maps can be easily found. However, the aim is for seeking an explicit expression with parameters m, g and s, preferably summation free or recursion available to realization on computers.

9 Asymmetrized Petal Bundles

- A petal bundle is a map that has only one vertex, or in other words, each edge of self-loop.
- From decomposing the set of rooted orientable petal bundles, a linear differential equation satisfied by the enumerating function with size as the parameter is discovered and then an explicit expression of the function is extracted.
- A quadratic equation of the enumerating function for rooted petal bundles on the surface of orientable genus 0 is discovered and then an explicit expression is extracted.
- From decomposing the set of rooted nonorientable petal bundles, a linear differential equation satisfied by the enumerating function with size as the parameter is discovered in company with the orientable case and then a favorable explicit expression of the function is also extracted.
- The numbers of orientable, nonorientable and total petal bundles with given size on all surfaces are, separately, obtained and then calculated for size not greater than 10.

9.1 Orientable petal bundles

A single vertex map is also called a *petal bundle*, its under graph is a *bouquet*. In this section, the orientable rooted petal bundles are investigated for determining their enumerating function with size as a parameter by a simple form.

Let $\mathscr{D}$ be the set of all non-isomorphic orientable rooted petal bundles. For convenience, the trivial map ϑ is assumed to be in $\mathscr{D}$.

Now, $\mathscr{D}$ is divided into two classes: $\mathscr{D}_{\mathrm{I}}$ and $\mathscr{D}_{\mathrm{II}}$, i.e.,

$$\mathscr{D} = \mathscr{D}_{\mathrm{I}} + \mathscr{D}_{\mathrm{II}} \tag{9.1}$$

where $(\mathscr{D})_{\mathrm{I}} = \{\vartheta\}$ and $\mathscr{D}_{\mathrm{II}}$, of course, consists of all petal bundles in $\mathscr{D}$ other than ϑ.

Lemma 9.1 Let $\mathscr{D}_{\langle \mathrm{II} \rangle} = \{D - a | \forall D \in \mathscr{D}_{\mathrm{II}}\}$. Then,

$$\mathscr{D}_{\langle \mathrm{II} \rangle} = \mathscr{D}. \tag{9.2}$$

Proof For any $D = (\mathscr{X}, \mathscr{P}) \in \mathscr{D}_{\langle \mathrm{II} \rangle}$, there exists a $D' = (\mathscr{X}', \mathscr{P}') \in \mathscr{D}_{\mathrm{II}}$ such that $D = D' - a'$. Because D' is orientable, group $\Psi' = \Psi_{\{\gamma, \mathscr{P}'\}}$ has two orbits

$$\{r'\}_{\mathscr{P}'} \text{ and } \{\alpha r'\}_{\mathscr{P}'}$$

DOI 10.1515/9783110481846-009

on $\mathscr{X}'$. Because $\gamma r' \in \{r'\}_{\mathscr{P}'}$, D has also two orbits

$$\{r\}_{\mathscr{P}} = \{r'\}_{\mathscr{P}'} - \{r', \gamma r'\} \text{ and } \{\alpha r\}_{\mathscr{P}} = \{\alpha r'\}_{\mathscr{P}'} - \{\alpha r', \beta r'\},$$

and hence D is orientable as well. From Theorem 3.4, petal bundle D' leads that D is a petal bundle. This implies that $\mathscr{D}_{\langle \mathrm{II} \rangle} \subseteq \mathscr{D}$.

Conversely, for any $D = (\mathscr{X}, \mathscr{P}) \in \mathscr{D}$, $\mathscr{P} = (r, \mathscr{P}r, \ldots, \mathscr{P}^{-1}r)$, $D' = (\mathscr{X} + Kr', \mathscr{P}')$, where

$$\mathscr{P}' = (r', \gamma r', r, \mathscr{P}r, \ldots, \mathscr{P}^{-1}r).$$

Because D is orientable, group $\Psi = \Psi_{\{\gamma, \mathscr{P}\}}$ has two orbits $\{r\}_\Psi$ and $\{\alpha r\}_\Psi$ on $\mathscr{X}$. Thus, group $\Psi' = \Psi_{\{\gamma, \mathscr{P}'\}}$ has two orbits

$$\{r'\}_{\Psi'} = \{r\}_\Psi + \{r', \gamma r'\} \text{ and } \{\alpha r'\}_{\Psi'} = \{\alpha r\}_\Psi + \{\alpha r', \beta r'\}$$

on $\mathscr{X}'$. This means that D' is also orientable. Because D' has only one vertex, $D' \in \mathscr{D}$. And from $D' \neq \vartheta$, it is only possible that $D' \in \mathscr{D}_{\mathrm{II}}$. Therefore, in view of $D = D' - a'$, $\mathscr{D} \subseteq \mathscr{D}_{\langle \mathrm{II} \rangle}$. □

From the last part in the proof of this lemma, for any $D = (r)_{\mathscr{J}} \in \mathscr{D}$, D' has $2m(D) + 1$ distinct choices such that $D' = D_i = D + e_i \in \mathscr{D}_{\mathrm{II}}$, $0 \leq i \leq 2m(D)$, and hence $D = D' - a'$ where $e_i = Kr'$ and

$$\begin{cases} D_0 = (r', \gamma r', r, \mathscr{J}r, \ldots, \mathscr{J}^{2m(D)-1}r), \ i = 0; \\ D_i = (r', r, \ldots, \mathscr{J}^{i-1}r, \gamma r', \mathscr{J}^i r, \ldots, \mathscr{J}^{2m(D)-1}), \\ \qquad 1 \leq i \leq 2m(D) - 1; \\ D_{2m(D)} = (r', r, \mathscr{J}r, \ldots, \mathscr{J}^{2m(D)-1}r, \gamma r'), i = 2m(D), \end{cases}$$

for $\gamma = \alpha\beta$.

Lemma 9.2 Let $\mathscr{H}(D) = \{D_i | i = 0, 1, 2, \ldots, 2m(D)\}$ for $D \in \mathscr{D}$. Then,

$$\mathscr{D}_{\mathrm{II}} = \sum_{D \in \mathscr{D}} \mathscr{H}(D). \tag{9.3}$$

Proof Because of Lemma 9.1, it is easily seen that the set on the left side of (9.3) is a subset of the set on the right.

Conversely, from $\mathscr{H}(D) \subseteq \mathscr{D}_{\mathrm{II}}$, for any $D \in \mathscr{D}$, the set on the right side of (9.3) is also a subset on the left. □

The importance of Lemma 9.2 is that (9.3) provides a one-to-one correspondence between the sets on its two sides. This is seen from the fact that for any two non-isomorphic petal bundles D_1 and D_2, $\mathscr{H}(D_1) \cap \mathscr{H}(D_2) = \emptyset$.

On the basis of Lemmas 9.1 and 9.2, the enumerating functions of sets $\mathscr{D}_{\mathrm{I}}$ and $\mathscr{D}_{\mathrm{II}}$ can be evaluated as a function of $\mathscr{D}$'s as

$$f_{\mathscr{D}}(x) = \sum_{D \in \mathscr{D}} x^{m(D)} \tag{9.4}$$

where $m(D)$ is the size of D.

Because $\mathscr{D}_{\mathrm{I}}$ consists of only the trivial map ϑ and ϑ has no edge,

$$f_{\mathscr{D}_{\mathrm{I}}}(x) = 1. \tag{9.5}$$

Lemma 9.3 For $\mathscr{D}_{\mathrm{II}}$,

$$f_{\mathscr{D}_{\mathrm{II}}}(x) = xf_{\mathscr{D}} + 2x^2 \frac{\mathrm{d}f_{\mathscr{D}}}{\mathrm{d}x}. \tag{9.6}$$

Proof From Lemma 9.2,

$$\begin{aligned} f_{\mathscr{D}_{\mathrm{II}}}(x) &= \sum_{D \in \mathscr{D}_{\mathrm{II}}} x^{m(D)} \\ &= x\left(\sum_{D \in \mathscr{D}} (2m(D)+1)x^{m(D)} \right) \\ &= x \sum_{D \in \mathscr{D}} x^{m(D)} + 2x \sum_{D \in \mathscr{D}} m(D) x^{m(D)} \\ &= xf_{\mathscr{D}} + 2x^2 \frac{\mathrm{d}f_{\mathscr{D}}}{\mathrm{d}x} \end{aligned}$$

where $f_{\mathscr{D}} = f_{\mathscr{D}}(x)$. This is (9.6). □

Theorem 9.1 The differential equation about h

$$\begin{cases} 2x^2 \dfrac{\mathrm{d}h}{\mathrm{d}x} = -1 + (1-x)h; \\ h_0 = h|_{x=0} = 1 \end{cases} \tag{9.7}$$

is well defined in the ring of infinite series with integral coefficients and finite terms of negative exponents. And, the solution is $h = f_{\mathscr{D}}(x)$.

Proof Suppose $h = H_0 + H_1 x + H_2 x^2 + \cdots + H_m x^m + \cdots$, for $H_i \in Z_+$, $i \geq 0$. According to the first relation of (9.7), via equating the coefficients of the terms with the same power of x on its two sides, the recursion

$$\begin{cases} -1 + H_0 = 0; \\ H_1 - H_0 = 0; \\ H_m = (2m-1)H_{m-1}, \ m \geq 2 \end{cases} \tag{9.8}$$

is soon found. From this, $H_0 = 1$(the initial condition!), $H_1 = 1, \ldots,$ and hence all the coefficients of h can be determined. Because only addition and multiplication are used in the evaluation, all H_m, $m \geq 1$, are integers from integrality of H_0. This is the first statement.

As for the last statement, from (9.1) and (9.5–6), it is seen that $h = f_{\mathscr{D}}(x)$ satisfies the first relation of (9.7). And from the initial condition $h_0 = f_{\mathscr{D}}(0) = 1$, we only have that $h = f_{\mathscr{D}}(x)$ by the first statement. □

In fact, from (9.8),

$$H_m = \prod_{i=1}^{m}(2i-1) = \frac{(2m)!}{2^m m!},$$

where $m \geq 0$.

Further, from Theorem 9.2,

$$f_{\mathscr{D}}(x) = 1 + \sum_{m\geq 1} \frac{(2m-1)!}{2^{m-1}(m-1)!}x^m. \tag{9.9}$$

Example 9.1 From (9.9), there are three orientable rooted petal bundles of size 2.

However, there are two orientable non-rooted petal bundles as shown in Figure 9.1(a) and (b).

In Figure 9.1(a), based on primal trail code (or dual trail code), there is only one rooted (r_1 as the root) element. In Figure 9.1(b), there are two rooted (r_2 and r_3 as the roots) elements.

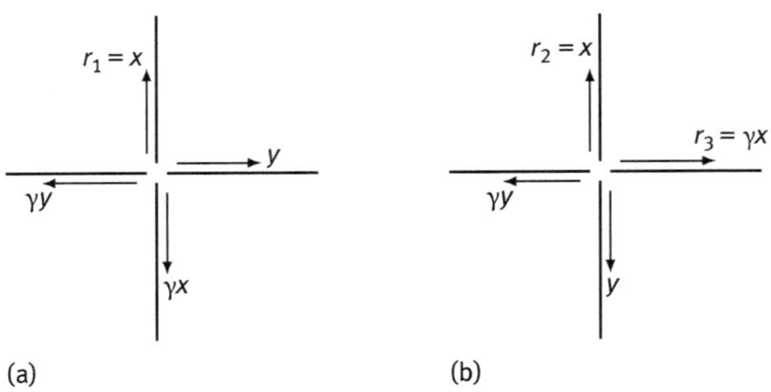

Figure 9.1: Petal bundles of size 2.

9.2 Planar pedal bundles

Petal bundles are here restricted to those of genus 0, i.e., *planar pedal bundle*. Rooted is still considered. Orientable petal bundles can be partitioned into classes by genus as

$$\mathscr{D} = \sum_{k\geq 0} \mathscr{D}_k, \tag{9.10}$$

where $\mathscr{D}_k$ is the set of rooted petal bundles with orientable genus k. What is discussed in this section is $\mathscr{D}_0$. For convenience, the trivial map ϑ is included in $\mathscr{D}_0$.

For this, $\mathscr{D}_0$ should be partitioned by the valency of root-face into classes as

$$\mathscr{D}_0 = \sum_{s\geq 0} \mathscr{F}_s, \tag{9.11}$$

where $\mathscr{F}_s$, $s \geq 0$, is planar rooted petal bundle with the root-face of valency s.

Lemma 9.4 Let $\mathscr{S}$ (the trivial map ϑ is included) and $\mathscr{T}$ (ϑ is excluded) be two sets of rooted maps. If for any $S = (\mathscr{X}, \mathscr{P}) \in \mathscr{S} - \vartheta$, there exist an integer $k \geq 1$ and maps $S_i = (\mathscr{X}_i, \mathscr{P}_i) \in \mathscr{T}$, $1 \leq i \leq k$, such that

$$\mathscr{X} = \sum_{i=1}^{k} \mathscr{X}_i \tag{9.12}$$

and $\mathscr{P}$ are different from $\mathscr{P}_i$, $1 \leq i \leq k$, only at vertex

$$(r)_{\mathscr{P}} = (\langle r_1\rangle_{\mathscr{P}_1}, \langle r_2\rangle_{\mathscr{P}_2}, \cdots, \langle r_k\rangle_{\mathscr{P}_k}), \tag{9.13}$$

where $r = r_1$. Then,

$$f_{\mathscr{S}}(x) = \frac{1}{1 - f_{\mathscr{T}}(x)}, \tag{9.14}$$

where $f_{\mathscr{S}}(x)$ and $f_{\mathscr{T}}(x)$ are the enumerating functions of, respectively, $\mathscr{S}$ and $\mathscr{T}$.

Proof First, $\mathscr{S}$ is classified based on k mentioned above, $k \geq 0$, i.e.,

$$\mathscr{S} = \sum_{k\geq 0} \mathscr{S}_k.$$

Naturally, $\mathscr{S}_0 = \{\vartheta\}$. Then, because any $M_k = (\mathscr{Y}_k, \mathscr{Q}_k) \in \mathscr{S}_k$, $k \geq 1$, has the form as shown in (9.12) and (9.13) ($\mathscr{X}$ and $\mathscr{P}$ are, respectively, replaced by $\mathscr{Y}_k$ and $\mathscr{Q}_k$), we have

$$\begin{aligned} f_{\mathscr{S}_k}(x) &= \sum_{M_k \in \mathscr{S}_k} x^{m(M_k)} \\ &= \sum_{\substack{(S_1, S_2, \ldots, S_k) \\ S_i \in \mathscr{T},\ 1 \le i \le k}} x^{m(S_1)+m(S_2)+\cdots+m(S_k)} \\ &= \left(f_{\mathscr{T}}(x)\right)^k. \end{aligned}$$

Therefore, by considering $f_{\mathscr{S}_0}(x) = 1$,

$$\begin{aligned} f_{\mathscr{S}}(x) &= \sum_{k \ge 0} f_{\mathscr{S}_k}(x) \\ &= 1 + \sum_{k \ge 1} \left(f_{\mathscr{T}}(x)\right)^k \\ &= \frac{1}{1 - f_{\mathscr{T}}(x)}. \end{aligned}$$

Notice that since x is an undeterminate, it can be considered for the values satisfying $|f_{\mathscr{T}}(x)| < 1$. This lemma is proved. □

If $\mathscr{S}$ and $\mathscr{T}$ are, respectively, seen as $\mathscr{D}_0$ and $\mathscr{F}_1$, it can be checked that the condition of Lemma 9.4 is satisfied, then

$$f_{\mathscr{D}_0}(x) = \frac{1}{1 - f_{\mathscr{F}_1}(x)}. \tag{9.15}$$

Further, another relation between $f_{\mathscr{D}_0}(x)$ and $f_{\mathscr{F}_1}(x)$ has to be found.

Lemma 9.5 Let $\mathscr{F}_{\langle 1 \rangle} = \{D - a | \forall D \in \mathscr{F}_1\}$. Then,

$$\mathscr{F}_{\langle 1 \rangle} = \mathscr{D}_0. \tag{9.16}$$

Proof Because $\vartheta \notin \mathscr{F}_1$, for any $D \in \mathscr{F}_1$, from the planarity of D, $D' = D - a$ is planar and from D with a single vertex, $D' = D - a$ is with a single vertex. Hence, $D' \in \mathscr{D}_0$. This implies that $\mathscr{F}_{\langle 1 \rangle} \subseteq \mathscr{D}_0$.

Conversely, for any $D' = (\mathscr{X}', \mathscr{P}') \in \mathscr{D}_0$, in view of a single vertex, $\mathscr{P}' = (r')_{\mathscr{P}'}$. Let

$$D = D' + a = (\mathscr{X}' + Kr, \mathscr{P}),$$

where $\mathscr{P} = (r, \langle r' \rangle_{\mathscr{P}'}, \gamma r)$. Naturally, D is of single vertex. Because D is obtained from D' by appending an edge, from Corollary 4.2 and Lemma 4.6, the planarity of D' shows that D is planar. And, from $(r)_{\mathscr{P}\gamma} = (r)$, $D \in \mathscr{F}_1$. Since $D' = D - a$, $D' \in \mathscr{F}_{\langle 1 \rangle}$. This implies that $\mathscr{D}_0 \subseteq \mathscr{F}_{\langle 1 \rangle}$. □

Because this lemma provides a one-to-one correspondence between $\mathscr{F}_1$ and $\mathscr{D}_0$, we obtained

$$\begin{aligned} f_{\mathscr{F}_1}(x) &= \sum_{D\in\mathscr{F}_1} x^{m(M)} \\ &= x\sum_{D\in\mathscr{D}_0} x^{m(D)} \\ &= xf_{\mathscr{D}_0}(x). \end{aligned} \tag{9.17}$$

In virtue of (9.17) and (9.15),

$$f_{\mathscr{D}_0}(x) = \frac{1}{1 - xf_{\mathscr{D}_0}(x)}. \tag{9.18}$$

Theorem 9.2 Let $h^{(0)} = f_{\mathscr{D}_0}(x)$ be the enumerating function of planar rooted petal bundles with the size as the parameter, then

$$h^{(0)} = \sum_{m\geq 0} \frac{(2m)!}{m!(m+1)!}. \tag{9.19}$$

Proof From (9.18), it is seen that $h^{(0)}$ satisfies the quadratic equation about h as

$$xh^2 - h + 1 = 0.$$

It can be checked that only one of its two solutions is in a power series with all coefficients nonnegative integers. That is,

$$h^{(0)} = \frac{1 - \sqrt{1-4x}}{2x}.$$

By expanding $\sqrt{1-4x}$ into a power series, (9.19) is soon found via rearrangement. □

From the quadratic equation, a non-linear recursion can be derived for determining the coefficients of h. However, a linear recursion can be extracted for getting a simple result. This is far from a universal way.

Example 9.2 From (9.19), the number of planar rooted petal bundles of size 3 is known to be five.

However, there are in all two planar non-rooted petal bundles, shown in Figure 9.2(a) and (b).

In Figure 9.2(a), by primal trail codes (or dual trail codes), there are three. Their roots are r_1, r_2 and r_3. In Figure 9.2(b), there are two. Their roots are r_4 and r_5.

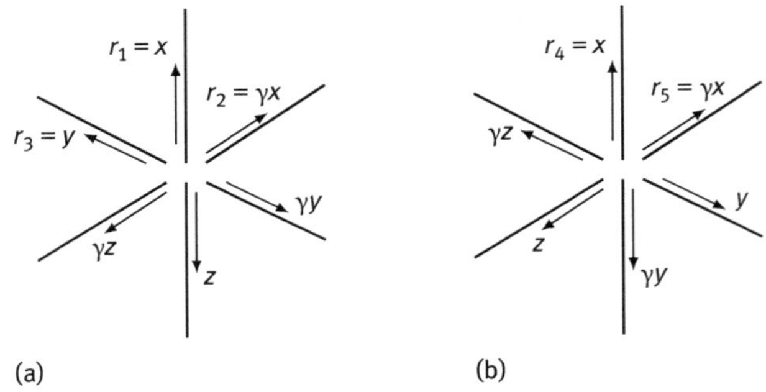

Figure 9.2: Planar petal bundles of size 3.

9.3 Nonorientable pedal bundles

The central task of this section is to determine the enumerating function of nonorientable rooted petal bundles with size as the parameter.

Let $\mathscr{U}$ be the set of all nonorientable rooted petal bundles. Because the trivial map is orientable, ϑ is never in $\mathscr{U}$. In other words, any map in $\mathscr{U}$ does have at least one edge. Now, $\mathscr{U}$ is partitioned into two classes: $\mathscr{U}_{\mathrm{I}} = \{M | \forall M \in \mathscr{U}, M - a \text{ orientable}\}$ and

$$\mathscr{U}_{\mathrm{II}} = \{M | \forall M \in \mathscr{U}, M - a \text{ nonorientable}\},$$

i.e.,

$$\mathscr{U} = \mathscr{U}_{\mathrm{I}} + \mathscr{U}_{\mathrm{II}}. \tag{9.20}$$

First, the decomposition of the two sets $\mathscr{U}_{\mathrm{I}}$ and $\mathscr{U}_{\mathrm{II}}$ should be investigated.

Lemma 9.6 Let $\mathscr{U}_{\langle \mathrm{I} \rangle} = \{M - a | \forall M \in \mathscr{U}_{\mathrm{I}}\}$. Then,

$$\mathscr{U}_{\langle \mathrm{I} \rangle} = \mathscr{D}, \tag{9.21}$$

where $\mathscr{D}$ is the set of all orientable rooted petal bundles given by (9.1).

Proof For $M = (\mathscr{X}, \mathscr{P}) \in \mathscr{U}_{\langle \mathrm{I} \rangle}$, if $M \neq \vartheta$, then $M' = (\mathscr{X}', \mathscr{P}')$, where $\mathscr{X}' = \mathscr{X} + Kr'$ and $\mathscr{P}'$ is different from $\mathscr{P}$ only at the vertex

$$(r')_{\mathscr{P}'} = (r', \beta r', r, \mathscr{P}r, \mathscr{P}^2 r, \ldots, \mathscr{P}^{-1} r)$$

such that $M = M' - a'$. From $M' \in \mathscr{M}_{\mathrm{I}}$, $M \in \mathscr{M}$. If $M = \vartheta$, then

$$M' = (Kr', (r', \beta r')) \in \mathscr{U}_{\mathrm{I}}$$

such that $M = M' - a'$. Meanwhile, $M \in \mathscr{D}$. Hence, $\mathscr{U}_{\langle \mathrm{I} \rangle} \subseteq \mathscr{D}$.

Conversely, for $M = (\mathscr{X}, \mathscr{P}) \in \mathscr{D}$, let $M' = (\mathscr{X}', \mathscr{P}')$ such that $\mathscr{X}' = \mathscr{X} + Kr'$. Because M has a single vertex,

$$\mathscr{P}' = (r')_{\mathscr{P}'} = (r', \beta r', r, \mathscr{P}r, \mathscr{P}^2 r, \ldots, \mathscr{P}^{-1} r).$$

Therefore, M' has a single vertex as well. And since $r', \beta r' \in \{r'\}_{\Psi'}$, $M' \in \mathscr{U}$. By reminding that $M = M' - a'$ is orientable, $M' \in \mathscr{U}_{\mathrm{I}}$. Thus, $M \in \mathscr{U}_{\langle \mathrm{I} \rangle}$. This implies that $\mathscr{D} \subseteq \mathscr{U}_{\langle \mathrm{I} \rangle}$. □

Lemma 9.7 For any $D = (\mathscr{X}, \mathscr{F}) \in \mathscr{D}$, $r = r(D)$, let $\mathscr{B}(D) = \{B_i | 0 \le i \le 2m(D)\}$, where $m(D)$ is the size of D, and

$$B_i(D) = \begin{cases} (r', \beta r', \langle r \rangle_{\mathscr{F}}), \ i = 0; \\ (r', r, \ldots, \beta r', \mathscr{F}^i r, \cdots), \\ \qquad 1 \le i \le 2m(D) - 1; \\ (r', r, \ldots, \mathscr{F}^i r, \ldots, \beta r'), \\ \qquad i = 2m(D). \end{cases} \tag{9.22}$$

Then,

$$\mathscr{U}_{\mathrm{I}} = \sum_{D \in \mathscr{D}} \mathscr{B}(D). \tag{9.23}$$

Proof For any $M = (\mathscr{X}, \mathscr{P}) \in \mathscr{U}_{\mathrm{I}}$, because

$$D = M - a \in \mathscr{D},$$

M is only some B_i, $1 \le i \le 2m(D)$ in (9.22) such that $\mathscr{P}r' = r$, or $\mathscr{P}^2 r' = r$ (here, r' and r are, respectively, the roots of M and D). Therefore, M is also an element of the set on the right side of (9.23).

Conversely, for an element M in the set on the right of (9.23), from Lemma 9.6, $M \in \mathscr{U}_{\mathrm{I}}$. This is to say that M is also an element of the set on the left side of (9.23). □

Example 9.3 Let $D = (Kx, (x, \gamma x)) \in \mathscr{D}$. Then, D is of size 1 i.e., $m(D) = 1$.

Three rooted petal bundles $B_0(D)$, $B_1(D)$ and $B_2(D) \in \mathscr{U}_{\mathrm{I}}$ are produced from D and shown in Figure 9.4(a-c), where $r' = y$ and $r = x$.

Lemma 9.8 Let $\mathscr{U}_{\langle \mathrm{II} \rangle} = \{M - a | \forall M \in \mathscr{U}_{\mathrm{II}}\}$. Then,

$$\mathscr{U}_{\langle \mathrm{II} \rangle} = \mathscr{U}. \tag{9.24}$$

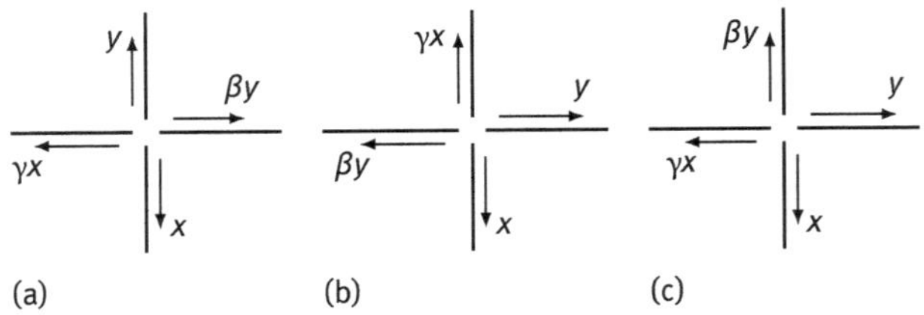

Figure 9.3: Nonorientable petal bundles from orientable ones.

Proof For $M = (\mathscr{X}, \mathscr{P}) \in \mathscr{U}_{\langle \mathrm{II}\rangle}$, let $M' \in \mathscr{U}_{\mathrm{II}}$ such that $M = M' - a'$. Because M' is a nonorientable petal bundle and $M' \in \mathscr{U}_{\mathrm{II}}$, M is a nonorientable petal bundle as well, i.e., $M \in \mathscr{U}$.

Conversely, for any $M = (\mathscr{X}, \mathscr{P}) \in \mathscr{U}$, there exists $M' = (\mathscr{X}', \mathscr{P}') \in \mathscr{U}_{\mathrm{II}}$ such that $M = M' - a'$, e.g., $\mathscr{X}' = \mathscr{X} + Kr'$, $\mathscr{P}' = (r', \gamma r', \langle r\rangle_{\mathscr{P}})$. Therefore, $M \in \mathscr{U}_{\langle \mathrm{II}\rangle}$. □

Further, observe for a map $M \in \mathscr{U}$ how many non-isomorphic maps $M' \in \mathscr{U}_{\mathrm{II}}$ are there such that $M = M' - a'$. Two cases should be considered: (1) r', $\gamma r'$ and r are in the same orbit of $\mathscr{P}'$; (2) r', $\beta r'$ and r are in the same orbit of $\mathscr{P}'$.

(1) Based on the rule of rooting, because has $\gamma r'$ only $2m(M) + 1$ possible positions, i.e., $\gamma r' = \mathscr{P}'r', \mathscr{P}'r, \mathscr{P}'(\mathscr{P}r), \cdots, \mathscr{P}'(\mathscr{P}^{2m(M)-1}r)$, so

$$I_i(M) = \begin{cases} (r', \gamma r', \langle r\rangle_{\mathscr{P}}), \ i = 0; \\ (r', r, \ldots, \gamma r', \mathscr{P}^i r, \cdots), \\ \qquad 1 \le i \le 2m(M) - 1; \\ (r', \langle r\rangle_{\mathscr{P}}, \gamma r'), \ i = 2m(M). \end{cases} \tag{9.25}$$

(2) Based on the rule of rooting, because $\beta r'$ also has $2m(M) + 1$ possible positions, i.e., $\beta r' = \mathscr{P}'r', \mathscr{P}'r, \mathscr{P}'(\mathscr{P}r), \cdots, \mathscr{P}'(\mathscr{P}^{2m(M)-1}r)$, so

$$J_i(M) = \begin{cases} (r', \beta r', \langle r\rangle_{\mathscr{P}}), \ i = 0; \\ (r', r, \ldots, \beta r', \mathscr{P}^i r, \cdots), \\ \qquad 1 \le i \le 2m(M) - 1; \\ (r', \langle r\rangle_{\mathscr{P}}, \beta r'), \ i = 2m(M). \end{cases} \tag{9.26}$$

Example 9.4 Let $M = (Kx, (x, \beta x)) \in \mathscr{U}$. The map M has only one edge, i.e., $m(M) = 1$.

Six nonorientable petal bundles $I_0(M)$, $I_1(M)$ and $I_2(M)$, with $J_0(M)$, $J_1(M)$ and $J_2(M) \in \mathscr{U}_{\mathrm{II}}$ are produced for M and shown as, respectively, in (a), (b) and (c), with (d), (e) and (f) of Figure 9.4, where $r' = y$ and $r = x$.

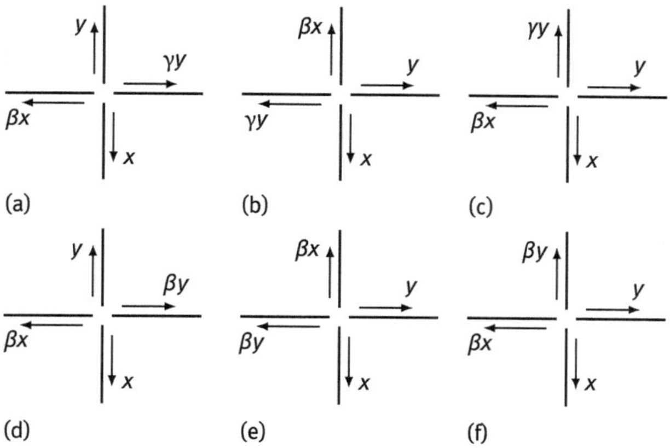

Figure 9.4: Nonorientable petal bundles from nonorientable ones.

Lemma 9.9 For any $M \in \mathscr{U}$, let

$$\begin{cases} \mathscr{I}(M) = \{I_i(M)|0 \le i \le 2m(M)\}; \\ \mathscr{J}(M) = \{J_j(M)|0 \le j \le 2m(M)\}. \end{cases} \tag{9.27}$$

Then,

$$\mathscr{U}_{\mathrm{II}} = \sum_{M \in \mathscr{U}} (\mathscr{I}(M) + \mathscr{J}(M)). \tag{9.28}$$

Proof For any $M' = (\mathscr{X}', \mathscr{P}') \in \mathscr{U}_{\mathrm{II}}$, because $M = M' - a' \in \mathscr{U}$, M' is only some I_i, $0 \le i \le 2m(M) - 1$ in (9.25), or some J_j, $0 \le j \le 2m(M) - 1$ in (9.26) such that $\mathscr{P}r' = r$, or $\mathscr{P}^2 r' = r$ (here, r' and r are, respectively, the roots of M' and M). Therefore, M is also an element of the set on the right side of (9.28).

Conversely, for an element M in the set on the right of (9.28), from Lemma 9.8, $M \in \mathscr{U}_{\mathrm{II}}$. This is, M is also an element of the set on the left side of (9.28). □

Lemma 9.10 Let $\mathscr{S}$ and $\mathscr{T}$ be two sets of maps. If for any $T \in \mathscr{T}$, there exists a set $\mathscr{L}(T) \subseteq \mathscr{S}$ such that

(i) for any $T \in \mathscr{T}$, $|\mathscr{L}(T)| = am(T) + b$ and for $S \in \mathscr{S}$, $m(T) = m(S) - c$, where a b and c are constants and $m(T)$ is an isomorphic invariant, e.g., the size; and
(ii) $\mathscr{S} = \sum_{T \in \mathscr{T}} \mathscr{L}(T)$, then

$$f_{\mathscr{S}}(x) = x^c\left(bf_{\mathscr{T}} + ax\frac{\mathrm{d}f_{\mathscr{T}}}{\mathrm{d}x}\right). \tag{9.29}$$

Proof Because $\mathscr{L}(T)$ provides a mapping from a map in $\mathscr{T}$ to a subset of $\mathscr{S}$ and the cardinality of the subset is only dependent on the parameter of the enumerating function (by (i)), and (ii) means that the mapping provides a partition on $\mathscr{S}$, then

$$\begin{aligned} x^{-c} f_{\mathscr{S}}(x) &= \sum_{T \in \mathscr{T}} (am(T) + b) x^{m(T)} \\ &= b \sum_{T \in \mathscr{T}} x^{m(T)} + ax \sum_{T \in \mathscr{T}} m(T) x^{m(T)-1} \\ &= b f_{\mathscr{T}} + ax \frac{\mathrm{d} f_{\mathscr{T}}}{\mathrm{d}x}. \end{aligned}$$

This is (9.29) by multiplying x^c to the two sides. □

Theorem 9.3 The enumerating function $g = f_{\mathscr{U}}(x)$ of nonorientable rooted petal bundles in the set $\mathscr{U}$ with size as the parameter satisfies the equation as

$$\begin{cases} 4x^2 \dfrac{\mathrm{d}g}{\mathrm{d}x} = (1 - 2x) g + h - 1; \\ \left.\dfrac{\mathrm{d}g}{\mathrm{d}x}\right|_{x=0} = 1, \end{cases} \tag{9.30}$$

where $h = f_{\mathscr{D}}(x)$ is the enumerating function of orientable rooted petal bundles given by (9.9).

Proof Because (9.23) and (9.28) provide the mappings from maps $D \in \mathscr{D}$ and $U \in \mathscr{U}$ to, respectively, a subset of $\mathscr{U}_{\mathrm{I}}$ with $2m(U) + 1$ elements and a subset of $\mathscr{U}_{\mathrm{II}}$ with $2(2m(U) + 1) = 4m(U) + 4$ elements, where D and U are one edge less than their images, from Lemma 9.10,

$$f_{\mathscr{U}_{\mathrm{I}}}(x) = x\left(h + 2x \frac{\mathrm{d}h}{\mathrm{d}x}\right) = xh + 2x^2 \frac{\mathrm{d}h}{\mathrm{d}x}$$

and

$$f_{\mathscr{U}_{\mathrm{II}}}(x) = x\left(2g + 4x \frac{\mathrm{d}g}{\mathrm{d}x}\right) = 2xg + 4x^2 \frac{\mathrm{d}g}{\mathrm{d}x}.$$

By (9.20) again,

$$g = xh + 2x^2 \frac{\mathrm{d}h}{\mathrm{d}x} + 2xg + 4x^2 \frac{\mathrm{d}g}{\mathrm{d}x}.$$

Via rearrangement and (9.7), (9.30) is soon obtained. □

9.4 The number of pedal bundles

Because (9.9) provides the number of orientable rooted petal bundles with size m, $m \geq 0$, i.e.,

$$H_m = \frac{(2m-1)!}{2^{m-1}(m-1)!}, \tag{9.31}$$

for $m \geq 1$. Of course, $H_0 = 1$.

Here, the number of nonorientable rooted petal bundles with size m is evaluated only by the equation shown in (9.30). Let G_m be the number of nonorientable rooted petal bundles with size m, $m \geq 1$.

In fact, G_m, $m \geq 1$, is determined by the recursion as

$$\begin{cases} G_m = (4m-2)G_{m-1} + H_m, & m \geq 2; \\ G_1 = 1. \end{cases} \tag{9.32}$$

The solution of the recursion (9.32) is obtained, i.e.,

$$\begin{aligned} G_m &= \frac{(2m-1)!}{2^{m-1}(m-1)!} + \prod_{i=2}^{m}(4i-2) \\ &\quad + \sum_{i=2}^{m-1} \frac{(2i-1)!}{2^{i-1}(i-1)!} \prod_{j=i+1}^{m}(4j-2). \end{aligned} \tag{9.33}$$

Example 9.5 When $m = 1$, there is one orientable rooted petal bundle of one edge, i.e., $M = (Kx, (x, \gamma x))$ and one nonorientable rooted petal bundle of one edge, i.e., $N = (Kx, (x, \beta x)) \in \mathscr{U}$.

By appending an edge Ky on M, three non-isomorphic nonorientable rooted petal bundles of two edges are produced and shown in Figure 9.3(a–c).

By appending an edge Ky on N, six non-isomorphic nonorientable rooted petal bundles of two edges are produced and shown in Figure 9.4(a–f). Then, $G_2 = 9$, which is in agreement with that provided by (9.32) or (9.33).

Now, H_m, G_m and $H_m^{(0)}$ for $m \leq 10$ are listed in Table 9.1.

Lemma 9.11 For an integer $m \geq 1$, the number of non-isomorphic nonorientable rooted petal bundles with size m is

$$G_m = (2^m - 1)H_m, \tag{9.34}$$

where H_m is given by (9.31).

Proof By induction. When $m = 1$, from $H_1 = 1$, $G_1 = 1$. (9.34) is true.

Table 9.1: Numbers of rooted petal bundles in 10 edges.

m	H_m	G_m	$H_m^{(0)}$
1	1	1	1
2	3	9	2
3	15	105	5
4	105	1575	14
5	945	29295	42
6	10395	654885	132
7	135135	17162145	429
8	2027025	516891375	1430
9	34459425	17608766175	4862
10	654729075	669787843725	16796

Assume G_k satisfies (9.34) for any $1 \le k \le m-1$, $m \ge 2$. Then, from (9.32),

$$\begin{aligned} G_m &= (4m-2)G_{m-1} + H_m \\ &= (4m-2)\Big((2^{m-1}-1)H_{m-1}\Big) + H_m. \end{aligned}$$

Since it can, from (9.31), be seen that

$$H_m = (2m-1)H_{m-1},$$

we have

$$\begin{aligned} G_m &= (4m-2)(2^{m-1}-1)\frac{H_m}{2m-1} + H_m \\ &= \left(\frac{4m-2}{2m-1}(2^{m-1}-1)+1\right)H_m \\ &= \Big(2(2^{m-1}-1)+1\Big)H_m \\ &= (2^m-1)H_m. \end{aligned}$$

This is (9.34). □

Theorem 9.4 For an integer $m \ge 1$, the number of non-isomorphic rooted petal bundles with size m is

$$2^m(2m-1)!!, \tag{9.35}$$

where

$$(2m-1)!! = \prod_{i=1}^{m}(2i-1). \tag{9.36}$$

Proof Because of (9.34), the number of non-isomorphic petal bundles with m edges is

$$H_m + G_m = 2^m H_m. \tag{9.37}$$

By substituting (9.31) into (9.37), we have

$$\begin{aligned} 2^m H_m &= 2^m \times \frac{(2m-1)!}{2^{m-1}(m-1)!} \\ &= 2 \times \frac{(2m-1)!}{(m-1)!} \\ &= 2^m(2m-1)!!. \end{aligned}$$

This is (9.35). □

The theorem above reminds that the number of embeddings of the bouquet of size m is $2^m(2m-1)!$, derived from (1.10) as a special case. This is $(m-1)!$ times the number of rooted petal bundles with m edges.

9.5 Notes

9.5.1 For two integers $m \geq 1$ and $p \geq 2$, determine the number of rooted petal bundles with size m on the surface of orientable genus p.

Notice 9.1 For the number of rooted petal bundles with size $m \geq 1$, in all, orientable surface has been done from Liu [103]. However, there are no new results for any integer $p \geq 2$.

9.5.2 For two integers $m \geq 1$ and $q \geq 3$, determine the number of rooted petal bundles with size m on the surface of nonorientable genus q.

Notice 9.2 For the number of rooted petal bundles with size $m \geq 1$, in all, nonorientable surface has been done from Liu [103]. For $q = 1$ and 2, seen in Yang and Liu [173]. However, there are no new results for any integer $q \geq 3$.

9.5.3 For two integers $m \geq 1$ and $p \geq 2$, determine the number of rooted unisheets with size m on the surface of orientable genus p.

Notice 9.3 Referred to as in Liu [90] (2010, p. 230). Because of a correspondence between unisheets and petal bundles, this problem is equivalent to IX.5.1.

9.5.4 For two integers $m \geq 1$ and $q \geq 3$, determine the number of rooted unisheets with size m on the surface of nonorientable genus q.

Notice 9.4 Referred to as in Liu [90] (2010, p. 230). Because of a correspondence between unisheets and petal bundles, this problem is equivalent to IX.5.2.

A map of order 2 is also called *bi-pole map*.

9.5.5 For a given integer $m \geq 1$, determine the number of all non-isomorphic orientable rooted bi-pole maps with size m

Notice 9.5 Referred to as in Liu [90] (2010, p. 232). Because of the triviality for $m = 1$ and 2, only $m \geq 3$ have to be considered. For $3 \leq m \leq 6$, the results are shown in AIII.2.

9.5.6 For a given integer $m \geq 1$, determine the number of all non-isomorphic nonorientable rooted bi-pole maps with size m.

Notice 9.6 Referred to as in Liu [90] (2010, p. 232). Because of the triviality for $m = 1$ and 2, only $m \geq 3$ have to be considered. For $3 \leq m \leq 6$, the results are shown in AIII.2.

9.5.7 For two integers $m \geq 1$ and $p \geq 0$, determine the number of all non-isomorphic orientable rooted bi-pole maps with size m of the surface of genus p.

Notice 9.7 Referred to as in Liu [90] (2010, p. 232). Because of the triviality for $m = 1$ and 2, only $m \geq 3$ have to be considered. For $3 \leq m \leq 6$, the results are shown in AIII.2.

9.5.8 For two integers $m \geq 1$ and $q \geq 1$, determine the number of all non-isomorphic nonorientable rooted bi-pole maps with size m on the surface of genus q.

Notice 9.8 Referred to as in Liu [90] (2010, p. 232). Because of the triviality for $m = 1$ and 2, only $m \geq 3$ have to be considered. For $3 \leq m \leq 6$, the results are shown in AIII.2.

9.5.9 For a given integer $m \geq 1$, determine the number of all non-isomorphic orientable rooted tri-pole maps with size m

Notice 9.9 Referred to as in Liu [90] (2010, p. 232). Because of the triviality for $m \leq 5$, only $m \geq 6$ have to be considered.

9.5.10 For a given integer $m \geq 1$, determine the number of all non-isomorphic nonorientable rooted tri-pole maps with size m.

Notice 9.10 Referred to as in Liu [90] (2010, p. 232). Because of the triviality for $m \leq 5$, only $m \geq 6$ have to be considered.

9.5.11 For two integers $m \geq 1$ and $p \geq 1$, determine the number of all non-isomorphic orientable rooted tri-pole maps with size m on the surface of genus p.

Notice 9.11 On the basis of joint trees sketched in Liu [72, topologically on pages: 14–18], detailed in Liu [73, combinatorically on page: 195–201] and clarified in Liu [77, algebraically on pages: 25–67], for a suitable $m \leq 10$, to do the problem for running in general.

9.5.12 For two integers $m \geq 1$ and $q \geq 1$, determine the number of all non-isomorphic nonorientable rooted tri-pole maps with size m on the surface of genus q.

Notice 9.12 On the basis of joint trees sketched in Liu [72, topologically on pages: 14–18], detailed in Liu [73, combinatorically on page: 195–201] and clarified in Liu [77, algebraically on pages: 25–67], for a suitable $m \leq 10$, to do the problem for running in general.

9.5.13 For two integers $n \geq 5$ and $p \geq s(n)$, where

$$s(n) = \left\lceil \frac{(n-3)(n-4)}{12} \right\rceil,$$

i.e., the least integer not less than the fractional $(n-3)(n-4)/12$ (or called the *up-integer* of the fractional), determine the number of rooted maps whose under graph is the complete graph of order n with orientable genus p.

Notice 9.13 Relevant references might be seen in Liu [70].

9.5.14 For two integers $n \geq 5$ and $p \geq t(n)$, where

$$t(n) = \left\lceil \frac{(n-3)(n-4)}{6} \right\rceil,$$

determine the number of rooted maps whose under graph is the complete graph of order n with nonorientable genus q.

Notice 9.14 Relevant references might be seen in Liu [70].

9.5.15 For two integers $n \geq 3$ and $p \geq c(n)$ where

$$c(n) = (n-4)2^{n-3} + 1,$$

determine the number of rooted maps whose under graph is the n-cube with orientable genus p.

Notice 9.15 Relevant references might be seen in Liu [70].

9.5.16 For two integers $n \geq 3$ and $q \geq d(n)$ where

$$d(n) = (n-4)2^{n-2} + 2,$$

determine the number of rooted maps whose under graph is the n-cube with nonorientable genus q.

Notice 9.16 Relevant references might be seen in Liu [70].

9.5.17 For three integers $m, n \geq 3$ and $p \geq r(n)$ where

$$r(n) = \left\lceil \frac{(m-2)(n-2)}{4} \right\rceil,$$

determine the number of rooted maps whose under graph is the complete bipartite graph of order $m + n$ with orientable genus p.

Notice 9.17 Relevant references might be seen in Liu [70] and Shao et al. [146].

9.5.18 For three integers $m, n \geq 3$ and $q \geq l(n)$ where

$$l(n) = \left\lceil \frac{(m-2)(n-2)}{2} \right\rceil,$$

determine the number of rooted maps whose under graph is the complete bipartite graph of order $m + n$ with nonorientable genus q.

Notice 9.18 Relevant references might be seen in Liu [70] and Shao et al. [146].

10 Asymmetrized Maps

- From decomposing the set of rooted orientable maps, a quadratic differential equation in Riccati type is discovered and then a recursion formula is extracted for determining a solution as the enumerating function with size as the parameter.
- A quadratic equation of the enumerating function in company with its partial values for rooted maps on the surface of orientable genus 0 is discovered with an extra parameter and then an explicit expression of the function with only size as a parameter is via characteristic parameters extracted for each term summation free.
- From decomposing the set of rooted nonorientable maps, a nonlinear differential equation satisfied by the enumerating function with size as the parameter is discovered in company with the orientable case and then a recursion formula is extracted for determining the function.
- The numbers of orientable, nonorientable and total maps with given size are, in all, obtained and then calculated for size not greater than 10.

10.1 Orientable equation

Corollary 8.7 shows that a map with symmetry becomes a map without symmetry whenever an element is chosen as the root. Such a map with a root is called a *rooted map*.

Rooting is, in fact, a kind of simplification in mathematics, particularly in recognizing distinct combinatorial configurations for reducing the complexity.

As soon as the rooted case is done, the general case can be recovered by considering the symmetry in a suitable way.

For maps, the estimation of the order of the automorphism group of a map in the previous chapter and the efficient algorithm for justifying and recognizing if two maps are isomorphic in Chapter 7 provide a theoretical foundation for transforming rooted maps into non-rooted maps. This will be seen in Chapter 11.

The main purpose of this chapter is to present some methods for investigating non-planar rooted maps as appendix to the monograph *Enumerative Theory of Maps* as in Liu [67], in which most pages are for planar maps, particularly rooted.

Let $\mathcal{M}$ be the set of all orientable rooted maps. For $M = (\mathcal{X}, \mathcal{P}) \in \mathcal{M}$, let

$$v_x = (x)_{\mathcal{P}} = (x, \mathcal{P}x, \dots, \mathcal{P}^{-1}x) \tag{10.1}$$

be the vertex incident with $x \in \mathcal{X}$. The root is always denoted by r. The rooted edge that is incident with r is denoted by $a = Kr$. The rooting of $M - a$ is taking $\mathcal{P}^{\delta}r$ as its root where

DOI 10.1515/9783110481846-010

$$\delta = \min\{i|\mathscr{P}^i r \notin Kr, i \geq 1\}. \tag{10.2}$$

In fact,

$$\delta = \begin{cases} 1, & \text{if } \mathscr{P}r \neq \gamma r; \\ 2, & \text{othewise.} \end{cases} \tag{10.3}$$

In virtue of Theorem 3.4, $M - a$ is a map if, and only if, a is not a harmonic loop except terminal link (or segmentation edge) of M.

Now, let us partition $\mathscr{M}$ into three parts: $\mathscr{M}_{\text{I}}$, $\mathscr{M}_{\text{II}}$ and $\mathscr{M}_{\text{III}}$, i.e.,

$$\mathscr{M} = \mathscr{M}_{\text{I}} + \mathscr{M}_{\text{II}} + \mathscr{M}_{\text{III}}, \tag{10.4}$$

where $\mathscr{M}_{\text{I}} = \{\vartheta\}$, i.e., consists of the trivial map, $\mathscr{M}_{\text{II}}$ and $\mathscr{M}_{\text{III}}$, respectively, consist of those with a as a segmentation edge and not.

Lemma 10.1 Let $\mathscr{M}_{\langle\text{II}\rangle} = \{M - a|\forall M \in \mathscr{M}_{\text{II}}\}$. Then,

$$\mathscr{M}_{\langle\text{II}\rangle} = \mathscr{M} \times \mathscr{M} \tag{10.5}$$

where $\times$ stands for the Cartesian product of two sets.

Proof For any $M = (\mathscr{X}, \mathscr{P}) \in \mathscr{M} \times \mathscr{M}$, let $M = M_1 + M_2$, $M_i = (\mathscr{X}_i, \mathscr{P}_i)$, $i = 1, 2$. Assume $M' = (\mathscr{X}', \mathscr{P}')$ such that $\mathscr{X}' = \mathscr{X} + Kr'$ and $\mathscr{P}'$ is different from $\mathscr{P}_2$ or $\mathscr{P}_1$ only at, respectively,

$$v_{r'} = (r')_{\mathscr{P}'} = (r', r_2, \mathscr{P}_2 r_2, \ldots, \mathscr{P}_2^{-1} r_2)$$

or

$$v_{\beta r'} = (\beta r')_{\mathscr{P}'} = (\alpha\beta r', r_1, \mathscr{P}_1 r_1, \ldots, \mathscr{P}_1^{-1} r_1).$$

Since $M' \in \mathscr{M}$ and its rooted edge $a' = Kr'$ is a segmentation edge, $M' \in \mathscr{M}_{\text{II}}$. It is checked that $M = M' - a'$. Therefore, $M \in \mathscr{M}_{\langle\text{II}\rangle}$.

Conversely, for any $M \in \mathscr{M}_{\langle\text{II}\rangle}$, we have $M' \in \mathscr{M}_{\text{II}}$ such that $M = M' - a'$ where $a' = Kr'$. From $M' \in \mathscr{M}_{\text{II}}$, $M = M_1 + M_2$ where $M_1, M_2 \in \mathscr{M}$. This implies that $M \in \mathscr{M} \times \mathscr{M}$. □

It is seen from this lemma that there is a one-to-one correspondence between $M(\in \mathscr{M}_{\langle\text{II}\rangle}$, or $\mathscr{M} \times \mathscr{M})$ and $M'(\in \mathscr{M}_{\text{II}})$. Hence,

$$\left|\mathscr{M}_{\text{II}}\right| = \left|\mathscr{M} \times \mathscr{M}\right|. \tag{10.6}$$

For $M = (\mathscr{X}, \mathscr{P}) \in \mathscr{M}_{\mathrm{III}}$, because $M - a$ is a map (Theorem 3.4), from (10.3), the root $r(M - a)$ of $M - a$ has two possibilities: when $\mathscr{P}r(M) \neq \gamma r(M)$, $r(M - a) = \mathscr{P}r(M)$; otherwise,

$$r(M - a) = \mathscr{P}^2 r(M).$$

Let $\tilde{M} = (\tilde{\mathscr{X}}, \tilde{\mathscr{P}}) = M - a$ where $\tilde{\mathscr{X}} = \mathscr{X} - Kr$ and $\tilde{\mathscr{P}}$ are different from $\mathscr{P}$ only at

$$\left.\begin{aligned}(\tilde{r})_{\tilde{\mathscr{P}}} &= (\mathscr{P}r, \mathscr{P}^2 r, \ldots, \mathscr{P}^{-1} r),\\ (\mathscr{P}\gamma r)_{\tilde{\mathscr{P}}} &= (\mathscr{P}\gamma r, \mathscr{P}^2 \gamma r, \ldots, \mathscr{P}^{-1}\gamma r)\end{aligned}\right\}\ \text{if } a \text{ is not a loop;}$$

otherwise, i.e., when a is a loop,

$$(\tilde{r})_{\tilde{\mathscr{P}}} = \begin{cases} (\mathscr{P}^2 r, \mathscr{P}^3 r, \ldots, \mathscr{P}^{-1} r), \ \gamma r = \mathscr{P}r; \\ (\mathscr{P}r, \ldots, \mathscr{P}^{s-1} r, \mathscr{P}^{s+1} r, \ldots, \mathscr{P}^{-1} r), \\ \qquad\qquad \text{if } \gamma r = \mathscr{P}^s r,\ s \geq 2. \end{cases}$$

Since $M(\mathscr{X}, \mathscr{P})$ is orientable, group $\Psi = \Psi_{\{\gamma, \mathscr{P}\}}$ has two orbits $\{r\}_\Psi$ and $\{\alpha r\}_\Psi$ on $\mathscr{X}$. For $\tilde{M} = (\tilde{\mathscr{X}}, \tilde{\mathscr{P}}) = M - a$, group $\tilde{\Psi} = \Psi_{\{\gamma, \tilde{\mathscr{P}}\}}$ also has two orbits

$$\{\tilde{r}\}_{\tilde{\Psi}} = \{r\}_\Psi - \{r, \gamma r\}$$

and

$$\{\alpha\tilde{r}\}_{\tilde{\Psi}} = \{\alpha r\}_\Psi - \{\alpha r, \beta r\}.$$

So, $\tilde{M}$ is also orientable.

Furthermore, for every element $y \in \{\tilde{r}\}_{\tilde{\Psi}}$, there exists exactly one position of $a = Kr$, i.e., γr is in the angle $\langle \alpha y, \tilde{\mathscr{P}} \rangle$, for $M \in \mathscr{M}_{\mathrm{III}}$ such that $\tilde{M} = M - a$. This means that in $\mathscr{M}_{\mathrm{III}}$, there are

$$|\{\tilde{r}\}_{\tilde{\Psi}}| = \frac{1}{2}|\tilde{\mathscr{X}}| = 2m(\tilde{M}),$$

where $m(\tilde{M})$ is the size of $\tilde{M}$, non-isomorphic maps for producing $\tilde{M}$. By considering for the case $\mathscr{P}r = \gamma r$, in $\mathscr{M}_{\mathrm{III}}$, there are in all $2m(\tilde{M}) + 1$ non-isomorphic maps for $\tilde{M}$.

Example 10.1 Let

$$\tilde{M} = (Kx + Ky + Kz, (x, y, \gamma z)(z, \gamma x, \gamma y)),$$

where $\tilde{r} = x$ is the root shown in Figure 10.1(a). Since it is orientable, the orbit of group $\tilde{\Psi}$ which $\tilde{r}$ is in can be written as a cyclic permutation as

$$(\tilde{r})_{\tilde{\Psi}} = (x, y, \gamma z, z, \gamma x, \gamma y).$$

Then, Figure 10.1(b–h) presents all the $2m(\tilde{M}) + 1 = 2 \times 3 + 1 = 7$ maps in $\mathscr{M}_{\text{III}}$, obtained by appending $a = Kr$ on map $\tilde{M}$ where (b) is for $\mathscr{P}r = \gamma r$ and (c–h) are for those obtained by appending $a = Kr$ in the order of $(\tilde{r})_{\tilde{\Psi}}$ from $\tilde{M}$.

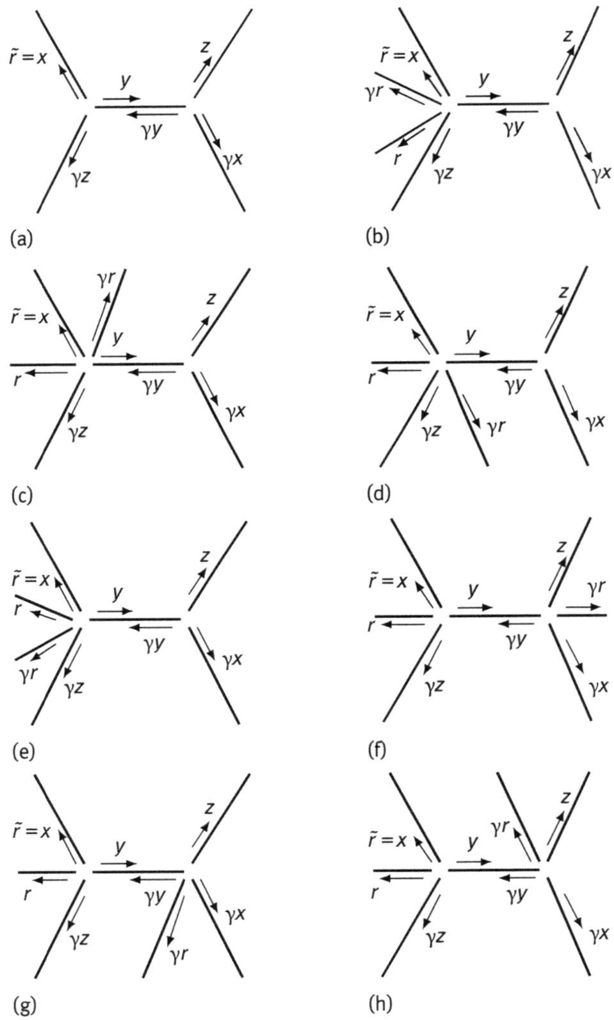

Figure 10.1: New maps obtained by appending an edge.

Lemma 10.2 Let $\mathscr{M}_{\langle\mathrm{III}\rangle} = \{M - a | \forall M \in \mathscr{M}_{\mathrm{III}}\}$. Then,

$$\mathscr{M}_{\langle\mathrm{III}\rangle} = \mathscr{M}. \tag{10.7}$$

Proof Because for any $M \in \mathscr{M}_{\mathrm{III}}$, $M - a$ is also a map (Theorem 3.4), so $\mathscr{M}_{\langle\mathrm{III}\rangle} \subseteq \mathscr{M}$.

Conversely, for any $M \in \mathscr{M}$, any one, e.g., M' of the $2m(M) + 1$ maps obtained by appending a' from M in the above way is with $M' \in \mathscr{M}_{\mathrm{III}}$. Because $M = M' - a'$, so $M \in \mathscr{M}_{\langle\mathrm{III}\rangle}$. □

For convenience, let $\mathscr{H}(M)$ be the set of all the $2m(\tilde{M}) + 1$ maps in $\mathscr{M}_{\mathrm{III}}$, obtained from M by appending an edge in the above way. From Theorem 8.1, they are all mutually non-isomorphic in the sense of rooting.

Lemma 10.3 For $\mathscr{M}_{\mathrm{III}}$,

$$\mathscr{M}_{\mathrm{III}} = \sum_{M \in \mathscr{M}} \mathscr{H}(M). \tag{10.8}$$

Proof For any $M = (\mathscr{X}, \mathscr{P}) \in \mathscr{M}_{\mathrm{III}}$, let $\tilde{M} = (\tilde{\mathscr{X}}, \tilde{\mathscr{P}}) = M - a$. Because a is not a segmentation edge, from Theorem 3.4 and Corollary 3.1, $\tilde{M} \in \mathscr{M}$. By orientability, because $r \in \{\tilde{r}\}_{\tilde{\Psi}}$ where $\tilde{\Psi} = \Psi_{\{\gamma, \tilde{\mathscr{P}}\}}$, there exists $y \in \{\tilde{r}\}_{\tilde{\Psi}}$ such that $\mathscr{P}y = \gamma r$, or $\mathscr{P}r = \gamma r$. Because $|\{\tilde{r}\}_{\tilde{\Psi}}| = 2m\tilde{M}$, the former has $2m(\tilde{M})$ possibilities and the latter, only one. This is the $2m(\tilde{M}) + 1$ possibilities in $\mathscr{H}(\tilde{M})$. Further, because $\tilde{M} \in \mathscr{M}$, M is an element of the set on the right side of (10.8).

Conversely, for any $M \in \mathscr{H}(\tilde{M})$, $\tilde{M} \in \mathscr{M}$, Since $\tilde{M} = M - a \in \mathscr{M}$, by Theorem 3.4 and Corollary 3.1 a is not a segmentation edge. Therefore, $M \in \mathscr{M}_{\mathrm{III}}$. □

Furthermore, (10.8) provides a one-to-one correspondence between the sets on its two sides. This enables us to construct all orientable maps with the rooted edge not a segmentation edge from general orientable maps with smaller size.

In order to determine the number of non-isomorphic orientable rooted maps in $\mathscr{M}$ with size $m \geq 0$, the *enumerating function* of set $\mathscr{M}$

$$f_{\mathscr{M}}(x) = \sum_{M \in \mathscr{M}} x^{m(M)} \tag{10.9}$$

has to be investigated for a simpler form in infinite power series where $m(M)$ is the size of M. In the series form of (10.9), the coefficient of the term with x^m, $m \geq 0$, is just the number of non-isomorphic orientable rooted maps with size m.

From (10.4),

$$f_{\mathscr{M}}(x) = f_{\mathscr{M}_{\mathrm{I}}}(x) + f_{\mathscr{M}_{\mathrm{II}}}(x) + f_{\mathscr{M}_{\mathrm{III}}}(x). \tag{10.10}$$

Lemmas above enable us to evaluate $f_{\mathcal{M}_{\mathrm{I}}}(x)$, $f_{\mathcal{M}_{\mathrm{II}}}(x)$ and $f_{\mathcal{M}_{\mathrm{III}}}(x)$ as functions of $f = f_{\mathcal{M}}(x)$.

First, because $\mathcal{M}_{\mathrm{I}}$ contains only one map ϑ and $m(\vartheta) = 0$, $f_{\mathcal{M}_{\mathrm{I}}}(x)$ contributes the constant term 1 of f, i.e.,

$$f_{\mathcal{M}_{\mathrm{I}}}(x) = 1. \tag{10.11}$$

Lemma 10.4 For $\mathcal{M}_{\mathrm{II}}$,

$$f_{\mathcal{M}_{\mathrm{II}}}(x) = xf^2. \tag{10.12}$$

Proof According to the one-to-one correspondence between $\mathcal{M}_{\mathrm{II}}$ and $\mathcal{M}_{\langle \mathrm{II} \rangle}$ and that the former is with its size 1 greater than the latter in the correspondence, by (10.6),

$$\begin{aligned} f_{\mathcal{M}_{\mathrm{II}}}(x) &= x \sum_{M \in \mathcal{M}_{\langle \mathrm{II} \rangle}} x^{m(M)} \\ &= x \sum_{M \in \mathcal{M} \times \mathcal{M}} x^{m(M)} \\ &= x\Big(\sum_{M \in \mathcal{M}} x^{m(M)} \Big)^2 \\ &= xf^2. \end{aligned}$$

This is (10.12). □

Lemma 10.5 For $\mathcal{M}_{\mathrm{III}}$,

$$f_{\mathcal{M}_{\mathrm{III}}}(x) = xf + 2x^2 \frac{\mathrm{d}f}{\mathrm{d}x}. \tag{10.13}$$

Proof From the one-to-one correspondence between $\mathcal{M}_{\mathrm{II}}$ and $\mathcal{M}_{\langle \mathrm{II} \rangle}$ and that the former is with its size 1 greater than the latter in the correspondence, and then by Lemmas 10.3 and 9.10,

$$\begin{aligned} f_{\mathcal{M}_{\mathrm{III}}}(x) &= x \sum_{M \in \mathcal{M}_{\langle \mathrm{III} \rangle}} x^{m(M)} \\ &= x\Big(f + 2x \frac{\mathrm{d}f}{\mathrm{d}x} \Big) \\ &= xf + 2x^2 \frac{\mathrm{d}f}{\mathrm{d}x}. \end{aligned}$$

This is (10.13). □

Theorem 10.1 The differential equation about f

$$\begin{cases} 2x^2\dfrac{\mathrm{d}f}{\mathrm{d}x} = -1 + (1-x)f - xf^2; \\ f_0 = f|_{x=0} = 1 \end{cases} \tag{10.14}$$

is well defined in the ring of infinite power series with all coefficients nonnegative integers and the terms of negative powers finite. And the solution is $f = f_{\mathcal{M}}(x)$.

Proof Suppose $f = F_0 + F_1x + F_2x^2 + \cdots + F_mx^m + \cdots$, $F_i \in Z_+$, $i \geq 0$. Based on the first relation of (10.14), by equating the coefficients on the two sits with the same power of x, the recursion

$$\begin{cases} -1 + F_0 = 0; \\ F_1 - F_0 - F_0^2 = 0; \\ F_m = (2m-1)F_{m-1} + \sum_{i=0}^{m-1} F_iF_{m-1-i}, \\ \qquad m \geq 2 \end{cases} \tag{10.15}$$

is soon extracted. Then, $F_0 = 1$ (the initial condition), $F_1 = 2, \ldots,$ all the coefficients of f can be uniquely found from this recursion. Because only addition and multiplication are used for evaluating all the coefficients from the initial condition that F_0 is an integer, F_m, $m \geq 1$, must all be integers. This is the first statement.

For the last statement, from (10.10) and (10.11–10.13), it is seen that $f = f_{\mathcal{M}}(x)$ satisfies the first relation of (10.14). And, $f_0 = f_{\mathcal{M}}(0) = 1$ is just the initial condition. By the uniqueness in the first statement, the only possibility is $f = f_{\mathcal{M}}(x)$. □

Although the form of the equation in Theorem 10.1 is rather simple, because of the occurrence of f^2, it is far from getting the solution directly. In fact, it is an equation in the Riccati's type. It has no analytic solution in general.

10.2 Planar rooted maps

Let $\mathcal{T}$ be the set of all planar rooted maps. It looks difficult to decompose $\mathcal{T}$ into some classes so that each class can be produced by $\mathcal{T}$ with only size as the parameter. Now, another parameter for a map M, i.e., the valency of the rooted vertex $n(M)$, is introduced. The enumerating function of $\mathcal{T}$ is

$$t(x,y) = f_{\mathcal{T}}(x,y) = \sum_{M \in \mathcal{T}} x^{m(M)}y^{n(M)}, \tag{10.16}$$

where $m(M)$ is still the size of M.

Assume that $\mathscr{T}$ is partitioned into three classes: $\mathscr{T}_0$, $\mathscr{T}_1$ and $\mathscr{T}_2$, i.e.,

$$\mathscr{T} = \mathscr{T}_0 + \mathscr{T}_1 + \mathscr{T}_2, \tag{10.17}$$

where $\mathscr{T}_0 = \{\vartheta\}$, $\mathscr{T}_1$ and $\mathscr{T}_2$ are the sets of planar rooted maps with the rooted edge, respective of a loop and a link (not loop).

For $M \in \mathscr{T}$, let $a = Kr$ be the rooted edge of M with the root $r = r(M)$. For maps $M_i \in \mathscr{T}$(i = 1 and 2), let $a_i = Kr_i$ be the rooted edge of M_i with the root $r_i = r(M_i)$.

The 1-*addition* of two maps $M_1 = (\mathscr{X}_1, \mathscr{P}_1)$ and $M_2 = (\mathscr{X}_2, \mathscr{P}_2)$ is to produce the map $M_1 +\cdot M_2 = M_1 \cup M_2$ with the root $r = r_1$ provided $M_1 \cap M_2 = \{v_r\}$ where $v_r = (\langle r_1\rangle_{\mathscr{P}_1}, \langle r_2\rangle_{\mathscr{P}_2})$.

Lemma 10.6 Let $\mathscr{T}_{\langle 1\rangle} = \{M - a|\forall M \in \mathscr{T}_1\}$, then

$$\mathscr{T}_{\langle 1\rangle} = \mathscr{T} \times\cdot \mathscr{T} \tag{10.18}$$

where $\mathscr{T} \times\cdot \mathscr{T} = \{M_1 + M_2|\forall M_1, M_2 \in \mathscr{T}\}$ is called the 1-*product* of $\mathscr{T}$ with itself.

Proof For any $M = (\mathscr{X}, \mathscr{P}) \in \mathscr{T}_{\langle 1\rangle}$, let $M' = (\mathscr{X}', \mathscr{P}') \in \mathscr{T}_1$ such that $M' - a' = M$. Because $a' = Kr'$ is a loop,

$$(r')_{\mathscr{P}'} = (r', \mathscr{P}'r', \ldots, \gamma r', \ldots, \mathscr{P}'^{-1}r').$$

From the planarity, $M' - a' = M_1 +\cdot M_2$, $M_i = (\mathscr{X}_i, \mathscr{P}_i)$, $i = 1, 2$, where $\mathscr{X} = \mathscr{X}_1 + \mathscr{X}_2 = \mathscr{X}' - Kr'$, $\mathscr{P}_1$ and $\mathscr{P}_2$ are different from $\mathscr{P}$ only at $(r)_{\mathscr{P}}$ becoming, respectively,

$$(r_1)_{\mathscr{P}_1} = (\mathscr{P}'r', \mathscr{P}(\mathscr{P}'r'), \ldots, \mathscr{P}'^{-1}\gamma r')$$

where $\gamma = \alpha\beta$ and

$$(r_2)_{\mathscr{P}_2} = (\mathscr{P}'\gamma r', \mathscr{P}(\mathscr{P}'\gamma r'), \ldots, \mathscr{P}'^{-1}r').$$

This implies $M \in \mathscr{T} \times\cdot \mathscr{T}$.

Conversely, for $M \in \mathscr{T} \times\cdot \mathscr{T}$, because $M = M_1 + M_2$, let $M' = M + a'$, $a' = Kr'$, such that

$$(r')_{\mathscr{P}'} = (r', \langle r_1\rangle_{\mathscr{P}_1}, \gamma r', \langle r_2\rangle_{\mathscr{P}_2}),$$

then $M' \in \mathscr{T}$ and $M = M' - a'$. Since a' is a loop, $M' \in \mathscr{T}_1$ and hence $M \in \mathscr{T}_{\langle 1\rangle}$. □

Because this lemma presents a one-to-one correspondence $M = M_1 + M_2$ between $M \in \mathscr{T}_{\langle 1\rangle}$ and $M_1, M_2 \in \mathscr{T}$ with $m(M) = m(M_1) + m(M_2)$ and $n(M) = n(M_1) + n(M_2)$, the enumerating function of $\mathscr{T}_{\langle 1\rangle}$

$$
\begin{aligned}
f_{\mathcal{T}_{\langle 1\rangle}}(x,y) &= \sum_{M \in \mathcal{T} \times \mathcal{T}} x^{m(M)} y^{n(M)} \\
&= \sum_{M_1, M_2 \in \mathcal{T}} x^{m(M_1)+m(M_2)} y^{n(M_1)+n(M_2)} \\
&= \Big(\sum_{M \in \mathcal{T}} x^{m(M)} y^{n(M)} \Big)^2 \\
&= t^2(x,y).
\end{aligned}
\tag{10.19}
$$

Then, from the one-to-one correspondence between $\mathcal{T}_1$ and $\mathcal{T}_{\langle 1\rangle}$ with the former of size 1 greater than the latter and the former of the rooted vertex valency two greater than the latter, the enumerating function of $\mathcal{T}_1$ is

$$
f_{\mathcal{T}_1}(x,y) = xy^2 f_{\mathcal{T}_{\langle 1\rangle}}(x,y) = xy^2 t^2(x,y). \tag{10.20}
$$

However, for $\mathcal{T}_2$, the correspondence between $\mathcal{T}_2$ and $\mathcal{T}_{\langle 2\rangle} = \{M \bullet a | \forall M \in \mathcal{T}_2\}$ with the former of size 1 greater than the latter is not of one-to-one where $M \bullet a$ is the contraction of the rooted edge a on map M. The root on $M \bullet a$ is defined to be $\mathcal{P}\gamma r$ when $\mathcal{P}\gamma r \neq \gamma r$; otherwise $(\mathcal{P}\gamma)^2 r$. Because a is a link, this is a basic transformation. According to Chapter 5, $M - a$ is planar if, and only if, M is. Hence,

$$
\mathcal{T}_{\langle 2\rangle} = \mathcal{T}. \tag{10.21}
$$

Further, observe what a correspondence between $\mathcal{T}_2$ and $\mathcal{T}$ is.

For $M = (\mathcal{X}, \mathcal{P}) \in \mathcal{T}$, let $(r)_{\mathcal{P}} = (r, \mathcal{P}r, \ldots, \mathcal{P}^{n(M)-1}r)$, where $n(M)$ is the valency of the rooted vertex v_r on M. By splitting a link at v_r, all those obtained are still planar because this operation is a basic transformation. For doing this, there are $n(M) + 1$ possibilities altogether.

Let $M_i = (\mathcal{X}_i, \mathcal{P}_i), 0 \leq i \leq n(M)$ be all the $n(M) + 1$ maps obtained from M by splitting an edge at the rooted vertex $v_r = (r)_{\mathcal{P}}$ as

$$
v_{r_i} = \begin{cases}
(r_0, \langle r \rangle_{\mathcal{P}}), \text{ while } v_{\beta r_0} = (\gamma r_0); \\
(r_1, \mathcal{P}r, \ldots, \mathcal{P}^{n(M)-1}r), \text{ while } v_{\beta r_1} = (\gamma r_1, r); \\
(r_2, \mathcal{P}^2 r, \ldots, \mathcal{P}^{n(M)-1}r), \text{ while } v_{\beta r_2} = (\gamma r_2, r, \mathcal{P}r); \\
\cdots\cdots \\
(r_{n(M)-1}, \mathcal{P}^{n(M)-1}r), \\
\quad \text{while } v_{\beta r_{n(M)-1}} = (\gamma r_{n(M)-1}, r, \ldots, \mathcal{P}^{n(M)-2}r); \\
(r_{n(M)}), \text{ while } v_{\beta r_{n(M)}} = (\gamma r_{n(M)}, \langle r \rangle_{\mathcal{P}}).
\end{cases}
\tag{10.22}
$$

Lemma 10.7 For a map $M \in \mathscr{T}$, let

$$\mathscr{K}(M) = \{M_i | i = 0, 1, 2, \ldots, n(M)\},$$

where M_i, $0 \le i \le n(M)$, is given by (10.22). Then,

$$\mathscr{T}_2 = \sum_{M \in \mathscr{T}} \mathscr{K}(M). \tag{10.23}$$

Proof For any $M \in \mathscr{T}_2$, because $a = Kr$ is a link, from (10.21), $M \bullet a \in \mathscr{T}$ and from (10.22), $M \in \mathscr{K}(M \bullet a)$. Therefore, M is an element of the set on the right side of (10.23).

Conversely, for M is an element of the set on the right of (10.23), there exists a map $M' \in \mathscr{T}$ such that $M \in \mathscr{K}(M')$. Because all maps in $\mathscr{K}(M)$ are planar if, and only if, M' is planar, $M \in \mathscr{T}$ as well. Moreover, from (10.22), the rooted edge a is a link, $M \in \mathscr{T}_2$. This meas that M is an element of the set on the left side of (10.23). □

Because (10.23) presents a one-to-one correspondence between maps with the same size on its two sides and the valency of rooted vertex of M_i $(0 \le i \le n(M))$ is $n(M) - i$ for any $M \in \mathscr{T}$, the enumerating function of $\mathscr{T}_2$ is

$$\begin{aligned} f_{\mathscr{T}_2}(x, y) &= \sum_{M \in \mathscr{T}_2} x^{m(M)} y^{n(M)} \\ &= xy \sum_{M \in \mathscr{T}} \left(\sum_{i=0}^{n(M)} y^i \right) x^{m(M)} \\ &= xy \sum_{M \in \mathscr{T}} \frac{1 - y^{n(M)+1}}{1 - y} x^{m(M)} \\ &= \frac{xy}{1 - y}(t_0 - yt), \end{aligned} \tag{10.24}$$

where $t = t(x, y)$ and $t_0 = t(x, 1)$.

Theorem 10.2 The enumerating function $t = t(x, y)$ of planar rooted maps satisfies the equation as

$$xy^2(1 - y)t^2 - (1 - y + xy^2)t + xyt_0 + (1 - y) = 0 \tag{10.25}$$

where $t_0 = t(x, 1)$.

Proof From (10.17),

$$t = f_{\mathscr{T}_0}(x, y) + f_{\mathscr{T}_1}(x, y) + f_{\mathscr{T}_2}(x, y).$$

Because $\mathcal{T}_0 = \{\vartheta\}$ and ϑ has no edge, $f_{\mathcal{T}_0}(x, y) = 1$. From (10.20) and (10.24),

$$t = 1 + xy^2t^2 + \frac{xy}{1-y}(t_0 - yt).$$

Via rearrangement of terms, (10.25) is soon found. □

Although (10.25) is a quadratic equation, because the unknown t_0 occurs and the equation becomes an identity when $y = 1$, complication courses to solve the equation directly are not easy.

The discriminant of the eq. (10.25), denoted by $D(x, y)$, is

$$\begin{aligned} D(x, y) &= (xy^2 - y + 1)^2 - 4(y-1)xy^2(y - 1 - xyt_0) \\ &= 1 - 2y + (1-2x)y^2 + \left(x^2 + 2x - H(x)\right)y^3 + H(x)y^4, \end{aligned} \tag{10.26}$$

where

$$H(x) = 4x^2t_0 + x^2 - 4x. \tag{10.27}$$

Assume that $D(x, y)$ has the form

$$\begin{aligned} D(x, y) &= (1-\theta y)^2(1 + ay + by^2) \\ &= 1 - (2\theta - a)y + (\theta^2 - 2a\theta + b)y^2 \\ &\quad + \theta(a\theta - 2b)y^3 + \theta^2 by^4. \end{aligned} \tag{10.28}$$

By comparing with (10.26),

$$\theta = 1 + \frac{a}{2}, \tag{10.29}$$

and

$$\begin{cases} 1 - 2x = \theta(4 - 3\theta) + b; \\ x^2 + 2x - H(x) = \theta\left(2(\theta - 1)\theta - 2b\right); \\ H(x) = \theta^2 b. \end{cases} \tag{10.30}$$

Then, an equation about b with θ as the parameter is found as

$$\begin{aligned} \frac{1}{4}\left(1 - 4\theta + 3\theta^2 - b\right)^2 &+ 1 - 4\theta + 3\theta^2 - b - \theta^2 b \\ &= 2(\theta - 1)\theta^2 - 2\theta b. \end{aligned}$$

By rearrangement, it becomes

$$b^2 - (10\theta^2 - 16\theta + 6)b + (9\theta^4 - 32\theta^3 + 42\theta^2 - 24\theta + 5) = 0. \quad (10.31)$$

The discriminant of (10.31) is

$$\begin{aligned}(10\theta^2 - 16\theta + 6)^2 &- 4(9\theta^4 - 32\theta^3 + 42\theta^2 - 24\theta + 5)\\ &= 64\theta^4 - 192\theta^3 + 208\theta^2 - 96\theta + 16\\ &= (8\theta^2 - 12\theta + 4)^2.\end{aligned}$$

Therefore, $b = (\theta - 1)^2$, or $9\theta^2 - 14\theta + 5 = (9\theta - 5)(\theta - 1)$.

The latter has to be chosen in our case. By the last relation of (10.30),

$$H(x) = \theta^2(9\theta - 5)(\theta - 1).$$

From the first relation of (10.30) and (10.27), the expressions for x and t_0 with parameter θ are extracted as

$$\begin{cases} x = (3\theta - 2)(1 - \theta);\\ t_0 = \frac{4\theta - 3}{(3\theta - 2)^2}. \end{cases} \quad (10.32)$$

This enables us to get b_0 as a power series of x as

$$t_0(x) = \sum_{m \geq 0} \frac{2 \times 3^m (2m)!}{m!(m+2)!} x^m. \quad (10.33)$$

by eliminating the parameter θ via Lagrangian inversion. More about this method can be seen in the monograph [67], or originally in Liu [105].

Example 10.2 For $m = 2$, it is known from (10.33) that the number of non-isomorphic planar rooted maps is the coefficient of x^2. That is 9.

Because there are four non-isomorphic planar maps of size 2 as shown in Figure 10.2. The arrows on the same map represent the roots of non-isomorphic rooted ones. As such there are, respectively, two, two, one and four non-isomorphic rooted maps in (a), (b), (c) and (d) of Figure 10.2 to get 9 altogether.

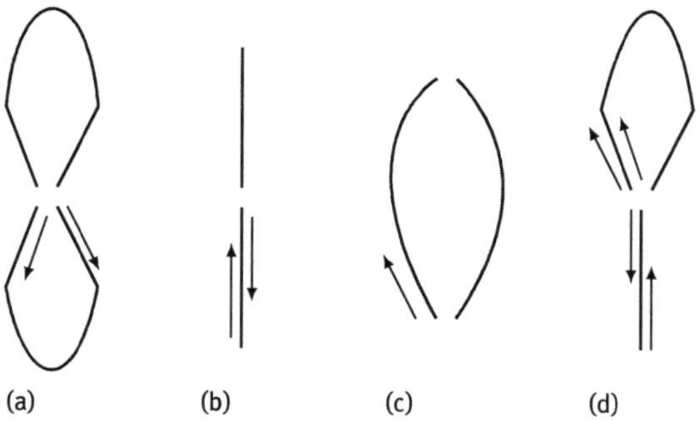

Figure 10.2: Planar rooted maps of two edges.

10.3 Nonorientable equation

Let $\mathcal{N}_m$ be the set of nonorientable rooted maps of size m. Of course, $m \geq 1$. And, let $\mathcal{N}_m$ be partitioned into $\mathcal{N}_m^{(\mathrm{I})}$ and $\mathcal{N}_m^{(\mathrm{II})}$, i.e.,

$$\mathcal{N}_m = \mathcal{N}_m^{(\mathrm{I})} + \mathcal{N}_m^{(\mathrm{II})} \tag{10.34}$$

where $\mathcal{N}_m^{(\mathrm{I})} = \{N | N \in \mathcal{N}_m, N - a \text{ orientable}\}$ and

$$\mathcal{N}_m^{(\mathrm{II})} = \mathcal{N}_m - \mathcal{N}_m^{(\mathrm{I})} = \{N | N \in \mathcal{N}_m, N - a \text{ nonorientable}\},$$

$a = e_r(N)$ is still the rooted edge.

Lemma 10.8 Let $\mathcal{N}_m^{\langle \mathrm{I} \rangle} = \{N - a | \forall N \in \mathcal{N}_m^{(\mathrm{I})}\}$, then

$$\mathcal{N}_m^{\langle \mathrm{I} \rangle} = \mathcal{M}_{m-1} \tag{10.35}$$

where $\mathcal{M}_{m-1}$ is the set of orientable rooted maps of size $m - 1$, $m \geq 1$.

Proof Because of the nonorientability of $N \in \mathcal{N}_m^{(\mathrm{I})}$ and the orientability of $N - a$, from Corollary 3.1, a is not a segmentation edge. Based on Theorem 3.4, $N - a$ is always an orientable map. So, for any $N \in \mathcal{N}_m^{\langle \mathrm{I} \rangle}$, $N \in \mathcal{M}_{m-1}$, $m \geq 1$. This implies $\mathcal{N}_m^{\langle \mathrm{I} \rangle} \subseteq \mathcal{M}_{m-1}$.

Conversely, for any $N = (\mathcal{X}, \mathcal{P}) \in \mathcal{M}_{m-1}$, by appending an edge a' on N, $N' = (\mathcal{X}', \mathcal{P}')$ is obtained where $\mathcal{X}' = \mathcal{X} + Kr'$ and $\mathcal{P}'$ is different from $\mathcal{P}$ only at the vertex

$$(r')_{\mathcal{P}'} = (r', \beta r', \langle r \rangle_{\mathcal{P}}).$$

Since Kr' is not a segmentation edge, from Theorem 3.7, N' is a map. And since $\beta r' \in \{r'\}_{\Psi'}$, $\Psi' = \Psi_{\{\mathcal{P}', \gamma\}}$, from Theorem 4.1, N' is nonorientable. Further, because $N = N' - a'$ is orientable and $m(N) + 1 = m(N') = m$, $N \in \mathcal{N}_m^{\langle \mathrm{I} \rangle}$. This implies $\mathcal{M}_{m-1} \subseteq \mathcal{N}_m^{\langle \mathrm{I} \rangle}$. □

For any $M = (\mathscr{X}, \mathscr{P}) \in \mathscr{M}_m$, since M is orientable, assume

$$\{r\}_\Psi = \{r, \psi_1 r, \ldots, \psi_{2m-1} r\},$$

$\psi_i \in \Psi = \Psi_{\{\gamma,\mathscr{P}\}}$, $i = 1, 2, \ldots, 2m(M) - 1 = 2m - 1$. By appending the edge r',

$$A(M) = \{A_0(M), A_1(M), \ldots, A_{2m}(M)\}$$

is obtained where $A_i(M) = M + e_{r_i} = (\mathscr{X}_i, \mathscr{P}_i)$ such that $\mathscr{X}_i = \mathscr{X} + Kr_i$ and $\mathscr{P}_i$ is determined in the following manner:

$$\begin{cases} \beta r_0 \text{ in the angle } \langle \alpha\mathscr{P}^{-1} r, r\rangle, \\ \beta r_i \ (i = 1, 2, \ldots, 2m(M) - 1) \text{ in the angle } \langle \alpha\mathscr{P}^{-1}\psi_i r, \psi_i r\rangle, \\ \beta r_{2m(M)} \text{ in the angle } \langle r, \alpha\mathscr{P}^{-1} r\rangle. \end{cases}$$

Because $\beta r_i \in \{r_i\}_{\Psi_i}$ where $\Psi_i = \Psi_{\{\gamma,\mathscr{P}_i\}}$, $i = 0, 1, \ldots, 2m(M)$, from Theorem 4.1, A_i are all nonorientable. Because $A_i(M) - e_{r_i} \in \mathscr{M}_m$, $A_i(M) \in \mathscr{N}_{m+1}^{(\mathrm{I})}$, $0 \le i \le 2m(M)$. From Lemma 10.8,

$$\mathscr{N}_{m+1}^{(\mathrm{I})} = \sum_{M \in \mathscr{M}_m} A(M) \tag{10.36}$$

$m \ge 0$. Of course, $\mathscr{M}_0$ consists of only the trivial map.

For $\mathscr{N}_m^{(\mathrm{II})}$, two cases should be considered: $\mathscr{N}_m^{(\mathrm{N})}$ and $\mathscr{N}_m^{(\mathrm{T})}$, i.e.,

$$\mathscr{N}_m^{(\mathrm{II})} = \mathscr{N}_m^{(\mathrm{N})} + \mathscr{N}_m^{(\mathrm{T})}, \tag{10.37}$$

where

$$\mathscr{N}_m^{(\mathrm{N})} = \{N \mid \forall N \in \mathscr{N}_m^{(\mathrm{II})}, a = e_r \text{ is a terminal, or segmentation edge}\}$$

and

$$\mathscr{N}_m^{(\mathrm{T})} = \{N \mid \forall N \in \mathscr{N}_m^{(\mathrm{II})}, a \text{ is neither terminal nor segmentation edge}\}.$$

Of course, $\mathscr{N}_m^{(\mathrm{T})} = \mathscr{N}_m^{(\mathrm{II})} - \mathscr{N}_m^{(\mathrm{N})}$.

Lemma 10.9 Let $\mathcal{N}_m^{\langle N\rangle} = \{N - a | \forall N \in \mathcal{N}_m^{(N)}\}$. Then,

$$\begin{aligned}\mathcal{N}_m^{\langle N\rangle} = & \sum_{\substack{n_1+n_2=m-1\\ n_1,n_2\geq 0}} \mathcal{M}_{n_1} \times \mathcal{N}_{n_2} + \sum_{\substack{n_1+n_2=m-1\\ n_1,n_2\geq 0}} \mathcal{N}_{n_1} \times \mathcal{M}_{n_2} \\ & + \sum_{\substack{n_1+n_2=m-1\\ n_1,n_2\geq 0}} \mathcal{N}_{n_1} \times \mathcal{N}_{n_2}\end{aligned} \tag{10.38}$$

where × represents the Cartesian product of sets.

Proof Easy to see except for noticing that $N - a$ has a transitive block which is the trivial map when a is a terminal edge. □

Lemma 10.10 Let $\mathcal{N}_m^{\langle T\rangle} = \{N - a | \forall N \in \mathcal{N}_m^{(T)}\}$. Then,

$$\mathcal{N}_m^{\langle T\rangle} = \mathcal{N}_{m-1} \tag{10.39}$$

where $m \geq 2$.

Proof Because $a = e_r$ is neither terminal nor segmentation edge, from Theorem 3.4, $N \in \mathcal{N}_m^{\langle T\rangle}$. By the nonorientability and the size $m-1$, $N \in \mathcal{N}_{m-1}$, i.e.,

$$\mathcal{N}_m^{\langle T\rangle} \subseteq \mathcal{N}_{m-1}.$$

On the other hand, for any $N = (\mathcal{X}, \mathcal{P}) \in \mathcal{N}_{m-1}$, we have $N' = (\mathcal{X}', \mathcal{P}')$ such that $\mathcal{X}' = \mathcal{X} + Kr'$ and $\mathcal{P}'$ is different from $\mathcal{P}$ only at the vertex $(r')_{\mathcal{P}'} = (r', \gamma r', \langle r\rangle_{\mathcal{P}})$. Since N is nonorientable and $a' = Kr'$ is neither terminal nor segmentation edge, $N' \in \mathcal{N}_m^{(T)}$. Thus, $N = N' - a' \in \mathcal{N}_m^{\langle T\rangle}$. This implies

$$\mathcal{N}_{m-1} \subseteq \mathcal{N}_m^{\langle T\rangle}.$$

In consequence, the lemma is proved. □

It should be noted that when $m = 1$, there is only one nonorientable map $(Kr, (r, \beta r))$, and $(Kr, (r, \beta r)) \in \mathcal{N}^{(I)}$. Thus, (10.39) is meaningful only for $m \geq 2$.

On the basis of this lemma, it is necessary to see how many $N' \in \mathcal{N}_{m+1}^{(T)}$ can be produced from one $N \in \mathcal{N}_m$ such that $N = N' - a'$.

Because N is nonorientable, let $I = \{r, \psi_1 r, \psi_2 r, \ldots, \psi_{2m-1} r\}$ consist of half the elements in $\{r\}_\Psi = \mathcal{X}$, $\Psi = \Psi_{\{\gamma, \mathcal{P}\}}$, such that for any $x \in I$, $Kx \cap I = \{x, \gamma x\}$. Two cases are now considered.

Case 1 For any $N = (\mathcal{X}, \mathcal{P}) \in \mathcal{N}_m$, let

$$B(N) = \{B_0(N), B_1(N), B_2(N), \ldots, B_{2m}(N)\}$$

where $B_j(N) = (\mathscr{X}_j, \mathscr{P}_j) = N + e_{r_j}, j = 0, 1, 2, \ldots, 2m$, have

$$\begin{cases} \beta r_0 \text{ in the angle } \langle \alpha\mathscr{P}^{-1}r, r\rangle, \\ \beta r_j (j = 1, 2, \ldots, 2m-1) \text{ in the angle } \langle \alpha\mathscr{P}^{-1}\psi_j r, jr\rangle, \\ \beta r_{2m} \text{ in the angle } \langle r, \alpha\mathscr{P}^{-1}r\rangle. \end{cases}$$

Case 2 For any $N = (\mathscr{X}, \mathscr{P}) \in \mathscr{N}_m$, let

$$C(N) = \{C_0(N), C_1(N), C_2(N), \ldots, C_{2m}(N)\}$$

where $C_j(N) = (\mathscr{Y}_j, \mathscr{Q}_j) = N + e_{r_j}, j = 0, 1, 2, \ldots, 2m$, have

$$\begin{cases} \gamma r_0 \text{ in the angle } \langle \alpha\mathscr{P}^{-1}r, r\rangle, \\ \gamma r_j (j = 1, 2, \ldots, 2m-1) \text{ in the angle } \langle \alpha\mathscr{P}^{-1}\psi_j r, jr\rangle, \\ \gamma r_{2m} \text{ in the angle } \langle r, \alpha\mathscr{P}^{-1}r\rangle. \end{cases}$$

On the basis of Lemma 10.10, from the conjugate axiom,

$$\mathscr{N}_{m+1}^{(\mathrm{T})} = \sum_{N \in \mathscr{N}_m} \big(B(N) + C(N)\big) \tag{10.40}$$

for $m \geq 1$.

Because $\mathscr{N} = \mathscr{N}_1 + \mathscr{N}_2 + \cdots$, the enumerating function

$$\begin{aligned} f_{\mathscr{N}}(x) &= \sum_{m \geq 1} \Big(\sum_{N \in \mathscr{N}_m} 1 \Big) x^m = f_{\mathscr{N}^{(\mathrm{I})}}(x) + f_{\mathscr{N}^{(\mathrm{II})}}(x) \\ &= f_{\mathscr{N}^{(\mathrm{I})}}(x) + f_{\mathscr{N}^{(\mathrm{N})}}(x) + f_{\mathscr{N}^{(\mathrm{T})}}(x). \end{aligned} \tag{10.41}$$

Lemma 10.11 For $\mathscr{N}^{(\mathrm{I})} = \mathscr{N}_1^{(\mathrm{I})} + \mathscr{N}_2^{(\mathrm{I})} + \cdots$,

$$f_{\mathscr{N}^{(\mathrm{I})}}(x) = x f_{\mathscr{M}} + 2x^2 \frac{\mathrm{d} f_{\mathscr{M}}}{\mathrm{d}x} \tag{10.42}$$

where $f_{\mathscr{M}} = f_{\mathscr{M}}(x)$ is the enumerating function of orientable rooted maps determined by eq. (10.14).

Proof On the basis of (10.35–10.36), from Lemma 9.10, the lemma is obtained. □

Lemma 10.12 For $\mathscr{N}^{(\mathrm{N})} = \mathscr{N}_1^{(\mathrm{N})} + \mathscr{N}_2^{(\mathrm{N})} + \cdots$,

$$f_{\mathscr{N}^{(\mathrm{N})}}(x) = 2x f_{\mathscr{M}} f_{\mathscr{N}} + x f_{\mathscr{N}}^2 \tag{10.43}$$

where $f_{\mathscr{M}} = f_{\mathscr{M}}(x)$ as in (10.10) and $f_{\mathscr{N}} = f_{\mathscr{N}}(x)$ as in (10.41).

Proof A direct result of Lemma 10.9. □

Lemma 10.13 For $\mathscr{N}^{(\mathrm{T})} = \mathscr{N}_1^{(\mathrm{T})} + \mathscr{N}_2^{(\mathrm{T})} + \cdots$,

$$f_{\mathscr{N}^{(\mathrm{T})}}(x) = 2xf_{\mathscr{N}} + 4x^2\frac{\mathrm{d}f_{\mathscr{N}}}{\mathrm{d}x}. \tag{10.44}$$

Proof On the basis of Lemma 10.10 with its extension (10.40), from Lemma 10.11, (10.44) is soon obtained. □

Theorem 10.3 The following equation about f

$$\begin{cases} 4x^2\dfrac{\mathrm{d}f}{\mathrm{d}x} = a(x)f - xf^2 - 2xb(x); \\ \left.\dfrac{\mathrm{d}f}{\mathrm{d}x}\right|_{x=0} = 1 \end{cases} \tag{10.45}$$

where

$$\begin{cases} a(x) = 1 - 2x - 2xf_{\mathscr{M}}; \\ b(x) = f_{\mathscr{M}} + 2x\dfrac{\mathrm{d}f_{\mathscr{M}}}{\mathrm{d}x} \end{cases}$$

is well defined in the ring of power series with all coefficients nonnegative integers and negative powers finite. And, the solution is $f = f_{\mathscr{N}}(x)$.

Proof Let $f = N_1x + N_2x^2 + N_3x^3 + \cdots$, then from (10.45) all the coefficients can be determined by the recursion

$$\begin{cases} N_m = (4m-2)N_{m-1} + (2m-1)F_{m-1} \\ \qquad +2\sum_{i=1}^{m-1} N_iF_{m-1-i} + \sum_{i=1}^{m-2} N_iN_{m-1-i}, \\ \qquad m \geq 2; \\ N_1 = 1, \end{cases} \tag{10.46}$$

where F_m, $m \geq 0$, are known in (10.15). Because all N_m, $m \geq 1$, determined by (10.46) are positive integers, the former statement is true. The latter is directly deduced from (10.41–10.44). □

10.4 Gross equation

Let $\mathscr{R}_m$ be the set of general (orientable and nonorientable) rooted maps with size m, $m \geq 0$. Of course, $\mathscr{R}_0$ consists of only the trivial map.

For $m \geq 1$, $\mathscr{R}_m$ is partitioned into two subsets $\mathscr{R}_m^{(\mathrm{N})}$ and $\mathscr{R}_m^{(\mathrm{T})}$, i.e.,

$$\mathscr{R}_m = \mathscr{R}_m^{(\mathrm{N})} + \mathscr{R}_m^{(\mathrm{T})} \tag{10.47}$$

where

$$\mathscr{R}_m^{(\mathrm{N})} = \{R | \forall R \in \mathscr{R}_m, e_r(R) \text{ is a terminal link or segmentation edge}\}$$

and

$$\mathscr{R}_m^{(\mathrm{T})} = \{R | \forall R \in \mathscr{R}_m, e_r(R) \text{ is neither terminal nor segmentation edge}\}.$$

Of course, $\mathscr{R}_m^{(\mathrm{T})} = \mathscr{R}_m - \mathscr{R}_m^{(\mathrm{N})}$.

Lemma 10.14 Let $\mathscr{R}_m^{\langle \mathrm{N}\rangle} = \{R - a | \forall R \in \mathscr{R}_m^{(\mathrm{N})}\}$, then

$$\mathscr{R}_m^{\langle \mathrm{N}\rangle} = \sum_{\substack{n_1+n_2=m-1\\ n_1,n_2\geq 0}} \mathscr{R}_{n_1} \times \mathscr{R}_{n_1}, \tag{10.48}$$

$m \geq 1$.

Proof For any $R \in \mathscr{R}_M^{(\mathrm{N})}$, because $a = e_r(R)$ is a terminal link or a segmentation edge, $R - a$ has two transitive blocks (when a is a terminal link, the trivial map is seen as a transitive block in its own right), $R - a = R_1 + R_2$ and $R_1 \in \mathscr{R}_{n_1}, R_2 \in \mathscr{R}_{n_2}$. In other words, the set on the left side of (10.48) is a subset of the set of its right.

Conversely, for any $R_1 = (\mathscr{X}_1, \mathscr{P}_1) \in \mathscr{R}_{n_1}$ and $R_2 = (\mathscr{X}_2, \mathscr{P}_2) \in \mathscr{R}_{n_2}$, by appending $a = e_r$, $R = (\mathscr{X}, \mathscr{P})$ is obtained where $\mathscr{X} = \mathscr{X}_1 + \mathscr{X}_2 + Kr$ and $\mathscr{P}$ is different from $\mathscr{P}_1$ and $\mathscr{P}_2$ only at the vertices $(r)_{\mathscr{P}} = (r, \langle r_1\rangle_{\mathscr{P}_1})$ and $(\gamma r)_{\mathscr{P}} = (\gamma r, \langle r_2\rangle_{\mathscr{P}_2})$. It is easily checked that $R \in \mathscr{R}_m$, $m = n_1 + n_2 + 1$. In other words, the set on the right side of (10.48) is a subset of the set on the left. □

Since $\mathscr{R} = \mathscr{R}_0 + \mathscr{R}_1 + \mathscr{R}_2 + \cdots$, the enumerating function

$$\begin{aligned} f_{\mathscr{R}}(x) &= \sum_{m\geq 0} \Big(\sum_{R\in\mathscr{R}_m} 1\Big) x^m \\ &= f_{\mathscr{R}_0}(x) + f_{\mathscr{R}^{(\mathrm{N})}}(x) + f_{\mathscr{R}^{(\mathrm{T})}}(x), \end{aligned} \tag{10.49}$$

where $\mathscr{R}^{(\mathrm{N})} = \mathscr{R}_1^{(\mathrm{N})} + \mathscr{R}_2^{(\mathrm{N})} + \cdots$ and $\mathscr{R}^{(\mathrm{T})} = \mathscr{R}_1^{(\mathrm{T})} + \mathscr{R}_2^{(\mathrm{T})} + \cdots$.

First, because $\mathscr{R}_0$ consists of only the trivial map,

$$f_{\mathscr{R}_0}(x) = 1. \tag{10.50}$$

Then, from Lemma 10.14,

$$f_{\mathscr{R}^{(\mathrm{N})}}(x) = x f_{\mathscr{R}}^2, \tag{10.51}$$

where $f_{\mathscr{R}} = f_{\mathscr{R}}(x)$.

In order to evaluate $f_{\mathscr{R}^{(\mathrm{T})}}(x)$, $\mathscr{R}^{(\mathrm{T})}$ has to be decomposed.

Lemma 10.15 Let $\mathscr{R}_m^{\langle T\rangle} = \{R - a | \forall R \in \mathscr{R}_m^{(T)}\}$, then

$$\mathscr{R}_m^{\langle T\rangle} = \mathscr{R}_{m-1}, \tag{10.52}$$

where $m \geq 1$.

Proof For any $R' \in \mathscr{R}_m^{(T)}$, because $a' = e_{r'}$ is neither terminal link nor segmentation edge, from Theorem 3.4, $R = R' - a' \in \mathscr{R}_{m-1}$. This implies

$$\mathscr{R}_m^{\langle T\rangle} \subseteq \mathscr{R}_{m-1}.$$

Conversely, for any $R = (\mathscr{X}, \mathscr{P}) \in R_{m-1}$, by appending the edge $a' = Kr'$, $R' = (\mathscr{X}', \mathscr{P}')$ is obtained where $\mathscr{X}' = \mathscr{X} + Kr'$ and $\mathscr{P}'$ is different from $\mathscr{P}$ only at the vertex $(r')_{\mathscr{P}'} = (r', \gamma r', \langle r\rangle_{\mathscr{P}})$. From Theorem 3.7, $R' \in \mathscr{R}_m$. Because a' is neither terminal link nor segmentation edge and $R = R' - a'$, $R \in \mathscr{R}_m^{\langle T\rangle}$. This implies

$$\mathscr{R}_{m-1} \subseteq \mathscr{R}_m^{\langle T\rangle}.$$

The lemma is proved. □

Based on this, what should be further considered for is how many $R' \in \mathscr{R}_{m+1}^{(T)}$ can be produced from one $R \in \mathscr{R}_m$ such that $R = R' - a'$.

Because R is a map (orientable, or nonorientable), let

$$I = \{r, \psi_1 r, \psi_2 r, \ldots, \psi_{2m-1} r\}$$

be the set of elements in correspondence with a primal trail code, or dual trail code. For any $x \in I$, $Kx \cap I = \{x, \gamma x\}$ has two possibilities as cases.

Case 1 For any $R = (\mathscr{X}, \mathscr{P}) \in \mathscr{R}_m$, let

$$D(R) = \{D_0(R), D_1(R), D_2(R), \ldots, D_{2m}(R)\}$$

where $D_j(R) = (\mathscr{X}_j, \mathscr{P}_j) = R + e_{r_j}$, $j = 0, 1, 2, \ldots, 2m$, have

$$\begin{cases} \beta r_0 \text{ in the angle } \langle \alpha\mathscr{P}^{-1} r, r\rangle, \\ \beta r_j (j = 1, 2, \ldots, 2m-1) \text{ in the angle } \langle \alpha\mathscr{P}^{-1}\psi_j r, jr\rangle, \\ \beta r_{2m} \text{ in the angle } \langle r, \alpha\mathscr{P}^{-1} r\rangle. \end{cases}$$

Case 2 For any $R = (\mathscr{X}, \mathscr{P}) \in \mathscr{R}_m$, let

$$E(R) = \{E_0(R), E_1(R), E_2(R), \ldots, E_{2m}(R)\}$$

where $E_j(R) = (\mathscr{Y}_j, \mathscr{Q}_j) = R + e_{r_j}$, $j = 0, 1, 2, \ldots, 2m$, have

$$\begin{cases} \gamma r_0 \text{ in the angle } \langle \alpha\mathscr{P}^{-1}r, r\rangle, \\ \gamma r_j (j = 1, 2, \ldots, 2m-1) \text{ in the angle } \langle \alpha\mathscr{P}^{-1}\psi_j r, jr\rangle, \\ \gamma r_{2m} \text{ in the angle } \langle r, \alpha\mathscr{P}^{-1}r\rangle. \end{cases}$$

Based on Lemma 10.15, from the conjugate axiom,

$$\mathscr{R}^{(\mathrm{T})}_{m+1} = \sum_{R\in\mathscr{R}_m} \big(D(R) + E(R)\big) \tag{10.53}$$

for $m \geq 1$.

Because $\mathscr{R}^{(\mathrm{T})} = \mathscr{R}^{(\mathrm{T})}_1 + \mathscr{R}^{(\mathrm{T})}_2 + \cdots$, from Lemma 10.15 with its extension (10.53) and Lemma 9.10, the enumerating function

$$f_{\mathscr{R}^{(\mathrm{T})}}(x) = 2xf_{\mathscr{R}} + 4x^2\frac{\mathrm{d}f_{\mathscr{R}}}{\mathrm{d}x} \tag{10.54}$$

Theorem 10.4 The equation about f

$$\begin{cases} 4x^2\dfrac{\mathrm{d}f}{\mathrm{d}x} = -1 + (1-2x)f - xf^2; \\ f_0 = f(0) = 1 \end{cases} \tag{10.55}$$

is well defined in the ring of power series with coefficients all nonnegative integers and terms of negative power finite. And the solution is $f = f_{\mathscr{R}}(x)$.

Proof In virtue of the initial condition of eq. (10.55), assume $f = R_0 + R_1x + R_2x^2 + \cdots$. Of course, $R_0 = f_0 = 1$. Further, from eq. (10.55), the recursion

$$\begin{cases} R_m = (4m-2)R_{m-1} + \sum_{i=0}^{m-1} R_iR_{m-1-i}, \\ \qquad m \geq 1; \\ R_0 = 1 \end{cases} \tag{10.56}$$

is soon found to determine all the coefficients R_m, $m \geq 0$. It is easily checked that all of them are positive integers and hence the former statement is true.

The latter is a direct result of (10.50–10.51) and (10.53). □

10.5 The number of rooted maps

First, let $\sigma_m = (F_0, F_1, \ldots, F_m)$, $m \geq 0$, be the $m+1$ dimensional vector where F_m, $m \geq 0$, is the number of non-isomorphic orientable rooted maps with size m. And

$\sigma_M^{\mathrm{R}} = (F_m, F_{m-1}, \ldots, F_0)$ is called the *reversed vector* of the vector σ_m. It is easy to check that

$$\sigma_m^{\mathrm{TR}} = ((\sigma_m)^{\mathrm{T}})^{\mathrm{R}} = ((\sigma_m)^{\mathrm{R}})^{\mathrm{T}} = \sigma_m^{\mathrm{RT}} \tag{10.57}$$

where T is the transposition of a matrix.

The recursion (10.15) for determining F_m, $m \geq 0$, becomes

$$\begin{cases} F_m = (2m-1)F_{m-1} + \sigma_{m-1}\sigma_{m-1}^{\mathrm{TR}}, \\ \qquad m \geq 1; \\ F_0 = 1. \end{cases} \tag{10.58}$$

By (10.58), the number of non-isomorphic orientable rooted maps with size m, $m \geq 1$, can be calculated. In the first column of Table 10.1, F_m, $m \leq 10$, are listed.

Then, let $\delta_m = (N_1, N_2, \ldots, N_1)$, where N_m is the number of non-isomorphic nonorientable rooted maps with size m for $m \geq 1$.

The recursion (10.46) for determining N_m, $m \geq 1$, becomes

$$\begin{cases} N_m = (4m-2)N_{m-1} + (2m-1)F_{m-1} \\ \qquad +2\delta_{m-1}\sigma_{m-2}^{\mathrm{TR}} + \delta_{m-2}\delta_{m-2}^{\mathrm{TR}}, \\ \qquad m \geq 2; \\ N_1 = 1 \end{cases} \tag{10.59}$$

where σ_{m-2} is given in (10.58).

By (10.59), the number N_m can be calculated for $m \geq 1$. In the second column of Table 10.1, the values of N_m, $m \leq 10$, are listed.

Finally, let $\rho_m = (R_0, R_1, \ldots, R_m)$, where R_m is the number of non-isomorphic general maps with size m for $m \geq 0$.

The recursion (10.56) becomes

$$\begin{cases} R_m = (4m-2)R_{m-1} + \rho_{m-1}\rho_{m-1}^{\mathrm{TR}}, \\ \qquad m \geq 1; \\ R_0 = 1. \end{cases} \tag{10.60}$$

By (10.60), the number R_m can be calculated for $m \geq 0$. In the third column of Table 10.1, the values of R_m, $m \leq 10$, are listed.

From those numbers in Table 10.1, it is also checked that the enumerating functions $f_{\mathscr{M}}(x)$, $f_{\mathscr{N}}(x)$ and $f_{\mathscr{R}}(x)$ of, respectively, non-isomorphic orientable, nonorientable and total (including both orientable and nonorientable) rooted general maps with size as the parameter satisfy the relation as

$$f_{\mathscr{R}}(x) = f_{\mathscr{M}}(x) + f_{\mathscr{N}}(x). \tag{10.61}$$

Table 10.1: Numbers of rooted maps with size less than 11.

m	F_m	N_m	R_m
0	1	0	1
1	2	1	3
2	10	14	24
3	74	223	297
4	706	4190	4896
5	8162	92116	100278
6	110410	2339894	2450304
7	1708394	67825003	69533397
8	29752066	2217740030	2247492096
9	576037442	80952028936	81528066378
10	12277827850	3268104785654	3280382613504

10.6 Notes

10.6.1 Given a relative genus $g \neq 0$ (the case $g = 0$ is solved), determine the number of rooted near triangulations of size $m \geq |g|$ on a surface of genus g.

Notice 10.1 The case of $g = 0$ is done in Dong and Liu [21], Cai and Liu [10, 11], and Ren and Liu [136]. Some investigation on $g = -1$ and $g = 1$ can be seen in Liu [78, pp. 67–90].

A rooted map with all vertices of the same valency except for probably one vertex is said to be *near regular*. Among them, near 3-regular and near 4-regular are often encountered in literature.

Although near triangulations or near quadrangulations are, respectively, the dual maps of near 3-regular, or near 4-regular maps, they are still considered for most convenience from a different point of view.

10.6.2 Given a relative genus $g \neq 0$ (the case $g = 0$ is solved), determine the number of rooted near 3-regular maps of size $m \geq |g|$ on a surface of genus g.

Notice 10.2 Relevant topics might be seen in Liu [67, pp. 117–141], or [76, pp. 108–131], and [107].

10.6.3 Given a relative genus $g \neq 0$ (the case $g = 0$ is solved), determine the number of rooted near quadrangulations of size $m \geq |g|$ on a surface of genus g.

Notice 10.3 The case of $g = -2, -1, 0, 1$ can be seen in Liu [78, pp. 97–119] with some corrections in [97, pp. 171–198]. Particularly, detailed in Ren and Liu [132, 133, 135].

10.6.4 Given a relative genus $g \neq 0$ (the case $g = 0$ is solved), determine the number of rooted near 4-regular maps of size $m \geq |g|$ on a surface of genus g.

Notice 10.4 The case of $g = -2, -1, 0, 1$ can be seen in Liu [78, pp. 97–119] with some corrections in [97, pp. 171–198]. Particularly, detailed in Ren and Liu [132, 133, 135].

For the problems above, another parameter $l \geq 1$ is absolutely necessary in almost all cases. It is the valency of the extra vertex or face accordingly as the regularity is for vertices or faces.

10.6.5 Given a relative genus $g \neq 0$ (the case $g = 0$ is solved), determine the number of non-separable rooted maps of size $m \geq |g|$ on a surface of genus g.

Notice 10.5 The case of $g = 0$ is done from Liu [108, 109].

10.6.6 Given a relative genus $g \neq 0$ (the case $g = 0$ is solved), determine the number of Eulerian rooted maps of size $m \geq |g|$ on a surface of genus g.

Notice 10.6 The case of $g = 0$ can be seen as in Liu [104] and Cai et al. [9].

10.6.7 Given a relative genus $g \neq 0$ (the case $g = 0$ is solved), determine the number of non-separable Eulerian rooted maps of size $m \geq |g|$ on a surface of genus g.

Notice 10.7 For the case of $g = 0$, referred to as in Liu [99]. For $g = -1$, referred to as in Hao et al. [34].

10.6.8 Given a relative genus $g \neq 0$ (the case $g = 0$ ia solved), determine the number of rooted general maps of size $m \geq |g|$ on a surface of genus g.

Notice 10.8 For the case of $g = 0$, seen as in Liu [106].

10.6.9 Given a relative genus $g \neq 0$ (the case $g = 0$ ia solved), determine the number of loopless rooted maps of size $m \geq |g|$ on a surface of genus g.

Notice 10.9 For the case of $g = 0$, seen as in Liu [111, 112].

10.6.10 Given a relative genus $g \neq 0$ (the case $g = 0$ is known), find a relation between general maps and quadrangulations on a surface of genus g.

Notice 10.10 For the case of $g = 0$, it is done from Liu [64, pp. 121–141], or [65, pp. 99–115]. In general, the theoretical basis is on 10.6.8 and 10.6.3.

10.6.11 Given a relative genus $g \neq 0$ (the case $g = 0$ is known), find a relation between general maps and triangulations on a surface of genus g.

Notice 10.11 For the case of $g = 0$, it is done from Liu [64, pp. 121–141], or [65, pp. 99–115]. In general, the theoretical basis is on 10.6.8 and 10.6.1.

10.6.12 Given a relative genus $g \neq 0$ (the case $g = 0$, a one-to-one correspondence between loopless planar rooted maps of size $m - 1$ and 2-connected planar rooted triangulations of $2m - 1$ unrooted faces should be found, but now unknown yet), find a relation between loopless rooted maps and triangulations on a surface of genus g.

Notice 10.12 The theoretical basis for the case of $g = 0$ is on Liu [67, pp. 199–208, 86–94], or [68, 108–118, 236–246].

10.6.13 Present an expression of the solution for eq. (10.14) by special functions, particularly the hyperbolic geometric function.

Notice 10.13 References are suggested to read Liu [67, pp. 41–45], or [68, pp. 40–44], particularly, in Wawtzynczyk [163].

11 Maps within Symmetry

- A relation between the number of rooted maps and the order of the automorphism group of a map is established.
- A general procedure is shown for determining the group order distribution of maps with given size via an example as an application of the relation.
- A principle for counting unrooted maps from rooted ones is provided.
- Based on the principle, a general procedure is shown for determining the genus distribution of unrooted maps with given size via two examples.
- Conversely, rooted maps can be also determined via unrooted maps.

11.1 Symmetric relation

First, observe how to derive the number of non-isomorphic unrooted maps from that of non-isomorphic rooted maps when the automorphism group is known, or in other words, how to transform results without symmetry to those with symmetry.

Theorem 11.1 Let $n_0(\mathscr{U};I)$ be the number of non-isomorphic rooted maps with a given set of invariants including the size in the set of maps $\mathscr{U}$ considered. If the order of automorphism group of each map M in $\mathscr{U}$ is independent of the map M itself, but only dependent on $\mathscr{U}$ and I, denoted by $\mathrm{aut}(\mathscr{U};I)$, then the number of non-isomorphic unrooted maps with I in $\mathscr{U}$ is

$$n_1(\mathscr{U};I) = \frac{\mathrm{aut}(\mathscr{U};I)n_0(\mathscr{U};I)}{4\epsilon} \tag{11.1}$$

where $\epsilon \in I$ is the size.

Proof Let map $M = (\mathscr{X}, \mathscr{P}) \in \mathscr{U}$. From Theorem 8.1, for any $x \in \mathscr{X}$,

$$|X_x| = |\{y| \, \exists \tau \in \mathrm{Aut}(M), y = \tau x\}| = \mathrm{aut}(M). \tag{11.2}$$

In view of Corollary 8.2, M itself produces

$$n_0(M) = \frac{|\mathscr{X}|}{|X_x|} = \frac{4\epsilon}{\mathrm{aut}(M)} \tag{11.3}$$

non-isomorphic rooted maps. Therefore, there are

$$\begin{aligned} n_0(\mathscr{U};I) &= \sum_{M \in \mathscr{U}} \frac{4\epsilon}{\mathrm{aut}(M)} \\ &= \frac{4\epsilon}{\mathrm{aut}(\mathscr{U};I)} n_1(\mathscr{U};I) \end{aligned}$$

non-isomorphic rooted maps in $\mathscr{U}$. Via rearrangement, (11.1) is soon obtained. □

DOI 10.1515/9783110481846-011

In Chapter 8, efficient algorithms are established for finding the automorphism group of a map; this enables us to know the number of non-isomorphic rooted maps from an unrooted map by (11.1).

However, from Chapters 9 and 10, it is unnecessary to know the automorphism group for counting rooted maps. This enables us to enumerate unrooted maps via automorphism groups by employing (11.1).

Problem of type 1 For a set of maps $\mathscr{M}$ known, the number of non-isomorphic rooted maps with a given size, determine the number of non-isomorphic unrooted maps with the given size according to the orders of their automorphism groups, or in other words, the distribution of unrooted maps on the orders of their automorphism groups.

Although this problem does not yet have general progress in present, a great number of results for rooted cases have already provided reachable conditions for the problem.

11.2 An application

In what follows, provide a general procedure for solving the problem of type 1 via the determination of the distribution of rooted petal bundles on the orders of the automorphism groups of corresponding unrooted maps on the basis of Chapter 9.

From Table 9.1, the number of non-isomorphic planar rooted petal bundles with size 4 is $H_4^{(0)} = 14$, shown in Figure 11.1(a–n).

In virtue of Corollary 8.2, the orders of their automorphism groups are possibly 1, 2, 4, 8 and 16, only five cases.

Case 1 $\mathrm{aut}(M) = 1$, $M = (Kx + Ky + Kz + Kt, \mathscr{J})$. None.

Case 2 $\mathrm{aut}(M) = 2$. Eight planar rooted petal bundles:

$$\mathscr{J}_1 = (x, y, \gamma y, \gamma x, z, \gamma z, t, \gamma t);\ \mathscr{J}_2 = (x, \gamma x, y, z, \gamma z, t, \gamma t, \gamma y);$$
$$\mathscr{J}_3 = (x, y, z, \gamma z, t, \gamma t, \gamma y, \gamma x);\ \mathscr{J}_4 = (x, y, \gamma y, z, \gamma z, \gamma x, t, \gamma t);$$
$$\mathscr{J}_5 = (x, y, z, \gamma z, \gamma y, t, \gamma t, \gamma x);\ \mathscr{J}_6 = (x, \gamma x, y, z, \gamma z, \gamma y, t, \gamma t);$$
$$\mathscr{J}_7 = (x, y, \gamma y, z, t, \gamma z, \gamma t, \gamma x);\ \mathscr{J}_8 = (x, \gamma x, y, \gamma y, z, t, \gamma t, \gamma z);$$

shown in Figure 11.1(a–h).

Case 3 $\mathrm{aut}(M) = 4$. Four planar rooted petal bundles:

$$\mathscr{J}_9 = (x, y, \gamma y, \gamma x, z, t, \gamma t, \gamma z);\ \mathscr{J}_{10} = (x, \gamma x, y, z, t, \gamma t, \gamma z, \gamma y);$$
$$\mathscr{J}_{11} = (x, y, z, t, \gamma t, \gamma z, \gamma y, \gamma x);\ \mathscr{J}_{12} = (x, y, z, \gamma z, \gamma y, \gamma x, t, \gamma t);$$

shown in Figure 11.1(i–l).

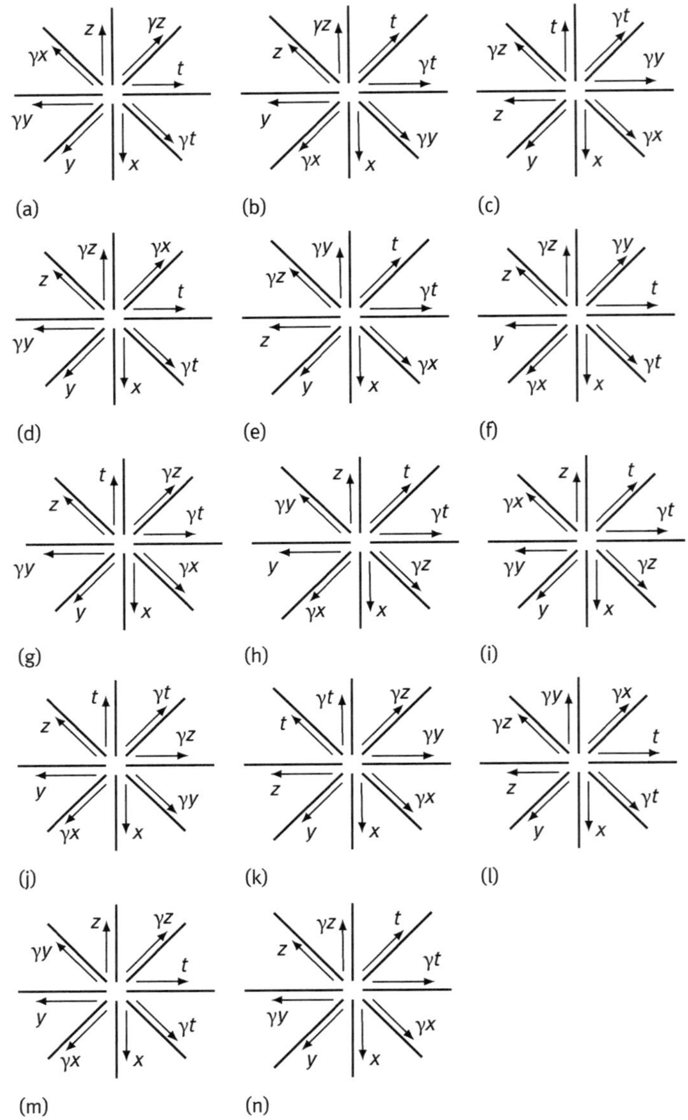

Figure 11.1: Planar petal bundles of size 4.

Case 4 $\mathrm{aut}(M) = 8$. Two planar rooted petal bundles:

$$\mathscr{I}_{13} = (x, \gamma x, y, \gamma y, z, \gamma z, t, \gamma t); \ \mathscr{I}_{14} = (x, y, \gamma y, z, \gamma z, t, \gamma t, \gamma x);$$

shown in Figure 11.1(m, n).

Case 5 aut(M) = 16. None.

This procedure can be done for determining the automorphism groups of unrooted maps via their primal trail codes, or dual trail codes, by computers and then via the collection of the same class of them according to the orders of the groups.

11.3 Symmetric principle

Whenever the distribution of rooted maps on the orders of automorphism groups is given for a set of maps, the number of non-isomorphic unrooted maps can soon be extracted

Theorem 11.2 Let $n_{0i}(\mathcal{M};I)$ be the number of non-isomorphic rooted maps with the set of invariants I and the order of their automorphism groups i in a set of maps $\mathcal{M}$ for $i|4\epsilon$, $1 \le i \le 4\epsilon$, where ϵ is the size, then the number of non-isomorphic unrooted maps in $\mathcal{M}$ is

$$n_1(\mathcal{M};I) = \sum_{\substack{i|4\epsilon \\ 1\le i\le 4\epsilon}} \frac{in_{0i}(\mathcal{M};I)}{4\epsilon}. \tag{11.4}$$

Proof Let $n_{1i}(\mathcal{M};I)$ be the number of non-isomorphic unrooted maps with the set of invariants I and the order of their automorphism groups i in the set of maps $\mathcal{M}$ for $i|4\epsilon$, $1 \le i \le 4\epsilon$, where ϵ is the size, then

$$n_1(\mathcal{M};I) = \sum_{\substack{i|4\epsilon \\ 1\le i\le 4\epsilon}} n_{1i}(\mathcal{M};I). \tag{11.5}$$

From Theorem 11.1, each unrooted map $M \in \mathcal{M}$, Aut(M) = i, produces

$$\frac{4\epsilon}{i}$$

non-isomorphic rooted maps. Therefore,

$$n_{1i}(\mathcal{M};I) = \frac{in_{0i}(\mathcal{M};I)}{4\epsilon}. \tag{11.6}$$

By substituting (11.6) into (11.5), (11.4) is soon obtained. □

On the choice of the set of invariants I, two types should be mentioned. One is that the set I consists of only the size and the genus for determining the genus distribution of non-isomorphic maps in a set of maps $\mathcal{M}$. The other is that the set I consists of only the size and the orders of automorphisms for determining the symmetric distribution of non-isomorphic maps in a set of maps $\mathcal{M}$.

Problem of type 2 For a set of maps $\mathscr{M}$ with the number of non-isomorphic maps given, determine the number of non-isomorphic under graphs of maps in $\mathscr{M}$.

Although the justification of whether or not two graphs are isomorphic is much far from easy, a feasible approach to it is presented from the above discussion. Because the under graphs are isomorphic if the two maps are isomorphic, the only thing we have to do is to classify non-isomorphic maps by their isomorphic under graphs.

On the other hand, for a graph, it is also possible to discuss how many non-isomorphic rooted maps are with the graph as their under graph, and then to discuss how many non-isomorphic unrooted maps are with the graph as their under graphs, and finally to classify maps according to the isomorphism of their under graphs.

11.4 General examples

On the basis of the 15 orientable rooted petal bundles of size 3 and the 9 nonorientable rooted petal bundles of size 2 (in Table 9.1), a general procedure is established for determining their genus distribution.

Orientable case Let $M = (Kx + Ky + Kz, \mathscr{J}_i)$, $1 \le i \le 15$.

Genus 0 Five orientable rooted petal bundles are shown in Figure 11.2(a–e). Here,

$$\mathscr{J}_1 = (x, \gamma x, y, \gamma y, z, \gamma z); \ \mathscr{J}_2 = (x, y, \gamma y, z, \gamma z, \gamma x)$$

with the order of its automorphism group aut$(M) = 6$ being one unrooted map;

$$\mathscr{J}_3 = (x, \gamma x, y, z, \gamma z, \gamma y); \ \mathscr{J}_4 = (x, y, \gamma y, \gamma x, z, \gamma z);$$

$$\mathscr{J}_5 = (x, y, z, \gamma z, \gamma y, \gamma x)$$

with the order of its automorphism group aut$(M) = 4$ being one unrooted map.

Genus 1 Ten orientable petal bundles are shown in Figure 11.2(f–o). Here,

$$\mathscr{J}_6 = (x, y, \gamma x, z, \gamma z, \gamma y); \ \mathscr{J}_7 = (x, y, z, \gamma z, \gamma x, \gamma y);$$

$$\mathscr{J}_8 = (x, y, \gamma y, z, \gamma x, \gamma z); \ \mathscr{J}_9 = (x, \gamma x, y, z, \gamma y, \gamma z);$$

$$\mathscr{J}_{10} = (x, y, z, \gamma y, \gamma z, \gamma x); \ \mathscr{J}_{11} = (x, y, \gamma x, \gamma y, z, \gamma z)$$

with the order of its automorphism group aut$(M) = 2$ being one unrooted map;

$$\mathscr{J}_{12} = (x, , y, z, \gamma y, \gamma x, \gamma z); \ \mathscr{J}_{13} = (x, y, \gamma x, z, \gamma y, \gamma z);$$

$$\mathscr{J}_{14} = (x, y, z, \gamma x, \gamma z, \gamma y)$$

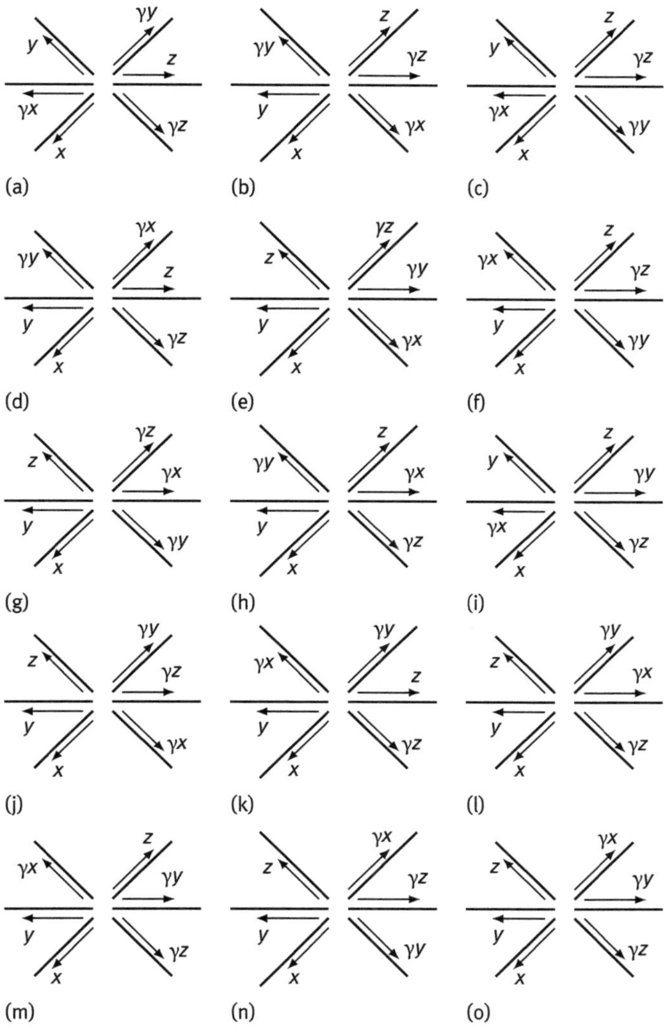

Figure 11.2: Orientable petal bundles of size 3.

with the order of its automorphism group $\mathrm{aut}(M) = 4$ being one unrooted map; and

$$\mathscr{I}_{15} = (x, y, z, \gamma x, \gamma y, \gamma z)$$

with the order of its automorphism group $\mathrm{aut}(M) = 12$ being one unrooted map itself.

Table 11.1 shows the genus distribution and group order distribution of orientable unrooted maps.

Table 11.1: Distributions of orientable petal bundles.

Genus	aut(M)						Dist.
	1	2	3	4	6	12	
0	0	0	0	1	1	0	2
1	0	1	0	1	0	1	3
Dist.	0	1	0	2	1	1	5

Nonorientable case Let $N = (Kx + Ky, \mathcal{J}_i)$, $1 \le i \le 9$.

Genus $\tilde{1}$ Five nonorientable rooted petal bundles are shown in Figure 9.4(e, a, c) and Figure 9.3(a,c). Here,

$$\mathcal{J}_1 = (x, \beta y, \beta x, y)$$

with the order of its automorphism group aut$(N) = 8$ being one unrooted map itself;

$$\mathcal{J}_2 = (x, \beta x, y, \gamma y); \ \mathcal{J}_3 = (x, \beta x, \gamma y, y);$$
$$\mathcal{J}_4 = (x, \gamma x, y, \beta y); \ \mathcal{J}_5 = (x, \gamma x, \beta y, y)$$

with the order of its automorphism group aut$(N) = 2$ being one unrooted map.

Genus $\tilde{2}$ Four nonorientable rooted petal bundles are shown in Figure 9.3(b) and Figure 9.4(b, d, f). Here,

$$\mathcal{J}_6 = (x, \beta y, \gamma x, y); \ \mathcal{J}_7 = (x, \gamma y, \beta x, y);$$
$$\mathcal{J}_8 = (x, \beta x, y, \beta y); \ \mathcal{J}_9 = (x, \beta x, \beta y, y)$$

with the order of its automorphism group aut$(N) = 4$ being two unrooted maps.

Table 11.2 shows the genus distribution and group order distribution of nonorientable unrooted maps.

Table 11.2: Distributions of nonorientable petal bundles.

Genus	aut(M)				Dist.
	1	2	4	8	
$\tilde{1}$	0	1	0	1	2
$\tilde{2}$	0	0	2	0	2
Dist.	0	1	2	1	4

11.5 Notes

11.5.1 Determine the number of non-isomorphic planar 4-regular unrooted maps of co-order $n + 1$ for $n \geq 1$.

Notice 11.1 On the basis of Liu [67, pp. 159–166] and the principles in Sections 11.1 and 11.3, find the distribution of rooted planar 4-regular maps by automorphism group order.

11.5.2 Determine the number of non-isomorphic planar loopless unrooted triangulations of size $3m$, $m \geq 2$.

Notice 11.2 On the basis of Liu [67, pp. 86–94] and the principles in Sections 11.1 and 11.3, find the distribution of planar rooted triangulations of size $3m$ by automorphism group order.

11.5.3 Determine the number of non-isomorphic planar Euler unrooted maps of size $m \geq 2$.

Notice 11.3 On the basis of Liu [67, pp. 142–148] and the principles in Sections 11.1 and 11.3, find the distribution of planar Eulerian rooted maps of size m by automorphism group order.

11.5.4 Determine the number of non-isomorphic planar nonseparable unrooted maps of size $m \geq 2$.

Notice 11.4 On the basis of Liu [67, pp. 184–191] and the principles in Sections 11.1 and 11.3, find the distribution of planar nonseparable rooted maps of size m by automorphism group order.

11.5.5 Prove, or disprove, the conjecture that almost all trees have the order of their automorphism group 1 when the size is greater enough.

Notice 11.5 References available can be seen in Liu [67, pp. 359–390], particularly in Yan and Liu [171].

11.5.6 Prove, or disprove, the conjecture that almost all maps with a given relative genus have the order of their automorphism group 1 when the size is large enough.

Notice 11.6 References available can be seen in Liu [67, pp. 359–390], particularly in [78, pp. 231–236].

11.5.7 Prove, or disprove, the conjecture that for a positive integer $t|\epsilon$, $t \geq 2$, almost no orientable map is with the order of automorphism group t when ϵ is large enough.

Notice 11.7 References needed can be seen in Liu [67, pp. 359–390], particularly in [78, pp. 231–236].

11.5.8 Prove, or disprove, the conjecture that for a positive integer $t|\epsilon$, $t \geq 2$, almost no nonorientable map is with the order of automorphism group t when ϵ is large enough.

Notice 11.8 References needed can be seen in Liu [67, pp. 359–390], particularly in [78, pp. 231–236].

11.5.9 Determine the genus distribution of 4-regular rooted maps of co-order $n + 1$, $n \geq 1$.

Notice 11.9 Two ways have to be considered. One is from the total number of 4-regular rooted maps of co-order $n + 1$ on all surfaces over relative genera. The other is for the number of the embeddings of co-order $n + 1$ on a surface of relative genus arbitrary given. For the second way, the starting point has been established in Liu [67, pp. 159–166].

11.5.10 Determine the genus distribution of loopless rooted triangulations of size $3m$, $m \geq 2$.

Notice 11.10 Two ways have to be considered. One is from the total number of loopless rooted triangulations with size $3m$ on all surfaces over relative genera. The other is for the number of the triangulations of size $3m$ on a surface of relative genus arbitrary given. For the second way, the starting point has been established in Liu [67, pp. 86–94].

11.5.11 Determine the genus distribution of rooted Eulerian map with size $m \geq 2$.

Notice 11.11 Two ways have to be considered. One is from the total number of rooted Eulerian maps with size m on all surfaces over relative genera. The other is for the number of the maps of size m on a surface of relative genus arbitrary given. For the second way, the starting point has been established in Liu [67, pp. 142–148]. For the first way, a general equation in orientable case is found so as getting the total number in Liu [78, pp. 139–142].

11.5.12 Determine the genus distribution of nonseparable rooted maps with size $m \geq 2$.

Notice 11.12 Two ways have to be considered. One is from the total number of nonseparable rooted maps with size m on all surfaces over relative genera. The other is for the number of the maps of size m on a surface of relative genus arbitrary given. For the first way, the starting point has been established in Liu [67, pp. 184–191].

Although corresponding problems about genus distribution can also be posed for unrooted case, they would be only suitable after the solution of rooted case in general.

Moreover, the genus distributions of maps with under graphs in some chosen classes can also be investigated.

12 Genus Polynomials

- The set of associate surfaces of a graph is constructed to determine all of its distinct embeddings, or its upper maps as well.
- A layer division of an associate surface of a graph is defined for establishing an operation to transform this surface into another associate surface. A procedure can be constructed for listing all other associate surfaces from an associate surface by this operation without repetition.
- A principle of determining the genus polynomial, called *handle polynomial*, of a graph is provided for the orientable case.
- The genus polynomial of a graph for nonorientable case, also called *crosscap polynomial*, is derived from the handle polynomial of the graph.

12.1 Associate surfaces

Given a graph $G = (V, E)$ and a spanning tree T, the edge set E is partitioned into E_T(tree edge) and $\bar{E}_T$(cotree edge), i.e., $E = E_T + \bar{E}_T$. Let $\bar{E}_T = \{i|i = 1, 2, \ldots, \beta\}$, $\beta = \beta(G)$ be the Betti number (or cyclic number) of G. If $i = (u[i], v[i])$, then i_u and i_v are, respectively, meant the semiedges of i incident with $u[i]$ and $v[i]$. Write $G' = (V+V_1, E_T+E_1)$, where $V_1 = \{v_i, \bar{v}_i|1 \leq i \leq \beta\}$ and $E_1 = \{(u[i], v_i), (v[i], \bar{v}_i)|1 \leq i \leq \beta\}$. Because G' is a tree itself, G' is called an *expanded tree* of T on G, and denoted by $\hat{T}_G$, or $\hat{T}$ in no confusion Liu [13, 14].

Let $\delta = (\delta_1, \delta_2, \ldots, \delta_\beta)$ be a binary vector, or as a binary number of β digits. Denoted by $\hat{T}^\delta$ that on $\hat{T}$, edges $(u[i], v_i)$ and $(v[i], \bar{v}_i)$ are labelled by i with indices: + (always omitted) or −, $1 \leq i \leq \beta$, where $\delta_i = 0$ means that the two indices are the same; otherwise, different. Then, δ is called an *assignment* of indices on $\hat{T}$.

For $v \in V$, let σ_v be a rotation at v and $\sigma_G = \{\sigma_v|\forall v \in V\}$, the rotation of G, then $\hat{T}_\sigma$ determines an embedding of $\hat{T}$ on the plane.

Theorem 12.1 For any σ as a rotation and δ as an assignment of indices, $\hat{T}_\sigma^\delta$ determines a joint tree shown in Section 1.3.

Proof By the definition of a joint tree, it is soon seen. □

According to the theory described in Chapter 1, the orientability and genus are naturally defined to be that of its corresponding embedding.

Lemma 12.1 Joint tree $\hat{T}_\sigma^\delta$ is orientable if, and only if, $\delta = 0$.

Proof Because $\delta = 0$ implies each label with its two occurrences of different indices, the lemma is true. □

DOI 10.1515/9783110481846-012

On a joint tree $\hat{T}_\sigma^\delta$, the surface determined by the boundary of the infinite face on the planar embedding of $\hat{T}_\sigma$ with δ on label indices is said to be an *associate surface*.

Lemma 12.2 The genus of a joint tree $\hat{T}_\sigma^\delta$ is that of its associate surface.

Proof Only from the definition of orientability of a joint tree. □

Two associate surfaces are the *same* if they have the same assignment with the same cyclic order. Otherwise, *distinct*. Let $\mathscr{F}(\beta)$ be the set of distinct surfaces on $I_\beta = \{1, 2, \ldots, \beta\}$.

For a surface $F \in \mathscr{F}(\beta)$ and a tree T on a graph G, if there exists a joint tree $\hat{T}_\sigma^\delta$ such that F is its associate surface, then F is said to be *admissible*. Let $\mathscr{F}_T(\beta)$ be the set of all distinct associate surfaces.

Given two integers p, $p \geq 0$, and q, $q \geq 1$, let $\mathscr{F}_T(\beta; p)$(or $\mathscr{F}_T(\beta; q)$, $q \geq 1$), $p \geq 0$, be all distinct admissible surfaces of orientable genus p (or nonorientable genus q).

Theorem 12.2 For any integer $p \geq 0$ (or $q \geq 1$), the cardinality $|\mathscr{F}_T(\beta; p)|$ (or $|\mathscr{F}_T(\beta; q)|$) is independent of the choice of tree T on G. Further, it is the number of distinct embeddings of G on a surface of orientable genus p (or nonorientable genus q).

Proof According to O1.14, a one-to-one correspondence between two sets of embeddings generated by two distinct spanning trees can be found such that same embeddings are in correspondence. This implies the theorem. □

Because of

$$|\mathscr{F}_T(\beta)| = \sum_{p \geq 0} |\mathscr{F}_T(\beta; p)| + \sum_{q \geq 1} |\mathscr{F}_T(\beta; q)|,$$

the following conclusion is found from the theorem.

Corollary 12.1 The cardinality $|\mathscr{F}_T(\beta)|$ is independent of the choice of tree T on G. Further, it is the number of distinct embeddings of G.

From Lemma 12.1, the nonorientability of an associate surface can be easily justified by only checking if it has a label i with the same index, i.e., $\delta(i) = 1$.

Theorem 12.3 There is a one-to-one correspondence between associate surfaces and embeddings of a graph.

Proof First, we can easily see that each embedding determines an associate surface. Then, we show that each associate surface is determined by an embedding. Because of Theorem 12.2, this statement is derived. □

From what is mentioned above, it is seen that the problem of determining the genus distribution of all embeddings for a graph is transformed into that of finding the number of all distinct admissible associate surfaces in each elementary equivalent class and the problem on minimum and maximum genus of a graph is that among all admissible associate surfaces of the graph. All of them are done on a polygon.

12.2 Layer division of a surface

Given a surface $S = (A)$. It is divided into segments layer by layer as in the following.

The zeroth layer contains only one segment, i.e., $A(= A_0)$.

The first layer is obtained by dividing the segment A_0 into l_1 segments, i.e., $S = (A_1, A_2, \ldots, A_{l_1})$, where $A_1, A_2, \ldots, A_{l_1}$ are called the *first layer segments*.

Suppose that on $k-1$st layer, the $k-1$st layer segments are $A_{\underline{n}_{(k-1)}}$, where $\underline{n}_{(k-1)}$ is an integral $k-1$-vector satisfied by

$$\underline{1}_{(k-1)} \le (n_1, n_2, \ldots, n_{k-1}) \le \underline{N}_{(k-1)}$$

with $\underline{1}_{(k-1)} = (1, 1, \ldots, 1)$,

$$\underline{N}_{(k-1)} = (N_1, N_2, \ldots, N_{k-1}),$$

$N_1 = l_1 = N_{(1)}$, $N_2 = l_{A_{\underline{N}_{(1)}}}$, $N_3 = l_{A_{\underline{N}_{(2)}}}$, …, $N_{k-1} = l_{A_{\underline{N}_{(k-2)}}}$, then the kth layer segments are obtained by dividing each $k-1$st layer segment as

$$A_{\underline{n}_{(k-1)},1}, A_{\underline{n}_{(k-1)},2}, \ldots, A_{\underline{n}_{(k-1)},l_{A_{\underline{n}_{(k-1)}}}}, \tag{12.1}$$

where

$$\underline{1}_{(k)} = (\underline{1}_{(k-1)}, 1) \le (\underline{n}_{(k-1)}, i) \le \underline{N}_{(k)} = (\underline{N}_{(k-1)}, N_k)$$

and $N_k = l_{A_{\underline{N}_{(k-1)}}}$, $1 \le i \le N_k$. Segments in (1.1) are called *sons* of $A_{\underline{n}_{(k-1)}}$. Conversely, $A_{\underline{n}_{(k-1)}}$ is the *father* of any one in (12.1).

A layer segment that has only one element is called an *end segment* and others are called *principle segments*.

For an example, let

$$S = (1, -7, 2, -5, 3, -1, 4, -6, 5, -2, 6, 7, -3, -4).$$

Figure 12.1 shows a layer division of S and Table 12.1, the principle segments in each layer.

For a layer division of a surface, if principle segments are dealt with vertices and edges are with the relationship between father and son, then what is obtained is a tree

Table 12.1: Layers and principle segments.

Layers	Principle segments
Zeroth layer	$A = \langle 1, -7, 2-5; 3, -1, 4, -6, 5; -2, 6, 7, -3-4\rangle$
First layer	$B = \langle 1; -7, 2; -5\rangle, C = \langle 3, -1; 4, -6; 5\rangle,$ $D = \langle -2, 6; 7; -3, -4\rangle$
Second layer	$E = \langle -7; 2\rangle, F = \langle 3; -1\rangle, G = \langle 4; -6\rangle,$ $H = \langle -2; 6\rangle, I = \langle -3; -4\rangle$

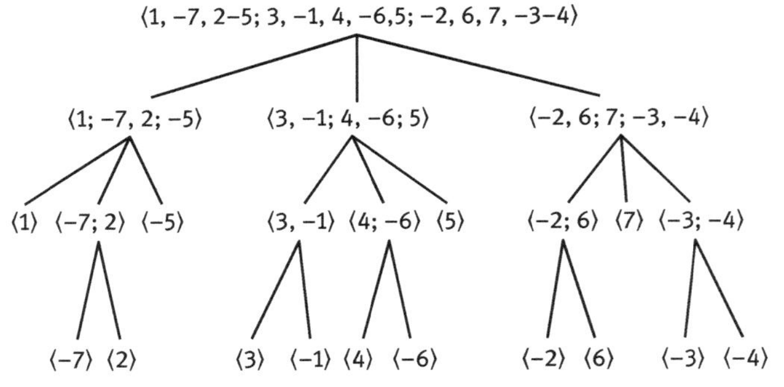

Figure 12.1: A layer division of S.

denoted by T. On T, by adding cotree edges as end segments, a graph $G = (V, E)$ is induced. For example, the graph induced from the layer division shown in Figure 12.1 is as

$$V = \{A, B, C, D, E, F, G, H, I\} \tag{12.2}$$

and

$$E = \{a, b, c, d, e, f, g, h, 1, 2, 3, 4, 5, 6, 7\}, \tag{12.3}$$

where

$$a = (A, B), b = (A, C), c = (A, D), d = (B, E),$$
$$e = (C, F), f = (C, G), g = (D, H), h = (D, I),$$

and

$$1 = (B, F), 2 = (E, H), 3 = (F, I), 4 = (G, I),$$
$$5 = (B, C), 6 = (G, H), 7 = (D, E).$$

By considering $E_T = \{a, b, c, d, e, f, g, h\}$, $\bar{E}_T = \{1, 2, 3, 4, 5, 6, 7\}$, $\delta_i = 0, i = 1, 2, \ldots, 7$, and the rotation σ implied in the layer division, a joint tree $\hat{T}_\sigma^\delta$ is produced.

Theorem 12.4 A layer division of a surface determines a joint tree. Conversely, a joint tree determines a layer division of its associate surface.

Proof From the procedure of constructing a layer division, a joint tree is determined. Conversely, it is natural. □

Then, an operation on a layer division is discussed for transforming an associate surface into another in order to visit all associate surfaces without repetition.

A layer segment with all its successors is called a *branch* in the layer division. The operation of interchanging the positions of two layer segments with the same father in a layer division is called an *exchanger*.

Lemma 12.3 A layer division of an associate surface of a graph under an exchanger is still a layer division of another associate surface. Conversely, the later under the same exchanger becomes the former.

Proof From the correspondence between layer divisions and associate surfaces, the lemma can be obtained. □

On the basis of this lemma, exchanger can be seen as an operation on the set of all associate surfaces of a graph.

Lemma 12.4 The exchanger is closed in the set of all associate surfaces of a graph.

Proof From the correspondence between joint trees and layer divisions, the conclusion of the lemma is seen. □

Lemma 12.5 Let $\mathscr{A}(G)$ be the set of all associate surfaces of a graph G, then for any S_1, $S_2 \in \mathscr{A}(G)$, there exist a sequence of exchangers on the set such that S_1 can be transformed into S_2.

Proof By considering the joint trees and layer divisions, the lemma is right. □

If $\mathscr{A}(G)$ is dealt as the vertex set and an edge as an exchanger, then what is obtained is called the *associate surface graph* of G, and denoted by $\mathscr{H}(G)$. From Theorem 12.3, it is also called the *surface embedding graph* of G.

Theorem 12.5 In $\mathscr{H}(G)$, there is a Hamilton path. Further, for any two vertices, $\mathscr{H}(G)$ has a Hamilton path with the two vertices as ends.

Proof By arranging an order, an Hamiltonian path can be extracted based on the procedure of the layer division. □

First, starting from a surface as marked in $\mathscr{A}(G)$, by doing exchangers at each principle in one layer to another for getting an unmarked surface to mark, by the FRP (Finite Recursion Principle), a Hamilton path can always be found in considering Theorem 12.3. This implies the first statement.

Further, for chosen $S_1, S_2 \in \mathscr{A}(G) = V(\mathscr{H}(G))$ adjective, starting from S_1, by doing exchangers avoid S_2 except the final step, on the basis of the strongly finite recursion principle, a Hamilton path between S_1 and S_2 can be obtained. This implies that $\mathscr{H}(G)$ has a Hamilton circuit and hence the last statement.

This theorem tells us that the problem of determining the minimum, or maximum genus of graph G has an algorithm in time linear on $\mathscr{H}(G)$.

12.3 Handle polynomials

Let $\mathscr{S}(G)$ be the set of associate surfaces of a graph G and $\mathscr{S}_g(G)$, the subset of $\mathscr{S}(G)$ with genus g. The the enumerating function

$$\gamma(G; z) = \sum_{g=g_{\min}}^{g_{\max}} |\mathscr{S}_g(G)| z^g \tag{12.4}$$

is called the *genus polynomial* of G, where $g_{\min}$ and $g_{\max}$ are, respectively, the minimum and maximum genus of G for orientable, or nonorientable case. In orientable case, $\mu(G; x) = \gamma(G; x)$ is called the *handle polynomial*. In nonorientable case, $\nu(G; y) = \gamma(G; y)$ is the *crosscap polynomial*.

On the basis of the theory described in Sections 12.1 and 12.2, (12.4) is, in fact, to determine the genus distribution of embeddings of G. Because the enumerating function of upper rooted maps of G is a constant times the genus polynomial $\gamma(G; z)$, for the enumeration of maps by genus it is enough only to discuss $\gamma(G; z)$.

Lemma 12.6 An orientable associate surface of a graph without two letters interlaced has a letter x such that xx^{-1} is a segment of the surface.

Proof Let $\langle x, x^{-1}\rangle$ be a segment of the surface with minimum of letters. If it does not contain only the letter x, then there is another letter y in it. Because x and y are noninterlaced, the segment $\langle y, y^{-1}\rangle$, or $\langle y^{-1}, y\rangle$, is a subsegment of $\langle x, x^{-1}\rangle$. However, it has at least one letter less than the minimum. □

Lemma 12.7 An orientable associate surface of a graph is with genus 0 if, and only if, no two letters are interlaced.

Proof On the basis of Lemma 12.6, by the FSP, the lemma can be easily found. □

Theorem 12.6 If an orientable associate surface of a graph has two letters interlaced, i.e., in form as $AxByCx^{-1}Dy^{-1}E$, then its genus is k, $k \geq 1$, if, and only if, the orientable genus of $ADCBE$ is $k-1$.

Proof On the basis of Relation 1 in Section 1.2, the theorem is easily proved. □

According to this theorem, a linear time algorithm can be designed for classifying the orientable associate surfaces of a graph G by their genus on $\mathscr{H}_O(G)$, the set of all orientable associate surfaces of G. Let $N_i(G)$ be the number of orientable associate surfaces of G with genus i, $i \geq 0$.

Theorem 12.7 The handle polynomial of G is

$$\mu(G;x) = \sum_{0 \leq i \leq \lfloor \frac{\beta}{2} \rfloor} N_i(G)x^i, \tag{12.5}$$

where β is the Betti number of G.

Proof From (12.4), the theorem follows. □

12.4 Crosscap polynomials

Let $\mathscr{F}^i_{2\beta} = \{S^i_{2\beta,j} | 1 \leq j \leq s_i\}$, where s_i is the number of orientable 2β-gons (surfaces) with genus i, then

$$\mathscr{F}_{2\beta} = \sum_{0 \leq i \leq s_i} \mathscr{F}^i_{2\beta}. \tag{12.6}$$

Given a surface $S \in \mathscr{F}_{2\beta}$, S induces $2^{2\beta} - 1$ nonorientable surfaces. Let $\mathscr{N}_S$ be the set of all nonorientable surfaces induced by S, then the polynomial

$$\delta_S(y) = \sum_{1 \leq j \leq \beta} |\mathscr{N}_j(S)| y^j \tag{12.7}$$

is called the *nonorientable form* of S where $\mathscr{N}_j(S)$ is the subset of $\mathscr{N}_S$, $1 \leq j \leq \beta$.

For a graph G with Betti number β, the set of all associate orientable surfaces determined by joint trees of G is denoted by $\mathscr{S}(G)$. Let $\mathscr{S}_\delta(G)$ for $\delta \in \Delta_{\mathscr{S}}$ be the subset of $\mathscr{S}(G)$ with nonorientable form δ, where $\Delta_{\mathscr{S}}$ is the set of all nonorientable forms of surfaces in $\mathscr{S}(G)$.

Theorem 12.8 The crosscap polynomial of a graph G is

$$\nu(G;y) = \sum_{\delta \in \Delta_{\mathscr{S}}} |\mathscr{S}_\delta(G)| \delta(y). \tag{12.8}$$

Proof From (12.4) and (12.7), the theorem is deduced. □

Theorem 12.9 If a nonorientable associate surface of a graph is in form as $AxBxC$, then its genus is k, $k \geq 1$, if, and only if, the genus of $AB^{-1}C$ is

$$\begin{cases} k-1, \text{ if } AB^{-1}C \text{ is nonorientable;} \\ \frac{k-1}{2}, \text{ otherwise.} \end{cases} \tag{12.9}$$

Proof From Relation 2 in 1.2, the theorem is easily proved. □

According to Theorem 12.9, a linear time algorithm can also be designed for determining the genus of a surface and then classify nonorientable associate surfaces of a graph by genus on $\mathscr{H}_N(G)$, the set of all nonorientable surfaces of G. Hence, the crosscap polynomial expressed by (12.8) can soon be found.

12.5 Notes

12.5.1 Find the handle polynomial of the bouquet B_n of size m, $m \geq 1$, by joint tree model.

Notice 12.1 In A2.2 of Appendix 2, one can see the handle polynomials of bouquets with size m, $1 \leq m \leq 5$, obtained by employing joint trees.

12.5.2 Find the crosscap polynomial of the bouquet of size m, $m \geq 1$, by joint tree model.

Notice 12.2 In A2.2 of Appendix 2, one can see the crosscap polynomials of bouquets with size m, $1 \leq m \leq 5$, obtained by employing joint trees. Relevant references can be seen in Yang and Liu [173] and Xu and Liu [169].

12.5.3 Find the handle polynomial of the wheel W_n of order n, $n \geq 4$ by joint tree model.

Notice 12.3 In A2.3 of Appendix 2, one can see the handle polynomials of wheels with order n, $4 \leq n \leq 8$, obtained by employing joint trees. Relevant references can be seen in Yang and Liu [175].

12.5.4 Find the crosscap polynomial of the wheel W_n of order n, $n \geq 4$, by joint tree model.

Notice 12.4 In A2.3 of Appendix 2, one can see the crosscap polynomials of wheels with order n, $4 \leq n \leq 7$, obtained by employing joint trees. Relevant references can be seen in Yang and Liu [175].

12.5.5 Find the handle polynomial of the complete graph K_n of order n, $n \geq 4$, by joint tree model.

Notice 12.5 In Liu [90, pp. 407–408], one can see the handle polynomials of complete graphs with order n, $4 \leq n \leq 6$, obtained by employing joint trees.

12.5.6 Find the crosscap polynomial of the complete graph K_n of order n, $n \geq 4$, by joint tree model.

Notice 12.6 In Liu [90, pp. 407–408], one can see the crosscap polynomials of complete graphs with order n, $4 \leq n \leq 6$, obtained by employing joint trees.

12.5.7 Find the handle polynomial of the complete bipartite graph $K_{m,n}$ of order $m+n$, $m, n \geq 3$, by joint tree model.

Notice 12.7 In A2.5 of Appendix 2, one can see the handle polynomials of complete bipartite graphs with order $m + n$, $6 \leq m + n \leq 8$, obtained by employing joint trees.

12.5.8 Find the crosscap polynomial of the complete bipartite graph $K_{m,n}$ of order $m + n$, $m, n \geq 3$, by joint tree model.

Notice 12.8 In A2.5 of Appendix 2, one can see the crosscap polynomials of complete bipartite graphs with order $m + n$, $6 \leq m + n \leq 7$, obtained by employing joint trees.

12.5.9 Find the handle polynomial of the n-cube Q_n of order n, $n \geq 3$, by joint tree model.

Notice 12.9 Referred to as in Wan and Liu [159] and Hao et al. [35].

12.5.10 Find the crosscap polynomial of the n-cube of order n, $n \geq 3$, by joint tree model.

Notice 12.10 Referred to as in Wan and Liu [159] and Hao et al. [35].

12.5.11 For the n-cube Q_n, $n \geq 3$, prove that the minimum genus $\gamma_n = g_{\min}(Q_n)$ with γ_{n-1} satisfies the relation

$$g_{\min}(Q_n) = 2^{n-4}(n-3) + g_{\min}(Q_{n-1})$$

from an associate surface of Q_{n-1} with genus γ_{n-1} to get an associate surface of Q_n with genus γ_n.

Notice 12.11 No progress turns out up to now. However, some references might be relevant as Shao and Liu [142] and Shao et al. [146].

12.5.12 For the complete bipartite graph $K_{m,n}$, $m \ge n \ge 4$, prove that the minimum genus $\gamma_{m,n} = g_{\min}(K_{m,n})$ with $\gamma_{m,n-1}$ satisfies the relation

$$g_{\min}(K_{m,n}) = \left\langle \frac{m-2}{4} \right\rangle + g_{\min}(K_{m,n-1}) - 1,$$

where

$$\left\langle \frac{m-2}{4} \right\rangle = \begin{cases} \left\lceil \frac{m-2}{4} \right\rceil, & m = 0(2 \nmid n), 1(2 \nmid n; 2|n, 2|\lfloor n/2 \rfloor), \\ & 3(2 \nmid n, 2 \nmid \lfloor n/2 \rfloor; 2|n, 2|\lfloor n/2 \rfloor); \\ \frac{m-2}{4}, & m = 2 (\text{mod } 4); \\ \left\lfloor \frac{m-2}{4} \right\rfloor, & m = 0(2|n), 1(2|n, 2 \nmid \lfloor n/2 \rfloor), \\ & 3(2 \nmid n, 2|\lfloor n/2 \rfloor; 2|n, 2 \nmid \lfloor n/2 \rfloor) \end{cases}$$

from an associate surface of $K_{m,n-1}$ with genus $\gamma_{m,n-1}$ to get an associate surface of $K_{m,n}$ with genus $\gamma_{m,n}$.

Notice 12.12 No progress turns out up to now. However, some references might be relevant as Shao and Liu [142] and Shao et al. [146].

12.5.13 For the complete graph K_n, $n \ge 5$, prove that the minimum genus $\gamma_n = g_{\min}(K_n)$ with γ_{n-1} satisfies the relation

$$g_{\min}(K_n) = \left\langle \frac{n-4}{6} \right\rangle + g_{\min}(K_{n-1}),$$

where

$$\left\langle \frac{n-4}{6} \right\rangle = \begin{cases} \left\lceil \frac{n-4}{6} \right\rceil, & n = 2, 1(2 \nmid \lfloor n/6 \rfloor), 3(2|\lfloor n/6 \rfloor), 5(2 \nmid \lfloor n/6 \rfloor); \\ \frac{n-4}{6}, & n = 4 (\text{mod } 6); \\ \left\lfloor \frac{n-4}{6} \right\rfloor, & n = 0, 1(2|\lfloor n/6 \rfloor), 3(2 \nmid \lfloor n/6 \rfloor), 5(2|\lfloor n/6 \rfloor) \end{cases}$$

from an associate surface of K_{n-1} with genus γ_{n-1} to get an associate surface of K_n with genus γ_n.

Notice 12.13 No progress turns out up to now. However, some references might be relevant as Shao and Liu [142] and Shao et al. [146].

13 Census with Partitions

- The planted trees are enumerated with vertex partition vectors in an elementary way instead of the methods used before.
- A summation free form of the number of outerplanar rooted maps is derived from the result on planted trees.
- On the basis of the result for planted outerplanar maps, the numbers of Hamilton cubic rooted maps is determined.
- The number of Halin rooted maps with vertex partition is obtained as a form without summation.
- Biboundary inner rooted maps on the sphere are counted by an explicit formula with vertex partitions.
- On the basis of joint tree model, the number of general rooted maps with vertex partition can also be expressed via planted trees in an indirected way.
- The pan-flowers, which have pan-Halin maps as a special case, are classified according to vertex partition and genus given.

13.1 Planted trees

A *plane tree* is such an upper planar rooted map of a tree. A *planted tree* is a plane tree of root-vertex valency 1. In Figure 13.1, (a) shows a plane tree and (b), a planted tree.

Let T be a planted tree of order n with vertices $v_0, v_1, v_2, \ldots, v_n$, $n \geq 1$, where v_0 is the rooted vertex. The segment recorded as travelling along the face boundary of T from v_0 back to itself and then v_0 left off is called a *V-code* of T when v_i is replaced by i for $i = 1, 2, \ldots, n$ as shown in Figures 13.2 and 13.3.

A sequence of numbers is said to be *polyhedral* if each adjacent pair of numbers occurs twice. It is easily seen that a V-code of a planted tree is a polyhedral segment.

The vector $\underline{n} = (n_1, n_2, \ldots, n_i, \cdots)$, where n_i $(i \geq 1)$ is the number of unrooted vertices of valency i, is called the *vertex partition* of a planted tree.

For a sequence of nonnegative integers n_1, n_2, …, n_i, $\cdots$ denoted by a vector $\underline{n} = (n_1, n_2, \ldots, n_i, \cdots)$, if

$$\sum_{i\geq 1}(2-i)n_i = 1, \tag{13.1}$$

then $\underline{n}$ is said to be *feasible*. Let

$$\underline{n}' = (n'_1, n'_2, \ldots, n'_i, \cdots),$$

DOI 10.1515/9783110481846-013

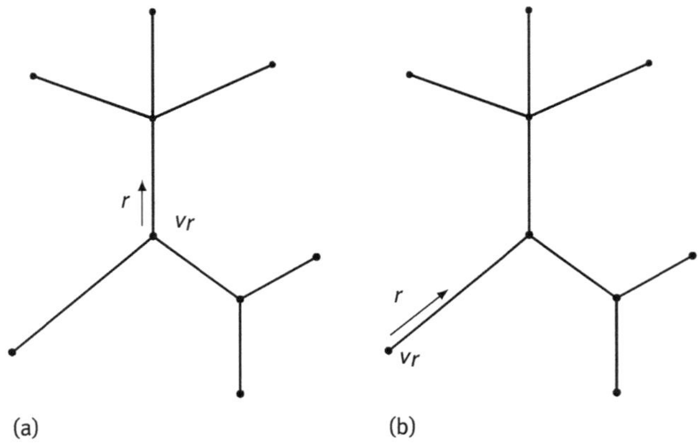

Figure 13.1: (a) Plane tree and (b) planted tree.

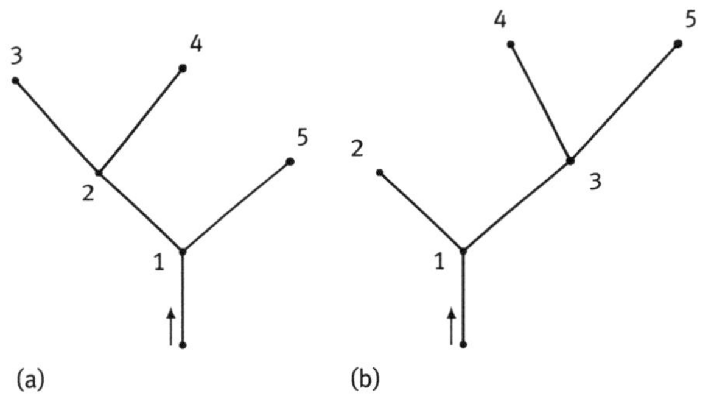

Figure 13.2: Trees with $\underline{n} = (3, 0, 2)$. (a) Sequence 123242151 and (b) Sequence 121343531.

where

$$n'_i = \begin{cases} n_1 - 1, \text{ when } i = 1; \\ n_{k-1} + 1, \text{ when } i = k - 1; \\ n_k - 1, \text{ when } i = k; \\ n_i, \text{ otherwise} \end{cases} \tag{13.2}$$

for a $k \geq 2$ and $n_1, n_k > 0$, then $\underline{n}'$ is called a *reduction* of $\underline{n}$.

Lemma 13.1 A reduction $\underline{n}'$ of a sequence of nonnegative integers $\underline{n}$ is feasible if, and only if, $\underline{n}$ is feasible.

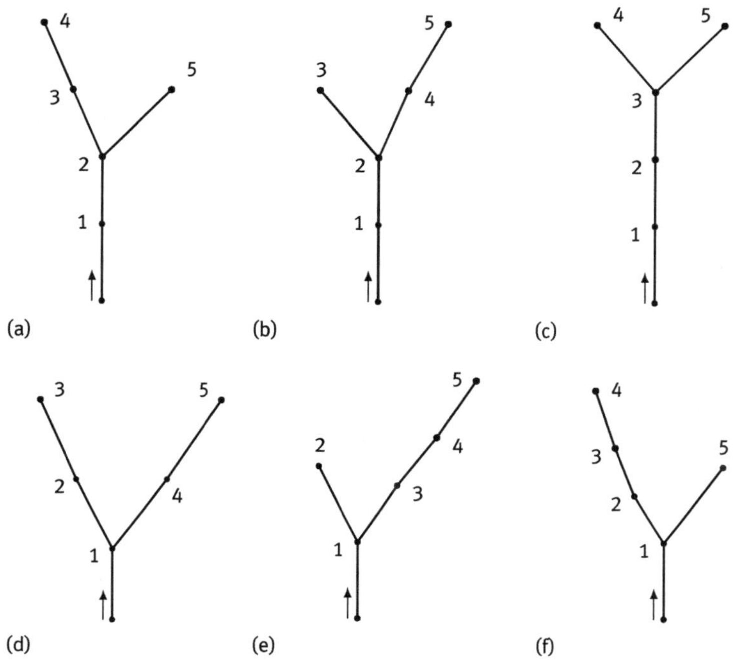

Figure 13.3: Trees with $\underline{n} = (2,2,1)$. (a) 123432521, (b) 1232345421, (c) 123435321, (d) 123214541, (e) 121345431 and (f) 123432151.

Proof By considering (13.2), we have the eqaulity as

$$\sum_{i\geq 1}(2-i)n_i' - \sum_{i\geq 1}(2-i)n_i = -1-(2-k)+(2-k+1) = 0.$$

This leads to the lemma. □

The sequence $\underline{n}_0 = (1,1)$ is feasible but no reduction can be done. So, it is called *irreducible*.

Lemma 13.2 Any feasible sequence $\underline{n}$ has $n_1 > 0$.

Proof By contradiction. Suppose $\underline{n}$ is feasible but $n_1 = 0$. Because

$$\sum_{i\geq 1}(2-i)n_i = \sum_{i\geq 2}(2-i)n_i \leq 0,$$

this contradicts (13.1), the feasibility. □

Lemma 13.3 Any feasible sequence $\underline{n} \neq \underline{n}_0$ can always be transformed into $\underline{n}_0$ only by reductions.

Proof Because of $\underline{n} \neq \underline{n}_0$, Lemma 13.2 enables us to get a reduction. Whenever the reduction is not $\underline{n}_0$, another reduction can also be done from Lemma 13.1. By the FRP (i.e., Finite Recursion Principle), the lemma is done. □

Theorem 13.1 For a nonnegative integer sequence $\underline{n} = (n_1, n_2, \ldots, n_i, \cdots)$, there exists such a planted tree that n_i is the number of unrooted vertices of valency i ($i \geq 1$) if, and only if, $\underline{n}$ is feasible.

Proof Necessity. Suppose T is such a planted tree. Because n_i is the number of unrooted vertices with valency i, $i \geq 1$ in T, the size of T is

$$\sum_{i\geq 1} n_i$$

and hence

$$1 + \sum_{i\geq 1} i n_i = 2\sum_{i\geq 1} n_i.$$

This means that $\underline{n}$ satisfies (13.1), i.e., $\underline{n}$ is feasible.

Sufficiency. First, it is seen that the irreducible sequence is the vertex partition of the planted tree whose under graph is a path of two edges. Then, by following the inversion of the procedure in the proof of Lemma 13.3, a planted tree with a given feasible sequence can be found. □

For a polyhedral segment L with 1 as both starting and ending numbers on the set $N = \{1, 2, 3, \ldots, n\}$, $n \geq 1$, let the vector $\underline{n}$ be the *point partition* of L where n_i is the number of occurrences of i in L, $i \geq 1$.

In a polyhedral segment L, if vuv is a subsegment of L, then u is said to be *contractible*. The operation of deleting u and then identifying v, or in other words vuv is replaced by v, is called *contraction*. If L can be transformed into a single point, then L is called a *celluliform*.

If the point partition of L satisfies (13.1), then L is said to be *feasible* as well.

It can be seen that any celluliform is a feasible segment but the converse is not necessarily true.

In what follows, the notation below is adopted as

$$\binom{n}{\underline{n}} = \binom{n}{n_1, n_2, \ldots, n_s} = \prod_{i=1}^{s-2} \binom{n - \sigma_{i-1}}{n_i} \tag{13.3}$$

where $s \geq 2$, $n_i \geq 0$ are all integers and

$$n = \sum_{i=1}^{s} n_i, \ \sigma_{i-1} = \sum_{j=1}^{i-1} n_j.$$

Notice that when $s = 2$, it becomes the combination of choosing n_1 from n.

Example 13.1 In Figure 13.2, two distinct planted trees of order 5 are with vertex partition $\underline{n} = (3, 0, 2)$ satisfying (13.1). Figure 13.2(a) is with sequence 123242151 and Figure 13.2(b) with sequence 121343531. Here,

$$\frac{1}{5}\binom{5}{3, 0, 2} = \frac{1}{5}\frac{5!}{3!0!2!} = 2.$$

In Figure 13.3, six distinct planted trees of order 5 shown by (a–f) are with vertex partition $\underline{n} = (2, 2, 1)$, satisfying (13.1). Here,

$$\frac{1}{5}\binom{5}{2, 2, 1} = \frac{1}{5}\frac{5!}{2!2!1!} = 6.$$

For a feasible segment of numbers on N, the occurrences of $i \in N$ divide the segment into sections in numbers equal to that of times of its occurrences. Each of the sections is called an *i-section*.

If a feasible segment on N is with the property that all numbers less than i have occurred before the first occurrence of i, $1 \leq i \leq n$, then it is called *favorable*. Denote by $1 \Leftrightarrow 1$ as a one-to-one correspondence between two sets.

Lemma 13.4 Let $\mathcal{T}_{\underline{n}}$ be the set of all planted trees of order $n + 1$ with vertex partition $\underline{n}$ and $\mathcal{L}_{\underline{n}}$, the set of all favorable celluliforms on N with point partition $\underline{n}$, then $\mathcal{T}_{\underline{n}}\ 1 \Leftrightarrow 1\ \mathcal{L}_{\underline{n}}$.

Proof Necessity. For $T \in \mathcal{T}_{\underline{n}}$, it is easy to check that its V-code $\mu(T)$ is uniquely a favorable celluliform, i.e., $\mu(T) \in \mathcal{L}_{\underline{n}}$.

Sufficiency. Let $\mu \in \mathcal{L}_{\underline{n}}$. Because of the uniqueness of the greatest point that is contractible, a point can be done by successfully contracting the greatest points. By reversing the procedure, a tree $T(\mu) \in \mathcal{T}_{\underline{n}}$ is done. □

Theorem 13.2 The number of non-isomorphic planted trees of order $n + 1$ with vertex partition $\underline{n}$ is

$$\frac{1}{n}\binom{n}{\underline{n}} = \frac{(n-1)!}{\underline{n}!} \tag{13.4}$$

where

$$\underline{n}! = \prod_{i\geq 1} n_i!. \tag{13.5}$$

Proof On the basis of Lemma 13.4, it suffices to discuss the set of all favorable celluliforms $\mathscr{L}_{\underline{n}}$. Since each favorable celluliform has n possibilities to choose the minimum point and different possibilities correspond to different ways of choosing $\underline{n}$ from n elements, the set of all ways is partitioned into

$$\frac{1}{n}\binom{n}{\underline{n}}$$

classes. A way is represented by the number sequence of length n with repetition as occurrence in the natural order. Two ways A and B are equivalent if, and only if, there exists a number $i \in N$ such that $A + i \pmod n$ is B in cyclic order. A way starting from 1 is said to be *standard*. Because of each class with n ways in which only the standard way enables us to form the V-code of a planted tree, the theorem is easily obtained. □

In Example 13.1, Figures 13.2 and 13.3 show two cases of (13.4). Take only $\underline{n} = (3, 0, 2)$ as an example. There are 10 ways of combinations of choosing 2 points with 3 occurrences each and 3 points with 1 occurrence each from 5 points numbered by 1, 2, 3, 4 and 5 as

(1) 111222345; (2) 111233345; (3) 111234445;
(4) 111234555; (5) 122233345; (6) 122234445;
(7) 122234555; (8) 123334445; (9) 123334555;
(10) 123444555

in which 2 classes are divided as $C_1 = \{(1), (5), (8), (10), (4)\}$ and $C_2 = \{(2), (6), (9), (3), (7)\}$ because of (5) = 222333451, (8) = 333444512, (10) = 444555123 and (4) = 555111234 as (1) = 111222345 for C_1, and the like for C_2.

For a general outerplanar rooted map $M = (\mathscr{X}_{\alpha,\beta}(X), \mathscr{P})$ with $(r)_{\mathscr{P}\gamma}$ on the specific circuit, where r is the root and $\gamma = \alpha\beta$ and its dual $M^* = (\mathscr{X}_{\beta,\alpha}(X), \mathscr{P}\gamma)$ with root r as well without loss of generality, let H_M be the map obtained from M^* by transforming the vertex $(r)_{\mathscr{P}\gamma}$ of M^* into vertices (r), $((\mathscr{P}\gamma)r)$, …, $((\mathscr{P}\gamma)^{-1}r)$. Such an operation is called *articulation*. The root r_H of H_M is taken as r, as shown in Figure 13.4, in which bold lines are on M and dashed lines on H_M. Here, multiedges are permitted.

Further, it is easily seen that H_M is a planted tree of size equal to the size of M.

Lemma 13.5 An outerplanar rooted map M of order $n + 1$ with face partition $\underline{s}$ is a one-to-one correspondence between a planted tree H_M and vertex partition $\underline{t} = \underline{s} + n\underline{1}_1$.

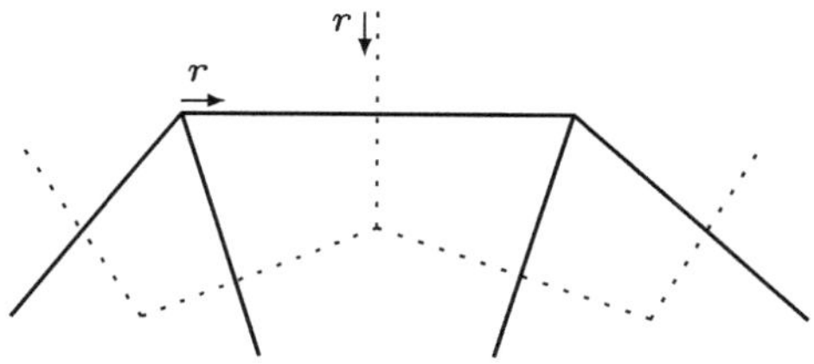

Figure 13.4: M and H_M.

Proof By the procedure of getting H_M from M, it is seen that the number of i-faces in M is the same as that in H_M for $i > 1$.

For $i = 1$, H_M has $n - 1$ articulate vertices greater than s_1. In virtue of the nonseparability of M, $s_1 = 0$.

Conversely, it is still true and hence the lemma. □

Note that all articulate edges in H_M have one-to-one correspondence with all edges on the root-face boundary of M.

Theorem 13.3 The number of non-isomorphic outerplanar rooted maps of order n with face partition $\underline{s}$ is

$$\frac{1}{n+s-1}\binom{n+s-1}{\underline{s}+(n-1)\underline{1}_1}, \tag{13.6}$$

where $s_1 + n = n$, i.e., $s_1 = 0$, because of no articulate vertex and s is the number of unrooted faces.

Proof On the basis of Lemma 13.5, the theorem is obtained from Theorem 13.2. □

Since a bipartite map has all of its faces of even valency, its face partition $\underline{s}$ is of all $s_i = 0$ when i is odd.

Corollary 13.1 The number of non-isomorphic outerplanar rooted bipartite maps of order $2m$ with face partition $\underline{s}$ is

$$\frac{1}{2m+s-1}\binom{n+s-1}{\underline{s}+(2m-1)\underline{1}_1}, \tag{13.7}$$

where $s_1 + n = n$, i.e., $s_1 = 0$, because of no articulate vertex and s is the number of unrooted faces.

A map is said to be *simple* if it has neither selfloop nor multiedge.

Corollary 13.2 The number of non-isomorphic outerplanar rooted simple maps of order n with face partition $\underline{s}$ is

$$\frac{1}{n+s-1}\binom{n+s-1}{\underline{s}+(n-1)\underline{1}_1}, \tag{13.8}$$

where $s_1 + n = n$, i.e., $s_1 = 0$, because of no articulate vertex and s is the number of unrooted faces.

Corollary 13.3 The number of non-isomorphic outerplanar rooted bipartite maps of order $2m$ with face partition $\underline{s}$ is

$$\frac{1}{2m+s-1}\binom{n+s-1}{\underline{s}+(2m-1)\underline{1}_1}, \tag{13.9}$$

where $s_1 + n = n$, i.e., $s_1 = 0$, because of no articulate vertex and s is the number of unrooted faces.

The method used in this section is from Liu [75].

13.2 Hamiltonian cubic map

Because of space constrain, this section concentrates on discussing about Hamiltonian planar rooted quadregular maps as upper maps of a Hamiltonian planar graph, and then provides the main idea for such general maps. A map is said to be *quadregular* if each of its vertices is of valency 4.

A Hamiltonian planar rooted quadregular map without two edges not on the Hamiltonian circuit, which are not able to rotate at each vertex, is called a *quaternity*.

Let $M_1 = (\mathscr{X}_{\alpha,be}(X_1), \mathscr{J}_1)$ and $M_2 = (\mathscr{X}_{\alpha,be}(X_2), \mathscr{J}_2)$ be two rooted maps with their roots, respectively, r_1 and r_2. Assume $(r_1)_{\mathscr{J}_1\gamma}$ and $(r_2)_{\mathscr{J}_2\gamma}$ are with the same length.

The map obtained by identifying r_1 and αr_2 with $Kr_1 = K\alpha r_2$ as well as $(\mathscr{J}_1\gamma)^i r_1$ and $(\alpha\mathscr{J}_2\gamma)^i r_2$ for $i \geq 1$ is called the *boundary identification* of M_1 and M_2, denoted by $I(M_1, M_2)$. The operation from M_1 and M_2 to $I(M_1, M_2)$ is called *boundary identifier*.

A boundary identification of two outerplanar cubic rooted maps is a quaternity because M_1 and M_2 are both outerplanar and cubic with its root $r_1 = r_2$.

Lemma 13.6 Let $\mathscr{Q}_{\underline{n}}$ and $\mathscr{I}_{\underline{n}}$ be the sets of all, respectively, quaternities and boundary identifiers with face partition $\underline{s}$, then there is a one-to-one correspondence between $\mathscr{Q}_{\underline{s}}$ and $\mathscr{I}_{\underline{s}}$.

Proof By considering the inverse of a boundary identifier, a quaternity becomes two cubic outerplanar maps whose boundary identification is just the quaternity with the same face partition $\underline{s}$. This is the lemma. □

From the proof of this lemma, we have

$$\mathcal{Q}_{\underline{s}} = \mathcal{I}_{\underline{s}}. \tag{13.10}$$

Lemma 13.7 The number of non-isomorphic outerplanar cubic rooted maps of order n with face partition $\underline{s}$ is

$$\frac{1}{n+s-1}\binom{n+s-1}{\underline{s}+(n-1)\underline{1}_1} \tag{13.11}$$

for $\underline{s} \in \mathcal{S}$cub, the set of all the vectors available as the face partition of an outerplanar cubic map.

Proof From Theorem 13.3, the conclusion of the lemma is true. □

Theorem 13.4 The number of non-isomorphic quaternities of order n with face partition $\underline{s}$ is

$$\sum_{\substack{\underline{s}_1,\underline{s}_2\in\mathcal{S}_{\text{cub}}\\ \underline{s}=\underline{s}_1+\underline{s}_2}} \frac{\binom{n+a_1-1}{\underline{s}_1+(n-1)\underline{1}_1}\binom{n+a_2-1}{\underline{s}_2+(n-1)\underline{1}_1}}{(n+a_1-1)(n+a_2-1)}, \tag{13.12}$$

where $a_i = |\underline{s}_i|$, called the ***absolute norm*** of $\underline{s}_i$, i.e., the sum of all the absolute values of entries in $\underline{s}_i$ for $i = 1, 2$.

Proof Since the set of all quaternities of order n is the Cartesian product of the set of all cubic outerplanar rooted maps and itself, the formula (13.12) is easily obtained. □

This method can be also employed for the case when the boundary is cubic and further for others with observing boundary combinatorics.

13.3 Halin maps

If a graph can be partitioned into a tree and a circuit whose vertex set consists of all articulate vertices of the tree, then it is called a *Halin graph*. A planar *Halin map* is an upper map of a Halin graph on the surface of genus 0 such that the circuit forms a face boundary.

Let $H = (\mathcal{X}_{\alpha,\beta}(X), \mathcal{J})$ be a planar Halin rooted map with $(r)_{\mathcal{J}\gamma}$, $\gamma = \alpha\beta$, as the face formed by the specific circuit where r is the root. The associate planted tree denoted by T_H is obtained by deleting all the edges $Kr, K(\mathcal{J}\gamma)r, \ldots, K(\mathcal{J}\gamma)^{-1}r$ on the circuit.

Lemma 13.8 A planar Halin rooted map with vertex partition $\underline{u}$ of the specific circuit with length n is a one-to-one correspondence with a planted tree with vertex partition $\underline{v} = \underline{u} + (n-1)(\underline{1}_1 - \underline{1}_3)$.

Proof By considering the procedure from a Halin map H to a tree T_H, carefully counting the numbers of vertices with the same valency and comparing them of H with those of T_H, the lemma is found. □

Theorem 13.5 The number of non-isomorphic planar Halin rooted map with vertex partition $\underline{u}$ of the specific circuit with length n is

$$A(\mathscr{H}_n^{\underline{u}}) = \frac{1}{|\underline{u}| + n - 1}\binom{|\underline{u}| + n - 1}{\underline{u} + (n-1)(\underline{1}_1 - \underline{1}_3)}, \tag{13.13}$$

where $|\underline{u}|$ is the absolute norm of $\underline{u}$.

Proof On the basis of Lemma 13.8, by Theorem 13.2, the conclusion of the theorem is done. □

Let $H_1 = (\mathscr{X}_{\alpha,\beta}(X_1), \mathscr{J}_1)$ and $H_2 = (\mathscr{X}_{\alpha,\beta}(X_2), \mathscr{J}_2)$ be two planar Halin rooted maps with $|\{r_1\}_{\mathscr{J}_1\gamma}| = |\{r_2\}_{\mathscr{J}_2\gamma}|$, the boundary identification of H_1 and H_2 is called a *double leaf*.

A graph with a specific circuit of all vertices of valency 4 is called a *quadcircularity*. An upper map of a quadcircularity is a *quadcirculation*.

Lemma 13.9 A planar rooted quadcirculation M is a double leaf if, and only if, the map obtained from M by deleting all edges on the specific circuit can be partitioned into two trees such that each of vertices on the circuit is articulate of both the trees.

Proof Since a double leaf is obtained by boundary identifier from two Halin maps, the conclusion of the lemma is directly deduced. □

Lemma 13.10 A planar rooted quadcirculation with vertex partition $\underline{u}$ of the specific circuit of length n is a one-to-one correspondence with a pair of planar Halin rooted maps H_1 and H_2 with vertex partitions, respectively, $\underline{s}$ and $\underline{t}$ such that

$$\underline{u} = \underline{s} + \underline{t} - (n-1)(2\underline{1}_3 - \underline{1}_4). \tag{13.14}$$

where $\underline{1}_i$ is the vector of all entries 0 but the ith 1 for $i = 3, 4$.

Proof By considering that $\underline{u}$ does not involve $n - 1$ unrooted 3-vertices in $\underline{s}$ and $\underline{t}$ each and involves $n - 1$ unrooted 4-vertices, the formula (13.14) holds. □

Theorem 13.6 The number of non-isomorphic double leafs with vertex partition $\underline{u}$ of the specific circuit of valency n is

$$\sum_{\substack{\underline{s}+\underline{t}=\underline{u}-(n-1)(2\underline{1}_3-\underline{1}_4) \\ \underline{s},\underline{t}\in\mathscr{S}_{\mathrm{dl}}}} A(\mathscr{H}_n^{\underline{s}})A(\mathscr{H}_n^{\underline{t}}), \tag{13.15}$$

where $\mathscr{S}_{\mathrm{dl}}$ is the set of vectors available as vertex partitions of planar Halin maps.

Proof On account of Lemma 13.10, the theorem is soon derived from Theorem 13.2 for the Cartesian product of two sets. □

Given a nonseparable graph G with a cocircuit C^* of an orientation defined, if G is planar in companion with such an orientation, then G is said to have the C^*-*oriented planarity*, or *cocircuit oriented planarity*. A planar upper map of such a graph is called a *cocircuit oriented map*. If each edge on the cocircuit is bisectioned and then snipped off each new 2-valent vertex as two articulate vertices in a cocircuit oriented map M so that what obtained is two disjoint plane trees, then M is called a *cocircular map*. The root is always chosen to be an element in an edge on the cocircuit in a cocircular map.

Lemma 13.11 A cocircular map with the oriented cocircuit of $n+1$ edges and the vertex partition $\underline{u}$ is a one-to-one correspondence with a pair of planted trees $\langle T_1, T_2\rangle$ with vertex partitions $\underline{u}_1$ and $\underline{u}_2$ such that $u_{11} = u_{21} = n$.

Proof By considering the uniqueness of a cocircular map composed of two planted trees, the conclusion is directly deduced. □

Let $\mathscr{U}_n$ be the set of all integer vectors feasible to a planted tree with n unrooted articulate vertices.

Theorem 13.7 The number of non-isomorphic cocircular maps with the oriented cocircuit of n edges and given vertex partition $\underline{u}$ is

$$\sum_{\substack{\underline{u}_1,\underline{u}_2\in\mathscr{U}_n\\ \underline{u}_1+\underline{u}_2=\underline{u}}} \frac{\binom{|\underline{u}_1|}{\underline{u}_1}\binom{|\underline{u}_2|}{\underline{u}_2}}{|\underline{u}_1||\underline{u}_2|}. \tag{13.16}$$

Proof Based on Lemma 13.11, (13.16) is derived from Theorem 13.2. □

A *cocirculation* is such a planar rooted map that has a cocircuit oriented. For this type of planar maps, the number of nonseparable ones can be determined from maps with cubic boundary of root-face.

More interestingly, maps with cubic boundary of root-face can be transformed into maps with root-vertex valency as a parameter.

In view of this, many types of planar maps with cubic boundary can be known from what have been done for counting maps with size and root-vertex valency as two parameters.

13.4 Biboundary inner rooted maps

A map is said to be *biboundary* if it has a circuit C from which two trees are obtained by deleting all the edges on C. In view of this, a Hamiltonian cubic map is a *uniboundary* map because it is not necessary to have two connected components as all the edges on the Hamiltonian circuit are deleted. Here, only planar case is considered.

Let $M = (\mathscr{X}, \mathscr{J})$ be a biboundary map, and $r = r(M)$ be its root. The length of boundary is m and the vertex partition vector of nonboundary vertices is $\underline{n} = (n_1, n_2, \cdots)$, n_i, $i \geq 1$, the number of i-vertices not on the boundary.

Assume $M_1 = (\mathscr{X}_1, \mathscr{J}_1)$ and $M_2 = (\mathscr{X}_2, \mathscr{J}_2)$ are two submaps of M. Denote by $C = (Kr, K\varphi r, \ldots, K\varphi^{m-1}r)$ the boundary circuit of M where

$$\varphi x_i = \begin{cases} \mathscr{J}\gamma x_i, & \text{if } \gamma x_i \text{ not incident} \\ & \quad \text{with an inner edge of } M_2; \\ \alpha\mathscr{J}\beta x_i, & \text{otherwise} \end{cases}$$

with $x_0 = x_m = r$ and $x_i = \varphi^i r$, $i = 1, 2, \ldots, m-1$.

Let $r_1 = r(M_1) = r$ and $r_2 = r(M_2) = \alpha\varphi^{m-1}r$. This means the root-vertex of M_1 adjacent to that of M_2 is M. Such a root is said to be *inner rooted.*

First, denoted by $\mathscr{B}_m$, $m \geq 6$, the set of all biboundary rooted maps with the boundary length m.

Lemma 13.12 Let $\mathscr{W}_{m_1}$ and $\mathscr{W}_{m_1}$ be, respectively, uniboundary maps of boundary lengths $m_1 \geq 3$ and $m_2 \geq 3$, then a pair of $\{W_1, W_2\}$, $W_1 \in \mathscr{W}_{m_1}$ and $W_2 \in \mathscr{W}_{m_2}$, $m = m_1 + m_2$, composes of

$$s_{m_2}(m_1) = \begin{bmatrix} m_2 \\ m_1 \end{bmatrix} \tag{13.17}$$

biboundary maps in $\mathscr{B}_m$. And, this combinatorial number is determined by the recursion as

$$\begin{cases} \begin{bmatrix} m_2 \\ m_1 \end{bmatrix} = \sum\limits_{i=0}^{m_1} \begin{bmatrix} m_2 - 1 \\ i \end{bmatrix}, & m_2 \geq 2 \\ \begin{bmatrix} m_2 \\ 0 \end{bmatrix} = 1, \; m_2 \geq 1; \; \begin{bmatrix} 1 \\ m_1 \end{bmatrix} = 1, \; m_1 \geq 0. \end{cases} \tag{13.18}$$

Proof By induction on m_2, $m_2 \geq 2$, for any $m_1 \geq 1$.

First, by checking the case of $m_2 = 2$, for $m_1 \geq 1$, we have

$$\begin{bmatrix} 2 \\ m_1 \end{bmatrix} = \begin{bmatrix} 2 \\ 0 \end{bmatrix} + \begin{bmatrix} 2 \\ 1 \end{bmatrix} + \cdots + \begin{bmatrix} 2 \\ m_1 \end{bmatrix} = m_1 + 1.$$

If there are 0 vertices in the first segment, then the second segment has to have m_1 vertices; if there is one vertex in the first segment, then the second segment has to have $m_1 - 1$ vertices; $\cdots$; if there are m_1 vertices in the first segment, then the second segment has to have 0 vertices. They are all together $m_1 + 1$. Thus, (13.18) is true for $m_2 = 2$.

Then, assume $t_{m_2-1}(m_1)$, $m_2 \geq 3$, have been determined by (13.18). To prove $t_{m_2}(m_1)$ is determined by (13.18). Because there are $m_1 + 1$ occurrences for putting m_1 vertices into m_2 segments as when m_1 vertices in the first segment, so no vertex in all other $m_2 - 1$ segments and hence $s_{m_2-1}(0)$ ways; when there are $m_1 - 1$ vertices in the first segment, then there is one vertex in all other $m_2 - 1$ segments and hence $s_{m_2-1}(1)$ ways; $\cdots$; when there are 0 vertices in the first segment, then there are m_1 vertices in all other $m_2 - 1$ segments and hence $s_{m_2-1}(m_1)$ ways. There are in all $s_{m_2}(m_1) = \sum_{i=0}^{m_1} s_{m_2-1}(i)$ ways. That is,

$$\begin{bmatrix} m_2 \\ m_1 \end{bmatrix} = \sum_{i=0}^{m_1} \begin{bmatrix} m_2 - 1 \\ i \end{bmatrix}, \quad m_2 \geq 2.$$

By the induction hypothesis, $s_{m_2}(m_1)$ is determined. The lemma is true. □

Then, denote by $\mathscr{D}_m$, $m \geq 6$, the set of all biboundary rooted maps with the boundary length m.

Lemma 13.13 Let $\mathscr{M}_{m_1}$ and $\mathscr{M}_{m_1}$ be the uniboundary rooted maps of boundary lengths, respectively, $m_1 \geq 3$ and $m_2 \geq 3$, then a pair $\{M_1, M_2\}$, $M_1 \in \mathscr{M}_{m_1}$ and $M_2 \in \mathscr{M}_{m_2}$, composes

$$t_{m_2}(m_1) = \left\langle \begin{matrix} m_2 \\ m_1 \end{matrix} \right\rangle \tag{13.19}$$

biboundary rooted maps in $\mathscr{D}_m$, $m = m_1 + m_2$. And, this combinatorial number is determined by

$$\left\langle \begin{matrix} m_2 \\ m_1 \end{matrix} \right\rangle = \sum_{i=0}^{m_1-1} \begin{bmatrix} m_2 - 1 \\ i \end{bmatrix}, \tag{13.20}$$

where the terms on the right side of (13.19) are given in Lemma 13.12.

Proof By induction on m_2 for any $m_1 \geq 1$.

First, when $m_2 = 2$, By considering for assigning m_1 vertices in M_2 edges on the boundary in the order determined as when $1, 2, \ldots, m_1$ vertices in the first edge (incident with the root), the second edge has to have, respectively, $m_1 - 1, m_1 - 2, \ldots, 0$ vertices, we have $t_2(m_1) = s_1(0) + s_1(1) + \cdots + s_1(m_1 - 1) = m_1$.

Then, assume $t_{m_2-1}(m_1)$, $m_2 \geq 3$, has been determined by (13.20). To prove that

$$t_{m_2}(m_1) = s_{m_2-1}(0) + s_{m_2-1}(1) + \cdots + s_{m_2-1}(m_1-1),$$

is determined by (13.20) as well. Because the first edge in M_2 edges is allowed to have $m_1, m_1-1, \ldots, 1$ vertex, the other m_2-1 edges are allowed to have only, respectively, $0, 1, \ldots, m_1-1$ vertex. This implies

$$\begin{bmatrix} m_2-1 \\ 0 \end{bmatrix} + \begin{bmatrix} m_2-1 \\ 1 \end{bmatrix} + \cdots + \begin{bmatrix} m_2-1 \\ m_1-1 \end{bmatrix}.$$

Hence, (13.20) is right. □

Denote by Q_i, $1 \leq i \leq t_{m_2}(m_1)$, the $t_{m_2}(m_1)$ biboundary inner rooted maps mentioned in this lemma with $\mathscr{M}(M_1, M_2) = \{Q_1, Q_2, \cdots, Q_{t_{m_2}(m_1)}\}$.

Lemma 13.14 Let $\mathscr{H}_m = \{\mathscr{M}(M_1, M_2) | \forall (M_1, M_2) \in \mathscr{M}_{m_1} \times \mathscr{M}_{m_2}, m_1 + m_2 = m\}$, then

$$\mathscr{D}_m = \sum_{H \in \mathscr{H}} H, \tag{13.21}$$

i.e., $\mathscr{H}_m$ is a partition of $\mathscr{D}_m$.

Proof For any $D \in \mathscr{D}_m$, it is known from biboundary maps that there exist m_1 and m_2, $m_1 + m_2 = m$, such that $M_1 \in \mathscr{M}_{m_1}$ and $M_2 \in \mathscr{M}_{m_2}$ compose of D. Thus, $D = (M_1, M_2) \in \mathscr{H}_m$.

Conversely, for any $Q \in H$, $H \in \mathscr{H}_m$, because Q is composed of two uniboundary maps $M_1 \in \mathscr{M}_{m_1}$ and $M_2 \in \mathscr{M}_{m_2}$, $m_1 + m_2 = m$, there exists $D \in \mathscr{D}_m$ such that $D = Q$.

In summary, the lemma is obtained. □

In what follows, observe the number of non-isomorphic uniboundary maps of boundary length m with vertex partition vector $\underline{n}$.

Lemma 13.15 The number of uniboundary rooted maps of boundary length m, $m \geq 3$, and nonboundary vertex partition vector $\underline{n}$ is

$$\eta(m, \underline{n}) = \frac{(m+n-1)!}{(m-1)!\underline{n}!}, \tag{13.22}$$

where $n = |\underline{n}| = n_1 + n_2 + \cdots$.

Proof Let $M = (\mathscr{X}, \mathscr{J})$ be a uniboundary rooted map of boundary length m, $m \geq 3$, and nonboundary vertex partition vector $\underline{n}$. Its root is r. Because of cubicness on the boundary, $\mathscr{J}r$ is incident with an articulate vertex of the tree. Let $\mathscr{J}r$ be the root to make the tree planted. Because of all 1-vertices of the planted tree on the boundary, its vertex partition vector is $\underline{n}+(m-1)\underline{1}_1$, where $\underline{1}_1$ is the vector of all entries 0 but the first entry 1. Since a planted tree with vertex partition is a one-to-one correspondence with a uniboundary rooted map of boundary length m, $m \geq 3$, and vertex partition vector $\underline{n}$, from Theorem 13.2, the number of non-isomorphic uniboundary rooted maps of boundary length m, $m \geq 3$, and vertex partition vector $\underline{n}$ is

$$\eta(m, \underline{n}) = \frac{(m+n-1)!}{(m+n_1-1)!(\underline{n}-n_1\underline{1}_1)!},$$

where $n = |\underline{n}| = n_1 + n_2 + \cdots$. By considering that $n_1 = 0$ in $\underline{n}$, the lemma is done. □

On the basis of the above tree lemmas, the main result of this section can be obtained.

Theorem 13.8 The number of biboundary inner rooted maps of boundary length $m \geq 6$ and nonboundary vertex partition is

$$\sum_{(m_1,m_2,\underline{n}^1,\underline{n}^2)\in\mathscr{L}} \left\langle \begin{matrix} m_2 \\ m_1 \end{matrix} \right\rangle \frac{(n^1+m_1-1)!}{(m_1-1)!\underline{n}^1!} \times \frac{(n^2+m_2-1)!}{(m_2-1)!\underline{n}^2!} \tag{13.23}$$

where $\mathscr{L} = \{(m_1, m_2, \underline{n}^1, \underline{n}^2) | m_1 + m_2 = m, \underline{n}^1 + \underline{n}^2 = \underline{n}, m_1, m_2 \geq 3\}$.

Proof For any given m_1 and m_2, $m_1 + m_2 = m$, with $\underline{n}^1$ and $\underline{n}^2$, $\underline{n}^1 + \underline{n}^2 = \underline{n}$, from Lemma 13.14–13.15, $\mathscr{D}_m$ can be classified into

$$\sum_{(m_1,m_2,\underline{n}^1,\underline{n}^2)\in\mathscr{L}} \frac{(n^1+m_1-1)!}{(m_1-1)!\underline{n}^1!} \frac{(n^2+m_2-1)!}{(m_2-1)!\underline{n}^2!}$$

classes. From Lemma 13.13, each class has

$$\left\langle \begin{matrix} m_2 \\ m_1 \end{matrix} \right\rangle$$

non-isomorphic biboundary inner rooted maps of boundary length m and nonboundary vertex partition vector $\underline{n}$. Thus, the theorem is proved. □

13.5 General maps

Based on the joint tree model shown in Chapter 12, it looks general maps on surfaces in a closed relation with joint trees. In this section, only the orientable case is considered as an instance.

Because of the independence with a tree chosen, general maps with a cotree marked are particularly investigated.

For the convenience of description, all maps are assumed to have no articulate edge.

Let $M = (\mathscr{X}, \mathscr{J})$ be a map with cotree edges $a_1 = Kx_1, a_2 = Kx_2, \ldots, a_l = Kx_l$ marked where $l = \beta(M)$ is the Betti number of M. The root of M is chosen on a cotree edge, assume $r = r(M) = x_1$.

Another map $H_M = (\mathscr{X}_H, \mathscr{J}_H)$ is constructed as

$$\mathscr{X}_H = \mathscr{X} + \sum_{i=1}^{l}(Ks_i + Kt_i - a_i), \tag{13.24}$$

where $Ks_i = \{x_i, \alpha x_i, \beta s_i, \gamma s_i\}$ and $Kt_i = \{\gamma x_i, \beta x_i, \beta t_i, \gamma t_i\}$, $1 \le i \le l$; $\mathscr{J}_H$ is defined as

$$(x)_{\mathscr{J}_H} = \begin{cases} (x)_{\mathscr{J}}, & \text{when } x \in \mathscr{X}; \\ (x), & \text{when } x \notin \mathscr{X}. \end{cases} \tag{13.25}$$

Lemma 13.16 For any rooted map M with a cotree marked, the map H_M is a planted tree with the number of articulate vertices two times the Betti number of M.

Proof Because it is connected without circuit on H_M, H_M is a tree in its own right. Since the number of cotree edges is the Betti number of M, from the construction of H_M and no articulate edge on M, the number of articulate vertices on H_M is two times the Betti number of M. □

Let $\mathscr{M}(l; \underline{n})$ be the set of all general rooted maps with a cotree of size l marked and vertex partition $\underline{n}$ including the root-vertex. And let $\mathscr{H}(\underline{n})$ be the set of all planted trees with articulate vertices two times the number of cotree edges and vertex partition $\underline{n}$ excluding the root-vertex.

Lemma 13.17 There is a one-to-one correspondence between $\mathscr{M}(l; \underline{n})$ and $\mathscr{H}(\underline{n} + (2l-1)\underline{1}_1)$ as the set of joint trees.

Proof For $M \in \mathscr{M}(l; \underline{n})$, it is easily seen that the corresponding H_M is just a joint tree of M and hence $H_M \in \mathscr{H}(\underline{n} + (2ł - 1)\underline{1}_1)$.

Conversely, for $H \in \mathscr{H}(\underline{n}+(2l-1)\underline{1}_1)$, in virtue of a joint tree its articulate vertices are pairwise marked as cotree edges of the corresponding map M as $H = H_M$, by counting the valencies of vertices, it is checked that $M \in \mathscr{M}(l; \underline{n})$.

Therefore, the lemma is true. □

This lemma enables us to determine the number of general rooted maps with a cotree marked with vertex partition given.

Theorem 13.9 The number of rooted general maps with a cotree marked for a vertex (root-vertex included) partition $\underline{n}$ given is

$$|\mathscr{M}(l;\underline{n})| = \frac{(n+2l-2)!}{(2l-1)!\underline{n}!}, \tag{13.26}$$

where l is the Betti number (the size of cotree) and $n = |\underline{n}|$.

Proof A direct result of Lemma 13.17 and Theorem 13.2. □

13.6 Pan-flowers

A map is called a *pan-flower* if it can be seen as a standard petal bundle added to a tree such that only all vertices are in the inner parts of edges on the petal bundle. The petal bundle is seen as the boundary of a pan-flower. In the absence of a circuit for the base graph of a petal bundle in general, a pan-flower is reasonably seen as a generalization of a map with a boundary, or a boundary map. A pan-Halin map is only a special case when the petal bundle is asymmetric.

For convenience, the petal bundle in a pan-flower is called the *base map*. If a edge of a fundamental circuit on the underlying graph of base graph is allowed to have no articulate vertex of the tree, then the pan-flower is said to be *pre-standard*. If the edge has at least one articulate vertex of the tree, then the pan-flower is *standard*.

This section is concerned with pan-flowers in the two classes for a vertex partition vector given.

Let $\mathscr{H}$psH be the set of all rooted pre-standard pan-flowers, where the root is chosen to be an element incident on the vertex of base map. For any $H = (\mathscr{X}, \mathscr{J}) \in \mathscr{H}$psH, the tree T_H is always seen as a planted tree whose root is first encountered on the rooted face of H starting from the root of H. Otherwise, the first encountered at the root-vertex of H from the root.

Lemma 13.18 Let $\mathscr{H}\mathrm{psH}(p;\underline{s})$ be the set of all rooted pre-standard pan-flowers with vertex partition vector $\underline{s} = (s_2, s_3, \cdots)$ on a surface of orientable genus p, then

$$|\mathscr{H}\mathrm{psH}(p;\underline{s})| = 2^{j_1-\delta_{p,1}+2}\binom{j_1+2p}{2p-1}|\mathscr{T}_1(\underline{j})|, \tag{13.27}$$

where $\mathscr{T}_1(\underline{j})$ is the set of planted trees with vertex partition vector $\underline{j} = (j_1, j_2, \cdots)$, such that $s_i = j_i$, $i \neq 3$; $s_3 = j_1 + j_3 + 1$, $p \geq 1$, $\underline{s} \geq \underline{0}$, $\underline{j} \geq \underline{0}$, but $\underline{s} \neq \underline{0}$ and $\underline{j} \neq \underline{0}$. Further, $\delta_{p,1}$ is the *Kronecker symbol*, i.e., $\delta_{p,1} = 1$, when $p = 1$; $\delta_{p,1} = 0$, otherwise.

Proof On the basis of pre-standardness, from the definition of pan-flowers it is seen that an element in the set on the left side of (13.27) has a corresponding element on the right side.

In what follows, it proves that each $T = (\mathscr{X}, \mathscr{J}) \in \mathscr{T}_1(\underline{j})$ produces

$$2^{j_1-\delta_{p,1}+2}\binom{j_1+2p}{2p-1}$$

maps in $\mathscr{H}$psH$(p;\underline{s})$.

Denoted by $(r), (x_1), (x_2), \ldots, (x_{j_1})$ all the articulate vertices of T, where $0 < l_1 < l_2 < \cdots < l_{j_1}$, such that $x_i = (\mathscr{J}\alpha\beta)^{l_i}r$, $i = 1, , 2, \ldots, j_1$, $r = r(T)$ is the root of T.

First, by considering that the underlying graph of base map has $2p$ loops, only one vertex, its embedding on the orientable surface of genus $2p$ has exactly one face. Because of the order of its automorphism group 8 when $p = 1$, there is only one possible way; $2p$ when $p \geq 2$, there are two possible ways.

Then, the assignment of the $j_1 + 1$ articulate vertices, (r), (x_i), $i = 1, 2, \ldots, j_1$, of T on the base map has the number of ways as choosing $2p-1$ from $j_1 + 2$ intervals with repetition allowable. That is,

$$\binom{j_1+2+(2p-1)-1}{2p-1} = \binom{j_1+2p}{2p-1}.$$

Finally, since each of elements r, x_i, $i = 1, 2, \ldots, j_1$, has two ways: one side $\{\alpha x, \alpha\beta x\}$, and the other $\{x, \beta x\}$; they have in all

$$2^{j_1+1}$$

ways

In summary of the three cases, the aim is achieved. □

On the basis of this lemma, by employing a result in Section 13.1, the following theorem can be deduced.

Theorem 13.10 The number of pre-standard rooted pan-flowers with vertex partition vector $\underline{s} = (s_2, s_3, \cdots)$ and its base involving m missing vertices on an orientable surface of genus p, $p \geq 1$, is

$$2^{m-\delta_{p,1}+1}\binom{m+2p-1}{2p-1}\binom{s_3}{m}\frac{n!m}{\underline{s}!}, \tag{13.28}$$

where $\underline{s}! = \prod_{i\geq 2} s_i! = s_2!s_3!\cdots$ and $n + 2 = \sum_{i\geq 2} s_i$.

Proof From Lemma 13.18, this number is

$$2^{m-\delta_{p,1}+1}\binom{m+2p-1}{2p-1}\tau_1(\underline{j}),$$

where $\underline{j} = (j_1, j_2, j_3, \cdots)$ such that $j_1 + 1 = m$, $j_3 = s_3 - m$ and $j_i = s_i$, $i \neq 3$, $i \geq 2$. Then, from Theorem 13.2, we have

$$\tau(\underline{j}) = \frac{(n'-1)!}{\underline{j}!} = \frac{n!}{(m-1)!s_2!(s_3-m)!s_4!\cdots} = \binom{s_3}{m}\frac{n!m}{\underline{s}!},$$

where $n = n' - 1 = \sum_{i\geq 1} j_i - 1 = \sum_{i\geq 2} s_i - 2$. By substituting this into the last, (13.28) is obtained. □

Let $\tilde{\mathscr{H}}$psH$(q;\underline{s})$ be the set of all pre-standard rooted pan-flowers with vertex partition vector $\underline{s} = (s_2, s_3, \cdots)$ on a nonorientable surface of genus q, $\underline{s} \geq \underline{0}$, but $\underline{s} \neq \underline{0}$.

Lemma 13.19 For $\tilde{\mathscr{H}}$psH$(q;\underline{s})$, $q > 0$, $\underline{s} \geq \underline{0}$, but $\underline{s} \neq \underline{0}$, we have

$$|\tilde{\mathscr{H}}\mathrm{psH}(q;\underline{s})| = 2^{m-\delta_{q,1}+1}\binom{m+q-1}{q-1}|\mathscr{T}_1(\underline{j})|, \tag{13.29}$$

where $\underline{j} = (m-1, s_2, s_3 - m, s_4, s_5, \cdots)$ and m is the number of trivalent vertices (i.e., missing vertices on its base).

Proof Similar to the proof of Lemma 13.18. However, note that the size of the base is q, $q \geq 1$, instead of $2p$, $p \geq 1$ and the order of automorphism group of the base is $2q$ when $q \geq 2$; 4 when $q = 1$. □

Similar to Theorem 13.11, we have

Theorem 13.11 The number of pre-standard rooted pan-flowers with vertex partition vector $\underline{s} = (s_2, s_3, \cdots)$ and its base involving m missing vertices on a nonorientable surface of genus q, $q \geq 1$, is

$$2^{m-\delta_{q,1}+1}\binom{m+q-1}{q-1}\binom{s_3}{m}\frac{n!m}{\underline{s}!}, \tag{13.30}$$

where $n + 2 = \sum_{i\geq 2} s_i$ and $\underline{s}! = \prod_{i\geq 2} s_i!$.

Proof Similar to the proof of Theorem 13.11, from Lemma 13.19 and Theorem 13.2, the theorem is done. □

For standard pan-flowers, let $\mathscr{H}$sH be the set of all such maps. From the definition, if the base map has m missing vertices, then it is not possible on an orientable surface of genus greater then $m/2$, or on a nonorientable surface of genus greater than m.

Lemma 13.20 Let $\mathscr{H}$sH$(p;\underline{s})$ be the set of all standard rooted pan-flowers with vertex partition vector $\underline{s} = (s_2, s_3, \cdots)$ on an orientable surface of genus p. If maps in

$\mathscr{H}$sH$(p;\underline{s})$ have their bases with $m \geq 2p$ missing vertices, then we have

$$|\mathscr{H}\mathrm{sH}(p;\underline{s})| = 2^{m-\delta_{p,1}+1}\binom{m-1}{2p-1}|\mathscr{T}_1(\underline{j})|, \tag{13.31}$$

where $\mathscr{T}_1(\underline{j})$, as above, is the set of panted trees with vertex partition vector $\underline{j} = (j_1, j_2, j_3, \cdots)$ for $j_1 = m-1$, $j_3 = s_3 - m$, $j_i = s_i$, $i \neq 3$, $i \geq 2$.

Proof For any $H \in \mathscr{H}$sH$(p;\underline{s})$, from the definitions of standardness and pan-flowers, it is seen that there exists a planted tree in $\mathscr{T}_1(\underline{j})$ corresponding to H. Thus, it suffices to prove that any planted tree in $T = (\mathscr{X}, \mathscr{J}) \in \mathscr{T}_1(\underline{j})$ produces

$$2^{m-\delta_{p,1}+1}\binom{m-1}{2p-1}$$

maps in $\mathscr{H}$sH$(p;\underline{s})$.

First, note that maps in $\mathscr{H}$sH$(p;\underline{s})$ are with their bases of size $2p$ on an orientable surface of genus p. Since the order of automorphism group of the base is $4p$ when $p \geq 2$, and 8 when $p = 1$, the base has, respectively, two ways when $p \geq 2$, and one way to choose its root.

Second, since the number of missing vertices on the base is m, T must have $m-1$ unrooted articulate vertices. Let them be incident with $(x_1), \ldots, (x_{m-1})$. From the standardness again, there are $m-1$ intervals for choice in the linear order $\langle (r), (x_1), (x_2), \cdots, (x_{m-1})\rangle$. Thus, any $2p-1$ points insertion divides the linear order into $2p$ nonempty segments. This has

$$\binom{m-1}{2p-1}$$

distinct ways.

Third, notice that each of the m articulate edges, including the root-edge in T, has two choices, and hence

$$2^m$$

distinct choices altogether.

In summary from the three cases, the lemma is easily found. □

Based on this, we have

Theorem 13.12 The number of standard rooted pan-flowers with vertex partition vector $\underline{s} = (s_2, s_3, \cdots)$ and their base map of m, $m \geq 2p$, unrooted vertices on an orientable surface S_p of genus $p \geq 1$, is

$$2^{m-\delta_{q,1}+1}\binom{m-1}{2p-1}\binom{s_3}{m}\frac{n!m}{\underline{s}!}, \tag{13.32}$$

where $n + 2 = \sum_{i\geq 2} s_i$.

Proof Similar to the proof of Theorem 13.11. However, by Lemma 13.20 instead of Lemma 13.8. □

A look again at the nonorientable case.

Lemma 13.21 Let $\tilde{\mathscr{H}}\mathrm{sH}(q;\underline{s})$ be the set of standard rooted pan-flowers with vertex partition vector $\underline{s} = (s_2, s_3, \cdots)$ on a nonorientable surface of genus $q \geq 1$. If each map in $\tilde{\mathscr{H}}\mathrm{sH}(q;\underline{s})$ has its base map of m unrooted vertices, then we have

$$|\tilde{\mathscr{H}}\mathrm{sH}(q;\underline{s})| = 2^{m-\delta_{q,1}+1}\binom{m-1}{q-1}\tau(\underline{j}), \tag{13.33}$$

where $\tau_1(\underline{j}) = |\mathscr{T}_1(\underline{j})|$, $\underline{j} = (j_1, j_2, j_3, \cdots)$, $j_1 = m-1$, $j_3 = s_3 - m$, $j_i = s_i$, $i \neq 3$, $i \geq 2$.

Proof Similar to the proof of Lemma 12.2.3. However, attention should be paid to the base of size q instead of $2p$ for a surface of genus q. □

Thus, we can also have

Theorem 13.13 The number of standard rooted pan-flowers with vertex partition vector $\underline{s} = (s_2, s_3, \cdots)$ and their bases of m unrooted vertices on a nonorientable surface of genus q, $q \geq 1$, is

$$2^{m-\delta_{q,1}+1}\binom{m-1}{q-1}\binom{s_3}{m}\frac{n!m}{\underline{s}!}, \tag{13.34}$$

where $\underline{s} \geq \underline{0}$, $\underline{s} \neq \underline{0}$, $m \geq q \geq 1$ and $n + 2 = \sum_{i\geq 2} s_i$.

Proof Similar to the proof of Theorem 13.13. However, by Lemma 13.21 instead of Lemma 13.20. □

13.7 Notes

13.7.1 Determine the vertex partition function of general maps rooted on the sphere.

Notice 13.1 On the basis of the theory illustrated in Liu [97, pp. 292–302], the function has been determined. This solution is particularly available for efficientization, and further for intelligentization.

Although an explicit expression can be obtained from Theorem 7.4.3 in Liu [64] and Theorem 15.9, a direct way is still necessary to get the explicit expression from the equation.

13.7.2 Determine the vertex partition function of general maps rooted on all surfaces.

Notice 13.2 On the basis of the theory illustrated in LiuP [97, pp. 418–430], the function has been determined. This solution is particularly available for efficientization, and further for intelligentization. Although an explicit expression can be obtained from Theorem 1.11 in Chapter 1 and Theorem 15.9 in Chapter 15, a direct way is still necessary to get the explicit expression from the equation.

13.7.3 Determine the vertex partition function of general Eulerian rooted maps on all surfaces.

Notice 13.3 On the basis of the theory illustrated in Liu [97, pp. 409–417], the function has been determined. This solution is particularly available for efficientization, and further for intelligentization. Although an explicit expression can be obtained from Theorem 7.4.3 in Liu [64] and Theorem 15.9 in Chapter 15, a direct way is still necessary to get the explicit expression from the equation.

13.7.4 Determine the vertex partition function of 2-edge connected rooted maps on the sphere.

Notice 13.4 Because there is no bridge equivalent to bi-edge (or 2-edge) connectedness, by the theory illustrated in Liu [97, pp. 389–398], the function can be determined. This solution is particularly available for efficientization, and further for intelligentization. Although an explicit expression can be obtained from Theorem 7.4.3 in Liu [64] and Theorem 15.9, a direct way is still necessary to get the explicit expression from the corresponding equation.

13.7.5 Determine the vertex partition function of 2-edge connected rooted maps on all surfaces.

Notice 13.5 On the basis of the theory illustrated in Liu [97, pp. 389–398], the function can be determined. This solution is particularly available for efficientization, and further for intelligentization. Although an explicit expression can be obtained from Theorem 1.11 in Chapter 1 and Theorem 15.9, a direct way is still necessary to get the explicit expression from the corresponding equation.

13.7.6 Determine the vertex partition function of the nonseparable rooted maps on the sphere.

Notice 13.6 On the basis of the theory illustrated in Liu [97, pp. 313–326], the function has been determined. This solution is particularly available for efficientization, and further for intelligentization. Although an explicit expression can be done from Theorem 7.4.3 in Liu [64] and Theorem 15.9 in Chapter 15, a direct way is still necessary to get the explicit expression from the equation.

13.7.7 Determine the vertex partition function of the nonseparable rooted maps on all surfaces.

Notice 13.7 On the basis of the theory illustrated in Liu [97, pp. 389–398], the function can be determined. This solution is particularly available for efficientization, and further for intelligentization. Although an explicit expression can be obtained from Theorem 1.11 in Chapter 1 and Theorem 15.9, a direct way is still necessary to get the explicit expression from the corresponding equation.

13.7.8 Determine the vertex partition function of loopless rooted maps on the sphere.

Notice 13.8 On the basis of the theory illustrated in Liu [97, pp. 303–312], the function has been determined. This solution is particularly available for efficientization, and further for intelligentization. Although an explicit expression can be done from Theorem 7.4.3 in Liu [64] and Theorem 15.9 in Chapter 15, a direct way is still necessary to get the explicit expression from the equation.

13.7.9 Determine the vertex partition function of loopless maps rooted on all surfaces.

Notice 13.9 On the basis of the theory illustrated in Liu [97, pp. 398–408], the function can be determined. This solution is particularly available for efficientization, and further for intelligentization. Although an explicit expression can be obtained from Theorem 1.11 in Chapter 1 and Theorem 15.9, a direct way is still necessary to get the explicit expression from the corresponding equation.

13.7.10 Determine the vertex partition function of Eulerian rooted triangulations on the sphere.

Notice 13.10 Whenever the equation is established, an available solution can be easily extracted for efficientization, and further for intelligentization. On the basis of (5.3.16), or Tutte formula, in Liu [78] (2009) and Theorem 15.9, a direct way is still necessary to get the explicit expression from the equation.

13.7.11 Determine the vertex partition function of Eulerian maps rooted on the sphere.

Notice 13.11 On the basis of the theory illustrated in Liu [78, pp. 120–124], the function has been determined. This solution is particularly available for efficientization, and further for intelligentization.

Although an explicit expression can be obtained from Theorem 5.2.2 in Liu [78] and Theorem 15.9, a direct way is still necessary to get the explicit expression from the equation.

13.7.12 Determine the vertex partition function of Eulerian maps rooted on all surfaces.

Notice 13.12 On the basis of the theory illustrated in Liu [97, pp. 409–417], the function has been determined. This solution is particularly available for efficientization, and further for intelligentization. Although an explicit expression can be obtained from Theorem 1.11 in Chapter 1 and Theorem 15.9 in Chapter 15, a direct way is still necessary to get the explicit expression from the equation.

14 Equations with Partitions

- The meson functional is used for describing equations discovered from census of maps via vertex, or face, partition as parameters.
- Functional equations are extracted from the census of general maps and nonseparable maps with the root-vertex valency and the vertex partition vector on the sphere.
- By observing maps without cut-edge on general surfaces, a functional equation has also be found with vertex partition.
- Functional equations satisfied by the vertex partition functions of Eulerian maps on the sphere and general surfaces are derived from suitable decompositions of related sets of maps.
- All these equations can be shown to be well defined. However, they are not yet solved in any way.

14.1 The meson functional

Let $f(\underline{y}) \in \mathscr{R}\{\underline{y}\}$, where $\underline{y} = (y_1, y_2, \cdots)$, be a function, and

$$V(f, y_i) \geq 0,\ i = 1, 2, \cdots .$$

A transformation is established as $\int_y : y^i \mapsto y_i,\ i = 1, , 2, \cdots$, convinced $y^0 = 1 \mapsto y_0$.

Since $\int_y$ is a function from the function space $\mathscr{F}$ with basis $\{1, y, y^2, \cdots\}$ to an abstract linear space $\mathscr{V}$ with basis $\{y_0, y_1, y_2, \cdots\}$, it is called the *meson functional*. i.e., the *Blissard operator*. For any

$$v_i = \sum_{j\geq 0} a_{ij} y^j,\ i = 1, 2,$$

it is easy to check that

$$\begin{aligned}\int_y (v_1 + v_2) &= \sum_{j\geq 0} (a_{1j} + a_{2j}) \int_y y^j \\ &= \sum_{j\geq 0} a_{1j} y_j + \sum_{j\geq 0} a_{2j} y_j \\ &= \int_y v_1 + \int_y v_2.\end{aligned}$$

Hence, the meson functional is linear.

DOI 10.1515/9783110481846-0114

The inverse of the meson functional $\int_y$ is denoted by $\int_y^{-1} : y_j \mapsto y^j,\ i = 1, 2, \cdots,$ convinced $\int_y^{-1} y_0 = 1$, or simply $y_0 = 1$. However, 1 is seen as a vector in $\mathscr{V}$.

Two linear operators called *left* and *right projection*, denoted by, respectively, $\mathbb{J}_y$ and $\mathbb{R}_y$, are defined in the space $\mathscr{V}$ as: let $v = \sum_{j\geq 0} a_j y_j \in \mathscr{V}$, then

$$\begin{cases} \mathbb{J}_y v = \sum\limits_{j\geq 0} (j+1) a_{j+1} y_j; \\ \mathbb{R}_y v = \sum\limits_{j\geq 1} \frac{1}{j} a_{j-1} y_j. \end{cases} \tag{14.1}$$

In other words, if y_i is considered as the vector with all entries 0 but only the ith 1, then the matrices corresponding to $\mathbb{J}_y$ and $\mathbb{R}_y$ are, respectively, as

$$L = (\underline{l}_1^{\mathrm{T}}, \underline{l}_2^{\mathrm{T}}, \underline{l}_3^{\mathrm{T}}, \cdots), \tag{14.2a}$$

where

$$\underline{l}_j = \begin{cases} \underline{0}, & \text{when } j = 1; \\ (j-1)\underline{1}_{j-1}, & \text{when } j \geq 2 \end{cases}$$

for $\underline{1}_j$ being the infinite vector of all entries 0 but only the $(j-1)$-st 1 and

$$R^{\mathrm{T}} = (\underline{r}_1^{\mathrm{T}}, \underline{r}_2^{\mathrm{T}}, \underline{r}_3^{\mathrm{T}}, \cdots), \tag{14.2b}$$

where

$$\underline{r}_j = \begin{cases} \underline{0}, & \text{when } j = 1; \\ \frac{1}{j-1}\underline{1}_{j-1}, & \text{when } j \geq 2 \end{cases}$$

for the super index "T" as the transpose.

It can be easily seen that

$$LR = \begin{pmatrix} I & \underline{0}^{\mathrm{T}} \\ \underline{0} & 0 \end{pmatrix}, \quad RL = \begin{pmatrix} 0 & \underline{0} \\ \underline{0}^{\mathrm{T}} & I \end{pmatrix}, \tag{14.3}$$

where I is the identity.

Theorem 14.1 For $v = v(y_0, y_1, \cdots) \in \mathscr{V}$, let $f(y) = \int_y^{-1} v$, then

$$\frac{\mathrm{d}}{\mathrm{d}y} f(y) = \int_y^{-1} \mathbb{J}_y v; \quad \int f(y)\mathrm{d}y = \int_y^{-1} \mathbb{R}_y v. \tag{14.4}$$

Proof By equating the coefficients of terms in same type on the two sides, the theorem is done. □

If $f(\underline{x}, \underline{y})$ is a function with two types of unknowns, and assuming $f(\underline{x}, \underline{y}) \in \mathscr{V}(\underline{x}, \underline{y})$, a bilinear space, it is easily seen that

$$\int_x^{-1} \int_y^{-1} f(\underline{x}, \underline{y}) = \int_y^{-1} \int_x^{-1} f(\underline{x}, \underline{y}). \tag{14.5}$$

Denote by $F(\underline{x}, \underline{y})$ the function in (14.5). Conversely, for $F(x, y) \in \mathscr{R}(x, y)$, we have

$$f(\underline{x}, \underline{y}) = \int_x \int_y F(x, y) \tag{14.6}$$

because of interchangeability between $\int_x$ and $\int_y$.

Let $f(z) \in \mathscr{R}\{z\}$. The following two operators on f as

$$\delta_{x,y} f = \frac{f(x) - f(y)}{x - y}, \tag{14.7}$$

and

$$\partial_{x,y} f = \frac{yf(x) - xf(y)}{x - y}. \tag{14.8}$$

are, respectively, called the (x, y)-*difference* and $\langle x, y\rangle$-*difference* of f with respect to z.

Lemma 14.1 For any function $f(z) \in \mathscr{R}\{x\}$, let $f = f(z)$, then

$$\partial_{x,y}(zf) = xy\delta_{x,y} f. \tag{14.9}$$

Proof Because of the linearity of the two operators $\partial_{x,y}$ and $\delta_{x,y}$, this enables us to discuss only $f(z) = z^n$, $n > 0$. Then, it is seen that

$$\begin{aligned}
\partial_{x,y} zf &= \partial_{x,y} z^{n+1} \\
&= \frac{yx^{n+1} - xy^{n+1}}{x - y} \\
&= xy\frac{x^n - y^n}{x - y} \\
&= xy\delta_{x,y} z^n \\
&= xy\delta_{x,y} f.
\end{aligned}$$

This is what we want to prove. □

Theorem 14.2 For any $f \in \mathscr{R}\{z\}$, we have

$$x^2y^2\delta^2_{x^2,y^2}(zf) - \partial^2_{x^2,y^2}(zf) = x^2y^2\delta_{x^2,y^2}(zf^2). \tag{14.10}$$

Proof From (14.7) and (14.8), the left side of (14.10) is

$$\frac{x^2y^2\Big((x^2f(x^2) - y^2f(y^2))^2 - x^2y^2(f(x^2) - f(y^2))^2\Big)}{x^2 - y^2}$$
$$= \frac{x^2y^2\Big(x^2f^2(x^2) - y^2f^2(y^2)\Big)}{x^2 - y^2}.$$

From (14.7), this is the right side of (1.6.10). □

For a set of maps $\mathscr{A}$, let

$$f_{\mathscr{A}}(x, \underline{y}) = \sum_{A\in\mathscr{A}} x^{m(A)}\underline{y}^{\underline{n}(A)}, \tag{14.11}$$

where $m(A)$ and $\underline{n}(A)$ are, respectively, the invariant parameter and vector on $\mathscr{A}$. Let $F_{\mathscr{A}}(x, y)$ be such a function of two unknowns such that

$$f_{\mathscr{A}}(x, \underline{y}) = \int_y F_{\mathscr{A}}(x, y). \tag{14.12}$$

The powers of x and y in $F_{\mathscr{A}}(x, y)$ are, respectively, called the *first parameter* and the *second parameter*.

Theorem 14.3 Let $\mathscr{S}$ and $\mathscr{T}$ be two sets of maps. If there is a mapping $\lambda(T) = \{S_1, S_2, \ldots, S_{m(T)+1}\}$ such that S_i and $\{i, m(T) + 2 - i\}$ are with a one-to-one correspondence from $\mathscr{T}$ to $\mathscr{S}$ for any $T \in \mathscr{T}$, where i and $m(T) + 2 - i$ are the contributions to, respectively, the first and the second parameters, $i = 1, 2, \ldots, m(T) + 1$, with the condition as

$$\mathscr{S} = \sum_{T\in\mathscr{T}} \lambda(T),$$

then

$$F_{\mathscr{S}}(x, y) = xy\delta_{x,y}(zf_{\mathscr{T}}), \tag{14.13}$$

where $f_{\mathscr{T}} = f_{\mathscr{T}}(z) = f_{\mathscr{T}}(z, \underline{y})$.

Proof From the definition of λ, we have

$$\begin{aligned}F_{\mathscr{S}}(x,y) &= \sum_{T\in\mathscr{T}}\sum_{i=1}^{m(T)+1} x^i y^{m(T)-i+2}\underline{y}^{n(T)}\\ &= xy\sum_{T\in\mathscr{T}}\frac{x^{m(T)+1}-y^{m(T)+1}}{x-y}\underline{y}^{n(T)}\\ &= xy\delta_{x,y}(zf_{\mathscr{T}}).\end{aligned}$$

This is (14.13). □

Theorem 14.4 Let $\mathscr{S}$ and $\mathscr{T}$ be two sets of maps. If there exists a mapping $\lambda(T) = \{S_1, S_2, \ldots, S_{m(T)-1}\}$ such that S_i and $\{i, m(T)-i\}$ are in a one-to-one correspondence for $T \in \mathscr{T}$, where i and $m(T)+2-i$ are the contributions to, respectively, the first and the second parameters, $i = 1, 2, \ldots, m(T)-1$, with the condition

$$\mathscr{S} = \sum_{T\in\mathscr{T}} \lambda(T),$$

then

$$F_{\mathscr{S}}(x,y) = \partial_{x,y}(f_{\mathscr{T}}), \tag{14.14}$$

where $f_{\mathscr{T}} = f_{\mathscr{T}}(z) = f_{\mathscr{T}}(z, \underline{y})$.

Proof From the definition of λ, we have

$$\begin{aligned}F_{\mathscr{S}}(x,y) &= \sum_{T\in\mathscr{T}}\sum_{i=1}^{m(T)-1} x^i y^{m(T)-i}\underline{y}^{n(T)}\\ &= xy\sum_{T\in\mathscr{T}}\frac{yx^{m(T)}-xy^{m(T)}}{x-y}\underline{y}^{n(T)}\\ &= \partial_{x,y}(f_{\mathscr{T}}).\end{aligned}$$

This is (14.14). □

14.2 General maps on the sphere

A map is said to be *general* if both loops and multi-edges are allowed. Of course, the vertex map ϑ is also treated as degenerate. Let $\mathscr{M}_{\rm gep}$ be the set of all rooted general planar maps. For any $M \in \mathscr{M}_{\rm gep}$, let $a = e_r(M)$ be the root-edge. Then, $\mathscr{M}_{\rm gep}$ can be divided into three classes: $\mathscr{M}_{\rm gep_0}$, $\mathscr{M}_{\rm gep_1}$ and $\mathscr{M}_{\rm gep_2}$, i.e.,

$$\mathscr{M}_{\rm gep} = \mathscr{M}_{\rm gep_0} + \mathscr{M}_{\rm gep_1} + \mathscr{M}_{\rm gep_2} \tag{14.15}$$

such that $\mathcal{M}_{\text{gep}_0}$ consists of a single map ϑ,

$$\mathcal{M}_{\text{gep}_1} = \{M | \forall M \in \mathcal{M}_{\text{gep}}, a \text{ is a loop}\}.$$

Of course,

$$\mathcal{M}_{\text{gep}_2} = \{M | \forall M \in \mathcal{M}_{\text{gep}}, a \text{ is a link}\}$$

in its own right.

Lemma 14.2 Let $\mathcal{M}_{\langle\text{gep}\rangle_1} = \{M - a | \forall M \in \mathcal{M}_{\text{gep}_1}\}$. Then, we have

$$\mathcal{M}_{\langle\text{gep}\rangle_1} = \mathcal{M}_{\text{gep}} \bigodot \mathcal{M}_{\text{gep}} \tag{14.16}$$

where $\odot$ is the 1-production as defined in Section 2.1.

Proof For a map $M \in \mathcal{M}_{\langle\text{gep}\rangle_1}$, because there is a map $\tilde{M} \in \mathcal{M}_{\langle\text{gep}\rangle_1}$, such that $M = \tilde{M} - \tilde{a}$, $\tilde{a} = e_r(\tilde{M})$, the root-edge of $\tilde{M}$, by considering the root-edge $\tilde{a}$ as a loop we see that $M = M_1 \dotplus M_2$, provided $M_1 \cap M_2 = o$, the common root-vertex of M and $\tilde{M}$. Since M_1 and M_2 are allowed to be any maps in $\mathcal{M}_{\text{gep}}$, including the vertex map ϑ, this implies that $M \in \mathcal{M}_{\text{gep}} \odot \mathcal{M}_{\text{gep}}$.

Conversely, for any $M \in \mathcal{M}_{\text{gep}} \odot \mathcal{M}_{\text{gep}}$, since $M = M_1 \dotplus M_2$, $M_1, M_2 \in \mathcal{M}_{\text{gep}}$, we can always construct a map $\tilde{M}$ by adding a loop $\tilde{a}$ at the common vertex of M_1 and M_2 as the root-edge of $\tilde{M}$ such that M_1 and M_2 are in different domains of the loop. Of course, $\tilde{M}$ is a general map. Because the root-edge of $\tilde{M}$ is a loop added, $\tilde{M} \in \mathcal{M}_{\text{gep}_1}$. However, it is easily seen that $M = \tilde{M} - \tilde{a}$. Therefore, $M \in \mathcal{M}_{\langle\text{gep}\rangle_1}$.

In consequence, the lemma is proved. □

For $\mathcal{M}_{\text{gep}_2}$, because the root-edges are all links, we consider the set $\mathcal{M}_{(\text{gep})_2} = \{M \bullet a | \forall M \in \mathcal{M}_{\text{gep}_2}\}$, $a = e_r(M)$, to be the root-edge as usual. The smallest map in $\mathcal{M}_{\text{gep}_2}$ is the link map $L = (Kr, (r)(\alpha\beta r))$ and it is seen that $L \bullet a = \vartheta$. Thus, $\vartheta \in \mathcal{M}_{(\text{gep})_2}$. For any $M \in \mathcal{M}_{\text{gep}_2}$, because the root-edge of M is not a loop, we know that $M \bullet a \in \mathcal{M}_{\text{gep}}$. Conversely, for any $M \in \mathcal{M}_{\text{gep}}$ we may always construct a map $\tilde{M} \in \mathcal{M}_{\text{gep}_2}$ by splitting the root-vertex of $\tilde{M}$ into two vertices with a new edge $\tilde{a}$ as the root-edge connecting them. This implies that $\tilde{M} \bullet \tilde{a} = M \in \mathcal{M}_{(\text{gep})_2}$. Therefore, we have

$$\mathcal{M}_{(\text{gep})_2} = \mathcal{M}_{\text{gep}}. \tag{14.17}$$

Lemma 14.3 For $\mathcal{M}_{\text{gep}_2}$ we have

$$\mathcal{M}_{\text{gep}_2} = \sum_{M \in \mathcal{M}_{\text{gep}}} \left\{\nabla_i M | \, 0 \le i \le m(M)\right\}, \tag{14.18}$$

where $m(M)$ is the valency of the root-vertex of M and ∇_i is the operator defined in Section 7.1.

Proof For any $M \in \mathscr{M}_{\text{gep}_2}$, because the root-edge a is a link, we may assume $a = (o_1, o_2)$ such that

$$o_1 = (r, S) \text{ and } o_2 = (\alpha\beta r, T).$$

Let $\tilde{M}$ be the map obtained by contracting the root-edge a into a vertex $\tilde{o} = (T, S)$ as the root-vertex of $\tilde{M}$. It is easily checked that $\tilde{M} \in \mathscr{M}_{\text{gep}}$ from (14.17) and that

$$M = \nabla_{|S|}\tilde{M},\ 0 \le |S| \le m(\tilde{M}),$$

where $m(\tilde{M}) = |S| + |T|$ and $|Z|, Z = S$ or T stand for the cardinality of Z. This implies that M is a member of the set on the right side of (14.18).

Conversely, for any M in the set on the right, because there exists a map $\tilde{M} \in \mathscr{M}_{\text{gep}}$ and an integer $i, 0 \le i \le m(\tilde{M})$, such that $M = \nabla_i\tilde{M}$, we may soon find that $M \in \mathscr{M}_{\text{gep}_2}$ by considering that the root-edge of M is always a link and that $M \in \mathscr{M}_{\text{gep}}$ as well. Thus, $M \in \mathscr{M}_{\text{gep}_2}$.

Therefore, the lemma follows. □

From the two lemmas above, we can now determine the contributions of $\mathscr{M}_{\text{gep}_i}$, $i = 0, 1, 2$, to the *enufunction*

$$g_{\mathscr{M}_{\text{gep}}}(x, \underline{y}) = \sum_{M \in \mathscr{M}_{\text{gep}}} x^{m(M)} \underline{y}^{\underline{n}(M)}, \tag{14.19}$$

where $\underline{n}(M) = (n_1(M), n_2(M), \ldots, n_i(M), \cdots)$, $n_i(m)$ is the number of vertices of valency i in M and $m(M)$, the valency of the root-vertex of M.

First, since ϑ has neither non-rooted vertex nor edge, we see that

$$g_{\mathscr{M}_{\text{gep}_0}} = 1. \tag{14.20}$$

Then, by Lemma 14.3,

$$g_{\mathscr{M}_{\text{gep}_1}} = x^2 g^2, \tag{14.21}$$

where $g = g_{\mathscr{M}_{\text{gep}}}(x, \underline{y})$, defined by (14.19).

Further, from Lemma 14.3

$$g_{\mathscr{M}_{\text{gep}_2}} = \int_y \sum_{M \in \mathscr{M}_{\text{gep}}} \left(\sum_{i=1}^{m(M)+1} x^i y^{m(M)-i+2} \right) \underline{y}^{\underline{n}(M)}.$$

By Theorem 14.3,

$$g_{\mathscr{M}_{\mathrm{gep}_2}} = x \int_y \left(y\delta_{x,y}(zg) \right). \tag{14.22}$$

Theorem 14.5 The enufunction g defined by (14..5) satisfies the following functional equation:

$$g = 1 + x^2 g^2 + x \int_y \left(y\delta_{x,y}(zg) \right). \tag{14.23}$$

Proof According to (14.15), from (14.20–14.22) the theorem is easily obtained. □

14.3 Nonseparable maps on the sphere

Let $\mathscr{M}_{\mathrm{ns}}$ be the set of all rooted nonseparable planar maps with the convention that the loop map $L_1 = (Kr, (r, \alpha\beta r))$ is included but the link map $L = (Kr, (r)(\alpha\beta r))$ is not for convenience.

Then, $\mathscr{M}_{\mathrm{ns}}$ is divided into two parts $\mathscr{M}_{\mathrm{ns}_0}$ and $\mathscr{M}_{\mathrm{ns}_1}$, i.e.,

$$\mathscr{M}_{\mathrm{ns}} = \mathscr{M}_{\mathrm{ns}_0} + \mathscr{M}_{\mathrm{ns}_1} \tag{14.24}$$

such that $\mathscr{M}_{\mathrm{ns}_0}$ consists of only the loop map L_1.

Lemma 14.4 A map $M \in \mathscr{M}_{\mathrm{ns}}$, $M \neq L_1$ if, and only if, its dual $M^* \in \mathscr{M}_{\mathrm{ns}}$.

Proof By contradiction. Assume $M = (\mathscr{X}_{\alpha,\beta}, \mathscr{P}) \in \mathscr{M}_{\mathrm{ns}}$, $M \neq L_1$ and its dual $M^* = (\mathscr{X}_{\beta,\alpha}, \mathscr{P}\alpha\beta) \notin \mathscr{M}_{\mathrm{ns}}$. Let

$$v^* = (x, \mathscr{P}\alpha\beta x, \ldots, (\mathscr{P}\alpha\beta)^m x)$$

be a cut-vertex of M^*. Then, we have a face $f^* = (x, \mathscr{P}x, \ldots, \mathscr{P}^n x)$ on M^* such that there exists an integer $j, 1 \leq j \leq n$, on f^* satisfying $\mathscr{P}^j x = (\mathscr{P}\alpha\beta)^i x$ for some $i, 1 \leq i \leq m$, i.e., $v_x^* = v_{\mathscr{P}^j x}^* = v^*$. However, f^* is a vertex of M that has the face v^* having the symmetry and hence f^* is a cut-vertex of M. A contradiction to the assumption appears. The necessity is true.

Conversely, from the duality the sufficiency is true as well. □

For any $M \in \mathscr{M}_{\mathrm{ns}}$, let $m(M)$ be the valency of the root-vertex and $\underline{n}(M) = (n_1(M), n_2(M), \cdots)$, $n_i(M)$ be the number of nonrooted vertices of valency $i, i \geq 1$.

From the nonimputability, the root-edge $a = (v_1, v_{\beta r})$ of any map M in $\mathcal{M}_{\mathrm{ns}_1}$ is always a link. The map $M \bullet a$ obtained by contracting the root-edge a in M has the same number of faces as M does.

Lemma 14.5 For any $M \in \mathcal{M}_{\mathrm{ns}_1}$ there is an integer $k \geq 1$ with

$$M \bullet a = \sum_{i=1}^{k} \cdot M_i \tag{14.25}$$

such that all M_i are allowed to be any map in $\mathcal{M}_{\mathrm{ns}}$ and that M_i, $i = 1, 2, \ldots, k$, does not have the form (14.25) for $k > 1$.

Proof In fact, from what was mentioned in Section 6.2, we see that k is the root-index of M and that all M_i, $1 \leq i \leq k$, do not have the form (14.2) for $k > 1$. From the nonseparability of M, by considering that all vertices of M_i except for the root-vertex are the same as those of M for $i = 1, 2, \ldots, k$, since M_i does not have the form (14.2) for $k > 1$, the root-vertex is not a cut-vertex for $i = 1, 2, \ldots, k$. This implies that all M_i, $1 \leq i \leq k$, are allowed to be any map in $\mathcal{M}_{\mathrm{ns}}$ including the loop map. The lemma follows. □

Now let us write

$$\mathcal{M}_k = \left\{ \sum_{i=1}^{k} \cdot M_i \,\middle|\, \forall M_i \in \mathcal{M}_{\mathrm{ns}}, 1 \leq i \leq k \right\}, \tag{14.26}$$

and

$$\mathcal{M}_{(\mathrm{ns})_1} = \{M \bullet a | \forall M \in \mathcal{M}_{\mathrm{ns}_1}\}, \tag{14.27}$$

where $a = e_r$, the root-edge of M.

Lemma 14.6 For $\mathcal{M}_{\mathrm{ns}_1}$, we have

$$\begin{aligned} \mathcal{M}_{(\mathrm{ns})_1} &= \sum_{k \geq 1} \mathcal{M}_k; \\ \mathcal{M}_k &= \mathcal{M}_{\mathrm{ns}}^{\times k} \end{aligned} \tag{14.28}$$

where $\times$ is the inner $1v$-production.

Proof By the definition of inner $1v$-product, the last form of (14.28) is easily seen.

From Lemma 14.5 we can find that

$$\mathcal{M}_{(\mathrm{ns})_1} = \bigcup_{k \geq 1} \mathcal{M}_k.$$

Moreover, for any $i, j, i \neq j$, we always have

$$\mathscr{M}_i \bigcap \mathscr{M}_j = \emptyset.$$

Therefore, the first form of (14.28) is true. □

Based on the two lemmas above, we can evaluate the contributions of $\mathscr{M}_{\mathrm{ns}_0}$ and $\mathscr{M}_{\mathrm{ns}_1}$ to the enufunction $f_{\mathscr{M}_{\mathrm{ns}}}$ of $\mathscr{M}_{\mathrm{ns}}$ with vertex partition, i.e.,

$$f = f_{\mathscr{M}_{\mathrm{ns}}} = \sum_{M \in \mathscr{M}_{\mathrm{ns}}} x^{m(M)} \underline{y}^{\underline{n}(M)}, \tag{14.29}$$

where $m(M)$ is the valency of root-vertex and

$$\underline{n}(M) = (n_1(M), n_2(M), \cdots)$$

with $n_i(M)$ being the number of nonroot-vertices of valency $i, i \geq 1$.

Since $\mathscr{M}_{\mathrm{ns}_0}$ consists of only the loop map, which has the root-vertex of valency 2 without nonrooted vertex, we have

$$f_{\mathscr{M}_{\mathrm{ns}_0}} = x^2. \tag{14.30}$$

For $\mathscr{M}_{\mathrm{ns}_1}$, we have to evaluate the function

$$\tilde{f}(x, z) = \sum_{M \in \mathscr{M}_{\mathrm{ns}_1}} x^{m(M)} z^{s(M)} \underline{y}^{\underline{n}(M)} \tag{14.31}$$

where $s(M)$ is the valency of the nonrooted vertex $v_{\beta r}$ incident with the root-edge e_r of M.

By considering that for $M \in \mathscr{M}_{\mathrm{ns}_1}$, $m(M \bullet a) = (m(M) - 1) + (s(M) - 1)$, we find

$$\tilde{f}(x, z) = xz \sum_{\tilde{M} \in \mathscr{M}_{\mathrm{ns}_1}} x^{m(\tilde{M}) - s(\tilde{M})} z^{s(\tilde{M})} \underline{y}^{\underline{n}(\tilde{M})},$$

where $s(\tilde{M})$ is the contribution of the valency of the nonrooted end of the root-edge of M to the valency of the root-vertex of $\tilde{M} = M \bullet a, M \in \mathscr{M}_{\mathrm{ns}_1}$. Because $s(\tilde{M})$ is allowed to be any number between 1 and $m(\tilde{M}) - 1$, from Lemma 14.2 we have

$$\tilde{f}(x, z) = xz \sum_{k \geq 1} \left(\sum_{M \in \mathscr{M}_{\mathrm{ns}}} x^{m(M)} \sum_{i=1}^{m(M)-1} \left(\frac{z}{x}\right)^i \underline{y}^{\underline{n}(M)} \right)^k .$$

By Theorem 14.4,

$$\tilde{f}(x,z) = xz \sum_{k\geq 1} (\partial_{x,z} f)^k = \frac{xz\partial_{x,z} f}{1 - \partial_{x,z} f},$$

where $f = f(u) = f_{\mathcal{M}_{\text{ns}}}(u, \underline{y})$, and hence

$$f_{\mathcal{M}_{\text{ns}_1}} = \int_y \tilde{f}(x,y) = x \int_y \frac{y\partial_{x,y} f}{1 - \partial_{x,y} f}. \tag{14.32}$$

Theorem 14.6 The enufunction of $\mathcal{M}_{\text{ns}}$ defined by (14.29) satisfies the following functional equation:

$$f = x^2 + x \int_y \frac{y\partial_{x,y} f}{1 - \partial_{x,y} f}. \tag{14.33}$$

Proof Since $f_{\mathcal{M}_{\text{ns}}} = f_{\mathcal{M}_{\text{ns}_0}} + f_{\mathcal{M}_{\text{ns}_1}}$, from (14.30) and (14.32) the theorem is obtained. □

14.4 Maps without cut-edge on surfaces

In this section, only maps without cut-edge (or, edge 2-connected) are considered. Let $\mathcal{M}$ be the set of all (including both orientable and nonorientable) rooted maps without cut-edge. Classify $\mathcal{M}$ into three classes as

$$\mathcal{M} = \mathcal{M}_0 + \mathcal{M}_1 + \mathcal{M}_2, \tag{14.34}$$

where $\mathcal{M}_0$ consists of only the vertex map ϑ, $\mathcal{M}_1$ is of all with the root-edge self-loop and, of course, $\mathcal{M}_2$ is of all with the root-edge not self-loop.

Lemma 14.7 The contribution of the set $\mathcal{M}_0$ to $f = f_{\mathcal{M}}(x, \underline{y})$ is

$$f_0 = 1, \tag{14.35}$$

where $f_0 = f_{\mathcal{M}_0}(x, \underline{y})$.

Proof Because of ϑ neither cut-edge nor nonrooted vertex, $m(\vartheta) = 0$ and $\underline{n}(\vartheta) = \underline{0}$. Thus, the lemma is obtained. □

In order to determine the enufunction of $\mathcal{M}_1$, how to decompose $\mathcal{M}_1$ should first be considered.

Lemma 14.8 For $\mathscr{M}_1$, we have

$$\mathscr{M}_{\langle 1\rangle} = \mathscr{M}, \tag{14.36}$$

where $\mathscr{M}_{\langle 1\rangle} = \{M - a|\forall M \in \mathscr{M}_1\}$, $a = Kr(M)$.

Proof Because of $L_1 = (r, \gamma r) \in \mathscr{M}_1$, $\gamma = \alpha\beta$, we have $L_1 - a = \vartheta \in \mathscr{M}$.

For any $S \in \mathscr{M}_{\langle 1\rangle}$, since there exists $M \in \mathscr{M}$ such that $S = M - a$, by considering the root-edge of M not cut-edge, it is seen that $S \in \mathscr{M}$. Thus, $\mathscr{M}_{\langle 1\rangle} \subseteq \mathscr{M}$.

Conversely, for any $M = (\mathscr{X}, \mathscr{J}) \in \mathscr{M}$, a new edge $a' = Kr'$ is added to the root-vertex $(r)_{\mathscr{J}}$ for getting S_i, whose root-vertex is

$$(r'r, \ldots, \mathscr{J}^i r, \gamma r', \mathscr{J}^{i+1} r, \ldots, \mathscr{J}^{m(M)-1} r),$$

where $0 \le i \le m(M) - 1$. Because $S_i - a' = M$, we have $S_i \in \mathscr{M}_1$. Hence, $\mathscr{M} \subseteq \mathscr{M}_{\langle 1\rangle}$. □

From this lemma, it is seen that each map $M = (\mathscr{X}, \mathscr{J})$ in $\mathscr{M}$ not only produces $S_i \in \mathscr{M}_1, 0 \le i \le m(M) - 1$ but also $S_m \in \mathscr{M}_1$ non-isomorphic to them. Its root-vertex is $(r', \langle r\rangle_{\mathscr{J}}, \gamma r')$. For $M \in \mathscr{M}$, let

$$\mathscr{S}_M = \{S_i|0 \le i \le m(M)\}. \tag{14.37}$$

Lemma 14.9 The set $\mathscr{M}_1$ has a decomposition as

$$\mathscr{M}_1 = \sum_{M\in\mathscr{M}} \mathscr{S}_M, \tag{14.38}$$

where $\mathscr{S}_M$ is given from (14.37).

Proof First, for $M \in \mathscr{M}_1$, because of $M' = M - a \in \mathscr{M}_{\langle 1\rangle}$, Lemma 14.8 enables us to have $M' \in \mathscr{M}$. From (14.37), $M \in \mathscr{S}_{M'}$ is obtained. Thus, the left side of (7.4.5) is a subset of that on the right side.

Conversely, for a map M on the left side of (14.38), because of the root-edge a self-loop, we have $M \in \mathscr{M}_1$. Thus, the set on the right side of (14.38) is a subset of that on the left side. □

On the basis of this lemma, we have

Lemma 14.10 For $g_1 = g_{\mathscr{M}^1}(x.\underline{y}) = f_{\mathscr{M}^1}(x^2, \underline{y})$, we have

$$f_1 = x^2\left(f + \frac{\partial f}{\partial x}\right), \tag{14.39}$$

where $f = f_{\mathscr{M}}(x, \underline{y})$.

Proof From Lemma 14.9,

$$f_1 = \sum_{M \in \mathscr{M}} (m(M) + 1) x^{m(M)} \underline{y}^{\underline{n}}.$$

By Lemma 9.10, we get

$$f_1 = x^2 \Big(f + x \frac{\partial f}{\partial x} \Big).$$

This is the conclusion of the lemma. □

In what follows, $\mathscr{M}_2$ is considered.

Lemma 14.11 For $\mathscr{M}_2$, let $\mathscr{M}_{(2)} = \{M \bullet a | \forall M \in \mathscr{M}_2\}$, then

$$\mathscr{M}_{(2)} = \mathscr{M} - \vartheta, \tag{14.40}$$

where ϑ is the vertex map.

Proof For any $M \in \mathscr{M}_{(2)}$, there is a map $M' \in \mathscr{M}_2$ such that $M = M' \bullet a'$. Because of a' neither cut-edge nor self-loop, $M \in \mathscr{M}$. And since the link map $L_0 = (Kr, (r)(\gamma r)) \notin \mathscr{M}_2$, $\mathscr{M}_{(2)} \subseteq \mathscr{M} - \vartheta$.

Conversely, for any $M = (\mathscr{X}, \mathscr{J}) \in \mathscr{M} - \vartheta$, let U_{i+1} be obtained by splitting the root-vertex $(r)_{\mathscr{J}}$ of M with an additional edge $a' = Kr'$ whose two ends are $(r', r, \ldots, \mathscr{J}^i r)$ and

$$(\gamma r', \mathscr{J}^{i+1} r, \cdots, \mathscr{J}^{m(M)-1} r), \ 1 \le i \le m(M).$$

Because a' is not a cut-edge, $U_i \in \mathscr{M}_2$, $1 \le i \le m(M)$. And because of $M = U_i \bullet a'$, $M \in \mathscr{M}_{(2)}$. Thus, $\mathscr{M} \subseteq \vartheta \subseteq \mathscr{M}_{(2)}$. □

For any $M = (\mathscr{X}, \mathscr{J}) \in \mathscr{M} - \vartheta$, let

$$\mathscr{U}_M = \{U_i | 1 \le i \le m(M)\}, \tag{14.41}$$

where U_i is the same as in the proof of Lemma 14.11.

Lemma 14.12 The set $\mathscr{M}^2$ has the following decomposition:

$$\mathscr{M}_2 = \sum_{M \in \mathscr{M} - \vartheta} \mathscr{U}_M, \tag{14.42}$$

where $\mathscr{U}_M$ is given from (14.41).

Proof First, for any $M \in \mathscr{M}_2$, from Lemma 14.5, $M' = M \bullet a \in \mathscr{M} - \vartheta$ and further $M \in \mathscr{U}_{M'}$. This implies that

$$\mathscr{M}_2 = \bigcup_{M \in \mathscr{M} - \vartheta} \mathscr{U}_M.$$

Then, for any $M_1, M_2 \in \mathscr{M} - \vartheta$, because of M_1 not isomorphic to M_2,

$$\mathscr{U}_{M_1} \bigcap \mathscr{U}_{M_2} = \emptyset.$$

Thus, (14.42) is right. The lemma is obtained. □

This lemma enables us to determine the contribution of $\mathscr{M}_2$ to $f_{\mathscr{M}}(x, \underline{y})$.

Lemma 14.13 For $f_2 = f_{\mathscr{M}_2}(x, \underline{y})$, we have

$$f_2 = x \int_y y \partial_{x,y}(f - 1), \tag{14.43}$$

where $f = f(z) = f_{\mathscr{M}}(z, \underline{y})$.

Proof From Lemma 14.12

$$f_2 = \int_y \sum_{M \in \mathscr{M} - \vartheta} \left(\sum_{i=1}^{m(M)} x^{i+1} y^{m(M)+2-i} \right) \underline{y}^{\underline{n}(M)}.$$

By employing Theorem 14.4, (14.43) is obtained. □

On the basis of those having been done, the main result can be deduced in what follows.

Theorem 14.7 The functional equation about f

$$x^2 \frac{\partial f}{\partial x} = -1 + (1 - x^2) f - x \int_y y \partial_{x,y}(f - 1) \tag{14.44}$$

is well defined on the field $\mathscr{L}\{\mathbb{R}; x, \underline{y}\}$. And its solution is $f = f(x) = f_{\mathscr{M}}(x, \underline{y})$.

Proof The first statement can be proved in a usual way except for involving a certain complication.

The second statement is derived from (14.34) in companion with (14.35), (14.39) and (14.43). □

14.5 Eulerian maps on the sphere

A map is called *Eulerian* if all the valencies of its vertices are even (or say, all vertices are *even*). Let $\mathcal{U}$ be the set of all the rooted planar Eulerian maps with the convention that the vertex map ϑ is in $\mathcal{U}$ for convenience.

Further, $\mathcal{U}$ is divided into three classes: $\mathcal{U}_0$, $\mathcal{U}_1$ and $\mathcal{U}_2$, i.e.,

$$\mathcal{U} = \mathcal{U}_0 + \mathcal{U}_1 + \mathcal{U}_2 \tag{14.45}$$

such that $\mathcal{U}_0 = \{\vartheta\}$, or simply write $\{\vartheta\} = \vartheta$, and

$$\mathcal{U}_1 = \{U | \forall U \in \mathcal{U} \text{ with } a = e_r(U) \text{ being a loop}\}.$$

Lemma 14.14 Any Eulerian map (not necessarily planar) has no cut-edge.

Proof By contradiction. Assume that a Eulerian map M has a cut-edge $e = (u, v)$ such that $M = M_1 \cup e \cup M_2$, $M_1 \cap M_2 = \emptyset$, where M_1 and M_2 are submaps of M with the property that M_1 is incident to u and M_2, to v. From the Eulerianity of M, u and v are the unique odd vertex in M_1 and M_2, respectively. This contradicts to that both M_1 and M_2 are a submap of M because the number of odd vertices in a map is even. □

Lemma 14.15 Let $\mathcal{U}_{\langle 1\rangle} = \{U - a | \forall U \in \mathcal{U}_1\}$, where $a = e_r(U)$ is the root-edge. Then, we have

$$\mathcal{U}_{\langle 1\rangle} = \mathcal{U} \bigodot \mathcal{U}, \tag{14.46}$$

where $\odot$ is the 1v-production.

Proof Because for $U \in \mathcal{U}_1$, the root-edge a is a loop, we see that $U - a = U_1 \dot{+} U_2$, where U_1 and U_2 are in the inner and outer domain of a, respectively. Of course, it can be checked that both U_1 and U_2 are maps in $\mathcal{U}$. Thus, the set on the left side of (14.46) is a subset of that on the right.

On the other hand, for any $U = U_1 \dot{+} U_2$, $U_1, U_2 \in \mathcal{U}$, we may uniquely construct a map U' by adding a loop at the common vertex of U_1 and U_2. The root-edge of U' is chosen to be the loop such that U_1 and U_2 are, respectively, in its inner and outer domains. It is easily seen that U' is an Eulerian map and hence $U' \in \mathcal{U}_1$. However, $U = U' - a \in \mathcal{U}_{\langle 1\rangle}$. This implies that the set on the right side of (14.46) is a subset of that on the left. □

For any map $U \in \mathcal{U}_2$, we see that the root-edge a of U has to be a link. From Lemma 14.14, if $U \bullet a = U_1 \dot{+} U_2$ such that the root-vertex is the common vertex, then the valencies of the vertices in both U_1 and U_2 are odd. Further, for any $U \in \mathcal{U}$, if $U = U_1 \dot{+} U_2$, then the valencies of the common vertex between U_1 and U_2 are both even as well.

Lemma 14.16 Let $\mathscr{U}_{(2)} = \{U \bullet a | \forall U \in \mathscr{U}_2\}$, where a is the root-edge of U. Then, we have

$$\mathscr{U}_{(2)} = \mathscr{U} - \vartheta, \tag{14.47}$$

where $\vartheta = \mathscr{U}_0$ for simplicity.

Proof From what has just been discussed, the set on the left side of (14.47) is a subset of that on the right.

Conversely, for any $U \in \mathscr{U} - \vartheta$, we may always construct a map U' by splitting the root-vertex into o_1 and o_2 with the new edge $a' = (o_1, o_2)$ as the root-edge of U' such that the valencies of o_1 and o_2 in U' are both even. However, we see that $U = U' \bullet a' \in \mathscr{U}_{(2)}$. This implies that the set on the right side of (14.47) is a subset of that on the left. □

For a map $U \in \mathscr{U} - \vartheta$, assume the valency of the root-vertex o is $2k$, $k \geq 1$,without loss of generality. The map U' is obtained by splitting the root-vertex into o_1 and o_2 with the new edge $a' = (o_1, o_2)$ such that the valency $\rho(o_1; U') = 2i$ and hence $\rho(o_2; U') = 2k - 2i + 2$ is denoted by $U_{[2i]}$, $i = 1, 2, \ldots, k$.

From Lemma 14.14, we see that the procedure works and that all the resultant maps $U_{[2i]}$ are also Eulerian maps.

Lemma 14.17 For $\mathscr{U}_2$, we have

$$|\mathscr{U}_2| = \sum_{U \in \mathscr{U} - \vartheta} \left| \left\{ U_{[2i]} \middle| i = 1, 2, \ldots, m(U) \right\} \right|, \tag{14.48}$$

where $2m(U)$ is the valency of the root-vertex of U.

Proof First, we show that for any $U \in \mathscr{U}_2$, it appears in the set on the right side of (14.48) only once. Assume that $a = (o, v)$ is the root-edge of U and that $\rho(o) = 2s$ and $\rho(v) = 2t$. Let U' be the map obtained by contracting the root-edge a, i.e., $U' = U \bullet a$. Then, there is the only possibility that $U = U'_{[2s]}$ in the set on the right side of (14.48).

Then, we show that for any map U in the set on the right side of (14.48), it appears also only once in $\mathscr{U}_2$. This is obvious from Lemma 14.16 because all elements are distinguished and they are all maps in $\mathscr{U}_2$ by considering the Eulerianity with the root-edges being links. □

In what follows, we see what kind of equation should be satisfied by the enufunction u of rooted planar Eulerian maps with vertex partition. Write

$$u = \sum_{U \in \mathscr{U}} x^{2m(U)} \underline{y}^{\underline{n}(U)}, \tag{14.49}$$

where $2m(U)$ is the valency of the root-vertex as mentioned above and $\underline{n}(U) = (n_2(U), \ldots, n_{2i}(U), \cdots)$, $n_{2i}(U)$ is the number of nonrooted vertices of valency $2i$, $i \geq 1$.

Theorem 14.8 The function u defined in (14.5) satisfies the following functional equation:

$$u = 1 + x^2u^2 + x^2 \int_y \left(y^2 \delta_{x^2,y^2}\left(u(\sqrt{z})\right)\right), \tag{14.50}$$

where $u(z) = u|_{x=z} = u(z, \underline{y})$.

Proof The contribution of $\mathscr{U}_0$ to u is

$$u_0 = 1 \tag{14.51}$$

since $2m(\vartheta) = 0$ and $\underline{n}(\vartheta) = \underline{0}$.

From Lemma 14.15, the contribution of $\mathscr{U}_1$ to u is

$$\begin{aligned} u_1 &= x^2 \sum_{U \in \mathscr{U}_{\langle 1 \rangle}} x^{2m(U)} \underline{y}^{\underline{n}(U)} \\ &= x^2u^2. \end{aligned} \tag{14.52}$$

The contribution of $\mathscr{U}_2$ to u is denoted by u_2. Let

$$\tilde{u}(z) = \sum_{U \in \mathscr{U}_2} x^{2m(U)} z^{2j(U)} \underline{y}^{\underline{\tilde{n}}(U)},$$

where $2j(U)$ is the valency of the nonrooted end of the root-edge and $\underline{\tilde{n}}(U) = (\tilde{n}_2(U), \tilde{n}_4(U), \ldots, \tilde{n}_{2i}(U), \cdots)$, $\tilde{n}_{2i}(U)$ is the number of vertices of valency $2i$ except for the two ends of the root-edge. It is easily seen that

$$\underline{\tilde{n}}(U) = \underline{n}(U) - e_{2j(U)},$$

where $e_{2j(U)}$ is the vector with all the components 0 except only for the $j(U)$th which is 1. In addition, it can be verified that

$$u_2 = \int_y \tilde{u}(y). \tag{14.53}$$

From Lemma 14.17, we have

$$\tilde{u}(z) = \sum_{U \in \mathscr{U} - \vartheta} \left(\sum_{i=1}^{m(U)} x^{2i} z^{2m(U)-2i+2} \right) \underline{y}^{\underline{n}(U)}.$$

By Theorem 14.3,

$$\tilde{u}(z) = x^2z^2\delta_{z^2,x^2}\left(u(\sqrt{t})\right).$$

Then by (14.53), we find that

$$u_2 = x^2 \int_y y^2 \delta_{y^2,x^2}\Big(u(\sqrt{t})\Big). \tag{14.54}$$

From (14.51), (14.52) and (14.54), the theorem is obtained. □

14.6 Eulerian maps on the surfaces

Let $\mathcal{M}_{\mathrm{Eul}}$ be the set of all orientable Euler rooted maps on surfaces. Because there is no cut-edge for any Eulerian map, Eulerian maps are classified into three classes as $\mathcal{M}^0_{\mathrm{Eul}}$, $\mathcal{M}^1_{\mathrm{Eul}}$ and $\mathcal{M}^2_{\mathrm{Eul}}$ such that $\mathcal{M}^0_{\mathrm{Eul}}$ consists of only the vertex map ϑ, $\mathcal{M}^1_{\mathrm{Eul}}$ has all its maps with the root-edge self-loop and

$$\mathcal{M}^2_{\mathrm{Eul}} = \mathcal{M}_{\mathrm{Eul}} - \mathcal{M}^0_{\mathrm{Eul}} - \mathcal{M}^1_{\mathrm{Eul}}. \tag{14.55}$$

Naturally, $\mathcal{M}^2_{\mathrm{Eul}}$ has all maps with the root-edge a link.

The enufunction $g = f_{\mathcal{M}_{\mathrm{Eul}}}(x, \underline{y})$ is of the powers $2m$ and $\underline{n} = (n_2, n_4, \cdots)$ of, respectively, x and $\underline{y}$ as the valency of root-vertex and the vertex partition operator.

Lemma 14.18 For $\mathcal{M}^0_{\mathrm{Eul}}$, we have

$$g_0 = 1, \tag{14.56}$$

where $g_0 = f_{\mathcal{M}^0_{\mathrm{Eul}}}(x^2, \underline{y})$.

Proof Because of ϑ with neither root-edge nor nonrooted vertex, $m(\vartheta) = 0$ and $\underline{n}(\vartheta) = \underline{0}$. The lemma is done. □

In order to determine the enufunction of $\mathcal{M}^1_{\mathrm{Eul}}$, a suitable decomposition of $\mathcal{M}^1_{\mathrm{Eul}}$ should be first considered.

Lemma 14.19 For $\mathcal{M}^1_{\mathrm{Eul}}$, we have

$$\mathcal{M}^{\langle 1\rangle}_{\mathrm{Eul}} = \mathcal{M}_{\mathrm{Eul}}, \tag{14.57}$$

where $\mathcal{M}^{\langle 1\rangle}_{\mathrm{Eul}} = \{M - a | \forall M \in \mathcal{M}^1_{\mathrm{Eul}}\}$, $a = Kr(M)$, the root-edge.

Proof Because of $L_1 = (r, \gamma r) \in \mathcal{M}^1_{\mathrm{Eul}}$, $\gamma\alpha\beta$, $L_1 - a = \vartheta \in \mathcal{M}_{\mathrm{Eul}}$ is seen.

For any $S \in \mathcal{M}^{\langle}_{\mathrm{Eul}}1\rangle$, since there is a map $M \in \mathcal{M}_{\mathrm{Eul}}$ such that $S = M - a$, from M as a Eulerian map, $S \in \mathcal{M}_{\mathrm{Eul}}$ is known. Thus, $\mathcal{M}^{\langle 1\rangle}_{\mathrm{Eul}} \subseteq \mathcal{M}_{\mathrm{Eul}}$.

Conversely, for any $M = (\mathscr{X}, \mathscr{J}) \in \mathscr{M}_{\text{Eul}}$, By adding a new root-edge $a' = Kr'$ at the vertex $(r)_{\mathscr{J}}$ to get S_i, whose root-vertex is $(r'r, \ldots, \mathscr{J}^{i}r, \gamma r', \mathscr{J}^{i+1}r, \cdots, \mathscr{J}^{2m(M)-1}r)$, $0 \le i \le 2m(M) - 1$. From $S_i - a' = M$, $S_i \in \mathscr{M}^{1}_{\text{Eul}}$. Thus, $\mathscr{M}_{\text{Eul}} \subseteq \mathscr{M}^{\langle 1\rangle}_{\text{Eul}}$. □

In the proof of this lemma, it is seen that each map $M = (\mathscr{X}, \mathscr{J})$ in $\mathscr{M}_{\text{Eul}}$ produces not only $S_i \in \mathscr{M}^{1}_{\text{Eul}}, 0 \le i \le 2m(M) - 1$, but also $S_{2m} \in \mathscr{M}^{1}_{\text{Eul}}$ non-isomorphic to them. Its root-vertex is $(r', \langle r\rangle_{\mathscr{J}}, \gamma r')$.

For $M \in \mathscr{M}_{\text{Eul}}$, let

$$\mathscr{S}_M = \{S_i | 0 \le i \le 2m(M)\}. \tag{14.58}$$

Lemma 14.20 Set $\mathscr{M}^{1}_{\text{Eul}}$ has the following decomposition:

$$\mathscr{M}^{1}_{\text{Eul}} = \sum_{M \in \mathscr{M}_{\text{Eul}}} \mathscr{S}_M, \tag{14.59}$$

where $\mathscr{S}_M$ is given from (14.58).

Proof First, for any $M \in \mathscr{M}^{1}_{\text{Eul}}$, from Lemma 14.19, $M' = M - a \in \mathscr{M}_{\text{Eul}}$, and hence $M \in \mathscr{S}_{M'}$. Thus,

$$\mathscr{M}^{1}_{\text{Eul}} = \bigcup_{M \in \mathscr{M}_{\text{Eul}}} \mathscr{S}_M.$$

Then, for any $M_1, M_2 \in \mathscr{M}_{\text{Eul}}$, because of the non-isomorphic between them,

$$\mathscr{S}_{M_1} \bigcap \mathscr{S}_{M_2} = \emptyset.$$

Therefore, the conclusion of the lemma is true. □

On the basis of this lemma, the following lemma can be seen.

Lemma 14.21 For $g_1 = g_{\mathscr{M}^{1}_{\text{Eul}}}(x.\underline{y}) = f_{\mathscr{M}^{1}_{\text{Eul}}}(x^2, \underline{y})$, we have

$$g_1 = x^2\left(g + 2x^2 \frac{\partial g}{\partial x^2}\right), \tag{14.60}$$

where $g = g_{\mathscr{M}_{\text{Eul}}}(x, \underline{y}) = f_{\mathscr{M}_{\text{Eul}}}(x^2, \underline{y})$.

Proof From (14.59),

$$g_1 = \sum_{M \in \mathscr{M}_{\text{Eul}}} (2m(M) + 1)x^{m(M)}\underline{y}^{\underline{n}}.$$

By employing Lemma 9.10, (14.60) is obtained. □

In what follows, $\mathscr{M}^{2}_{\text{Eul}}$ is investigated.

Lemma 14.22 For $\mathcal{M}^2_{\mathrm{Eul}}$, let $\mathcal{M}^{(2)}_{\mathrm{Eul}} = \{M \bullet a | \forall M \in \mathcal{M}^2_{\mathrm{Eul}}\}$, then

$$\mathcal{M}^{(2)}_{\mathrm{Eul}} = \mathcal{M}_{\mathrm{Eul}} - \vartheta, \tag{14.61}$$

where ϑ is the vertex map.

Proof Because of $L_1 \notin \mathcal{M}^2_{\mathrm{Eul}}$, $\vartheta \notin \mathcal{M}^{(2)}_{\mathrm{Eul}}$. Then, $\mathcal{M}^2_{\mathrm{Eul}} \subseteq \mathcal{M}_{\mathrm{Eul}} - \vartheta$.

Conversely, for any $M = (\mathcal{X}, \mathcal{P}) \in \mathcal{M}_{\mathrm{Eul}} - \vartheta$, since $M_{2j} = (\mathcal{X} + Kr_{2j}, \mathcal{P}_{2j}) \in \mathcal{M}^2_{\mathrm{Eul}}$, where the two ends of $a_{2j} = Kr_{2j}$ are obtained by splitting the root-vertex $(r)_{\mathcal{P}}$ of M, i.e.,

$$(r_{2j})_{\mathcal{P}_{2j}} = (r_{2j}, r, \mathcal{P}r, \ldots, (\mathcal{P})^{2j-2})$$

and

$$(\gamma r_{2j})_{\mathcal{P}_{2j}} = (\gamma r_{2j}, \mathcal{P})^{2j-1} r, \ldots, (\mathcal{P})^{2m-1}),$$

$1 \le i \le m-1$. Because $M = M_{2j} \bullet a_{2j}$, $M_{2j} \in \mathcal{M}^2_{\mathrm{Eul}}$. Thus, $\mathcal{M}_{\mathrm{Eul}} - \vartheta \subseteq \mathcal{M}^2_{\mathrm{Eul}}$. □

For any $M = (\mathcal{X}, \mathcal{J}) \in \mathcal{M}_{\mathrm{Eul}} - \vartheta$, let

$$\mathcal{M}_M = \{M_{2j} | 1 \le j \le m(M)\}, \tag{14.62}$$

where M_{2j}, $1 \le j \le m(M) - 1$, has appeared in the proof of Lemma 14.22.

Lemma 14.23 Set ${\mathcal{M}_{\mathrm{Eul}}}^2$ has the following decomposition:

$${\mathcal{M}_{\mathrm{Eul}}}^2 = \sum_{M \in \mathcal{M}_{\mathrm{Eul}} - \vartheta} \mathcal{M}_M, \tag{14.63}$$

where $\mathcal{M}_M$ is given from (14.62).

Proof First, for any $M \in \mathcal{M}^2_{\mathrm{Eul}}$, because $M' = M \bullet a \in \mathcal{M}_{\mathrm{Eul}} - \vartheta$, Lemma 14.22 tells us that $M \in \mathcal{M}_(M)$. Thus,

$${\mathcal{M}_{\mathrm{Eul}}}^2 = \bigcup_{M \in \mathcal{M}_{\mathrm{Eul}} - \vartheta} \mathcal{M}_M.$$

Then, for any $M_1, M_2 \in \mathcal{M}_{\mathrm{Eul}} - \vartheta$, because of the non-isomorphic between M_1 and M_2,

$$\mathcal{M}_{M_1} \bigcap \mathcal{M}_{M_2} \neq \emptyset.$$

This implies (14.63). □

On the basis of this lemma, the following conclusion can be drawn.

Lemma 14.24 For $g_2 = f_{\mathcal{M}^2_{\mathrm{Eul}}}(x, \underline{y})$, we have

$$g_2 = x^2 \int_y y^2 \delta_{x^2,y^2} g(\sqrt{z}), \tag{14.64}$$

where $g = g(x) = f_{\mathcal{M}_{\mathrm{Eul}}}(x^2, \underline{y})$.

Proof From Lemma 14.23,

$$g_2 = \sum_{M \in \mathcal{M}_{\mathrm{Eul}} - \vartheta} \int_y \left(\sum_{j=1}^{m(M)} x^{2j} y^{2m(M)+2-2j} \right) \underline{y}^{\underline{n}}.$$

By employing Theorem 14.3,

$$g_2 = x^2 \int_y y^2 \delta_{x^2,y^2} g(\sqrt{z}).$$

This is the lemma. □

Now, the main result of this section can be described.

Theorem 14.9 The functional equation about g

$$2x^4 \frac{\partial g}{\partial x^2} = -1 + (1 - x^2) g - x^2 \int_y y^2 \delta_{x^2,y^2} g(\sqrt{z}) \tag{14.65}$$

is well defined on the field $\mathcal{L}\{\mathbb{R}; x, \underline{y}\}$. Further, its solution is $g = g_{\mathcal{M}_{\mathrm{Eul}}}(x, \underline{y}) = f_{\mathcal{M}_{\mathrm{Eul}}}(x, \underline{y})$.

Proof The last conclusion is drawn from (14.55), in companion with (14.56), (14.60) and (14.64).

The former conclusion is a result of the well definedness for the equation system obtained by equating the coefficients on the two sides of (14.65). □

14.7 Notes

14.7.1 For given orientable genus $p \neq 0$, determine a functional equation satisfied by the vertex function of the set of all general rooted maps on the surface of genus p.

Notice 14.1 For $p = 0$, the functional equation has been established from Liu [113]. Its solution can be seen in [97, pp. 292–300]. No general result for $p \neq 0$ is obtained yet.

14.7.2 For given nonorientable genus $q \geq 1$, determine a functional equation satisfied by the vertex function of the set of rooted maps on the surface of genus q.

Notice 14.2 For general rooted maps of nonorientable genus $q \geq 1$, it seems no functional equation has been extracted yet.

14.7.3 Determine a functional equation satisfied by the vertex partition function of nonseparable rooted maps on the Klein bottle.

Notice 14.3 For nonseparable rooted maps, no such functional equation has been extracted on Klein bottle yet.

14.7.4 Determine a functional equation satisfied by the vertex partition function of nonseparable rooted maps on the torus.

Notice 14.4 For nonseparable rooted maps, no such functional equation has been extracted on torus yet.

14.7.5 Determine a functional equation satisfied by the vertex partition function of bipartite rooted maps on the torus.

Notice 14.5 For bipartite rooted maps, no such functional equation is found on torus yet.

14.7.6 Solve the functional equation about f as

$$(1 - x^2 f)f = 1 + x \int_y y\delta_{x,y}(zf_z). \tag{14.66}$$

Notice 14.6 The equation has been solved theoretically in the systematization stage from Liu [97, pp. 292–230]. The solution is available for efficientization and further intelligentization.

14.7.7 Solve the functional equation about f as

$$\Big(- x^2 f + \int_y f_y \Big) f = x^2 \int_y \Big(f_y + y^2 \delta_{x^2,y^2} f_{\sqrt{z}} \Big). \tag{14.67}$$

Notice 14.7 This is the eq. (9.9.1) in Liu [97, pp. 368–376]. Hence, the problem is solved at the systematization stage. The solution of the equation is also available for efficientization and further intelligentization.

14.7.8 Solve the functional equation about f as

$$\Big(\int_y (1 + xy) f_y - x^2 f \Big) f = \int_y \Big(f_y + xy\delta_{x,y}(zf_z) \Big). \tag{14.68}$$

Notice 14.8 This is eq. (9.5.1) in Liu [97, pp. 327–340]. Hence, the problem is solved at the systematization stage. The solution of the equation is also available for efficientization and further intelligentization.

14.7.9 Solve the functional equation about f as

$$f = \int_y \frac{1}{1 - \partial_{x,y}(z^2 f_z)}. \tag{14.69}$$

Notice 14.9 This is eq. (9.3.1) in Liu [97, pp. 303–313]. Hence, the problem is solved at the systematization stage. The solution of the equation is also available for efficientization and further intelligentization.

14.7.10 Solve the functional equation about f as

$$f = \int_y \frac{1 - \partial_{x^2,y^2}(zf_{\sqrt{z}})}{1 - 2\partial_{x^2,y^2}(zf_{\sqrt{z}}) - x^2y^2\delta_{x^2,y^2}(zf^2_{\sqrt{z}})}. \tag{14.70}$$

Notice 14.10 This is eq. (9.3.1) in Liu [97, pp. 358–368]. Hence, the problem is solved at the systematization stage. The solution of the equation is also available for efficientization and further intelligentization.

14.7.11 Solve the functional equation about f as

$$f = x^2 + x \int_y \frac{y\partial_{x,y} f_z}{1 - \partial_{x,y} f_z}. \tag{14.71}$$

Notice 14.11 This is eq. (9.4.1) in Liu [97, pp. 313–327]. Hence, the problem is solved at the systematization stage. The solution of the equation is also available for efficientization and further intelligentization.

14.7.12 Solve the functional equation about f as

$$f = x^2 + x^2 \int_y \frac{y^2 \delta_{x^2,y^2} f_{\sqrt{z}}}{(1 - \partial_{x^2,y^2} f_{\sqrt{z}})^2 - (xy\delta_{x^2,y^2} f_{\sqrt{z}})^2}. \tag{14.72}$$

Notice 14.12 This is eq. (9.7.1) in Liu [97, pp. 348–358]. Hence, the problem is done in the systematization stage. The solution of the equation is also available for efficientization and further intelligentization.

14.7.13 Solve the functional equation about f as

$$(1 - x^2 f)f = 1 + x^2 \int_y y^2 \delta_{x^2,y^2} f_{\sqrt{z}}. \tag{14.73}$$

Notice 14.13 This is eq. (9.6.1) in Liu [97, pp. 341–348]. Hence, the problem is solved at the systematization stage. The solution of the equation is also available for efficientization and further intelligentization.

15 Upper Maps of a Graph

- A semi-automorphism of a graph is a bijection from its semiedge set to itself generated by the binary group sticking on all edges similar to the partitions in correspondence.
- An automorphism of a graph is a bijection from the edge set to itself similar to the adjacency on edges in correspondence.
- The semi-automorphism group of a graph is different from its automorphism group if, and only if, a loop occurs.
- Non-isomorphic upper rooted and unrooted maps of a graph can be done from the embeddings of the graph via its automorphism group or semi-automorphism group of the graph.

15.1 Semi-automorphisms on a graph

A pregraph is considered as a partition on the set of all semiedges as shown in Chapter 1. Let $G = (\mathscr{X}, \delta; \pi)$ be a pregraph where $\mathscr{X}$, δ and π are, respectively, the set of all semiedges, the permutation determined by edges and the partition on $\mathscr{X}$.

Two pregraphs $G_1 = (\mathscr{X}_1, \delta_1; \pi_1)$ and $G_2 = (\mathscr{X}_2, \delta_2; \pi_2)$ are said to be *seme-isomorphic* if there is a bijection $\tau : \mathscr{X}_1 \longrightarrow \mathscr{X}_2$ such that

$$\begin{array}{ccc} \mathscr{X}_1 & \xrightarrow{\quad\tau\quad} & \mathscr{X}_2 \\ \gamma_1 \big\downarrow & & \big\downarrow \gamma_2 \\ \gamma_1(\mathscr{X}_1) & \xrightarrow{\quad\tau_{\gamma_1}\quad} & \gamma_2(\mathscr{X}_2) \end{array} \qquad (15.1)$$

are commutative for $\gamma = \delta$ and π, where τ_{γ_1} is induced from τ on $\gamma_1(\mathscr{X}_1)$. The bijection τ is called a *semi-automorphism* between G_1 and G_2.

Example 15.1 Given two pregraphs $G = (\mathscr{X}, \delta_1; \pi_1)$ where

$$\begin{cases} \mathscr{X} = \sum_{i=1}^{8}\{x_i(0), x_i(1)\}, \ \delta_1 = \prod_{i=1}^{8}(x_i(0), x_i(1)); \\ \pi_1 = \{X_i | \, 1 \le i \le 8\} \end{cases}$$

with

$$\begin{cases} X_1 = \{x_1(0), x_6(0), x_6(1)\}, \\ X_2 = \{x_1(1), x_2(0), x_3(0)\}, \\ X_3 = \{x_2(1), x_5(0), x_7(0)\}, \\ X_4 = \{x_4(0), x_5(1), x_7(1), x_8(0)\}, \\ X_5 = \{x_4(1), x_5(1), x_8(1)\} \end{cases}$$

DOI 10.1515/9783110481846-015

and $H = (\mathscr{Y}, \delta_2; \pi_2)$ where

$$\begin{cases} \mathscr{Y} = \sum_{i=1}^{8}\{y_i(0), y_i(1)\}, \ \delta_2 = \prod_{i=1}^{8}(y_i(0), y_i(1)); \\ \pi_2 = \{Y_i | \ 1 \le i \le 8\} \end{cases}$$

with

$$\begin{cases} Y_1 = \{y_4(0), y_6(0), y_7(0)\}, \\ Y_2 = \{y_5(0), y_6(1), y_7(1), y_8(0)\}, \\ Y_3 = \{y_3(0), y_5(1), y_8(1)\}, \\ Y_4 = \{y_2(0), y_3(1), y_4(1)\}, \\ Y_5 = \{y_1(0), y_1(1), y_2(1)\}. \end{cases}$$

Let $\tau : \mathscr{X} \longrightarrow \mathscr{Y}$ be a bijection with (15.1) commutative for $\gamma_i = \delta_i$, $i = 1$ and 2, as

$$\begin{pmatrix} x_1 & x_2 & x_3 & x_4 & x_5 & x_6 & x_7 & x_8 \\ \delta_2 y_2 & \delta_2 y_4 & y_3 & y_8 & y_6 & y_1 & y_7 & y_5 \end{pmatrix}.$$

Because of

$$\begin{cases} \tau X_1 = \{\tau x_1(0), \tau x_6(0), \tau x_6(1)\} \\ \qquad = \{y_2(1), y_1(0), y_1(1)\} = Y_5, \\ \tau X_2 = \{\tau x_1(1), \tau x_2(0), \tau x_3(0)\} \\ \qquad = \{y_2(0), y_4(1), y_3(1)\} = Y_4, \\ \tau X_3 = \{\tau x_2(1), \tau x_5(0), \tau x_7(0)\} \\ \qquad = \{y_4(0), y_6(0), y_7(0)\} = Y_1, \\ \tau X_4 = \tau\{x_4(0), x_5(1), x_7(1), x_8(0)\} \\ \qquad = \{y_6(1), y_7(1), y_5(0), y_8(0)\} = Y_2, \\ \tau X_5 = \{\tau x_3(1), \tau x_4(1), \tau x_8(1)\} \\ \qquad = \{y_3(0), y_8(1), y_5(1)\} = Y_3, \end{cases}$$

we have $\tau_{\pi_1}\pi_1 = \pi_2\tau$, i.e., (15.1) is commutative for $\gamma_i = \pi_i$, $i = 1$ and 2. Therefore, τ is a semi-isomorphism between G and H.

Lemma 15.1 If two pregraphs G and H are semi-isomorphic, then they have the same number of connected components provided omission of isolated vertex.

Proof By contradiction. Suppose G and H are semi-isomorphic with a semi-isomorphism $\tau : G \to H$ but $G = G_1 + G_2$ with two components: G_1 and G_2, and H a component itself. From the the commutativity of (15.1), H as well has two components. This contradicts the assumption that H is a component itself. □

If $G = H$, then a semi-isomorphism between G and H is called a *semi-automorphism* of G. Lemma 15.1 enables us to discuss semi-automorphism of only a graph instead of a pregraph without loss generality.

Lemma 15.1 allows us to consider only graphs instead of pregraphs for semi-automorphisms.

Moreover, for the sake of brevity, only graphs of order greater than 4 are considered as the general case in what follows.

Theorem 15.1 The set of all semi-automorphisms of a graph forms a group.

Proof Because of all semi-automorphisms as permutations acting on the set of semiedges, the commutativity leads to the closedness in the set of all semi-automorphisms under composition with the associate law. Moreover, it is easy to check that the identity permutation is a semi-automorphisms and the inverse of a semi-automorphism is still a semi-automorphism. This theorem holds. □

This group in Theorem 15.1 is called the *semi-automorphism group* of the graph.

Example 15.2 In Example 15.1, the pregraph $G = (\mathscr{X}, \delta_1; \pi_1)$ is a graph. It can be easily seen that

$$\begin{aligned}\tau_1 &= \begin{pmatrix} x_1 & x_2 & x_3 & x_4 & x_5 & x_6 & x_7 & x_8 \\ x_1 & x_2 & x_3 & x_4 & x_5 & \delta_1 x_6 & x_7 & x_8 \end{pmatrix} \\ &= (x_6(0), x_6(1))\end{aligned}$$

is a semi-automorphism on G. It is also seen that the semi-automorphism group of G is $\mathrm{Authf}(G) = \{\tau_i | 0 \le i \le 11\}$, where

$$\begin{cases} \tau_0 &= 1, \text{ the identity;} \\ \tau_1 &= (x_5, x_7); \\ \tau_2 &= (x_4, x_8); \\ \tau_3 &= (x_5, x_7)(x_4, x_8); \\ \tau_4 &= (x_2, x_3)(x_4, \delta_1 x_5)(x_7, \delta_1 x_8); \\ \tau_5 &= (x_2, x_3)(x_5, \delta_1 x_8)(x_7, \delta_1 x_4); \\ \tau_i &= (x_6(0), x_6(1))\tau_{i-6}, \ 6 \le i \le 11. \end{cases}$$

15.2 Automorphisms on a graph

Now, let us go back to the usual form of a graph $G = (V, E)$ where V and E are, respectively, the vertex and edge sets. In fact, if X_i as described in Section 15.1 is denoted by v_i, then $V = \{v_0 | i = 0.1.2. \cdots\}$ and $E = \{x_j | j = 0, 1, 2, \cdots\}$.

An *edge-isomorphism* of two pregraphs $G_i = (V_i, E_i)$, $i = 1$ and 2, is defined as a bijection $\tau : E_1 \longrightarrow E_2$ with diagram

$$\begin{array}{ccc} E_1 & \xrightarrow{\tau} & E_2 \\ \eta_1 \downarrow & & \downarrow \eta_2 \\ V_1 & \xrightarrow{\tau_{\eta_1}} & V_2 \end{array} \tag{15.2}$$

commutative, where η_i, $i = 1$ and 2, is seen as a mapping $2^{E_i} \longrightarrow V_i$.

When $G = G_1 = G_2$, an edge-isomorphism between G_1 and G_2 becomes an *edge-automorphism* on G.

Lemma 15.2 If two pregraphs G and H are edge-isomorphic, then they have the same number of components provided omission of isolated vertex.

Proof Similar to the proof of Lemma 15.1. □

This lemma enables us to discuss only graphs instead of pregraphs for edge-isomorphisms or edge-automorphisms.

Theorem 15.2 All edge-automorphisms of a graph G form a group, denoted by $\mathrm{Aut}_{\mathrm{ee}}(G)$.

Proof Similar to the proof of Theorem 15.1. □

Example 15.3 The graph G in Example 15.2 has its $\mathrm{Aut}_{\mathrm{ee}}(G) = \{\tau_i | 0 \le i \le 5\}$, where

$$\begin{cases} \tau_0 &= 1, \text{ the identity;} \\ \tau_1 &= (x_5, x_7); \\ \tau_2 &= (x_4, x_8); \\ \tau_3 &= (x_5, x_7)(x_4, x_8); \\ \tau_4 &= (x_2, x_3)(x_4, x_5)(x_7, x_8); \\ \tau_5 &= (x_2, x_3)(x_5, x_8)(x_7, x_4). \end{cases}$$

An *isomorphism*, or in the sense above *vertex-isomorphism*, between two pregraphs $G_i = (V_i, E_i)$, $i = 1$ and 2, is defined as a bijection $\tau : V_1 \longrightarrow V_2$, which satisfies that the diagram

$$\begin{array}{ccc} V_1 & \xrightarrow{\tau} & V_2 \\ \Big\downarrow{\scriptstyle \xi_1} & & \Big\downarrow{\scriptstyle \xi_2} \\ E_1 \subseteq V_1 \times V_1 & \xrightarrow{\tau_{\xi_1}} & E_2 \subseteq V_2 \times V_2 \end{array} \tag{15.3}$$

is commutative, where $\xi_i(v_i) = E_{iv_i}$ for $v_i \in V_i$, $i = 1$ and 2.

When $G = G_1 = G_2$, an isomorphism between G_1 and G_2 is called an *automorphism* of G.

Lemma 15.3 If two pregraphs G and H are isomorphic, then they have the same number of components.

Proof Similar to the proof of Lemma 15.2. □

This lemma enables us to discuss only graphs instead of pregraphs for isomorphisms or automorphisms.

Theorem 15.3 All automorphisms of a graph G form a group denoted by Aut(G).

Proof Similar to the proof of Theorem 15.2. □

The group mentioned in this theorem is called the *automorphism group* of G.

Example 15.4 The graph G in Example 15.2 has its $\mathrm{Aut}(G) = \{\tau_i | 0 \le i \le 1\}$, where $\tau_0 = 1$, the identity; $\tau_1 = (X_3, X_5)$.

Because there is no influence on the automorphism group of a graph when deleting loops, or replacing multiedge by a single edge, τ_i, $i = 1, 2$ and 3, in Example 15.3 is to the identity τ_0, and τ_4 and τ_5 in Example 15.3 to τ_1 here.

15.3 Relationships

Fundamental relationships among those groups mentioned in the previous section are then explained for the coming usages.

Theorem 15.4 $\mathrm{Aut}_{\mathrm{hf}}(G) \sim \mathrm{Aut}_{\mathrm{ee}}(G)$ if, and only if, G is loopless.

Proof Necessity. By contradiction. Suppose $\mathrm{Aut}_{\mathrm{hf}}(G) \sim \mathrm{Aut}_{\mathrm{ee}}(G)$ with an edge-automorphism τ but G has a loop denoted by $z = (z(0), z(1))$. Assume $\tau(l) = l$ without loss of generality. However, both the semi-automorphisms: τ_1 and τ_2 corresponding to τ are found as

$$\tau_1(x) = \begin{cases} \tau(x), & \text{when } x \neq z; \\ z(0), & \text{when } x = z(0); \\ z(1), & \text{when } x = z(1) \end{cases}$$

and

$$\tau_2(x) = \begin{cases} \tau(x), & \text{when } x \neq z; \\ z(0), & \text{when } x = z(1); \\ z(1), & \text{when } x = z(0). \end{cases}$$

This implies $\mathrm{Aut}_{\mathrm{hf}}(G) \nsim \mathrm{Aut}_{\mathrm{ee}}(G)$, a contradiction.

Sufficiency. Because there is no loop in G, the symmetry between two ends of a link leads to $\mathrm{Aut}_{\mathrm{hf}}(G) \sim \mathrm{Aut}_{\mathrm{ee}}(G)$. □

From the proof of Theorem 15.4, the following corollary can be done.

Corollary 15.1 Let l be the number of loops in G, then

$$\mathrm{Aut}_{\mathrm{hf}}(G) \sim S_2^l \times \mathrm{Aut}_{\mathrm{ee}}(G),$$

where S_2 is the symmetric group of order 2.

Proof Because of exact two semi-automorphisms deduced from an edge-automorphism and a loop, the conclusion is done. □

From this corollary, we can easily find

$$\mathrm{aut}_{\mathrm{hf}}(G) = 2^l \times \mathrm{aut}_{\mathrm{ee}}(G). \tag{15.4}$$

Because there is no contribution of a loop to the automorphism group of G, the graph G has its automorphism group $\mathrm{Aut}(G)$ always for that obtained by deleting all loops on G.

Theorem 15.5 $\mathrm{Aut}_{\mathrm{ee}}(G) \sim \mathrm{Aut}(G)$ if, and only if, G is simple.

Proof Because there is no contribution of either loops or multiedges to the automorphism group $\mathrm{Aut}(G)$, the theorem holds. □

In virtue of the proof of Theorem 12.5, a graph with multiedges G has its automorphism group $\mathrm{Aut}(G)$ always for its underlying simple graph, i.e., one obtained by substituting a link for each multiedge on G.

Lemma 15.4 Let G be a graph with i-edges of number m_i, $i \geq 2$. Then, its edge-automorphism group

$$\mathrm{Aut}_{\mathrm{ee}}(G) = \sum_{i\geq 2} m_i S_i \times \mathrm{Aut}(G),$$

where S_i is the symmetric group of order i, $i \geq 2$.

Proof In virtue of S_m as the edge-automorphism group of link bundle P_m of size m, $m \geq 2$, the lemma is done. □

On the basis of Lemma 15.4, we can obtain

Corollary 15.2 Let l and m_i be, respectively, the number of loops and i-edges, $l \geq 1$, $i \geq 2$ in G, then

$$\mathrm{aut}_{\mathrm{hf}}(G) = 2^l n_{\mathrm{me}} \mathrm{aut}(G) \tag{15.5}$$

where

$$n_{\mathrm{me}} \equiv n_{\mathrm{me}}(G) = \sum_{i\geq 1} i! m_i$$

which is called the *multiplicity* of G.

Proof By considering Corollary 15.1, the conclusion is done. □

15.4 Upper maps with symmetry

For map $M = (\mathscr{X}_{\alpha,\delta}, \mathscr{P})$, its automorphisms are discussed with asymmetrization in Chapter 8. Let $\mathscr{M}(G)$ be the set of all non-isomorphic maps with underlying graph G.

Lemma 15.5 For an automorphism ζ on map $M = (\mathscr{X}_{\alpha,\delta}(X), \mathscr{P})$, we have exhaustively

$$\zeta|_{\delta(X)} \in \mathrm{Aut}_{\mathrm{hf}}(G) \text{ and } \zeta\alpha|_{\delta(X)} \in \mathrm{Aut}_{\mathrm{hf}}(G),$$

where $G = G(M)$, the under graph of M, and $\delta(X) = X + \delta X$.

Proof Because $\mathscr{X}_{\alpha,\delta}(X) = (X+\delta X)+(\alpha X+\alpha\delta X) = \delta(X)+\alpha\delta(X)$, by Conjugate Axiom each $\zeta \in \mathrm{Aut}(M)$ has exhaustively two possibilities: $\zeta|_{\delta(X)} \in \mathrm{Aut}_{\mathrm{hf}}(G)$ and $\zeta\alpha|_{\delta(X)} \in \mathrm{Aut}_{\mathrm{hf}}(G)$. □

On the basis of Lemma 15.5, we can find

Theorem 15.6 Let $\mathscr{E}_g(G)$ be the set of all embeddings of a graph G on a surface of genus g (orientable or nonorientable), then the number of non-isomorphic maps in $\mathscr{E}_g(G)$ is

$$m_g(G) = \frac{1}{2 \times \mathrm{aut}_{\mathrm{hf}}(G)} \sum_{\tau \in \mathrm{Aut}_{\mathrm{hf}}(G)} \left|\Phi(\tau)\right| \tag{15.6}$$

where $\Phi(\tau) = \{M \in \mathscr{E}_g(G) | \tau(M) = M \text{ or } \tau\alpha(M) = M\}$.

Proof Suppose $X_1, X_2, \ldots, X_m$ are all the equivalent classes of $X = \mathscr{E}_g(G)$ under the group $\mathrm{Aut}_{\mathrm{hf}}(G) \times \langle\alpha\rangle$, then $m = m_g(G)$. Let

$$S(x) = \{\tau \in \mathrm{Aut}_{\mathrm{hf}}(G) \times \langle\alpha\rangle | \ \tau(x) = x\}$$

be the stabilizer at x, a subgroup of $\mathrm{Aut}_{\mathrm{hf}}(G) \times \langle\alpha\rangle$. Because

$$|\mathrm{Aut}_{\mathrm{hf}}(G) \times \langle\alpha\rangle| = |S(x_i)||X_i|,$$

$x_i \in X_i$, $i = 1, 2, \ldots, m$, we have

$$m|\mathrm{Aut}_{\mathrm{hf}}(G) \times \langle\alpha\rangle| = \sum_{i=1}^{m} |S(x_i)||X_i|. \tag{1}$$

By observing $|S(x_i)|$ independent of the choice of x_i in the class X_i, the right side of (1) is

$$\begin{aligned} \sum_{x \in X} |S(x)| &= \sum_{x \in X} \sum_{\tau \in S(x)} 1 \\ &= \sum_{\tau \in \mathrm{Aut}_{\mathrm{hf}}(G) \times \langle\alpha\rangle} \sum_{x = \tau(x)} 1 \\ &= \sum_{\tau \in \mathrm{Aut}_{\mathrm{hf}}(G) \times \langle\alpha\rangle} \left|\Phi(\tau)\right|. \end{aligned} \tag{2}$$

From (1) and (2), the theorem can be easily derived. □

The theorem above shows how to find non-isomorphic upper maps of a graph when the semi-automorphism group of the graph is known.

Theorem 15.7 Let G be a graph with l loops and m_i multi-edges of multiplier i and $\mathscr{E}_g(G)$, the set of all embeddings of G on a surface of genus g (orientable or nonorientable), then the number of non-isomorphic maps in $\mathscr{E}_g(G)$ is

$$m_g(G) = \frac{1}{2^{l+1} n_{\mathrm{me}} \mathrm{aut}(G)} \sum_{\tau \in \mathrm{Aut}_{\mathrm{hf}}(G)} \left|\Phi(\tau)\right| \tag{15.7}$$

where $\Phi(\tau) = \{M \in \mathscr{E}_g(G) | \tau(M) = M \text{ or } \tau\alpha(M) = M\}$ and n_{me} is the multiplicity of G.

Proof On the basis of Theorem 15.6, the conclusion is easily drawn from Corollary 15.2. □

Corollary 15.3 Let G be a simple graph. Then, the number of nonisomorphic maps in $\mathscr{E}_g(G)$ is

$$m_g(G) = \frac{1}{2\mathrm{aut}(G)} \sum_{\tau \in \mathrm{Aut}_{\mathrm{hf}}(G)} \left|\Phi(\tau)\right| \tag{15.8}$$

where $\Phi(\tau) = \{M \in \mathscr{E}_g(G) | \tau(M) = M \text{ or } \tau\alpha(M) = M\}$ and n_{me} is the multiplicity of G.

Proof A direct result of Theorem 15.7 by considering G with neither loop nor multiedge. □

15.5 Via asymmetrized upper maps

Another approach for determining non-isomorphic upper maps of a graph is via rooted ones whenever its distinct embeddings are known.

Theorem 15.8 For a graph G, let $\mathscr{R}_g(G)$ and $\mathscr{E}_g(G)$ be, respectively, the sets of all non-isomorphic rooted upper maps and all distinct embeddings of G with size $\epsilon(G)$ on a surface of genus g (orientable or nonorientable). Then,

$$|\mathscr{R}_g(G)| = \frac{2\epsilon(G)}{\mathrm{aut}_{\mathrm{hf}}(G)} |\mathscr{E}_g(G)|. \tag{15.9}$$

Proof Let $\mathscr{M}_g(G)$ be the set of all non-isomorphic upper maps of G. By (11.3), we have

$$|\mathscr{R}_g(G)| = \sum_{M \in \mathscr{M}_g(G)} \frac{4\epsilon(G)}{\mathrm{aut}(M)}$$

i.e.,

$$\frac{4\epsilon(G)}{2 \times \mathrm{aut}_{\mathrm{hf}}(G)} \sum_{M \in \mathscr{M}_g(G)} \frac{2 \times \mathrm{aut}_{\mathrm{hf}}(G)}{\mathrm{aut}(M)}.$$

By considering that $2 \times \mathrm{aut}_{\mathrm{hf}}(G) = |\mathrm{Aut}_{\mathrm{hf}}(G) \times \langle\alpha\rangle|$ is

$$\left|(\mathrm{Aut}_{\mathrm{hf}}(G) \times \langle\alpha\rangle)|_M\right| \times \left|\mathrm{Aut}_{\mathrm{hf}}(G) \times \langle\alpha\rangle(M)\right|$$

and $(\mathrm{Aut_{hf}}(G) \times \langle\alpha\rangle)|_M = \mathrm{Aut}(M)$ from Lemma 15.5, we have $|\mathscr{R}_g(G)|$ is

$$\frac{4\epsilon(G)}{|\mathrm{Aut_{hf}}(G) \times \langle\alpha\rangle|} \sum_{M\in\mathscr{M}_g(G)} \left|\mathrm{Aut_{hf}}(G) \times \langle\alpha\rangle(M)\right|$$
$$= \frac{2\epsilon(G)}{\mathrm{aut_{hf}}(G)}|\mathscr{E}_g(G)|.$$

This is (15.7). □

This theorem enables us to determine all the superrooted maps of a graph when the semi-automorphism group of the graph is known.

Theorem 15.9 For a graph G with $l, l \geq 1$, loops and m_i multiedges of multiplier $i, i \geq 2$, let $\mathscr{R}_g(G)$ and $\mathscr{E}_g(G)$ be, respectively, the sets of all non-isomorphic rooted upper maps and all distinct embeddings of G with size $\epsilon(G)$ on a surface of genus g (orientable or nonorientable). Then,

$$|\mathscr{R}_g(G)| = \frac{\epsilon(G)}{2^{l-1}n_{\mathrm{me}}\mathrm{aut}(G)}|\mathscr{E}_g(G)|, \tag{15.10}$$

where n_{me} is the multiplicity of G.

Proof A direct result of Theorem 15.8 from Corollary 15.2. □

Corollary 15.4 For a simple graph G, let $\mathscr{R}_g(G)$ and $\mathscr{E}_g(G)$ be, respectively, the sets of all non-isomorphic rooted upper maps and all distinct embeddings of G with size $\epsilon(G)$ on a surface of genus g (orientable or nonorientable). Then,

$$|\mathscr{R}_g(G)| = \frac{2\epsilon(G)}{\mathrm{aut}(G)}|\mathscr{E}_g(G)|. \tag{15.11}$$

Proof The case of $l = 0$ and $m_i = 0, i \geq 2$, of Theorem 15.9. □

Corollary 15.5 The number of rooted upper maps of a simple graph G with n_i vertices of valency $i, i \geq 1$, on orientable surfaces is

$$\frac{2\epsilon}{\mathrm{aut}(G)} \prod_{i\geq 2}((i-1)!)^{n_i} \tag{15.12}$$

where ϵ is the size of G.

Proof Because the number of distinct embeddings on orientable surfaces

$$\sum_{g\geq 0} \mathscr{E}_g(G) = \prod_{i\geq 2}((i-1)!)^{n_i} \tag{15.13}$$

is known, Corollary 15.4 leads to the conclusion. □

Corollary 15.6 The number of rooted upper maps of bouquet B_m, $m \geq 1$, on orientable surfaces is

$$\frac{(2m)!}{2^m m!}. \tag{15.14}$$

Proof Because the number of all distinct embeddings of B_m on orientable surfaces $(2m_1)!$ and the order of its semi-automorphism group $2^m m!$ are known, the conclusion is deduced from Theorem 15.8. □

In virtue of petal bundles, which are all upper maps of bouquets, (15.14) is in coincidence with (9.9).

The number of non-isomorphic upper maps of a graph can also be derived from rooted ones.

Theorem 15.10 For a given graph G, let $\mathscr{E}_k(G)$ be the set of all its nonequivalent embeddings with automorphism group order k. Then the number of all non-isomorphic unrooted supper maps of G is

$$n_{\mathrm{ur}}(G) = \frac{1}{2^{l(G)+1}\mathrm{aut}(G)}\Big(\sum_{\substack{i|4\epsilon\\ 1\leq i\leq 4\epsilon}} i|\mathscr{E}_i(G)|\Big) \tag{15.15}$$

where $\epsilon = \epsilon(G)$ is the size of G and $l(G)$ is the number of loops in G.

Proof On the basis of Theorem 15.8, we have

$$|\mathscr{R}_i(G)| = \frac{2\epsilon(G)}{\mathrm{aut}_{1/2}(G)}|\mathscr{E}_i(G)|$$

where $\mathscr{R}_i(G)$ is rooted upper maps of G with automorphism group order i. By Theorem 15.4 and Corollary 15.1,

$$|\mathscr{R}_i(G)| = \frac{\epsilon(G)}{2^{l(G)-1}\mathrm{aut}(G)}|\mathscr{E}_i(G)|.$$

Because the $4\epsilon(G)/i$ rooted maps produced by an unrooted map in $\mathscr{R}_i(G)$ are known in the proof of Theorem 11.1, we have

$$\begin{aligned}\frac{i}{4\epsilon(G)}|\mathscr{R}_i(G)| &= \frac{i}{4\epsilon(G)}\frac{\epsilon(G)}{2^{l(G)-1}\mathrm{aut}(G)}|\mathscr{E}_i(G)|\\ &= \frac{i}{2^{l(G)+1}\mathrm{aut}(G)}|\mathscr{E}_i(G)|.\end{aligned}$$

Overall, possibly $i|4\epsilon(G)$ is the conclusion of the theorem. □

Further, this theorem can be generalized for any types of graphs.

Theorem 15.11 For a set of graphs $\mathscr{G}$, the number of non-isomorphic unrooted upper maps of all graphs in $\mathscr{G}$ is

$$nur(\mathscr{G}) = \sum_{G\in\mathscr{G}} \frac{1}{2^{l(G)+1}\mathrm{aut}(G)} \sum_{\substack{i|4\epsilon(G)\\1\le i\le 4\epsilon(G)}} i|\mathscr{E}_i(G)|. \tag{15.16}$$

Proof From Theorem 15.10, overall $G \in \mathscr{G}$, the the conclusion is easily drawn. □

For a given genus g of an orientable or nonorientable surface, let $\mathscr{E}_k(G; g)$ be the set of all nonequivalent embeddings of a graph G on the surface with automorphism group order k.

Theorem 15.12 For a given genus g of an orientable or nonorientable surface, the number of all non-isomorphic unrooted supper maps of a graph G on the surface is

$$nur(G; g) = \frac{1}{2^{l(G)+1}\mathrm{aut}(G)}\Big(\sum_{\substack{i|4\epsilon\\1\le i\le 4\epsilon}} i|\mathscr{E}_i(G; g)|\Big) \tag{15.17}$$

where $\epsilon = \epsilon(G)$ is the size of G.

Proof By classification of maps and embeddings as well with genus, from Theorem 15.11 the theorem is done. □

Furthermore, this theorem can also be generalized for any types of graphs.

Theorem 15.13 For a given genus g of an orientable or nonorientable surface, the number of non-isomorphic unrooted upper maps of all graphs in a set of graphs $\mathscr{G}_{\mathrm{P}}$ with a given property P is

$$nur(\mathscr{G}; g) = \sum_{G\in\mathscr{G}_{\mathrm{P}}} \frac{1}{4\epsilon(G)} \sum_{\substack{i|4\epsilon(G)\\1\le i\le 4\epsilon(G)}} i|\mathscr{E}_i(G; g)|. \tag{15.18}$$

Proof A particular case of Theorem 15.11. □

15.6 Notes

15.6.1 For a given integer $n \geq 1$, determine the embedding distribution of outer planar graphs of order n by the order of semi-automorphism groups.

Notice 15.1 Referred to as in Liu [114], Dong and Yan [22].

15.6.2 For a given integer $n \geq 1$, determine the embedding distribution of Eulerian planar graphs of order n by the order of semi-automorphism groups.

Notice 15.2 Referred to as in Cai et al. [9], Cai and Liu [12], Liu [99–102], particularly, [97, pp. 409–418] etc.

15.6.3 For a given integer $n \geq 1$, determine the embedding distribution of general planar graphs of order n by the order of semi-automorphism groups.

Notice 15.3 Referred to as in Liu [88], or [97, pp. 418–430].

15.6.4 For a given integer $n \geq 1$, determine the embedding distribution of nonseparable planar graphs of order n by the order of semi-automorphism groups.

Notice 15.4 Referred to as in Liu [105, 108–110] and [97, pp. 313–326] etc.

15.6.5 For a given integer $n \geq 1$, determine the embedding distribution of cubic planar graphs of order n by the order of semi-automorphism groups.

Notice 15.5 Referred to as in Cai and Liu [10, 11, 67, pp. 117–141], etc.

15.6.6 For a given integer $n \geq 1$, determine the embedding distribution of 4-regular planar graphs of order n by the order of semi-automorphism groups.

Notice 15.6 Referred to as in Ren and Liu [132, 133, 135], Liu [78, pp. 97–119, 86, 87], Pan and Liu [131], and Zhang et al. [182], etc.

15.6.7 For a given integer $n \geq 1$, determine the embedding distribution of nonseparable graphs of order n by the order of semi-automorphism groups.

Notice 15.7 Referred to as in Liu [67, pp. 191–198], [76, pp. 173–180], [78, pp. 163–171], Li and Liu [53, 54], etc.

15.6.8 For a given integer $n \geq 1$, determine the embedding distribution of general graphs of order n by the order of semi-automorphism groups.

Notice 15.8 Referred to as in Liu [88, 97, pp. 418–430], etc.

15.6.9 For a given integer $n \geq 1$, determine the embedding distribution of Elerian graphs of order n by the order of their semi-automorphism groups.

Notice 15.9 Referred to as in Liu [85, 97, pp. 409–417], etc.

15.6.10 For a given integer $n \geq 1$, determine the embedding distribution of end-cut free graphs of order n by the order of semi-automorphism groups.

Notice 15.10 Referred to as in Liu [84, 97, pp. 379–388], etc.

15.6.11 For a given integer $n \geq 1$, determine the embedding distribution of cubic graphs of order n by the order of semi-automorphism groups.

Notice 15.11 Referred to as in Liu [78, pp. 78–94], Li and Liu [52, 54], etc.

15.6.12 For a given integer $n \geq 1$, determine the embedding distribution of 4-regular graphs of order n by the embedding order of semi-automorphism groups.

Notice 15.12 Referred to as in Liu [78, pp. 110–118, 86, 87, 97; pp. 171–197], etc.

15.6.13 For a given integer $n \geq 1$, determine the embedding distribution of 5-regular graphs of order n by the order of semi-automorphism groups.

Notice 15.13 No observation on 5-regular graphs are considered in this aspect yet.

16 Genera of a Graph

- A principle for reducing two adjacent edges with maximum genus decreased by exact 1 on a graph is established. An efficient theorem of characterizing the maximum genus of a graph is found on the basis of a spanning tree arbitrarily chosen.
- Recursion formulae are posed for determining the minimum genus of a graph not necessary to be with symmetry.
- A base for estimating bounds of the average genus of a graph is established via joint trees.
- Certain ways involving genus are proposed to access the thickness of a graph.
- Certain ways involving genus are discussed for accessing the interlacedness of a graph.

16.1 Recursion theorem

The theoretical idea of this section is to establish reductions initiated from Liu [61] in 1979. A *reducible operation* (or in short, a *reduction*) was defined as follows. Let G be a connected graph of valencies not less than 3, and x and y be two adjacent edges. Such a transformation, denoted by $\pi(v'; x, y)$, at vertex v' for x and y from G into G' of valencies not less than 3 is called a reducible operation whenever $G' = \pi(v'; x, y)G$ is connected as well, where v' is the common vertex v of x and y when its valency is greater than 3; or the vertex adjacent to the common vertex otherwise and

$$G' = \begin{cases} G - \{x, y\}, & \text{if } v' \text{ is } v; \\ G - \{v\}, & \text{otherwise.} \end{cases} \tag{16.1}$$

A pair of adjacent edges satisfying (16.1) in a graph is said to be *available*.

In fact, for a spanning tree T on a graph $G = (V, E)$, let a and b be an adjacent pair of cotree edges, or a adj b, the operation from G to G' such that $G' \sim_{\text{top}} G - \{a, b\}$ when their common vertex v is of valency greater than 3; or $G' = G - \{v\}$ otherwise is just the reduction.

Because of the connectedness of G', each time of reduction in deleting a vertex has to be with an edge deleted in a spanning tree of G. This enables us to consider reduction $\pi(v'; x, y)$ as deleting a pair of cotree edges denoted simply by $\pi_{x,y}$ for a spanning tree.

Let $n_G(\pi)$ be the maximum number of reductions (or in short, the *reduction number*) that can be done on G and write

$$\Pi_G = \{(x_i, y_i) | 1 \le i \le n_G(\pi), \pi(v_i'; x_i, y_i) \text{ available}\}. \tag{16.2}$$

DOI 10.1515/9783110481846-016

A graph is called a *cascade* if all circuits are independent, i.e., pairwise vertex-disjoint (or V-disjoint). Because there is no circuit on a tree, trees are all cascades as degenerate.

On the basis of the theory described in Liu [61], for a Eulerian circuit $C(v, e, e^{-1})$ with edges and vertices (or $C(e, e^{-1})$ as an edge set) obtained by the E-C procedure on double graph GG, a *maximum genus Eulerian circuit* is defined as a regular Eulerian circuit with γ_{Max} instead of β.

Theorem 16.1 Given a graph $G = (V, E)$ connected and not a cascade. Let $G' = \pi(v' : x, y)G$, then for any $(x, y) \in \Pi_G$,

$$\max_{S\in\mathscr{S}} g_S(G) = \max_{S\in\mathscr{S}'} g_S(G') + 1, \tag{16.3}$$

where $\mathscr{S}$ and $\mathscr{S}'$ are the sets of orientable surfaces on which G and G' can be embedded respectively and g_S is the genus of S.

Proof Because at most one genus is reduced by a reduction, we have

$$\gamma_{\text{Max}}(G') \leq \gamma_{\text{Max}}(G) - 1.$$

On the other hand, from the maximality of Π_G, a maximum genus Eulerian circuit $C(v, e, e^{-1})$ can be constructed and $C(e, e^{-1})$ is in the form $(Axx^{-1}yy^{-1}BCD)$, where C is the segment other than $x^{-1}y$ between the two reflective vertices. By the operation $\Delta_{x,y}$, for the sake of brevity, instead of $\Delta^p_{\xi,\eta,\xi',\eta'}$ in Liu [61] on $C(e, e^{-1})$, we have

$$\begin{aligned}\Delta_{x,y}C(e, e^{-1}) &= (AxCy^{-1}Bx^{-1}yD)\\ &\sim_{\text{top}} (ABCDxyx^{-1}y^{-1}).\end{aligned}$$

This shows that genus is reduced by at least 1 from a reduction and hence

$$\gamma_{\text{Max}}(G') \geq \gamma_{\text{Max}}(G) - 1.$$

Thus, the lemma is done. □

16.2 Maximum genus

A graph obtained from a number of reductions on G that cannot yet do the reduction is called an *irreducible* graph of G.

Lemma 16.1 All irreducible graphs of a given graph G are cascades.

Proof First, it is seen that for a reduction $\pi(v'; x, y)$ on G, in virtue of the connectedness of $G' = \pi(v'; x, y)G$, both x and y are not a cut-edge. This implies that each of x and y is on a circuit and the two circuits have a common vertex.

Then, by considering that a graph can always be reduced by a reduction whenever it has two circuits with a vertex in common, each irreducible graph of G is a cascade.

This is the lemma. □

Lemma 16.2 The maximum genus of a cascade is 0.

Proof Because any Eulerian circuit obtained from a cascade by the E-C procedure in Liu [61] cannot do the operation $\Delta_{x,y}$ for any pair of edges x and y, the cascade is with its maximum genus 0 on the basis of Theorem 16.1. □

Theorem 16.2 For any connected graph G, its orientable maximum genus

$$\gamma_{\mathrm{Max}}(G) = n_G(\pi),$$

the reduction number of G for $\pi(v'; x, y)$.

Proof From Theorem 16.1,

$$\gamma_{\mathrm{Max}}(G) = n_G(\pi) + \gamma_{\mathrm{Max}}(G_0),$$

where G_0 is an irreducible graph of G.

On account of Lemma 16.1, G_0 is a cascade. Then, from Lemma 16.2,

$$\gamma_{\mathrm{Max}}(G_0) = 0.$$

Therefore, the theorem is obtained. □

This theorem has an advantage in simplifying the procedure of determining the maximum genus particularly for a cubic graph because a reduction in this case is corresponding only to delete a vertex.

In Xuong [170], the Betti deficiency $\xi(G)$ is defined to be the minimum number of odd size components among all cotrees for determining the orientable maximum genus of a graph.

A spanning tree T of G, which has its cotree with ξ odd size components, is called *reasonable*. A graph whose edge set can be partitioned into a spanning tree and a matching is called a *snuff*.

Lemma 16.3 All irreducible graphs of G for a reasonable tree are snuffs.

Proof Because each even size component is pairwise reduced and each odd size component is pairwise reduced except exactly for one edge, the irreducible graph has its cotree edges pairwise vertex-disjoint, and hence a matching. Thus, the irreducible graph is a snuff. The proof is done. □

Attention 16.1 The irreducible graph is not independent of the choice of reductions but is independent of their order. Figure 16.1(a) shows a graph H with a tree represented by bold lines, Figure 16.1(b) shows the irreducible graph H_1 for the reductions $\Pi_1 = \{\pi_{x_1,z_1}, \pi_{x_2,y_2}, \pi_{x_3,y_3}\}$ and Figure 16.1(c) shows the irreducible graph H_2 for the reductions $\Pi_2 = \{\pi_{x_1,y_1}, \pi_{x_2,z_2}, \pi_{y_3,z_3}\}$. In H_1, there is no way to reduce the number of odd size components. However, H_2 shows that the number of odd size components can be reduced by two because the existence of two fundamental circuits with a vertex in common. Although both H_1 and H_2 are with three odd size components, only H_2 can transform to have one odd size component by employing the following method. For two fundamental circuits C_x and C_y with a vertex in common, where x and y are cotree edges, suppose x' and y' at a common vertex are on, respectively, C_x and C_y without loss of generality. By putting x' as a cotree edge and x as a tree edge if $x \neq x'$ while for y' and y as well, the resultant tree T' has its cotree with at least two odd size components less than T has. This is because the new cut-edge except for articulate edge occurs in H_1.

Lemma 16.4 Let T be a spanning tree of G such that the number of odd size components in its cotree is the Betti deficiency $\xi(G)$, then the reduction number $n'_G(\pi)$ for $\pi(x,y)$, where both x and y are in the cotree, is $n_G(\pi)$.

Proof First, because a reduction $\pi_{x,y}$ can always correspond to a reduction $\pi(v'; x, y)$, we have $n'_G(\pi) \le n_G(\pi)$.

Then, to show $n'_G(\pi) \nless n_G(\pi)$. By contradiction, assume $n'_G(\pi) < n_G(\pi)$. Let G'_0 be the graph obtained by doing $\pi_{x,y}$ for $n'_G(\pi)$ times on G. Because $n'_G(\pi) < n_G(\pi)$, we can construct $n_G(\pi)$ reductions $\pi_{x,y}$ for T' to find G'_1, which has at least two odd size components less than G'_0 has. A contradiction to the minimality of T.

Therefore, the lemma is true. □

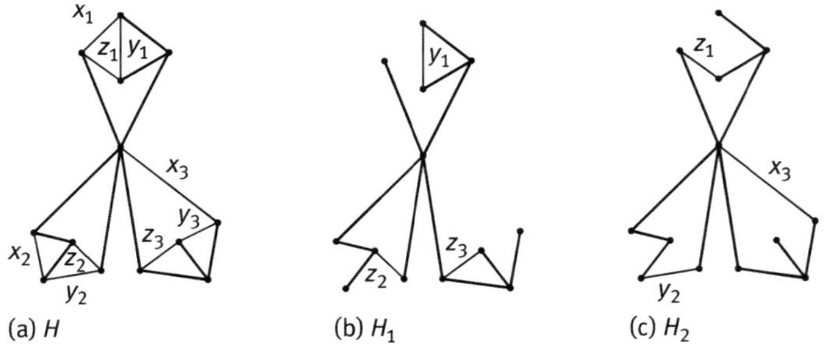

Figure 16.1: Irreducible graphs.

Theorem 16.3 For a connected graph G, its Betti deficiency

$$\xi(G) = \beta(G) - 2n_G(\pi). \tag{16.4}$$

Proof From Lemma 16.4, the Betti number, i.e., the number of cotree edges,

$$\beta(G) = 2n_G(\pi) + \xi(G).$$

This is the same quality as (16.3). □

Corollary 16.1 The orientable maximum genus of a connected graph G is

$$\gamma_{\mathrm{Max}}(G) = \frac{\beta(G) - \xi(G)}{2}. \tag{16.5}$$

Proof A direct result of Theorems 16.2 and 16.3. □

The nonorientable maximum genus of a graph was completely determined in Liu [60].

16.3 Minimum genus

Two ways are considered for determining the minimum genus of a general graph and a specific graph. This section concerns only orientables cases because nonorientable cases can be done similarly with additional choices of cotree edges.

Because of the difficulty, only a procedure of accessing the minimum genus of a general graph is first provided in decreasing the upper bounds of the minimum step by step until the minimum genus based on the theory described in Section 12.2.

Let $G = (V, E)$ be a graph and T be a spanning tree of G. For a rotation system σ on G, $\hat{T}_\sigma$ is the expanded tree of T. Let x_i, $1 \le i \le \beta$, be all the cotree edges where β is the Betti number of G.

Suppose the associate surface S, a polyhegon on $X = \{x_i, x_i^{-1} | 1 \le i \le \beta\}$, is divided into k layers: $0, 1, \ldots, k-1$.

Let Π_T^{+j} be the set of all exchangers on the layers at least j, $0 \le j \le k-1$, for the given associate surface S on T of G. The value

$$\gamma_T^{+j} = \min_{\pi \in \Pi_T^{+j}} g(\pi S)$$

is called a *minimal j-genus* of G.

Lemma 16.5 The sets of Π^{+j} for $1 \le j \le k-1$ have the following relation

$$\Pi_T^{+0} \supset \Pi_T^{+1} \supset \cdots \supset \Pi_T^{+(k-1)}. \tag{16.6}$$

Proof It is apparently seen from Section 9.3. □

Lemma 16.6 All the minimal j-genera of G satisfy the following inequality as

$$\gamma_T^{+0} \leq \gamma_T^{+1} \cdots \leq \gamma_T^{+(k-1)}. \tag{16.7}$$

Proof A direct result of Lemma 16.5. □

Lemma 16.7 For any associate surface S of G, γ_T^{+0} is a constant, i.e., independent of S.

Proof In virtue of Theorem 9.1.5 in Liu [77], Π^{+0} is independent of the choice of a spanning tree on G and hence the lemma. □

This lemma enables us to adopt γ^{+0} instead of γ_T^{+0}.

Theorem 16.4 The minimum genus of G is $\gamma_{\min}(G) = \gamma^{+0}$.

Proof A direct result of Lemmas 16.6 and 16.7. □

Another way is to determe the minimum genus of a graph G_n when G_n can be constructed from G_{n-1} by a local operation such as addition of a few of vertices, or edges for $n \geq 2$ and the minimum genus of G_1 is easily determined. Take only Q_n, the n-cube; $K_{m,n}$, the complete bipartite graph of order $m + n$; and K_n, the complete graph of order n as examples. Of course, the genera of Q_1, $K_{1,1}$ and K_1 are all known to be 0.

For the n-cube Q_n, $n \geq 3$, show that the minimum genus $\gamma_n = g_{\min}(Q_n)$ with γ_{n-1} satisfies the relation

$$g_{\min}(Q_n) = 2^{n-4}(n-3) + g_{\min}(Q_{n-1}) \tag{16.8}$$

from an associate surface of Q_{n-1} with genus γ_{n-1} to get an associate surface of Q_n with genus γ_n.

Theorem 16.5 The minimum genus of the n-cube Q_n determined by (16.8) is

$$\gamma_{\min}(Q_n) = 2^{n-3}(n-4) + 1$$

for $n \geq 3$.

Proof By induction on n. For $n = 3$, it is easy to see that $\gamma_{\min}(Q_3) = 0 = 2^0(-1) + 1$ because Q_3 is the cube, which is planar. Then, by the hypothesis,

$$\gamma_{\min}(Q_{n-1}) = 2^{n-4}(n-5) + 1$$

for $n > 3$. From (16.8), we have

$$\gamma_{\min}(Q_n) = 2^{n-4}(n-3) + 2^{n-4}(n-5) + 1 = 2^{n-3}(n-4) + 1.$$

This is the conclusion of the theorem. □

For the complete bipartite graph $K_{m,n}$, $m \geq n \geq 4$, show that the minimum genus $\gamma_{m,n} = g_{\min}(K_{m,n})$ with $\gamma_{m,n-1}$ satisfies the relation

$$g_{\min}(K_{m,n}) = \left\langle \frac{m-2}{4} \right\rangle + g_{\min}(K_{m,n-1}) - 1, \tag{16.9}$$

where

$$\left\langle \frac{m-2}{4} \right\rangle = \begin{cases} \left\lceil \frac{m-2}{4} \right\rceil, & m = 0(2 \nmid n), 1(2 \nmid n; 2|n, 2|\lfloor n/2 \rfloor), \\ & \quad 3(2 \nmid n, 2 \nmid \lfloor n/2 \rfloor; 2|n, 2|\lfloor n/2 \rfloor); \\ \frac{m-2}{4}, & m = 2(\mathrm{mod}\ 4); \\ \left\lfloor \frac{m-2}{4} \right\rfloor, & m = 0(2|n), 1(2|n, 2 \nmid \lfloor n/2 \rfloor), \\ & \quad 3(2 \nmid n, 2|\lfloor n/2 \rfloor; 2|n, 2 \nmid \lfloor n/2 \rfloor) \end{cases}$$

from an associate surface of $K_{m,n-1}$ with genus $\gamma_{m,n-1}$ to get an associate surface of $K_{m,n}$ with genus $\gamma_{m,n}$.

Theorem 16.6 The minimum genus of the complete bipartite graph $K_{m,n}$ determined by (16.9) is

$$\gamma_{\min}(K_{m,n}) = \left\lceil \frac{(m-2)(n-2)}{4} \right\rceil.$$

for $m \geq n \geq 2$.

Proof By induction on n for any $m \geq n$. When $n = 2$, it is easy to see that for any $m \geq n = 2$, $\gamma_{\min}(K_{m,2}) = 0$ because of $K_{m,2}$, a graph of m parallel paths with same ends, which is planar. Then, from (16.9), by the hypothesis

$$\gamma_{\min}(K_{m,n}) = \left\langle \frac{m-2}{4} \right\rangle + \left\lceil \frac{(m-2)(n-3)}{4} \right\rceil - 1$$

for $m \geq n > 2$. By checking each case, the theorem can be done. □

For the complete graph K_n, $n \geq 5$, show that the minimum genus $\gamma_n = g_{\min}(K_n)$ with γ_{n-1} satisfies the relation

$$g_{\min}(K_n) = \left\langle \frac{n-4}{6} \right\rangle + g_{\min}(K_{n-1}). \tag{16.10}$$

where

$$\left\langle \frac{n-4}{6} \right\rangle = \begin{cases} \left\lceil \frac{n-4}{6} \right\rceil, & n = 2, 1(2 \nmid \lfloor n/6 \rfloor), 3(2 \mid \lfloor n/6 \rfloor), \\ & 5(2 \nmid \lfloor n/6 \rfloor); \\ \frac{n-4}{6}, & n = 4 (\mathrm{mod}\ 6); \\ \left\lfloor \frac{n-4}{6} \right\rfloor, & n = 0, 1(2 \mid \lfloor n/6 \rfloor), 3(2 \nmid \lfloor n/6 \rfloor), \\ & 5(2 \mid \lfloor n/6 \rfloor) \end{cases}$$

from an associate surface of K_{n-1} with genus γ_{n-1} to get an associate surface of K_n with genus γ_n.

Theorem 16.7 The minimum genus of the complete graph K_n determined by (16.9) is

$$\gamma_{\min}(K_n) = \left\lceil \frac{(n-3)(n-4)}{12} \right\rceil$$

for $n \geq 4$.

Proof By induction on n for any $n \geq 4$. When $n = 4$, it is easily seen that for any $n = 4$, $\gamma_{\min}(K_4) = 0$ because of K_4, the tetrahedron, which is planar. Then, from (16.10), by the hypothesis

$$\gamma_{\min}(K_n) = \left\langle \frac{n-4}{6} \right\rangle + \left\lceil \frac{(n-4)(n-5)}{12} \right\rceil$$

for $n > 4$. By checking each case, the theorem can be obtained. □

16.4 Average genus

Let ξ_G be a variable that represents the genus of an embedding randomly chosen from all possible embeddings of a graph G. The problem considered now is to determine the *expectation* of ξ_G as

$$\mathbf{E}(\xi_G) = \sum_i i p_i(G), \tag{16.11}$$

where $p_i(G)$ is the probability of $\xi_G = i$, an integer (negative when nonorientable and positive in orientable).

Attention 16.2 Because of Section 12.4, nonorientable embeddings of a graph can be deduced from orientanle ones, only orientable embeddings are considered in what follows without loss of generality. The range of the summation in (16.11) is always restricted to be $i \geq 0$.

Assume that the genus polynomial of a graph G denoted by

$$g_G(x) = \sum_{i\geq 0} A_i(G)x^i \tag{16.12}$$

is known, where $A_i(G)$ is the number of distinct embeddings of G on a surface of genus i.

In virtue of (1.10),

$$\begin{aligned} g_G(1) &= \sum_{i\geq 0} A_i(G) = n_O(G) \\ &= \prod_{j\geq 1} \left((j-1)!\right)^{n_i} \end{aligned}$$

where n_j is the number of vertices of valency j, $j \geq 1$, in G. It is seen that the probability of $\xi_G = i$, $i \geq 0$, is

$$p_(G) = \frac{A_i(G)}{n_O(G)} \tag{16.13}$$

and hence,

$$E(\xi_G) = \frac{1}{g_G(1)} \frac{\mathrm{d}}{\mathrm{d}x} g_G(x)\bigg|_{x=1}, \tag{16.14}$$

which can then be called the *average genus* of G and denoted by $\gamma_{\mathrm{avg}}(G)$.

On the basis of (16.14), whenever the genus polynomial $g_G(x)$ is known as an explicit form, the average genus $\gamma_{\mathrm{avg}}(G)$ is determined only by routine calculation.

So, all graphs in Appendix II have their average genus known.

However, only few graphs are known to have their genus polynomial with an explicit form. Most attention to our work is concentrated on a better bound estimation of average genus of a graph, or a certain class of graphs.

Let G be a graph connected and simple. The former is in our convention and the latter is without loss of generality within subdivision on some edges.

For a vertex u of valency $s+1 (s \geq 3)$ and an integer k, $2 \leq k \leq s-1$, in G, a basic splitting at u such that the end called the *first end* of the new edge with the marked edge is of valency $k+2$ and the other end called the *second end* is apparently of valency $(s+1)-(k+1)+1 = s-k+1$, as described in Sections 3.3 and 3.4 is called a fit k-splitting. The graph obtained by such a fit k-splitting on G is called a *fit k-split*.

For a vertex u of valency $s+1$, assume its rotation as $\sigma(u) = (e_0, e_1, \cdots, e_s)$, where e_0 is rooted in an embedding of G. Then, there are exactly $k+1$ fit k-split embeddings with the rotation at the fist end of the new edge e as

$$(e, e_0, e_1, e_2, \ldots, e_k), (e_0, e, e_1, e_2, \ldots, e_k), \ldots, (e_0, e_1, e_2, \ldots, e, e_k).$$

Denote by G_0,G_1, …, G_k the $k+1$ under graphs of these $k+1$ fit k-split embeddings.

Theorem 16.8 For G with $G_0, G_1, \ldots, G_k$ described above, we have the genus polynomial of G

$$g_G(x) = \frac{1}{k+1}\sum_{i=0}^{s} g_{G_i}(x). \tag{16.15}$$

Proof On the basis of Section 2.4, from the basicness of fit k-splitting, the lemma is done. □

Because of $g_{G_i}(x) = (k+1)g_G(x)$ for all $0 \le i \le k$, we have

$$\gamma_{\text{avg}}(G) = \sum_{i=0}^{k} \gamma_{\text{avg}}(G_i). \tag{16.16}$$

This enables us to denote by $G_{\langle 1,2,\ldots,k\rangle}$ all the G_i, $0 \le i \le k$.

Further, for any $i_1, i_2, \cdots$ and $i_k \in \{1, 2, 3, \ldots, s\}$ instead of, respectively, $1, 2, \cdots$ and k in the above fit k-splitting at u with its rotation $(e_0, e_{i_1}, e_{i_2}, \ldots, e_{i_s})$, let $G_{\langle i_1, i_2, \ldots, i_k\rangle}$ be the corresponding fit k-split graph.

Theorem 16.9 Let $\mathscr{H}_k$ be the set of fit k-split graphs over all

$$\{i_1, i_2, \ldots, i_k\} \subseteq \{1, 2, \ldots, s\}$$

at a vertex u of valency $s+1$, then we have

$$\gamma_{\text{avg}}(G) = \frac{k!(s-k)!}{s!} \sum_{H \in \mathscr{H}_k} \gamma_{\text{avg}}(H). \tag{16.17}$$

Proof Because of choosing $\{i_1, i_2, \ldots, i_k\}$ from $\{1, 2, \ldots, s\}$ with

$$\binom{s}{k} = \frac{s!}{k!(s-k)!}$$

ways, from (16.16) the theorem is easily deduced. □

For estimating a bound of the average genus of a graph, lemmas not mentioned earlier have to be exploited first. Before the lemma is stated, let us see what happens on the genus when two letters x and y are added to the polyhegon $P = (aba^{-1}b^{-1})$ for getting another polyhegon. When x and y are interlaced, by Reduced rule 2 with z^{-1} instead of $\gamma z (z = x, y)$ in Section 5.3,

$$(xyax^{-1}by^{-1}a^{-1}b^{-1}) \sim_{\text{el}} (xyx^{-1}y^{-1}),$$

whose genus is not greater than the genus of P. When x and y are not interlaced, we have

$$(xyay^{-1}bx^{-1}a^{-1}b^{-1}) \sim_{\mathrm{el}} (xyx^{-1}y^{-1}aba^{-1}b^{-1}),$$

whose genus is greater than the genus of P.

Lemma 16.8 Let $P = (ABC)$ be a polyhegon with its genus $g(P) = k$, $k \geq 0$. Then, two polyhegons

$$P_1 = (xyAx^{-1}By^{-1}C) \text{ and } P_2 = (xyAy^{-1}Bx^{-1}C)$$

satisfy at least one of the following two inequalities:

$$g(P_1) \geq k + 1 \text{ and } g(P_2) \geq k + 1.$$

Proof By induction on the genus k. When $k = 0$, on account of Reduced rule 2 in Section 5.3, $g(P_1) = g(BACxyx^{-1}y^{-1}) \geq 1$ is known.

Assume the conclusion is true for $k - 1$, $k \geq 1$. Hence, show that it is also true for k. Because $g(P) = k$ and $k \geq 1$, Theorem 5.3 gives

$$(ABC) = A_0aB_0bC_0a^{-1}D_0b^{-1}E_0)$$

and hence

$$\begin{cases} (xyAx^{-1}By^{-1}C) = xyA_1aB_1bC_1a^{-1}D_1b^{-1}E_1; \\ (xyAy^{-1}Bx^{-1}C) = xyA_2aB_2bC_2a^{-1}D_2b^{-1}E_2. \end{cases}$$

By employing Reduced rule 2 in Section 5.3,

$$\begin{cases} P \sim_{\mathrm{el}} (A_0D_0C_0B_0E_0aba^{-1}b^{-1}); \\ P_1 \sim_{\mathrm{el}} (A_1D_1C_1B_1E_1aba^{-1}b^{-1}); \\ P_2 \sim_{\mathrm{el}} (A_2D_2C_2B_2E_2aba^{-1}b^{-1}). \end{cases} \tag{*}$$

Denote $P' = (A_0D_0C_0B_0E_0) = (A'B'C')$, $P_1' = (A_1D_1C_1B_1E_1)$ and $P_2' = (A_2D_2C_2B_2E_2)$. Then, we have

$$\{P_1', P_2'\} = \{(xyA'x^{-1}B'y^{-1}C'), (xyA'y^{-1}B'x^{-1}C')\}.$$

Because of $g(P') = k - 1$ from $g(P) = k$, by the hypothesis of induction,

$$g(P_1') \geq k \text{ or } g(P_2') \geq k.$$

From (*), we have

$$g(P_1) \geq k+1 \text{ or } g(P_2) \geq k+1.$$

This is just the lemma. □

Lemma 16.9 Let G be a graph (simple as above) with its maximum genus $\gamma_{\max}(G) \geq 1$. Then, there exists a pair of adjacent edges $\{\xi, \eta\}$ such that the graph $G_{\langle\xi,\eta\rangle}$ obtained by deleting the edges ξ and η from G is still connected and

$$\gamma_{\max}(G_{\langle\xi,\eta\rangle}) = \gamma_{\max}(G) - 1.$$

Proof A directed result of Theorem 16.1. □

Theorem 16.10 For any graph G of valency at most d, we have its average genus

$$\gamma_{\text{avg}}(G) \geq \frac{1}{d-1}\gamma_{\max}(G). \tag{16.18}$$

Proof By induction on $\gamma_{\max}(G)$. When $\gamma_{\max}(G) = 0$, because $\gamma_{\text{avg}}(G) = 0$, the theorem is true.

Assume that the theorem is true for any graph H of $\gamma_{\max}(G) = k-1$, $k \geq 1$. To show that for graph G, $\gamma_{\max}(G) = k$. On the basis of Lemma 18.9, $\gamma_{\max}(G_{\langle\xi,\eta\rangle}) = k-1$ from $\gamma_{\max}(G) = k$. Moreover, the vertex valencies are still at most d as those of G. By the inductive hypothesis,

$$\begin{aligned}\gamma_{\text{avg}}(G_{\langle\xi,\eta\rangle}) &\geq \frac{1}{d+1}\gamma_{\max}(G_{\langle\xi,\eta\rangle}) \\ &= \frac{1}{d-1}(\gamma_{\max}(G)-1) \\ &= \frac{\gamma_{\max}(G)}{d-1} - \frac{1}{d-1}.\end{aligned} \tag{1}$$

In virtue of joint tree principles shown in Section 12.1, because $G' = G_{\langle\xi,\eta\rangle}$, G and G' have a common spanning tree T. Then, relationships between joint trees $\hat{T}^{\delta}_{\sigma}$ and $\hat{T}^{\delta'}_{\sigma'}$, $\delta = \delta' = 0$ (orientable!), are observed.

Let $\mathscr{P}$ and $\mathscr{P}'$ be the sets of *associate polyhegons* (surfaces) in, respectively, G and G'. For each $P' \in \mathscr{P}'$, denote by $\mathscr{P}_{P'}$ the set of polyhegons in $\mathscr{P}$ induced from P', then by Theorem 1.11, we have

$$|\mathscr{P}_{P'}| = (d_0-1)(d_0-2)(d_1-1)(d_2-1), \tag{2}$$

where d_0, d_1 and d_2 are the valencies of vertices that are, respectively, the common end of ξ and η, the other end of ξ and the other end of η. Because $|\mathscr{P}_{P'}|$ is independent of the choice of P', we can write it as $\Delta(\mathscr{P}')$. Since $\{\mathscr{P}_{P'} | \forall P' \in \mathscr{P}'\}$ is a partition of $\mathscr{P}$, we have

$$|\mathscr{P}| = \Delta(\mathscr{P}')|\mathscr{P}'|. \tag{3}$$

Because of the adjacency between ξ and η, there are

$$\lambda = (d_0 - 2)(d_1 - 1)(d_2 - 1)$$

pairs $\{(\xi\eta A\xi^{-1}B\eta^{-1}C), (\xi\eta A\eta^{-1}B\xi^{-1}C)\}$ of polyhegons in $\mathscr{P}_{P'}$ for each $P' = (ABC) \in \mathscr{P}'$. From Lemma 16.8, one of $(\xi\eta A\xi^{-1}B\eta^{-1}C)$ and $(\xi\eta A\eta^{-1}B\xi^{-1}C)$ has its genus at least $g(ABC) + 1$. Then, the average increment of genus is at least

$$\frac{\lambda}{|\mathscr{P}_{P'}|} = \frac{1}{d_0 - 1} \tag{4}$$

and for polyhegons in $\mathscr{P}_{P'}$ genus is at least $g(P') + 1$, and none with genus less than $g(P')$. In consequence,

$$\sum_{P \in \mathscr{P}_{P'}} g(P) \geq |\mathscr{P}_{P'}| \left(g(P') + \frac{1}{d_0 - 1} \right). \tag{5}$$

Now, we have

$$\begin{aligned}
\gamma_{\mathrm{avg}}(G) &= \frac{\sum\limits_{P \in \mathscr{P}} g(P)}{|\mathscr{P}|}, \text{ by (3)}, \\
&= \frac{\sum\limits_{P' \in \mathscr{P}'} \sum\limits_{P \in \mathscr{P}_{P'} g(P)}}{|\mathscr{P}|} g(P), \text{ by (5)}, \\
&\geq \frac{\sum\limits_{P' \in \mathscr{P}'} |\mathscr{P}_{P'}| \left(g(P') + \frac{1}{d_0 - 1} \right)}{|\mathscr{P}|}, \text{ by (3)}, \\
&= \frac{\sum\limits_{P' \in \mathscr{P}'} \left(g(P') + \frac{1}{d_0 - 1} \right)}{|\mathscr{P}'|} \\
&= \gamma_{\mathrm{avg}}(G') + \frac{1}{d_0 - 1}, \text{ by } d_0 \leq d, \\
&\geq \gamma_{\mathrm{avg}}(G') + \frac{1}{d - 1}, \text{ by (1)}, \\
&\geq \frac{\gamma_{\mathrm{avg}}(G)}{d - 1}.
\end{aligned}$$

This is the theorem. □

Corollary 16.2 For a cubic graph G, we have

$$\gamma_{\text{avg}}(G) \geq \frac{\gamma_{\max}(G)}{2}. \tag{16.19}$$

Proof The case of $d = 3$ in Theorem 16.10. □

16.5 Thickness

In mid-twentieth century, most papers on layouts and routings of VLSI circuit design concerned with *planar decomposition* of a graph $G = (V, E)$, i.e., partition of the edge set as $E = E_1 + E_2 + \cdots, +E_k$ such that all the edges induced subgraphs $G_i = G[E_i]$, $1 \leq i \leq k$, are planar. The minimum of k among all possible planar decompositions is called the *thickness* of G, denoted by $t(G)$.

Because of the difficulty to determine $t(G)$ for a graph G, several approaches are used to try to access the problem. A common way is to fix a spanning tree as oriented. However, this section is working on a circuit oriented, particularly a Hamiltonian circuit oriented.

Let G be a graph and C be one of its circuits. For an embedding $\mu(G)$ of G in a space, let B_i, $1 \leq i \leq k$, be all connected components of $\mu(G) - \mu(C)$, then $[B_i]$ is called a *bridge* (mod C). This definition is from the lectures in Academia Sinica in Liu [89]. However, the origin of such a bridge is in combinatorial version with some sophistication in Tutte [152].

Given a bridge $B \in \mathscr{B}_C$, the set of all bridges (mod C) of G, C is divided into pieces by its boundary points. Each piece is called a *B-section* of C; otherwise, *alternative*. Two bridges B_1 and B_2 are said to be *parallel* if all boundary points of B_1 are in one of B_2-sections.

It is easily seen that the parallelism between bridges is an equivalence on $\mathscr{B}$. Let $\mathscr{B}_i$, $1 \leq i \leq s$, be all classes under this equivalence.

Theorem 16.11 A graph G is planar if, and only if, for each circuit C, $\mathscr{B}_C$ has at most two classes.

Proof Necessity. It is observed that every circuit C has all of its bridges in at most two classes from the Jordan closed curve axiom.

Sufficiency. Because of the given condition, a procedure can be established by following the fundamental circuits one by one to embed the bridges in the manner that one class is put in one domain of the circuit, and the other class, if any, in other domain. Finally, by the Finite recursion principle, an planar embedding is obtained. □

This theorem can be used to deduce a number of corollaries among which the following two are available for our case here.

Corollary 16.3 A Hamiltonian graph G is planar if, and only if, $\mathscr{B}_C$ has at most two classes where C is a given Hamiltonian circuit.

Proof Because each bridge is an edge, the fact that C has all its bridges in at most two classes leads to the fact that every circuit has all its bridges in at most two classes. From Theorem 16.11, the corollary follows. □

In fact, one class of the bridges is put in one domain of the Hamiltonian circuit, while the other, in the other domain, is in a planar embedding.

Corollary 16.4 A Hamiltonian graph G is outerplanar if, and only if, $\mathscr{B}_C$ has at most one class where C is a given Hamiltonian circuit.

Proof A particular case of Corollary 16.3. □

For a given subgraph H of a graph $G = (V, E)$, the *H-thickness* of G is the least number k such that $E = E_1 + E_2 + \cdots + E_k$ with $G[E_i]$ planar and $E_H \subseteq E_i$ for all $1 \le i \le k$.

Theorem 16.12 The C-thickness of a Hamiltonian graph with such a circuit C is

$$t_C(G) = \left\lceil \frac{s}{2} \right\rceil \tag{16.20}$$

where s is the number of classes in $\mathscr{B}_C$.

Proof On account of Corollary 16.3, the theorem is deduced. □

Let $\mu(G)$ be an embedding of graph G determined by a joint tree $\hat{T}_\delta$. Two cotree edges are said to be *interlaced* if the two occurrences of one are alternative with the two occurrences of the other. A partition of the set of all cotree edges as $E(\bar{T}) = E_1 + E_2 + \cdots + E_k$ with all E_i, $1 \le i \le k$, no pair of edges interlaced is called a *cotree decomposition*, or precisely cotree k-decomposition.

For an embedding $\mu(G)$ of graph G with a spanning tree T given on a surface S, the *embedding thickness* of G is

$$t_\mu(G) = \min\{k | \text{for all } k\text{-compositions}\}.$$

And, the *surface thickness* is $t_S(G) = \min t_\mu(G)$ among all possible surface embeddings.

Given an embedding $\mu(G)$ of graph G by a joint tree $\hat{T}_\delta$, let s_μ be the number of classes on the set of cotree edges on the basis of the equivalence that two cotree edges are equivalent if, and only if, without interlace.

Theorem 16.13 For an embedding $\mu(G)$ of graph G with a spanning tree T given on a surface S, the embedding thickness is determined by

$$t_\mu(G) = \left\lceil \frac{s_\mu}{2} \right\rceil. \tag{16.21}$$

Proof Let $A(G)$ be the *associate graph* produced from the polyhegon of a joint tree in such a way that the vertices are all occurrences of letters (or numbers) and edges, all pairs of occurrences of a letter on the polyhegon, then the embedding thickness is the C-thickness of the Hamiltonian graph $A(G)$. In virtue of Theorem 13.4.2, the theorem follows. □

Although both (16.20) and (16.21) provide an upper bound of $t_S(G)$, the exact value of $t_S(G)$ for a graph is still less known.

16.6 Interlacedness

Although the determination of the crossing number of a graph is difficult without general result obtained up to now except for some specific type of graphs (see in Liu and Liu [57], Hao and Liu [38], Lu et al. [117], this section concerns with another type of crossing numbers for a general graph for accessing its crossing number.

First, the *interlace number*, denoted by $\theta(P)$, of a polyhegon P is defined to be the cardinality of the set $\Theta(P) = \{(a,b)|\forall a, b \in \mathscr{A}, a \text{ Int } b\}$, where $\mathscr{A}$ is the number of all letters in P.

Theorem 16.14 For an orientable polyhegon P, its genus is

$$g(P) = \min_{\pi\in\Pi} \theta(\pi P), \tag{16.22}$$

where Π is the set of all possible operations determined by Reduced Rule 2 in Section 5.3 on P.

Proof On the basis of Section 5.3, the theorem can be easily done. □

Let $G = (V, E)$ be a graph and T a spanning tree of G. For a rotation system σ on G, $\hat{T}_\sigma$ is the expanded tree of T. Let x_i, $1 \le i \le \beta$, be all the cotree edges where β is the Betti number of G.

Suppose the associate surface S, a polyhegon on $X = \{x_i, x_i^{-1}|1 \le i \le \beta\}$, is divided into k layers: $0, 1, \ldots, k-1$.

Let Ω^{+j} be the set of all exchangers on the layers at least j, $0 \leq j \leq k-1$, for the given associate surface S of G. The value

$$\theta_T^{+j} = \min_{\omega \in \Omega^{+j}} \theta(\omega S)$$

is called a *minimal j-crossing number* of G.

Lemma 16.10 All the minimal j-crossing numbers of G satisfy the following inequality as

$$\theta_T^{+0} \leq \theta_T^{+1} \cdots \leq \theta_T^{+(k-1)}.$$

Proof A direct result of Lemma 16.5. □

For a spanning tree T, the crossing number over all T-immersions is called the crossing number of T-immersions. Because there is a one-to-one correspondence between the set of all T immersions and the set of all joint trees on T, it is seen that the crossing number of T-immersion $\mu_T(G)$ is just the interlace number of its corresponding polyhedron written as $\theta(\mu_T(G))$ as well.

Lemma 16.11 For a given spanning tree T of a graph G, the T-crossing number of G is

$$\mathrm{cr}_T(G) = \theta_T^{+0}.$$

Proof By considering the relationship between T-immersions and joint trees, the lemma is deduced from Lemma 16.10. □

The *cotree crossing number* of a graph is the minimum among the T-crossing numbers over all possible spanning tree T.

Lemma 16.12 θ_T^{+0} is independent of the choice of spanning tree T.

Proof Let $\mathscr{U}_T$ be the set of all T immersions of a graph. Because $\mathscr{U}_T$ is seen to be Π^{+0}. In virtue of Theorem 9.1.5 (from LiuP [77], Π^{+0} is independent of the choice of a spanning tree on G and hence the lemma. □

In virtue of this lemma, θ_T^{+0} can replace θ^{+0}.

Theorem 16.15 The cotree crossing number of a graph is θ^{+0}.

Proof A direct result of Lemmas 16.11–16.12. □

16.7 Notes

16.7.1 To look for a constant number of configurations forbidden to characterize the up-embeddability of a graph.

Notice 16.1 There is no constant number of forbidden minors for up-embeddability. Referred to as in Liu [61], Cai et al. [8], Chai and Liu [13], Dong et al. [25], Dong et al. [26], Lv and Liu [118–120].

16.7.2 On the basis of joint trees, to seek for a new characterization of the maximum genus of a graph.

Notice 16.2 Referred to as in Liu [91], Dong et al. [26], Lv and Liu [121, 122].

16.7.3 By employing the joint tree model, to show (16.8) for hypercubes Q_n, $n \geq 4$.

Notice 16.3 It is clear that symmetries of Q_n have to be considered on the basis of Liu [91]. Referred to as in Shao and Liu[141, 143, 145].

16.7.4 By employing the joint tree model, to show (16.9) for complete bipartite graphs $K_{m,n}$ of order $m + n$, $m, n \geq 4$.

Notice 16.4 It is clear that symmetries of $K_{m,n}$ have to be considered on the basis of Liu [91]. Referred to as in Shao and Liu[141, 143, 145].

16.7.5 By employing the joint tree model, to show (16.10) for complete graphs K_n of order n, $n \geq 6$.

Notice 16.5 It is clear that symmetries of K_n have to be considered on the basis of Liu [91]. Referred to as in Shao and Liu [141, 143, 145].

16.7.6 By employing the joint tree model, to determine the genus of complete *eq-tripartite* graphs $K_{n,n,n}$ of order $3n$ for $n \geq 4$.

Notice 16.6 It is clear that symmetries of $K_{n,n,n}$ have to be considered on the basis of Liu [91]. Referred to as in Shao and Liu[141, 143, 145].

16.7.7 By employing the joint tree model, to determine the genus of complete tripartite graphs $K_{m,n,s}$ of order $m + n + s$ for $m, n, s \geq 4$.

Notice 16.7 It is clear that symmetries of $K_{m,n,s}$ have to be considered on the basis of Liu [91]. Referred to as in Shao and Liu[141, 143, 145].

16.7.8 By employing the joint tree model, to determine the genus of complete s-partite graphs $K_{n_1,n_2,\ldots,n_s}$ of order $n_1 + n_2 + \cdots + n_s$ for $n_i \geq 3$, $1 \leq i \leq s$, $s \geq 4$.

Notice 16.8 It is clear that symmetries of $K_{n_1,n_2,\ldots,n_s}$ have to be considered on the basis of Liu [91]. Referred to as in Shao and Liu[141, 143, 145].

16.7.9 Determine the exact value of average genus of bouquets of size m for $m \geq 8$.

Notice 16.9 Referred to as in Chen and Liu [15].

16.7.10 Determine the exact value of average genus of wheels of order n for $m \geq 7$.

Notice 16.10 Referred to as in Chen and Liu [15].

16.7.11 Determine the thickness of the hypercube Q_n, $n \geq 4$.

Notice 16.11 No general result has been found yet.

16.7.12 Determine the crossing number of the hypercube Q_n, $n \geq 4$.

Notice 16.12 No general result is known yet. Referred to as in Hao and Liu [38], Liu and Liu [57], Lu et al. [117], etc.

17 Isogemial Graphs

- A graph is said to be a set of all its embeddings on surfaces for any integer as genus.
- Two operations are established on a graph to find conditions for getting the same genus distribution.
- An isogemial theorem is extracted for constructing an infinite family of isogemial graphs.
- Two particular types of non-isomorphic isogemial graphs are presented.

17.1 Basic concepts

The embedding genus distribution (or polynomial) of a graph used to be most possibly considered as complete combinatorial invariants of graphs as in GrossJL and Furst [33]. However, two types of graph sequences were shown in Wan and Liu [158] to have pairwise the same embedding genus distribution but not isomorphic. As a matter of fact, a graph can be represented by the set of its embeddings on all surfaces. Because of nonorientable embeddings of a graph determined by its orientable ones, we are allowed only to observe orientable embeddings in Liu [74].

Denote by $\mathscr{E}_g$ the set of all embeddings on a surface of genus g, an integer.

Theorem 17.1 Two graphs G and H are isomorphic if, and only if, for a non negative integer g there are $\mu \in \mathscr{E}_g(G)$ and $\nu \in \mathscr{E}_g(H)$ such that $\mu \simeq \nu$ (i.e., μ and ν are isomorphic).

For convenience, two graphs with the same embedding genus distribution are said to be *isogemial*.

In this chapter, a general principle is established to extract isogemial graphs that are not isomorphic.

17.2 Two operations

In order to extract the genus polynomial of a graph from a smaller one, some operations have to be introduced for getting a graph enlarged. This section discusses only two types of operations as typical example.

One is called *parataxis*, i.e., adding two parallel edges $a = (a_1, a_2)$ and $b = (b_1, b_2)$ by inserting two vertices a_i, and b_i to edge (u_i, v_i), $i = 1, 2$, assumed on a spanning tree without loss of generality from the principle of joint trees. The other is called *cascade*, i.e., adding a vertex v with tree edges (v, a_1), (v, b_1) and (v, u) by inserting a vertex u

DOI 10.1515/9783110481846-017

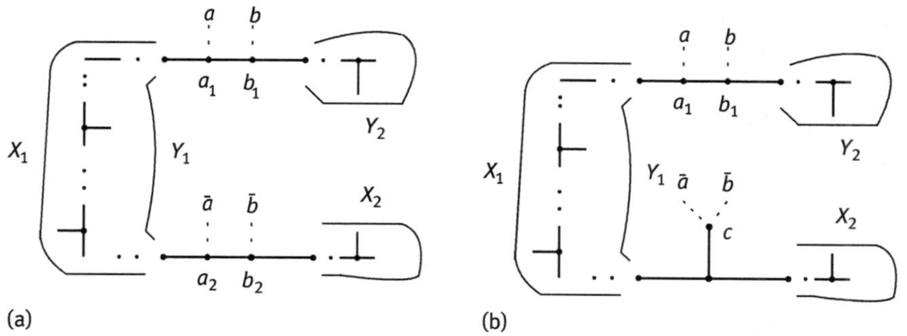

Figure 17.1: Two ways to get a graph enlarged. (a) Parataxis and (b) Cascade.

to (u_2, v_2) and two vertices a_1 and b_1 to (u_1, v_1), assumed on a spanning tree, as well, without loss of generality. They are shown in Figure 17.1, where $\bar{x} = x^{-1}$ for $x = a$ and b.

On the basis of Theorem 13.1.3 in Liu [74], it suffices to discuss the associate surfaces. For parataxis, it is seen that each associate surface, e.g., $(X_1Y_2Y_1X_2)$, of G produces 16 associate surfaces of $G^{\{a,b\}}$ enlarged from G by parataxis as follows (see Figure 17.1):

$$\begin{aligned}
A_0 &= (X_1abY_2Y_1a^{-1}b^{-1}X_2),\ A_1 = (X_1abY_2Y_1a^{-1}X_2b^{-1}),\\
A_2 &= (X_1abY_2Y_1b^{-1}X_2a^{-1}),\ A_3 = (X_1abY_2Y_1X_2b^{-1}a^{-1}),\\
A_4 &= (X_1aY_2bY_1a^{-1}b^{-1}X_2),\ A_5 = (X_1aY_2bY_1a^{-1}X_2b^{-1}),\\
A_6 &= (X_1aY_2bY_1b^{-1}X_2a^{-1}),\ A_7 = (X_1aY_2bY_1X_2b^{-1}a^{-1}),
\end{aligned}$$

$$\begin{aligned}
B_0 &= (X_1bY_2aY_1a^{-1}b^{-1}X_2),\ B_1 = (X_1bY_2aY_1a^{-1}X_2b^{-1}),\\
B_2 &= (X_1bY_2aY_1b^{-1}X_2a^{-1}),\ B_3 = (X_1bY_2aY_1X_2b^{-1}a^{-1}),\\
B_4 &= (X_1Y_2baY_1a^{-1}b^{-1}X_2),\ B_5 = (X_1Y_2baY_1a^{-1}X_2b^{-1}),\\
B_6 &= (X_1Y_2baY_1b^{-1}X_2a^{-1}),\ B_7 = (X_1Y_2baY_1X_2b^{-1}a^{-1}).
\end{aligned}$$

Then, it is seen that each associate surface, e.g., $(X_1Y_2Y_1X_2)$, of G produces 16 associate surfaces of $G^{\{ab\}}$ enlarged from G by cascade as follows (see Figure 17.1):

$$\begin{aligned}
C_0 &= (X_1abY_2Y_1a^{-1}b^{-1}X_2),\ C_1 = (X_1abY_2Y_1b^{-1}a^{-1}X_2),\\
C_2 &= (X_1abY_2Y_1X_2a^{-1}b^{-1}),\ C_3 = (X_1abY_2Y_1X_2b^{-1}a^{-1}),\\
C_4 &= (X_1aY_2bY_1a^{-1}b^{-1}X_2),\ C_5 = (X_1aY_2bY_1b^{-1}a^{-1}X_2),\\
C_6 &= (X_1aY_2bY_1X_2a^{-1}b^{-1}),\ C_7 = (X_1aY_2bY_1X_2b^{-1}a^{-1}),
\end{aligned}$$

$$\begin{aligned}
D_0 &= (X_1bY_2aY_1a^{-1}b^{-1}X_2),\ D_1 = (X_1bY_2aY_1b^{-1}a^{-1}X_2),\\
D_2 &= (X_1bY_2aY_1X_2a^{-1}b^{-1}),\ D_3 = (X_1bY_2aY_1X_2b^{-1}a^{-1}),\\
D_4 &= (X_1Y_2baY_1a^{-1}b^{-1}X_2),\ D_5 = (X_1Y_2baY_1b^{-1}a^{-1}X_2),\\
D_6 &= (X_1Y_2baY_1X_2a^{-1}b^{-1}),\ D_7 = (X_1Y_2baY_1X_2b^{-1}a^{-1}).
\end{aligned}$$

From the theory mentioned in Liu [65, 69], it is seen that two associate surfaces are in the same equivalent class if one can be transformed into the other by the following four principles:

Principle 0 $(Axx^{-1}B) \sim (AB)$ without $A = B = \emptyset$.

Principle 1 For A, B, C, D and E allowed to be the empty $\emptyset$,

$$\begin{aligned}(AxByCx^{-1}Dy^{-1}E) &\sim (ADCBExyx^{-1}y^{-1}) \\ &\sim ((ADCBE)(xyx^{-1}y^{-1})).\end{aligned}$$

Principle 2 For any A, B and C but $AB \neq \emptyset$,

$$(xABx^{-1}C) \sim (xBAx^{-1}C) \sim (x(AB)x^{-1}C).$$

Principle 3 For any $A \neq \emptyset$, $E \neq \emptyset$, $C \neq \emptyset$, B and $D \neq \emptyset$,

$$\begin{aligned}(ExAx^{-1}ByCy^{-1}D) &\sim (AxEx^{-1}ByCy^{-1}D) \\ &\sim (xEx^{-1}AByCy^{-1}D).\end{aligned}$$

17.3 Isogemial theorem

According to these principles, it is seen that

$$\begin{aligned}A_0 &\sim (X_1Y_2Y_1X_2aba^{-1}b^{-1}) \sim A_5, \text{ by Principle 1,} \\ A_1 &\sim (X_1X_2Y_2Y_1aba^{-1}b^{-1}) \sim A_4, \text{ by Principle 1} \\ &\qquad \text{and inverse of } (X_1Y_1Y_2X_2),\end{aligned}$$

$$\begin{aligned}B_2 &\sim (X_1X_2Y_1Y_2aba^{-1}b^{-1}) \sim B_7, \text{ by Principle 1,} \\ &\qquad \text{and inverse of } (X_1Y_2Y_1X_2), \\ B_3 &\sim (X_1Y_1X_2Y_2aba^{-1}b^{-1}) \sim B_6, \text{ by Principle 1} \\ &\qquad \text{and inverse of } (X_1Y_2X_2Y_1),\end{aligned}$$

$$\begin{aligned}A_2 &\sim (X_1abY_2Y_1b^{-1}X_2a^{-1}) \sim B_0, \text{ by Principle 3,} \\ &\qquad \text{inverse and interchange of } a \text{ and } b, \\ A_3 &\sim (X_1aY_2Y_1X_2a^{-1}) \sim B_4, \text{ if } X_1 \text{ and } Y_1 \text{ are} \\ &\qquad \text{interchangeable,}\end{aligned}$$

$$\begin{aligned}A_6 &\sim (X_1abY_2Y_1b^{-1}X_2a^{-1}) \sim B_1, \text{ by interchange} \\ &\qquad \text{of } a \text{ and } b, \\ A_7 &\sim (X_1aY_2bY_1b^{-1}X_2a^{-1}) \sim B_5, \text{ by Principle 3} \\ &\qquad \text{and interchange of } a \text{ and } b.\end{aligned}$$

Similarly, it is also seen that

$$
\begin{aligned}
C_0 &\sim (X_1Y_2Y_1X_2aba^{-1}b^{-1}) \sim C_2, \text{ by Principle 1,}\\
C_4 &\sim (X_1Y_1Y_2X_2aba^{-1}b^{-1}) \sim D_1, \text{ by Principle 1,}
\end{aligned}
$$

$$
\begin{aligned}
C_6 &\sim (X_1X_2Y_1Y_2aba^{-1}b^{-1}) \sim D_3, \text{ by Principle 1,}\\
D_5 &\sim (X_1Y_2Y_1X_2aba^{-1}b^{-1}) \sim D_7, \text{ by Principle 1,}
\end{aligned}
$$

$$
\begin{aligned}
C_1 &\sim (X_1abY_2Y_1b^{-1}X_2a^{-1}) \sim D_6, \text{ if } X_1 \text{ and } Y_1 \text{ are interchangeable,}\\
C_3 &\sim (X_1aY_2Y_1X_2a^{-1}) \sim D_4, \text{ if } X_1 \text{ and } Y_1 \text{ are interchangeable,}
\end{aligned}
$$

$$
\begin{aligned}
C_5 &\sim (X_1aY_2bY_1b^{-1}X_2a^{-1}) \sim D_0, \text{ by interchange of } a \text{ and } b,\\
C_7 &\sim (X_1aY_2bY_1X_2b^{-1}a^{-1}) \sim D_2, \text{ by interchange of } a \text{ and } b.
\end{aligned}
$$

For parataxis on a graph, let

$$
\begin{aligned}
&\mathscr{A}_1 = \{A_0, A_5\},\ \mathscr{A}_2 = \{A_1, A_4\},\ \mathscr{A}_3 = \{B_2, B_7\},\\
&\mathscr{A}_4 = \{B_3, B_6\},\ \mathscr{A}_5 = \{A_2, B_0\},\ \mathscr{A}_6 = \{A_3, B_4\},\\
&\mathscr{A}_7 = \{A_6, B_1\} \text{ and } \mathscr{A}_8 = \{A_7, B_5\}.
\end{aligned}
$$

On the other hand, for cascade on a graph let

$$
\begin{aligned}
&\mathscr{D}_1 = \{C_0, D_2\},\ \mathscr{D}_2 = \{C_4, D_1\},\ \mathscr{D}_3 = \{C_6, D_3\},\\
&\mathscr{D}_4 = \{D_5, D_7\},\ \mathscr{D}_5 = \{C_1, D_6\},\ \mathscr{D}_6 = \{C_3, D_4\},\\
&\mathscr{D}_7 = \{C_5, D_0\} \text{ and } \mathscr{D}_8 = \{C_7, D_2\}.
\end{aligned}
$$

Lemma 17.1 If X_1 and Y_1 are interchangeable, then there is a bijection between the sets of associate surfaces of $G^{\{a,b\}}$ and $G^{\{ab\}}$ such that corresponding surfaces are in the same equivalent class.

Proof From what is discussed above, we can see that $\mathscr{A}_i$ and $\mathscr{D}_i$ are in the same equivalent class for $i = 1, 2, 3, 4, 5, 6, 7$ and 8. This just presents a bijection between the sets of associate surfaces of $G^{\{a,b\}}$ and $G^{\{ab\}}$. □

Theorem 17.2 If X_1 and Y_1 are interchangeable, then the orientable embedding distributions of $G^{\{a,b\}}$ and $G^{\{ab\}}$ by genus are the same.

Proof From Lemma 17.1, the theorem is easily found. □

Because $G^{\{a,b\}}$ and $G^{\{ab\}}$ are non-isomorphic in general, this theorem enables us to construct many families of graphs non-isomorphic with same genus distribution for embeddings and even same handle polynomial for upper maps if their automorphism groups of same order from Theorem 1.

17.4 Non-isomorphic isogemial graphs

One thing that should be mentioned is that the condition about X_1 and Y_1 is reasonable. For example, let X_2 and Y_2 be connected only by a cotree edge x, then $X_2 = x$ and $Y_2 = x^{-1}$, or vice versa, seen as a degenerate case. It is easy to check that X_2 and Y_2 are interchangeable in this case, as shown in Figure 17.2 where bold and thin lines stand for tree and cotree edges, respectively.

Now, we have the following corollaries for two pairs of graph sequences: $X_G(n)$ and $Y_G(n)$; $X_n(G; x_i)$ and $Y_n(G; x_i)$ for $n \geq 1$ and $1 \leq i \leq m$. The former is for doing parataxis and cascade iteratively on cotree edge x as shown in Figure 17.2(b and c); the later is for distinct cotree edges $x_1, x_2, \ldots, x_m$.

Lemma 17.2 For doing parataxis and cascade iteratively on a cotree edge x, X_1 and Y_1 are interchangeable.

Proof As shown in Figure 17.1, we see that $X_2 = x$ and $Y_2 = x^{-1}$ (or *vise versa*, $X_2 = x^{-1}$ and $Y_2 = x$). For parataxis,

$$\begin{aligned}(X_1aY_2Y_1Y_2a^{-1}) &= (X_1ax^{-1}Y_1xa^{-1}) = (X_1yY_1y^{-1}) \\ &\sim (Y_1yX_1y^{-1})\end{aligned}$$

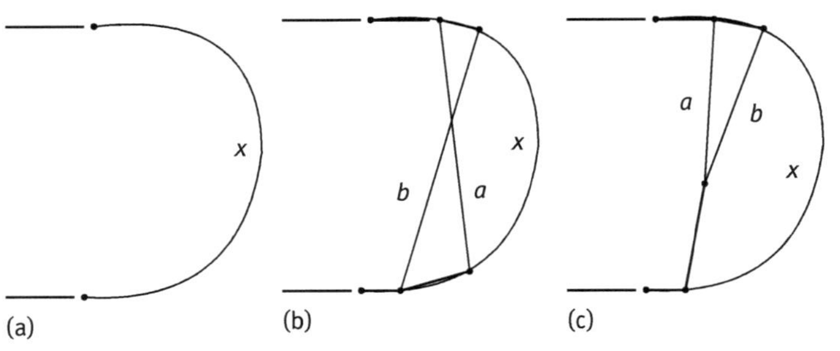

Figure 17.2: Two ways to get a graph enlarged. (a) Part in a graph, (b) 0-parataxis and (c) 0-cascade.

where $y = ax^{-1}$. For cascade, we have

$$\begin{aligned}(X_1abY_2Y_1b^{-1}X_2a^{-1}) &= (X_1abx^{-1}Y_1b^{-1}xa^{-1})\\ &\sim ((X_1aY_1a^{-1})(bxb^{-1}x^{-1}))\ \text{(by Principle 1)}\\ &\sim ((Y_1aX_1a^{-1})(bxb^{-1}x^{-1}))\end{aligned}$$

and $(X_1aY_2Y_1Y_2a^{-1})$ as shown in the case for parataxis.

Therefore, X_1 and Y_1 are interchangeable. □

Theorem 17.3 Graphs $X_G(n)$ and $Y_G(n)$, $n \geq 1$, are with the same genus distribution. Graphs $X_G(n)$ and $Y_G(n)$, $n \geq 2$, are non-isomorphic.

Proof The first statement is from Lemma 2 and Theorem 1. The second statement is from the the observation that $X_G(n)$ and $Y_G(n)$ have different numbers of independent triangles for $n \geq 2$ and hence they are not isomorphic because x is not a loop as shown in Figure 17.3. □

For G as a loop obtained by contracting X_1 and Y_1 into the vertex a_1, shown in Figure 17.1, $X_G(n)$ and $Y_G(n)$ are isomorphic when $n = 1$ and 2 in Theorem 17.3, shown as follows.

The starting graph is taken to be G, as shown in Figure 17.4 which is a self-loop.

By doing 0-parataxis and 0-cascade on G, the results $X_G(1)$ and $Y_G(1)$ are, respectively, shown in Figure 17.5(a and b). It is seen that $X_G(1) \sim Y_G(1)$ in topological equivalence.

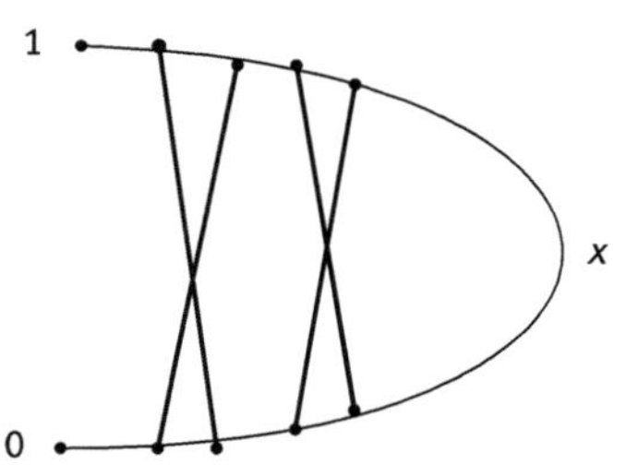

(a)

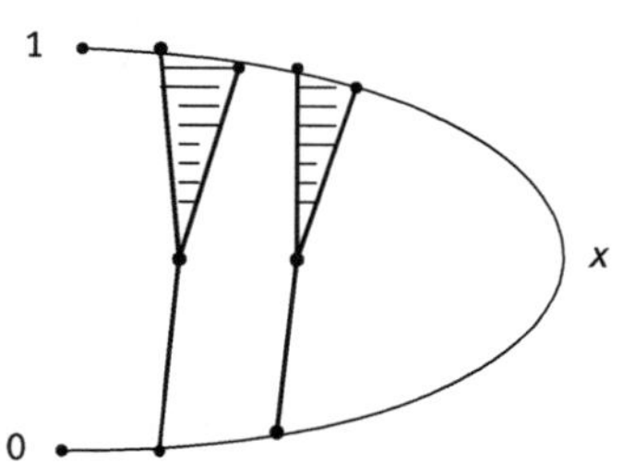

(b)

Figure 17.3: Non-isomorphic isogemial graphs. (a) With one triangle and (b) With two independent triangles.

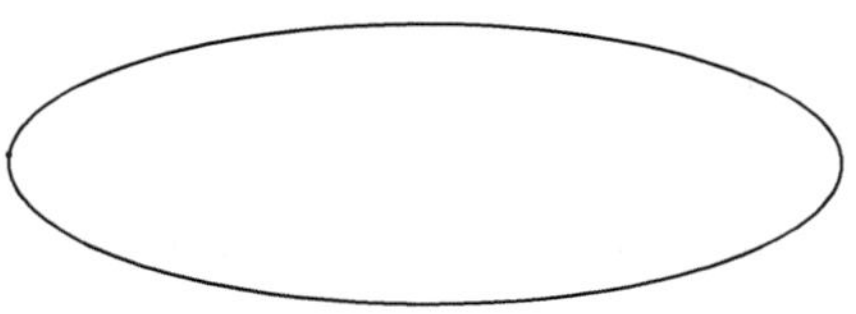

Figure 17.4: Graph G.

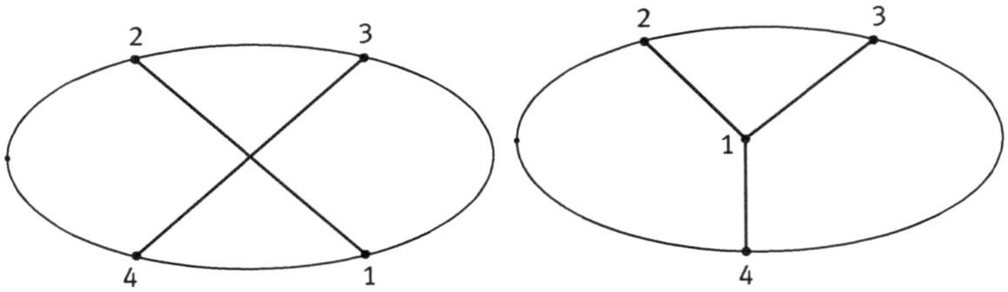

Figure 17.5: (a) $X_G(1)$ and (b) $Y_G(1)$.

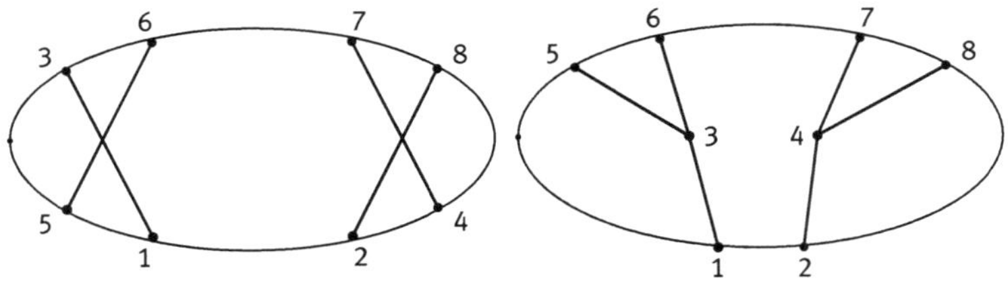

Figure 17.6: (a) $X_G(2)$ and (b) $Y_G(2)$.

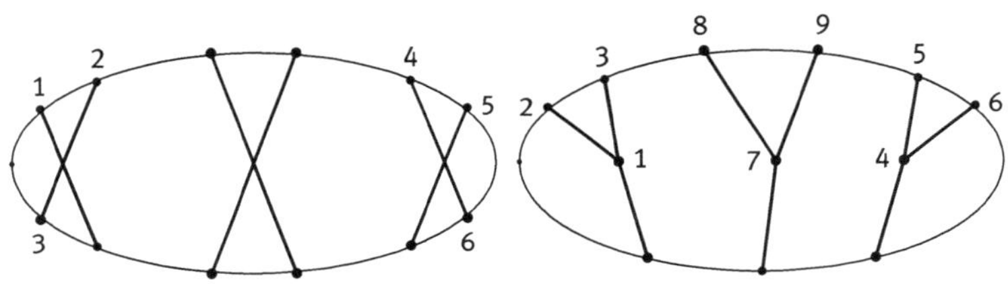

Figure 17.7: (a) $X_G(3)$ and (b) $Y_G(3)$.

By doing 0-parataxis and 0-cascade on, respectively, $X_G(1)$ and $Y_G(1)$, shown in Figure 17.5, the results $X_G(2)$ and $Y_G(2)$ are shown in Figure 17.6(a and b). It is also seen that $X_G(2) \sim Y_G(2)$.

By doing 0-parataxis and 0-cascade on, respectively, $X_G(2)$ and $Y_G(2)$, shown in Figure 17.6, the results $X_G(3)$ and $Y_G(3)$ are shown in Figure 17.7(a and b). It is seen that $X_G(3) \nsim Y_G(3)$ because of $X_G(3)$ with two independent triangles and $Y_G(3)$ with three independent triangles.

From then on, we have known that $X_G(n) \nsim Y_G(n)$ for all $n \geq 3$.

Corollary 17.1 For G as a self-loop, graphs $X_G(n)$ and $Y_G(n)$, $n \geq 3$, are non-isomorphic isogemial.

Proof Because $Y_G(3)$ has three independent triangles but $X_G(3)$ does not have, $X_G(3)$ and $Y_G(3)$ are non-isomorphic. similarly, $X_G(3)$ and $Y_G(3)$ $(n \geq 4)$ are non-isomorphic. □

Now, let G be an arbitrary graph and x_i, $i = 1, 2, \ldots, m$, its m edges. The results from G by, respectively, doing 0-parataxis and 0-cascade for n times on x_i, $i = 1, 2, \ldots, m$, without a nonempty set as a cocircuit are denoted by $X_n(G; x_i)$ and $Y_n(G; x_i)$.

Corollary 17.2 All pairs $\{X_n(G; e_i), Y_n(G; e_i)\}$ are isogemial but nonisomorphic for $n \geq 2$ and $1 \leq i \leq m$. Further, the genus distributions of $\{X_n(G; e_i), Y_n(G; e_i)\}$ and $\{X_n(G; e_j), Y_n(G; e_j)\}$ are different when e_i and e_j are not transitive on G.

Proof The first statement is from Lemma 17.2 and Theorem 17.1 in virtue of x_i, $i=1, 2, \ldots, m$, as cotree edges from no nonempty set as a cocircuit. Because of the transitivity as $x_i = x_j$, $1 \leq i, j \leq m$, the second statement is true on its own right. □

An explanation of Corollary 17.2 can be seen in what follows.

The starting graph is taken to be G as shown in Figure 17.8, which is a cubic graph of order 6.

By doing 0-parataxis and 0-cascade twice on the edge $e_1 = (0, 1)$ of G, the results $X_2(G; e_1)$ and $Y_2(G; e_1)$ are, respectively, shown in Figure 17.1(a and b). It is seen that $X_2(G; e_1)$ and $Y_2(G; e_1)$, then $X_n(G; e_1)$ and $Y_n(G; e_1)$ for $n \geq 3$ as well are isogemial but non-isomorphic as shown in Figure 17.9.

By doing 0-parataxis and 0-cascade twice on edge $e_2 = (1, 4)$ of G, the results $X_2(G; e_2)$ and $Y_2(G; e_2)$ are, respectively, shown in Figure 17.1(a and b). It is seen that $X_2(G; e_2)$ and $Y_2(G; e_2)$, then $X_n(G; e_2)$ and $Y_n(G; e_2)$ for $n \geq 3$ as well are isogemial but non-isomorphic as shown in Figure 17.10.

Because G is arbitrarily chosen, Corollary 17.2 enables us to find a great number of graph pairs $\{X_n, Y_n\}$ that are isogemial but non-isomorphic for $n \geq 2$.

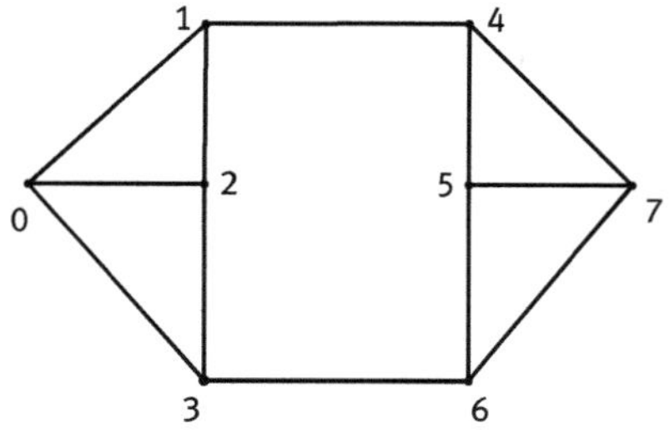

Figure 17.8: A cubic graph G.

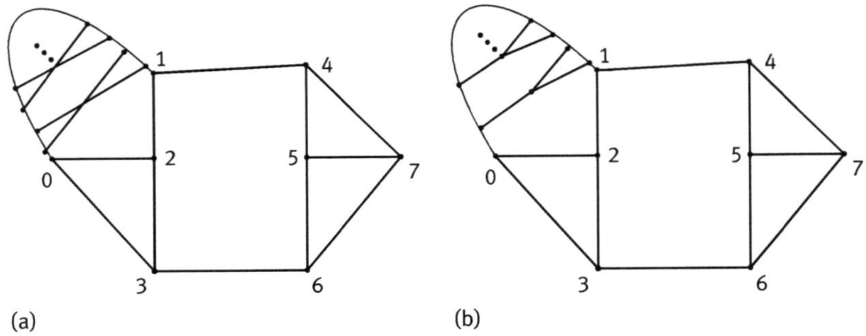

Figure 17.9: (a) $X_2(G; e_1)$ and (b) $Y_2(G; e_1)$.

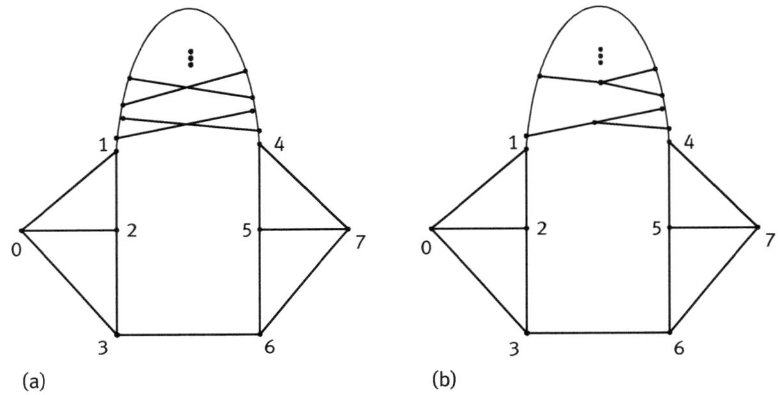

Figure 17.10: (a) $X_2(G; e_2)$ and (b) $Y_2(G, e_2)$.

17.5 Notes

17.5.1 Because of the complexity of checking if an embedding of a graph is isomorphic to one of embeddings of another graph, Theorem 17.2 is not efficient although the recognition of two embeddings isomorphic can be efficiently done.

Notice 17.1 Our purpose is to establish an algorithm of recognizing two graphs isomorphic with high efficiency via their embeddings.

17.5.2 Some other operations are necessary to be discovered for finding new types of non-isomorphic isogemial graphs.

Notice 17.2 Referred to as in Wan and Liu [158].

17.5.3 As shown in Theorem 17.1, one might like to seek more general conditions for X_1 and Y_1 shown in Figure 17.1 interchangible.

Notice 17.3 No more general condition has been found yet.

17.5.4 Can you find a type of 3-regular graph isogimial with a type of 4-regular graphs?

Notice 17.4 No graph in such type is found yet.

17.5.5 Do any two non-isomorphic isogemial graphs have the same thickness?

Notice 17.5 No observation is done yet.

17.5.6 Do any two non-isomorphic isogemial graphs have the same crossing number?

Notice 17.6 No observation is done yet.

The *vertex independency* of a graph is the minimum cardinality over all vertex independent sets in the graph.

17.5.7 Do two non-isomorphic isogemial graphs have the same vertex independency?

Notice 17.7 No observation is done yet.

17.5.8 Do two non-isomorphic isogemial graphs have the same vertex partition?

Notice 17.8 No observation is done yet.

The *rank polynomial* of a graph $G = (V, E)$ is given as

$$\rho(x, y) = \sum_{S\subseteq E} x^{r([S])} y^{|E|-r([S])},$$

where $r([S])$ is the rank(i.e., $|V[S]| - 1$) of the subgraph of G induced by S.

17.5.9 What types of graphs with same rank polynomial are isogemial?

Notice 17.9 No observation is done yet.

The *matching polynomial* of a graph is known as

$$\phi(x) = \sum_{i=0}^{\lfloor n/2 \rfloor} \Phi_i x^i,$$

where Φ_i is the number of all matchings of size i in G with the convention that $\Phi_0 = 1$

17.5.10 What types of graphs with same matching polynomial are isogemial?

Notice 17.10 No observation is done yet.

The *characteristic polynomial* of a graph $G = (V, E)$ is given as

$$\kappa(x) = \det(xI - A),$$

where A is the adjacent matrix of G and I is the unit matrix of $|V| \times |V|$.

17.5.11 What types of graphs with same characteristic polynomial are isogemial?

Notice 17.11 No observation is done yet.
The *chromatic polynomial* of a graph $G = (V, E)$ is defined by

$$\chi(x) = \sum_{i=0}^{|V|} \Psi_i x^i,$$

where Ψ_i is the number of ways for coloring vertices by i distinct colors such that no pair of adjacent vertices are with same color.

17.5.12 What types of graphs with same polynomial are isogemal?
Notice 17.12 No observation is done yet.

18 Surface Embeddability

- From a tree-travel as starting point, a criterion for a graph embeddable on a surface of genus given is established.
- On the basis of homology and cohomology theorems for genus 0, a criterion for a graph embeddable on a surface of genus ($\neq 0$) given is established.
- On the basis of joint trees, a criterion for a graph embeddable on a surface of genus given is established.
- From a forbidden configuration for a surface of genus g as a startng point, a criterion for a graph embeddable on a surface of $g + 1$ is established.

18.1 Via tree-travels

Along the Kurotowski research line for determining the embeddability of a graph on a surface of genus not zero, the number of forbidden minors is greater than a hundred even for the projective plane, a nonorientable surface of genus 1 in Glover et al. [32] and further in Archdeacon [2].

However, this chapter extends the results in Liu [60], which is on the basis of the method established in Liu [60, 61] for dealing with the problem on the maximum genus of a graph since 1979.

Although the principle idea looks like from the joint trees, a main difference of a tree used here is not corresponding to an embedding of the graph considered.

Given a graph $G = (V, E)$, let T be a spanning tree of G. If each cotree edge is added to T as an articulate edge, what obtained is called a *protracted tree* of G, denoted by $\breve{T}$.

A protracted tree $\breve{T}$ is oriented via an orientation of T or its fundamental circuits. In order to guarantee the well definedness of the orientation for a given rotation at all vertices on G and a selected vertex of T, the direction of a cotree edge is always chosen in coincidence with its direction first appeared along the the face boundary of $\breve{T}$. For convenience, vertices on the boundary are marked by the ordinary natural numbers as the root vertex, the starting vertex, by 0. Of course, the boundary is a travel on G, called a *tree-travel.*

Figure 18.1(a) shows a spanning tree T of K_5 (i.e., the complete graph of order 5), shown by bold lines; (b) the protracted tree $\breve{T}$ of T.

Let $C = C(V; e)$ be the tree-travel obtained from the boundary of $\breve{T}$ with 0 as the starting vertex. Apparently, the travel as an edge sequence $C = C(e)$ provides a double covering of $G = (V, E)$, denoted by

$$C(V; e) = 0P_{0,i_1}v_{i_1}P_{i_1,i_2}v_{i_2}P_{i_2,i'_1}v_{i'_1}P_{i'_1,i'_2}v_{i'_2}P_{i'_2,2\epsilon}0, \tag{18.1}$$

where $\epsilon = |E|$.

DOI 10.1515/9783110481846-018

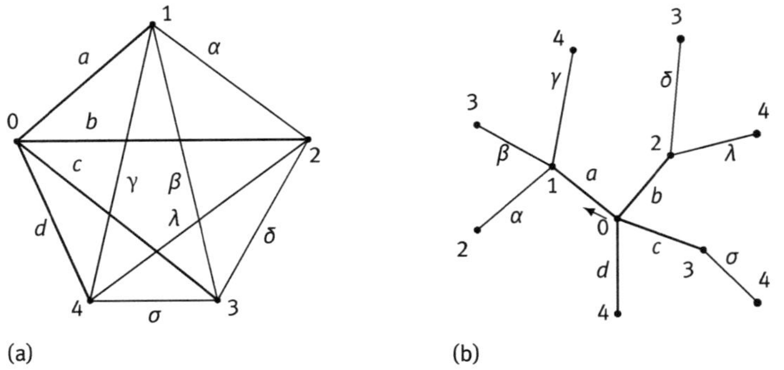

Figure 18.1: (a) T and (b) $\check{T}$.

For a vertex-edge sequence Q as a tree-travel, denote by $[Q]$eg the edge sequence induced from Q missing vertices, then $C_{eg} = [C(V;e)]$eg is a polyhegon (i.e., a polyhedron with only one face).

Example 18.1 From $\check{T}$ in Figure 18.1(b), obtain the thee-travel

$$C(V;e) = 0P_{0,8}0P_{8,14}0P_{14,18}0P_{18,20}0,$$

where $v_0 = v_8 = v_{14} = v_{18} = v_{20} = 0$ and

$$\begin{cases} P_{0,8} = a1\alpha2\alpha^{-1}1\beta3\beta^{-1}1\gamma4\gamma^{-1}1a^{-1}; \\ P_{8,14} = b2\delta3\delta^{-1}2\lambda4\lambda^{-1}2b^{-1}; \\ P_{14,18} = c3\sigma4\sigma^{-1}3c^{-1}; \\ P_{18,20} = d4d^{-1}. \end{cases}$$

For natural number i, if av_ia^{-1} is a segment in C, then a is called a *reflective edge* and v_i the *reflective vertex* of a.

Because there is nothing important for articulate vertices (1-valent vertices) and 2-valent vertices in an embedding, we are allowed to restrict ourselves to discussing graphs with neither 1-valent nor 2-valent vertices without loss of generality. From vertices of all greater than or equal to 3, we are allowed only to consider all reflective edges as on the cotree.

If v_{i_1} and v_{i_2} are both reflective vertices in (18.1), their reflective edges are adjacent in G and $v_{i_1'} = v_{i_1}$ and $v_{i_2'} = v_{i_2}$, $[P_{v_{i_1,i_2}}]$eg $\cap$ $[P_{v_{i_1',i_2'}}]$eg $= \emptyset$, but neither $v_{i_1'}$ nor $v_{i_2'}$ is a reflective vertex, then the transformation from C to

$$\triangle_{v_{i_1},v_{i_2}} C(V;e) = 0P_{0,i_1}v_{i_1}P_{i_1',i_2'}v_{i_2}P_{i_2,i_1'}v_{i_1'}P_{i_1,i_2}v_{i_2'}P_{i_2',0}0 \tag{18.2}$$

is called an operation of *interchange segments* for $\{v_{i_1}, v_{i_2}\}$.

Example 18.2 In $C = C(V;e))$ of Example 18.1, $v_2 = 2$ and $v_4 = 3$ are two reflective vertices, α and β their reflective edges, $v_9 = 2$ and $v_15 = 3$. For interchange of segments once on C, we have

$$\triangle_{2,3}C = 0P_{0,2}2P_{9,15}3P_{4,9}2P_{2,4}3P_{15,20}0\ (= C_1),$$

where

$$\begin{cases} P_{0,2} = a1\alpha\ (= P_{1;0,2}); \\ P_{9,15} = \delta 3\delta^{-1}2\lambda 4\lambda^{-1}2b^{-1}0c3\ (= P_{1;2,8}); \\ P_{4,9} = \beta^{-1}1\gamma 4\gamma^{-1}1a^{-1}0b2\ (= P_{1;8,13}); \\ P_{2,4} = \alpha^{-1}1\beta\ (= P_{1;13,15}); \\ P_{15,20} = \sigma 4\sigma^{-1}3c^{-1}0d4d^{-1}\ (= P_{1;15,20}). \end{cases}$$

Lemma 18.1 Polyhegon $\triangle_{v_i,v_j}C_{\mathrm{eg}}$ is orientable if, and only if, C_{eg} is orientable and the genus of $\triangle_{v_{i_1},v_{i_2}}C_{\mathrm{eg}}$ is exactly 1 greater than that of C_{eg}.

Proof Because of the invariance of orientability for $\triangle$-operation on a polyhegon, the first statement is true. □

In order to prove the second statement, assume that cotree edges α and β are reflective edges at vertices, respectively,v_{i_1} and v_{i_2}. Because

$$C_{\mathrm{eg}} = A\alpha\alpha^{-1}B\beta\beta^{-1}CDE,$$

where

$$\begin{cases} A\alpha = [P_{0,i_1}]_{\mathrm{eg}};\ \alpha^{-1}B\beta = [P_{i_1,i_2}]_{\mathrm{eg}}; \\ \beta^{-1}C = [P_{i_2,i_1'}]_{\mathrm{eg}};\ D = [P_{i_1',i_2'}]_{\mathrm{eg}}; \\ E = [P_{i_2',i_\epsilon}]_{\mathrm{eg}}, \end{cases}$$

we have

$$\begin{aligned} \triangle_{v_{i_1},v_{i_2}}C_{\mathrm{eg}} &= A\alpha D\beta^{-1}C\alpha^{-1}B\beta E \\ &\sim_{\mathrm{top}} ABCDE\alpha\beta\alpha^{-1}\beta^{-1},\ \text{(Theorem 3.3.3 in Liu [77])} \\ &= C_{\mathrm{eg}}\alpha\beta\alpha^{-1}\beta^{-1}\ \text{(Transform 1, in Section 3.1 of Liu [77]).} \end{aligned}$$

Therefore, the second statement is true.

If interchange segments can be done on C successively for k times, then C is called a *k-tree travel*. Since one reflective edge is reduced for each interchange of segments on C and C has at most $m = \lfloor \beta/2 \rfloor$ reflective edges, we have $0 \le k \le m$, where $\beta = \beta(G)$ is the Betti number (or corank) of G. When $k = m$, C is also called *normal*.

For a k-tree travel $C_k(V; e, e^{-1})$ of G, graph G_k is defined as

$$G_k = T \bigcup \left[\mathrm{Eref} \bigcap E_{\bar{T}} - \sum_{j=1}^{k} \{e_j, e_j'\} \right] \tag{18.3}$$

where T is a spanning tree, $[X]$ represents the edge-induced subgraph by edge subset X and $e \in \mathrm{Eref}$, $e \in E_{\bar{T}}$, $\{e_j, e_j'\}$ are, respectively, reflective edge, cotree edge, pair of reflective edges for interchange segments.

Example 18.3 On C_1 in Example 18.2, $v_{1;3} = 3$ and $v_{1;5} = 4$ are two reflective vertices, $v_{1;8} = 3$ and $v_{1;10} = 4$. By doing interchange segments on C_1, we obtain

$$\triangle_{3,4} C_1 = 0P_{1;0,10} 3P_{1;17,19} 4P_{1;12,15} 3P_{1;10,12} 4P_{1;19,20} 0 \ (= C_2),$$

where

$$\begin{cases} P_{1;0,10} = a1\alpha 2b^{-1}0c3\beta^{-1}1\gamma 4\gamma^{-1}1a^{-1}0b2\delta(= P_{2;0,10}); \\ P_{1;17,19} = c^{-1}0d(= P_{2;10,12}); \\ P_{1;12,17} = \lambda^{-1}2\alpha^{-1}1\beta 3\sigma 4\sigma^{-1}(= P_{2;12,17}); \\ P_{1;10,12} = \delta^{-1}2\lambda(= P_{2;17,19}); \\ P_{1;19,20} = d^{-1}(= P_{2;19,20}). \end{cases}$$

Because $[P_{2;6,16}]\mathrm{eg} \cap [P_{2;12,19}]\mathrm{eg} \neq \emptyset$ for $v_{2;12} = 4$ and $v_{2;19} = 4$, only $v_{2;6} = 4$ and $v_{2;16} = 4$ with their reflective edges γ and σ are allowed for doing interchange segments on C_2. The protracted tree $\breve{T}$ in Figure 18.1(b) provides a 2-tree travel C, and then a 1-tree travel as well.

However, if interchange segments are done for pairs of cotree edges as $\{\beta, \gamma\}$, $\{\delta, \lambda\}$ and $\{\alpha, \sigma\}$ in this order, it is known that C is also a 3-tree travel.

For C in Example 18.1, the reflective vertices of cotree edges β and γ are, respectively, $v_4 = 3$ and $v_6 = 4$, choose $4' = 15$ and $6' = 19$, so we have

$$\triangle_{4,6} C = 0P_{1;0,4} 3P_{1;4,8} 4P_{1;8,17} 3P_{1;17,19} 4P_{1;19,20} 0(= C_1),$$

where

$$\begin{cases} P_{1;0,4} = P_{0,4};\ P_{1;4,8} = P_{15,19};\ P_{1;8,17} = P_{6,15}; \\ P_{1;17,19} = P_{4,6};\ P_{1;19,20} = P_{19,20}. \end{cases}$$

For C_1, subindices of the reflective vertices for reflective edges δ and λ are 5 and 8, choose $5' = 17$ and $8' = 19$, and find

$$\triangle_{5,8} C_1 = 0P_{2;0,5} 3P_{2;5,7} 4P_{2;7,16} 3P_{2;16,19} 4P_{2;19,20} 0(= C_2),$$

where

$$\begin{cases} P_{2;0,12} = P_{1;0,12};\ P_{2;12,14} = P_{1;17,19};\ P_{2;14,17} = P_{1;14,17}; \\ P_{2;17,19} = P_{1;12,14};\ P_{2;19,20} = P_{1;19,20}. \end{cases}$$

For C_2, subindices of the reflective vertices for reflective edges α and σ are 2 and 5, choose $2' = 18$ and $5' = 19$, and find

$$\triangle_{5,8}C_2 = 0P_{3;0,2}3P_{3;2,3}4P_{3;3,16}3P_{3;16,19}4P_{3;19,20}0(= C_3),$$

where

$$\begin{cases} P_{3;0,2} = P_{2;0,2};\ P_{2;2,3} = P_{2;18,19};\ P_{3;3,16} = P_{2;5,18}; \\ P_{3;16,19} = P_{2;2,5};\ P_{3;19,20} = P_{2;19,20}. \end{cases}$$

Because $\beta(K_5) = 6$, so $m = 3 = \lfloor \beta/2 \rfloor$. Thus, tree-travel C is normal.

This example shows that the problem of determining the maximum orientable genus of a graph can be transformed into that of determining a k-tree travel of a graph with k maximum in Liu [61].

Lemma 18.2 Among all k-tree travel of a graph G, the maximum of k is the maximum orientable genus $\gamma_{\max}(G)$ of G.

Proof In order to prove this lemma, the following two facts have to be known (both of them can be done via the finite recursion principle in Section 1.3 of Liu [77, pp. 11–17].

Fact 1 In a connected graph G considered, there exists a spanning tree such that any pair of cotree edges whose fundamental circuits with vertex in common are adjacent in G.

Fact 2 For a spanning tree T with Fact 1, there exists an orientation such that on the protracted tree $\check{T}$, no two articulate subvertices (articulate vertices of T) with odd out-degree of cotree have a path in the cotree.

Because of that if two cotree edges for a tree are with their fundamental circuits without vertex in common, then they for any other tree are with their fundamental circuits without vertex in common as well. Fact 1 enables us to find a spanning tree with a number of pairs of adjacent cotree edges as much as possible and Fact 2 enables us to find an orientation such that the number of times for doing successive interchange of segments is as much as possible. From Lemma 18.1, the lemma can be done. □

The purpose of what follows is for characterizing the embeddability of a graph on a surface of genus not necessary to be zero via k-tree travels.

Theorem 18.1 A graph G can be embedded into an orientable surface of genus k if, and only if, there exists a k-tree travel $C_k(V; e)$ such that G_k is planar.

Proof Necessity. Let $\mu(G)$ be an embedding of G on an orientable surface of genus k. From Lemma 18.2, $\mu(G)$ has a spanning tree T with its edge subsets E_0, $|E_0| = \beta(G) - 2k$, such that $\hat{G} = G - E_0$ is with exactly one face. By successively doing the inverse of interchange segments for k times, a k-tree travel is obtained on $\hat{G}$. Let K be consisted of the k pairs of cotree edge subsets. Thus, from Operation 2 in Section 3.3 of Liu [77, pp. 50–55], $G_k = G - K = \hat{G} - K + E_0$ is planar.

Sufficiency. Because of G with a k-tree travel $C_k(V; e)$, let K consist of the k pairs of cotree edge subsets in successively interchanging segments for k times. Since $G_k = G - K$ is planar, by successively doing the inverse of interchange segments for k times on $C_k(V; e)$ in its planar embedding, an embedding of G on an orientable surface of genus k is easily obtained. □

Example 18.4 In Example 18.1, for $G = K_5$, C is a 1-tree travel for the pair of cotree edges α and β. And, $G_1 = K_5 - \{\alpha, \beta\}$ is planar. Its planar embedding is

$$\begin{cases} [4\sigma^{-1}3c^{-1}0d4]\text{eg} = (\sigma^{-1}c^{-1}d);\ [4d^{-1}0a1\gamma4]\text{eg} = (d^{-1}a\gamma); \\ [3\sigma4\lambda^{-1}2\delta3]\text{eg} = (\sigma\lambda^{-1}\delta);\ [0c3\delta^{-1}2b^{-1}0]\text{eg} = (c\delta^{-1}b^{-1}); \\ [2\lambda4\gamma^{-1}1a^{-1}0b2]\text{eg} = (\lambda\gamma^{-1}a^{-1}b). \end{cases}$$

By recovering $\{\alpha, \beta\}$ from G and then interchanging segments once on C, obtain C_1. From C_1, on the basis of a planar embedding of G_1, an embedding of G on an orientable surface of genus 1 (the torus) is produced as

$$\begin{cases} [4\sigma^{-1}3c^{-1}0d4]\text{eg} = (\sigma^{-1}c^{-1}d);\ [4d^{-1}0a1\gamma4]\text{eg} = (d^{-1}a\gamma); \\ [3\sigma4\lambda^{-1}2\delta3\beta^{-1}1a^{-1}0b2\alpha^{-1}1\beta3]\text{eg} = (\sigma\lambda^{-1}\delta\beta^{-1}a^{-1}b2\alpha^{-1}\beta); \\ [0c3\delta^{-1}2b^{-1}0]\text{eg} = (c\delta^{-1}b^{-1});\ [2\lambda4\gamma^{-1}1\alpha2]\text{eg} = (\lambda\gamma^{-1}\alpha). \end{cases}$$

Similarly, we further discuss the nonorientable case. Let $G = (V, E)$, T a spanning tree and

$$C(V; e) = 0P_{0,i}v_iP_{i,j}v_jP_{j,2\epsilon}0 \tag{18.4}$$

be the travel obtained from 0 along the boundary of protracted tree $\check{T}$. If v_i is a reflective vertex and $v_j = v_i$, then

$$\tilde{\Delta}_\xi C(V; e) = 0P_{0,i}v_iP_{i,j}^{-1}v_jP_{j,2\epsilon}0 \tag{18.5}$$

is obtained by doing a *reverse segment* for the reflective vertex v_i on $C(V; e)$.

If reverse segment can be done successively for k on C, then C is called a $\tilde{k}$-*tree travel*. Because one reflective edge is reduced for each reverse segment and at most β reflective edges on C, we have $0 \le k \le \beta$, where $\beta = \beta(G)$ is the Betti number of G (or corank). When $k = \beta$, C(or G) is called *twist normal*.

Lemma 18.3 A connected graph is twist normal if, and only if, the graph is not a tree.

Proof Because of trees no cotree edge themselves, the reverse segment cannot be done, this leads to the necessity. Conversely, because a graph is not a tree, the graph has to be with a circuit, a tree-travel has at least one reflective edge. Because there is no effect on other reflective edges after doing reverse segment once for a reflective edge, reverse segment can always be done for successive $\beta = \beta(G)$ times, and hence this tree-travel is twist normal. Therefore, sufficiency holds. □

Lemma 18.4 Let C be obtained by doing reverse segment at least once on a tree-travel of a graph. Then, the polyhegon $[\triangle_i C]_{\text{eg}}$ is nonorientable and its genus

$$\tilde{g}([\triangle_\xi C]_{\text{eg}}) = \begin{cases} 2g(C) + 1, & \text{when } C \text{ orientable;} \\ \tilde{g}(C) + 1, & \text{when } C \text{ nonorientable.} \end{cases} \tag{18.6}$$

Proof Although a tree-travel is orientable with genus 0 itself, after the first reverse segment on what are obtained, the nonorientability is always kept unchanged. This leads to the first conclusion. Assume C_{eg} is orientable with genus $g(C)$ (in fact, only $g(C) = 0$ will be used!). Because of

$$[\triangle_i C]_{\text{eg}} = A\xi B^{-1}\xi C$$

where $[P_{0,i}]_{\text{eg}} = A\xi$, $[P_{i,j}]_{\text{eg}} = \xi^{-1}B$ and $[P_{j,\epsilon}]_{\text{eg}} = C$, from (3.1.2) in Liu [77],

$$[\triangle_i C]_{\text{eg}} \sim_{\text{top}} ABC\xi\xi.$$

Noticing that from Operation 0 in Section 3.3 of Liu [77], $C_{rseg} \sim_{\text{top}} ABC$, Lemma 3.1.1 in Liu [77] leads to

$$\tilde{g}([\triangle_\xi C]_{\text{eg}}) = 2g([C]_{\text{eg}}) + 1 = 2g(C) + 1.$$

Assume C_{eg} is nonorientable with genus $g(C)$. Because of

$$C_{\text{eg}} = A\xi\xi^{-1}BC \sim_{\text{top}} ABC,$$

$\tilde{g}([\triangle_\xi C]_{\text{eg}}) = \tilde{g}(C) + 1$. Thus, this implies the second conclusion. □

As a matter of fact, only reverse segment is enough on a tree-travel for determining the nonorientable maximum genus of a graph.

Lemma 18.5 Any connected graph, except only for trees, has its Betti number as the nonorientable maximum gemus.

Proof From Lemmas 18.3 and 18.4, the conclusion can easily be drawn. □

For a $\tilde{k}$-tree travel $C_{\tilde{k}}(V;e)$ on G, the graph $G_{\tilde{k}}$ is defined as

$$G_{\tilde{k}} = T \bigcup \left[\mathrm{Eref} - \sum_{j=1}^{k} \{e_j\} \right] \tag{18.7}$$

where T is a spanning tree, $[X]$ the induced graph of edge subset X and $e \in$ Eref and $\{e_j, e_j'\}$, respectively, a reflective edge and that used for reverse segment.

Theorem 18.2 A graph G can be embedded into a nonorientable surface of genus k if, and only if, G has a $\tilde{k}$-tree travel $C_{\tilde{k}}(V;e)$ such that $G_{\tilde{k}}$ is planar.

Proof From Lemma 18.3, for k, $1 \le k \le \beta(G)$, any connected graph G but tree has a $\tilde{k}$-tree travel.

Necessity. Because G is embeddable on a nonorientable surface $S_{\tilde{k}}$ of genus k, let $\tilde{\mu}(G)$ be an embedding of G on $S_{\tilde{k}}$. From Lemma 5, $\tilde{\mu}(G)$ has a spanning tree T with cotree edge set E_0, $|E_0| = \beta(G) - k$, such that $\tilde{G} = G - E_0$ has exactly one face. By doing the inverse of reverse segment for k times, a $\tilde{k}$-tree travel of $\tilde{G}$ is obtained. Let K be a set consisting k cotree edges. From Operation 2 in Section 3.3 of Liu [77, pp. 50–55], $G_{\tilde{k}} = G - K = \tilde{G} - K + E_0$ is planar.

Sufficiency. Because of G with a $\tilde{k}$-tree travel $C_{\tilde{k}}(V;e)$, let K be the set of k cotree edges used for successively doing reverse segment. Since $G_{\tilde{k}} = G - K$ is planar, by successively doing reverse segment for k times on $C_{\tilde{k}}(V;e)$ in a planar embedding of $G_{\tilde{k}}$, an embedding of G on a nonorientable surface $S_{\tilde{k}}$ of genus k is then extracted. □

Example 18.5 On $K_{3,3}$, take a spanning tree T, as shown in Figure 18.2(a) by bold lines. Figure 18.2(b) shows a protracted tree $\breve{T}$ of T. From $\breve{T}$, we get a tree-travel

$$C = 0P_{0,11}2P_{11,15}2P_{15,0}0\ (= C_0),$$

where $v_0 = v_{18}$ and

$$\begin{cases} P_{0,11} = c4\delta5\delta^{-1}4\gamma3\gamma^{-1}4c^{-1}0d2e3\beta1\beta^{-1}3e^{-1}; \\ P_{11,15} = d^{-1}0a1b5\alpha; \\ P_{15,0} = \alpha^{-1}5b^{-1}1a^{-1}. \end{cases}$$

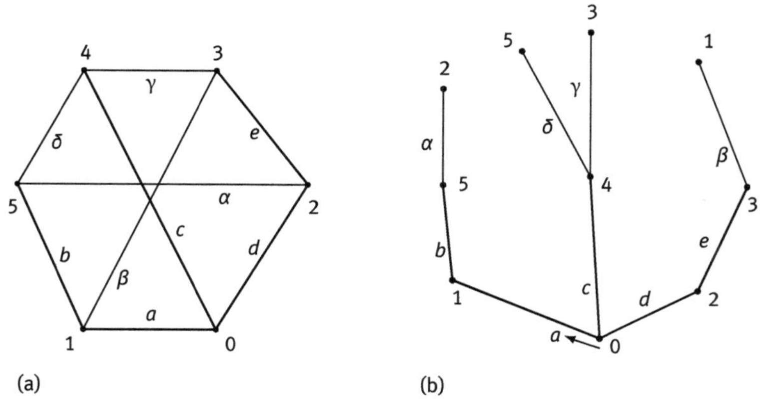

Figure 18.2: (a) T and (b) $\check{T}$.

Because $v_{15} = 2$ is the reflective vertex of cotree edge α and $v_{11} = v_{15}$,

$$\triangle_3 C_0 = 0P_{1;0,11}2P_{1;11,15}2P_{1;15,0}0\ (= C_1),$$

where

$$\begin{cases} P_{1;0,11} = P_{0,11} = c4\delta 5\delta^{-1}4\gamma 3\gamma^{-1}4c^{-1}0d2e3\beta 1\beta^{-1}3e^{-1}; \\ P_{1;11,15} = P_{11,15}^{-1} = \alpha^{-1}5b^{-1}1a^{-1}0d; \\ P_{1;15,0} = P_{15,0} = \alpha^{-1}5b^{-1}1a^{-1}. \end{cases}$$

Since $G_{\tilde{1}} = K_{3,3} - \alpha$ is planar, from C_0 its planar embedding is given by

$$\begin{cases} f_1 = [5P_{16,0}0P_{0,2}0]_{\text{eg}} = (b^{-1}a^{-1}c\delta); \\ f_2 = [3P_{4,8}3]_{\text{eg}} = (\gamma^{-1}c^{-1}de); \\ f_3 = [1P_{13,14}5P_{2,4}3P_{8,9}1]_{\text{eg}} = (\delta^{-1}\gamma\beta b); \\ f_4 = [1P_{9,13}1]_{\text{eg}} = (\beta^{-1}e^{-1}d^{-1}a). \end{cases}$$

By doing reverse segment on C_0, we get C_1. On this basis, an embedding of $K_{3,3}$ on the projective plane (i.e., nonorientable surface $S_{\tilde{1}}$ of genus 1) is obtained as

$$\begin{cases} \tilde{f}_1 = [5P_{1;16,0}0P_{1;0,2}0]_{\text{eg}} = f_1 = (b^{-1}a^{-1}c\delta); \\ \tilde{f}_2 = [3P_{1;4,8}3]_{\text{eg}} = f_2 = (\gamma^{-1}c^{-1}de); \\ \tilde{f}_3 = [1P_{1;9,11}2P_{1;11,13}1]_{\text{eg}} = be^{-1}e^{-1}\alpha^{-1}b^{-1}); \\ \tilde{f}_4 = [0P_{1;14,15}2P_{1;15,16}5P_{1;2,4}3P_{1;8,9}1P_{1;13,14}0]_{\text{eg}} \\ \quad = (d\alpha^{-1}\delta^{-1}\gamma\beta a^{-1}). \end{cases}$$

18.2 Via homology

On a surface of genus 0, the embeddability of a graph is the planarity, which is well known in literature. Here, we only mention some classic theorems in topology related to the present topic in our language.

What is obtained by deleting an edge from the edge set and/or identifying the two ends of an edge deleted on a graph is called a *minor* of the graph.

Theorem A (Kuratowski [46]) A graph can be embedded into the plane if, and only if, neither K_5 nor $K_{3,3}$ is a minor of the graph.

In this theorem, K_5 and $K_{3,3}$ are the complete graphs of order 5 and the complete bipartite of order 6 with each part of three vertices.

Theorem B (Whitney [166]) A graph $G = (V, E)$ can be embedded into the plane if, and only if, there exists another graph $G^* = (V^*, E^*)$ with a bijection $\tau : E \Longrightarrow E^*$ such that for any circuit $C \subseteq E$, $\tau(C) = C^* \subseteq E^*$ is a cocircuit.

The statement of this theorem is in different form but equivalent to the original of Whitney's for the convenience of coming usage.

Theorem C (MacLane [123]) A graph G can be embedded into the plane if, and only if, there exists a set $\mathscr{C}$ of $\beta(G)$ circuits such that each edge of G occurs either once or twice in $\mathscr{C}$.

In this theorem, $\beta(G)$ is the corank of G, i.e., the dimension of the cycle space of G.

Theorem D (LefschetzS [48]) A graph can be embedded into the plane if, and only if, there exists a set $\mathscr{C}$ of $\beta(G) + 1$ circuits such that each edge of G occurs exactly twice in $\mathscr{C}$.

The set $\mathscr{C}$ in this theorem is called a *double covering* of G.

Although the above four theorems have been employed as a basis for investigating the embeddability of a graph on surfaces of genus not zero, only those with genus 1 (orientable or nonorientable) get results rather complicated as in Archdeacon [2] on Theorem A and Abrams and Slilaty [1] on Theorem B for the projective plane.

However, in the monographs Liu [64, 65], a graph was considered as a set of polyhedra and then two mutually dual planarity characterizations were established on the basis of homological and cohomological spaces. From these results, the above four theorems were naturally reduced.

The purpose of this chapter is to generalize the above three theorems B, C and D for characterizing the surface embeddability (planarity is only a special case of genus

0) of a graph. Theorem A is not discussed in this chapter because of the limitation in length.

Let $\mathscr{C}$ be a set of travels as a double covering of $G = (V, E)$. Given a vertex $v \in V$ for a pair of edges at v, if there exists a travel $C \in \mathscr{C}$ such that they are successive on C, then the pair of edges are called *sharing* vertex v. This relation is extended as an equivalence on the set of edges at a vertex. If all edges at v are in the same class under this equivalence, then $\mathscr{C}$ is called *transitive* at v.

Lemma 18.6 A double covering of a graph determines an embedding if, and only if, it is transitive at all vertices.

Proof If double covering $\mathscr{C}$ is an embedding of G, then its skeleton is G, as in Section 2.2 of Liu [77, pp. 29–31]. Because of edges at each vertex with a rotation, i.e., a cyclic permutation, this implies the necessity.

Conversely, assume $\mathscr{C}$ to be transitive at all vertices of G, then all edges at each vertex form a cyclic permutation. If this cyclic permutation is taken as a rotation, on the basis described in Section 3.5 of Liu [77, pp. 57–61], $\mathscr{C}$ is an embedding of G. This is the sufficiency. □

If a double covering $\mathscr{C}$ of G by travels is transitive at each vertex, then $\mathscr{C}$ is said to be *consistent*. $\mathscr{C}$ is called *orientable* when there exists an orientation of each travel such that the two occurrences of each edge have different directions; *nonorientable*, otherwise.

For convenience, all graphs in what follows are nonseparable with neither loop nor isthmus without loss of generality.

Lemma 18.7 Let $\mathscr{C}$ be an orientable double covering of G with $\beta(G) - 2p + 1$ travels, where p is a nonnegative integer, then $\mathscr{C}$ is consistent if, and only if, the dimension of the space generated by travels in $\mathscr{C}$ is $\dim\langle\mathscr{C}\rangle = \beta(G) - 2p$.

Proof Denote $k = \beta(G) - 2p + 1$, then $\dim\langle\mathscr{C}\rangle = k - 1$.

Sufficiency. Because $\beta(G) = \epsilon(G) - \nu(G) + 1$, so $k = \epsilon(G) - \nu(G) - 2p + 2$. Assume $\mathscr{C}$ is not consistent, then $\mathscr{C}$ is a polygon Σ with at least $\nu(G) + 1$ vertices. Because of $\epsilon(\Sigma) = \epsilon(G)$,

$$\begin{aligned}\dim(\Sigma) &= \beta(\Sigma) - 2p = \epsilon(\Sigma) - \nu(\Sigma) + 1 - 2p\\ &\le \epsilon(G) - (\nu(G) + 1) - 2p + 1\\ &= \dim\langle\mathscr{C}\rangle - 1.\end{aligned}$$

This is a contradiction to $\dim(\Sigma) = \dim\langle\mathscr{C}\rangle$. Therefore, $\mathscr{C}$ is consistent.

Necessity. Because $\mathscr{C}$ is transitive at each vertex of G, $\mathscr{C}$ as a polyhedron is an embedding of G and hence $\dim\langle\mathscr{C}\rangle = k - 1 = \beta(G) - 2p$. □

Lemma 18.8 Let $\mathscr{C}$ be an embedding of G on an orientable surface of genus p, then $\dim\langle\mathscr{C}\rangle = \beta(G) - 2p$.

Proof Because $\mathscr{C}$ is an embedding of G on an orientable surface of genus p, by Eulerian characteristic, this embedding has $\phi = |\mathscr{C}| = \beta(G) - 2p + 1$ faces. Assume ∂f_i is the boundary of face f_i, $1 \le i \le \phi$. Because $\sum_{1\le i\le\phi} \partial f_i = 0$,

$$\dim\langle\mathscr{C}\rangle \le \beta(G) - 2p = \phi - 1.$$

From the connectedness of G, $\mathscr{C}$ has its $\phi - 1$ vectors independent (otherwise, $\mathscr{C}$ has that these $\phi - 1$ travels form a polyhedron. However, the skeleton of this polyhedron is a proper subgraph of G. A contradiction to G connected occurs). Therefore, $\dim\langle\mathscr{C}\rangle = \beta(G) - 2p$. □

Theorem 18.3 A graph G can be embedded on an orientable surface of genus p if, and only if, there exists an orientable double covering $\mathscr{C}$ with $\beta(G) - 2p + 1$ travels such that $\dim\langle\mathscr{C}\rangle = \beta(G) - 2p$.

Proof Necessity. If Σ is an embedding of G on an orientable surface of genus p, let $\mathscr{C}$ as polyhedron Σ be the set of all boundaries of faces of Σ. From Eulerian characteristic, $|\mathscr{C}| = \beta(G) - 2p + 1$. Naturally, $\mathscr{C}$ is an orientable double covering of G by travels. From Lemma 3, $\dim\langle\mathscr{C}\rangle = \beta(G) - 2p$.

Sufficiency. Assume $\mathscr{C}$ is a double covering of G with $\beta(G) - 2p + 1$ travels such that $\dim\langle\mathscr{C}\rangle = \beta(G) - 2p$, then from Lemma 18.6, and 18.7, $\mathscr{C}$ is an embedding of G on an orientable surface of genus p. □

Since $\dim\langle\mathscr{C}\rangle = \beta(G) - 2p$ leads to that $\mathscr{C}$ has $\beta(G) - 2p + 1$ travels, it is allowed to employ the former instead of the latter.

Corollary 18.1 (Theorem D above) A nonseparable G is planar if, and only if, there exists a set $\mathscr{C}$ of $\beta(G) + 1$ circuits, which is a double covering of G.

Proof The case of $p = 0$ in Theorem 18.3 because of $\dim\langle\mathscr{C}\rangle = \beta(G)$ and $|\mathscr{C}| = \beta(G) + 1$. □

Corollary 18.2 (Theorem C above) A nonseparable graph $G = (V, E)$ is planar if, and only if, there exists a set $\mathscr{C}$ of $\beta(G) + 1$ circuits in G such that each $e \in E$ occurs in either one or two circuits of $\mathscr{C}$.

Proof Because $p = 0$, we have $\dim\langle\mathscr{C}\rangle = \beta(G)$ and $\mathscr{C} + C_0$ is a double covering of G where

$$C_0 = \sum_{C\in\mathscr{C}} C.$$

From Theorem 18.3, the corollary holds. □

Corollary 18.3 A graph $G = (V, E)$ can be embedded on an orientable surface of genus $p \geq 0$ if, and only if, there exist another graph $G_p^* = (V^*, E^*)$ and a bijection $\tau : E \longrightarrow E^*$ such that $\dim\langle \tau E_v | \forall v \in V\rangle = \beta(G^*) - 2p$.

Proof In fact, the dual form of Theorem 18.3. □

Corollary 18.4 (Theorem B above) A connected graph $G = (V, E)$ is planar if, and only if, there exist another graph $G_0^* = (V_0^*, E_0^*)$ and a bijection $\tau : E \longrightarrow E_0^*$ such that $\dim\langle \tau E_v | \forall v \in V\rangle = \beta(G^*)$.

Proof The case of $p = 0$ in Corollary 18.3. □

In this corollary, G_0^* is the algebraic dual of Whitney.

Lemma 18.9 Let $\mathscr{C}$ be a nonorientable double covering of G with $\beta(G) - q + 1$ travels where q is a positive integer; if the dimension of the space generated by $\mathscr{C}$ is $\dim\langle\mathscr{C}\rangle = \beta(G) - q$, then $\mathscr{C}$ is consistent.

Proof Denote $k = \beta(G) - q + 1$, then $\dim\langle\mathscr{C}\rangle = k - 1$. Because $\beta(G) = \epsilon(G) - \nu(G) + 1$, $k = \epsilon(G) - \nu(G) - q + 2$. Assume $\mathscr{C}$ is not consistent, then the polyhedron Σ represented by $\mathscr{C}$ has at least $\nu(G) + 1$ vertices. Because $\epsilon(\Sigma) = \epsilon(G)$,

$$\begin{aligned}\dim(\Sigma) &= \epsilon(\Sigma) - \nu(\Sigma) + 1 - q \\ &\leq \epsilon(G) - (\nu(G) + 1) - q + 1 \\ &= \dim\langle\mathscr{C}\rangle - 1.\end{aligned}$$

This is a contradiction to $\dim(\Sigma) = \dim\langle\mathscr{C}\rangle$. Thus, $\mathscr{C}$ is consistent. □

Lemma 18.10 Let $\mathscr{C}$ represent an embedding of G on a nonorientable surface of genus q, then

$$\dim\langle\mathscr{C}\rangle = \beta(G) - q.$$

Proof Since $\mathscr{C}$ is an embedding of G on a nonorientable surface of genus q, from Eulerin characteristic, this embedding has $\phi = |\mathscr{C}| = \beta(G) - q + 1$ faces. Assume ∂f_i is the boundary of face f_i, $1 \leq i \leq \phi$. Because

$$\sum_{1\leq i\leq \phi} \partial f_i = 0,$$

we have

$$\dim\langle\mathscr{C}\rangle \leq \beta(G) - 2p = \phi - 1.$$

In virtue of the connectedness of G, any $\phi-1$ elements of $\mathscr{C}$ as vectors are independent (otherwise, the $\phi-1$ travels in $\mathscr{C}$ form a polyhedron themselves. However, the skeleton of this polyhedron is a proper subgraph of G. This contracts to the connectedness of G). Therefore, $\dim\langle\mathscr{C}\rangle = \beta(G) - q$. □

Theorem 18.4 A graph G can be embedded on a nonorientable surface of genus q if, and only if, there exists a nonorientable double covering of G with $\beta(G) - q + 1$ travels such that $\dim\langle\mathscr{C}\rangle = \beta(G) - q$.

Proof Necessity. If Σ is an embedding of G on a nonorientable surface of genus q, let $\mathscr{C}$ be the set of all face boundaries of polyhedron Σ. From Eulerin characteristic, $|\mathscr{C}| = \beta(G) - 2p + 1$. In virtue of Lemma 5, $\dim\langle\mathscr{C}\rangle = \beta(G) - q$.

Sufficiency. Assume $\mathscr{C}$ is a nonorientable double covering of G with $\beta(G) - q + 1$ travels and $\dim\langle\mathscr{C}\rangle = \beta(G) - q$, then from Lemmas 18.1 and 18.4, we have $\mathscr{C}$ is an embedding of G on the nonorientable surface of genus q. □

18.3 Via joint trees

The term "joint three" seems first appeared in Liu [72] and then in Liu [73] in a certain detail and in Liu [79] first in English. However, its theoretical idea was initiated in early articles of the author Liu [60, 61] from which maximum genus of a graph in both orientable and nonorientable cases were investigated.

The central idea is to transform a problem related to embeddings of a graph on surfaces (i.e., compact 2-manifolds without boundary in topology) into that on polyhegons (or polygons of even size with binary boundaries). The following two principles can be seen in Liu [79].

Principle A Joint trees of a graph have a one-to-one correspondence between embeddings of the graph with the same orientability and genus (i.e., on the same surfaces).

Principle B Associate polyhegons (as surfaces) of a graph have a one-to-one correspondence between joint trees of the graph with the same orientability and genus (i.e., on the same surfaces).

The two principles above are employed in this paper as the theoretical foundation. They enable us to discuss in any way among associate polyhegons, joint trees and embeddings of a graph considered.

Given a graph $G = (V, E)$, let T be a spanning tree of G and $\bar{T}$, its cotree. If each cotree edge $a \in \bar{T}$ is replaced by its two semiedges as new edges marked by $a = a^{+1}$(or a) and a^{-1}(or $\bar{a}$), then G becomes a tree, called an *extended tree* T, denoted by $\check{T}$.

If each vertex on $\check{T}$ is assigned with a rotation σ, then $\check{T}_\Sigma$ is, by Jordan closed curve axiom, seen as a planar embedding of $\check{T}$ where

$$\Sigma = \{\sigma(v) | \forall v \in V\}.$$

Let $a_1, a_2, \ldots, a_\beta, \beta = \beta(G)$ be the sequence of all cotree edges of G in agreement with the order of first occurrences around the face boundary of $\breve{T}_\Sigma$. Given a binary number of β digits as $\delta = \delta_1\delta_2 \cdots \delta_\beta$ where

$$\delta_i = \begin{cases} 1, & \text{when the two occurrences of } a_i \text{ are with same powers;} \\ 0, & \text{otherwise} \end{cases}$$

for $1 \le i \le \beta$. Thus, $\breve{T}_\Sigma^\delta$ is called a *joint tree* of G. The polyhegon of length 2β determined by the boundary of $\breve{T}_\Sigma^\delta$ is called an *associate polyhegon* of G (or joint tree $\breve{T}_\Sigma^\delta$).

From Liu [74, pp. 46–48], the following facts are known.

Fact 1 Given a graph $G = (V, E)$. For any δ chosen, the number of joint trees $\breve{T}_\Sigma^\delta$ is independent of the choice of T, i.e.,

$$\prod_{v \in V} (\rho(v) - 1)!$$

where $\rho(v)$ is the valency of $v \in V$.

Fact 2 Given a graph $G = (V, E)$. For any δ chosen, the set of associate polyhegons is independent of the choice of T. There is a one-to-one correspondence between the set of joint trees and the set of associate polyhegons.

Fact 3 Between the set of embeddings and the set of associate polyhegons of a graph, there is a one-to-one correspondence.

Fact 4 The general genera of a joint tree and its associate polyhegon of a graph are the same as that of their corresponding embedding.

The last fact enables us to call an associate polyhegon as an *associate surface* and then from the facts, an embedding of graph G can be represented by $\{\Sigma(G), \delta\}$, which is so called the *joint tree model.*

A *partition* of a set X is such a set of subsets of X that any two subsets are without common element and the union of all the subsets is X.

Theorem 18.5 A partition $P(X)$ of a set X determines an equivalence on X such that the subsets in $P(X)$ are the equivalent classes.

Let $P(X) = \{p_1, p_2, \ldots, p_{k_1}\}$ and $Q(X) = \{q_1, q_2, \ldots, q_{k_2}\}$ be two partitions of X. If for any q_j, $1 \le j \le k_1$, there exists a p_i, $1 \le i \le k_2$, such that $q_j \subseteq p_i$, then $Q(X)$ is called a *refinement* of $P(X)$, and $P(X)$ an *enlargement* of $Q(X)$ except only for $P(X) = Q(X)$. The partition of X with each subset, of a single element, or only one subset, which is X in its own right is, respectively, called the 1-*partition*, or 0-*partition* and denoted by $1(X)$, or $0(X)$.

Theorem 18.6 For a set X and its partition $P(X)$, the 0-partition $0(X)$ (or 1-partition $1(X)$) can be obtained by refinements (or enlargements) for at most $O(\log|X|)$ times in the worst case.

Proof In the worst case, it suffices to consider $P(X) = 1(X)$(or $0(X)$) and only one more subset produced in a refinement. This induces

$$1 + 2 + 2^2 + \cdots + 2^{\log|X|} = \frac{2^{1+\log|X|} - 1}{2 - 1} = O(|X|) \tag{18.8}$$

times of refinements (or enlargements) needed for getting $0(X)$ (or $1(X)$). The theorem is obtained. □

For two partitions $P = \{p_1, p_2, \ldots, p_s\}$ and $Q = \{q_1, q_2, \ldots, q_t\}$ of a set X, the *family intersection* of P and Q is defined to be

$$P \cap Q = \bigcup_{i=1}^{s} \{p_i \cap q_1, p_i \cap q_2, \ldots, p_i \cap q_t\}. \tag{18.9}$$

Actually, $\{p_i \cap q_1, p_i \cap q_2, \ldots, p_i \cap q_t\}$ for $i = 1, 2, \ldots, l$ are partitions of p_i.

Theorem 18.7 The family intersection satisfies the commutative and associate laws. And further, $P \cap Q$ is a refinement of both P and Q.

A sequence of refinements of polyhegon P from $0(P)$ to $1(P)$, written as $\mathbf{P} = P^{\langle 0\rangle} P^{\langle 1\rangle} P^{\langle 2\rangle} \cdots P^{\langle s\rangle}$, is called a *scheme*, where $P^{\langle 0\rangle} = 0(P)$, $P^{\langle s\rangle} = 1(P)$ and $P^{\langle i\rangle} \prec P^{\langle i+1\rangle}$, $1 \le i \le s-1$.

Because a polyhegon is a cyclic order, each subset of a refinement in a scheme is dealt with a linear order.

Assume $X = \{x_1, x_2, \ldots, x_l\}$ is a subset of $P^{\langle i\rangle}$ and $\langle x_1, x_2, \cdots, x_l\rangle$ is a segment of subsets of $P^{\langle i+1\rangle}$, $0 \le i \le s-1$. x_j, $1 \le j \le l$, is called a *son* of X in $P^{\langle i\rangle}$.

If a segment as

$$\langle x_1 \cup \cdots \cup x_h \cup \cdots \cup x_k \cup \cdots \cup x_l\rangle$$

in $P^{\langle i\rangle}$ is replaced by

$$\langle x_1 \cup \cdots \cup x_k \cup \cdots \cup x_h \cup \cdots \cup x_l\rangle$$

to get $P(x_h, x_k)$ from $P = P_0$, it is easy to check that $P(x_h, x_k)$ is a polyhegon as well. Such a transformation from P_0 to $P\ (x_h, x_k)$ is called an *interchanger* of two sons. A subset with at least two elements in **P** is called a *principler*. Specifically, P is a principler in its own right. Of course, the interchanger is meaningful only for principlers. The cardinality of a subset is called its *order*.

Attention 18.1 The scheme **P** of polyhegon P is transformed into the scheme $\mathbf{P}(x_h, x_k)$ of polyhegon $P(x_h, x_k)$ in company with the interchanger between x_h and x_k on P.

Let $\mathscr{P}$ and $\widetilde{\mathscr{P}}$ be the sets of, respectively, all polyhegons and their schemes generated by interchangers on P in company with its scheme **P**.

The graph $G_P = (V_P, E_P)$ where $V_P = \mathscr{P}$ and for any $R, S \in V_P$,

$$(R, S) \in E_P \Longleftrightarrow S \text{ can be done by an interchanger from } R,$$

which is called an *associate graph* of P.

Theorem 18.8 For any associate graph G_P of polyhegon P, we have $G_P = K_{|\mathscr{P}|}$, i.e., the complete graph of order $|\mathscr{P}|$.

Proof Because of $|\mathscr{P}| - 1$ schemes gotten from **P** by one interchanger, each vertex in G_P is of valency $|\mathscr{P}| - 1$. However, G_P is of order $|\mathscr{P}|$ and hence the conclusion. □

Corollary 18.5 In $G_P = (V_P, E_P)$, there is a Hamiltonian path.

Corollary 18.6 For any two vertices, $G_P = (V_P, E_P)$ has a Hamiltonian path between the two vertices.

For $v_0 = P_0$ arbitrarily chosen, if all interchangers are arranged in the order as π_1, $\pi_2, \pi_3, \ldots$, then by FRP (the Finite Recursion Principle in Section 1.1), $(v_0, v_1, v_2, v_3 \cdots)$ where for $i \geq 1$, $v_i = \pi_i \pi_{i-1} \cdots \pi_1 P_0$ is a Hamiltonian circuit on $G_P = (V_P, E_P)$. In virtue of the arbitrariness chosen for an order on the set of all interchangers, such a procedure is done on $G_P = (V_P, E_P)$ in time linear.

Two schemes with trees isomorphic are said to be *isomorphic*. Of course, all schemes of a polyhegon generated by all non-articulate vertices as root are *congruent*.

Theorem 18.9 A joint tree determines an isomorphic class of its schemes.

Proof In virtue of the procedure above, the theorem is done. □

Theorem 18.10 Associate polyhegon of graph G under no distinction have 2-tuple $\{\Sigma(G), \lambda\}$ as complete invariants where $\Sigma(G)$ and λ are, respectively, a set of rotations at vertices and a binary number of $\beta(G)$ digits.

Proof In virtue of the change of joint tree via an interchanger, by Facts 1–4 and Theorem 18.9, the conclusion can be drawn. □

A graph G can be seen as the set $\tilde{\mathscr{P}}_G$ of all classes of schemes that are with G as their associate graph. Then, the interchanger on $\tilde{\mathscr{P}}_G$ becomes the exchanger on the set of all associate surfaces of G.

Theorem 18.11 A class of schemes determines a joint tree. Conversely, a joint tree determines a class of schemes.

Proof Known from the description above. □

Then, an operation on $\tilde{\mathscr{P}}_G$ is discussed for transforming an associate polyhegon into another in order to visit all associate polyhegon without repetition.

A subset in a scheme with all its successors is also called a *branch*. The operation of interchanging the positions of two subsets with the same predecessor in the scheme is called an *exchanger*.

Lemma 18.11 An associate polyhegon of a graph under an exchanger is still another associate polyhegon. Conversely, the later under the same exchanger becomes the former.

Proof Similar to Lemma 12.3. □

On the basis of this lemma, the exchanger can be seen as an operation on the set of all associate polyhegons of a graph.

Lemma 18.12 The exchanger is closed in the set of all associate polyhegons of a graph.

Proof From Theorem 18.11, the lemma is a direct conclusion of Lemma 18.1. □

Lemma 18.13 Let $\mathscr{A}(G)$ be the set of all associate polyhegons of a graph G, then for any $S_1, S_2 \in \mathscr{A}(G)$, there exist a sequence of exchangers on the set such that S_1 can be transformed into S_2.

Proof Because of exchange corresponding to transposition of two elements in a rotation at a vertex, in virtue of permutation principle that any two rotations can be transformed from one into another by transpositions, from Theorem 18.11 and Lemma 18.11, the conclusion is drawn. □

If $\mathscr{A}(G)$ is dealt as the vertex set and an edge as an exchanger, then what is obtained is called the *associate polyhegon graph* of G, and denoted by $\mathscr{H}(G)$. From Principle A, it is also called the *surface embedding graph* of G.

Theorem 18.12 In $\mathscr{H}(G)$, there is a Hamilton path. Further, for any two vertices, $\mathscr{H}(G)$ has a Hamilton path with the two vertices as ends.

Proof Since a rotation at each vertex is a cyclic permutation (or in short a cycle) on the set of semiedges with the vertex, an exchanger of layer segments is corresponding to a transposition on the set at a vertex.

Any two cycles at a vertex v can be transformed from one into another by $\rho(v)$ transpositions, where $\rho(v)$ is the valency of v, i.e., the order of cycle(rotation). This enables us to do exchangers from the first layer according to the order from left to right at one vertex to the other.

Because of the finiteness, an associate polyhegon can always be transformed into another by $|\mathscr{A}(G)|$ exchangers.

From Theorem 1 with Principles 1–2, the conclusion is drawn. □

First, starting from a surface in $\mathscr{A}(G)$, by doing exchangers at each principle segments in one layer to another, a Hamilton path can always be found in considering Theorems 18.1 and 18.2. Then, a Hamilton path can be found on $\mathscr{H}(G)$.

Further, for chosen $S_1, S_2 \in \mathscr{A}(G) = V(\mathscr{H}(G))$ adjective, starting from S_1, by doing exchangers avoid S_2 except the final step, on the basis of the strongly finite recursion principle, a Hamilton path between S_1 and S_2 can be obtained. In consequence, a Hamilton circuit can be found on $\mathscr{H}(G)$.

Corollary 18.7 In $\mathscr{H}(G)$, there exists a Hamilton circuit.

Theorem 18.12 tells us that the problem of determining the minimum, or maximum genus of graph G has an algorithm in time linear on $\mathscr{H}(G)$.

For a graph G, let $\mathscr{S}(G)$ be the the associate polehegons (or surfaces) of G, and $\mathbf{S}_p$ and $\mathbf{S}_{\tilde{q}}$ the subsets of, respectively, orientable and nonorientable polyhegons of genus $p \geq 0$ and $q \geq 1$.

Then, we have

$$\mathscr{S}(G) = \sum_{p\geq 0} \mathbf{S}_p + \sum_{q\geq 1} \mathbf{S}_{\tilde{q}}.$$

Theorem 18.13 A graph G can be embedded on an orientable surface of genus p if, and only if, $\mathscr{S}(G)$ has a polyhegon in $\mathbf{S}_p$, $p \geq 0$. Moreover, for an embedding of G, there exist a sequence of exchangers by which the corresponding polyhegon of the embedding can be transformed into one in $\mathbf{S}_p$.

Proof For an embedding of G on an orientable surface of genus p, there is an associate polyhegon in $\mathbf{S}_p$, $p \geq 0$. This is the necessity of the first statement.

Conversely, given an associate polyhegen in $\mathbf{S}_p$, $p \geq 0$, from Theorems 18.11–18.12 with Principles A and B, an embedding of G on an orientable surface of genus p can be done. This is the sufficiency of the first statement.

The last statement of the theorem is directly seen from the proof of Theorem 18.12. □

For an orientable embedding $\mu(G)$ of G, denote by $\tilde{\mathbf{S}}_\mu$ the set of all nonorientable associate polyhegons induced from $\mu(G)$.

Theorem 18.14 A graph G can be embedded on a nonorientable surface of genus $q(\geq 1)$ if, and only if, $\mathscr{S}(G)$ has a polyhegon in $\tilde{\mathbf{S}}_q$, $q \geq 1$. Moreover, if G has an embedding $\tilde{\mu}$ on a nonorientable surface of genus q, then it can always be done from an orientable embedding μ arbitrarily given to another orientable embedding μ' by a sequence of exchangers such that the associate polyhegon of $\tilde{\mu}$ is in $\tilde{\mathbf{S}}_{\mu'}$.

Proof For an embedding of G on a nonorientable surface of genus q, Principle B leads to that its associate polyhegon is in $\mathbf{S}_q$, $q \geq 1$. This is the necessity of the first statement.

Conversely, let $S_{\tilde{q}}$ be an associate polyhegon of G in $\tilde{\mathbf{S}}_q$, $q \geq 1$. From Principles A and B, an embedding of G on a nonorientable surface of genus q can be found from $S_{\tilde{q}}$. This is the sufficiency of the first statement.

Since a non orientable embedding of G has exactly one under orientable embedding of G by Principle A, Theorem 2 directly leads to the second statement. □

18.4 Via configurations

The classical version of Jordan curve theorem in topology states that a single closed curve C separates the sphere into two connected components of which C is their common boundary. In this section, we investigate the polyhedral statements and proofs of the Jordan curve theorem.

Let $\Sigma = \Sigma(G; F)$ be a polyhedron whose underlying graph is $G = (V, E)$ with F as the set of faces. If any circuit C of G not a face boundary of Σ has the property that there exist two proper subgraphs In and Ou of G such that

$$In \bigcup Ou = G; \; In \bigcap Ou = C, \tag{18.10}$$

then Σ is said to have the *first Jordan curve property*, or simply written as 1-JCP. For a graph G, if there is a polyhedron $\Sigma = \Sigma(G; F)$, which has the 1-JCP, then G is said to have the 1-JCP as well.

Of course, in order to make sense for the problems discussed in this section, we always suppose that all the members of F in the polyhedron $\Sigma = \Sigma(G; F)$ are circuits of G.

Theorem A (First Jordan curve theorem) G has the 1-JCP if, and only if, G is planar.

Proof Because the homology $\mathscr{H}_1(\Sigma) = 0, \Sigma = \Sigma(G; F)$, from Theorem 4.2.5 in Liu [77], we know that $\text{Im}\,\partial_2 = \text{Ker}\,\partial_1 = \mathscr{C}$, the cycle space of G and hence $\text{Im}\,\partial_2 \supseteq F$, which contains a basis of $\mathscr{C}$. Thus, for any circuit $C \notin F$, there exists a subset D of F such that

$$C = \sum_{f \in D} \partial_2 f; \; C = \sum_{f \in F \setminus D} \partial_2 f. \tag{18.11}$$

Moreover, if we write

$$Ou = G[\bigcup_{f \in D} f]; \; In = G[\bigcup_{f \in F \setminus D} f],$$

then Ou and In satisfy the relations in (A) since any edge of G appears exactly twice in the members of F. This is the sufficiency.

Conversely, if G is not planar, then G has embedding only on surfaces of genus not 0. Because of the existence of non contractible circuit, such a circuit does not satisfy the 1-JCP and hence G is without 1-JCP. This is the necessity. □

Let $\Sigma^* = \Sigma(G^*; F^*)$ be a dual polyhedron of $\Sigma = \Sigma(G; F)$. For a circuit C in G, let $C^* = \{e^* \mid \forall e \in C\}$, or say the corresponding vector in $\mathscr{G}_1^*$, of $C \in \mathscr{G}_1$.

Lemma 18.14 Let C be a circuit in Σ. Then, $G^* \setminus C^*$ has at most two connected components.

Proof Assume H^* to be a connected component of $G^* \setminus C^*$ but not the only one. Let D be the subset of F corresponding to $V(H^*)$. Then,

$$C' = \sum_{f \in D} \partial_2 f \subseteq C.$$

However, if $\emptyset \neq C' \subset C$, then C itself is not a circuit. This is a contradiction to the condition of the lemma. Because any edge appears twice in the members of F, there is only one possibility, namely,

$$C = \sum_{f \in F \setminus D} \partial_2 f.$$

Hence, $F \setminus D$ determines the other connected component of $G^* \setminus C^*$ when $C' = C$. □

Any circuit C in G, which is the underlying graph of a polyhedron $\Sigma = \Sigma(G; F)$, is said to have the *second Jordan curve property*, or simply written as 2-JCP for Σ with its dual $\Sigma^* = \Sigma(G^*; F^*)$ if $G^* \setminus C^*$ has exactly two connected components. A graph G is said to have the 2- JCP if all the circuits in G have the property.

Theorem B (Second Jordan curve theorem) A graph G has the 2-JCP if, and only if, G is planar.

Proof To prove the necessity. Because for any circuit C in G, $G^*\backslash C^*$ has exactly two connected components, any C^* that corresponds to a circuit C in G is a cocircuit. Since any edge in G^* appears exactly twice in the elements of V^*, which are all cocircuits, from Lemma 1, V^* contains a basis of Ker δ_1^*. Moreover, V^* is a subset of Im δ_0^*. Hence, Ker $\delta_1 \subseteq$ Im δ_0. From Lemma 4.3.2 in Liu [77], Im $\delta_0^* \subseteq$ Ker δ_1^*. Then, we have Ker $\delta_1^* =$ Im δ_0^*, i.e., the cohomology $\mathscr{H}_1(\Sigma^*) = 0$. From the dual case of Theorem 4.3.2 in Liu [77], G^* is planar and hence so is G.

Conversely, to prove the sufficiency. From the planar duality, for any circuit C in G, C^* is a cocircuit in G^*. Then, $G^*\backslash C^*$ has two connected components and hence C has the 2-JCP. □

For a graph G, of course, connected without loop, associated with a polyhedron $\Sigma = \Sigma(G;F)$, let C be a circuit and E_C the set of edges incident to, but not on C. We may define an equivalence on E_C, denoted by $\sim_C$ as the transitive closure of that $\forall a, b \in E_C$,

$$\begin{aligned} a \sim_C b \Leftrightarrow \exists f \in F,\ & (a^\alpha C(a,b) b^\beta \subset f) \\ & \vee (b^{-\beta} C(b,a) a^{-\alpha} \subset f), \end{aligned} \tag{18.12}$$

where $C(a,b)$, or $C(b,a)$, is the common path from a to b, or from b to a in $C \cap f$, respectively. It can be seen that $|E_C/\sim_C| \leq 2$ and the equality holds for any C not in F only if Σ is orientable.

In this case, the two equivalent classes are denoted by $E_{\mathscr{L}} = E_{\mathscr{L}}(C)$ and $E_{\mathscr{R}} = E_{\mathscr{R}}(C)$. Further, let $V_{\mathscr{L}}$ and $V_{\mathscr{R}}$ be the subsets of vertices by which a path between the two ends of two edges in $E_{\mathscr{L}}$ and $E_{\mathscr{R}}$ without common vertex with C passes, respectively.

From the connectedness of G, it is clear that $V_{\mathscr{L}} \cup V_{\mathscr{R}} = V\backslash V(C)$. If $V_{\mathscr{L}} \cap V_{\mathscr{R}} = \emptyset$, then C is said to have the *third Jordan curve property*, or simply written as 3-JCP. In particular, if C has the 3-JCP, then every path from $V_{\mathscr{L}}$ to $V_{\mathscr{R}}$ (or vice versa) crosses C and hence C has the 1-JCP. If every circuit that is not the boundary of a face f of $\Sigma(G)$, one of the underlain polyhedra of G has the 3-JCP, then G is said to have the 3-JCP as well.

Lemma 18.15 Let C be a circuit of G that is associated with an orientable polyhedron $\Sigma = \Sigma(G;F)$. If C has the 2-JCP, then C has the 3-JCP. Conversely, if $V_{\mathscr{L}}(C) \neq \emptyset$, $V_{\mathscr{R}}(C) \neq \emptyset$ and C has the 3-JCP, then C has the 2-JCP.

Proof For a vertex $v^* \in V^* = V(G^*)$, let $f(v^*) \in F$ be the corresponding face of Σ. Suppose In^* and Ou^* are the two connected components of $G^*\backslash C^*$ by the 2-JCP of C. Then,

$$In = \bigcup_{v^* \in In^*} f(v^*) \quad \text{and} \quad Ou = \bigcup_{v^* \in Ou^*} f(v^*)$$

are subgraphs of G such that $In \cup Ou = G$ and $In \cap Ou = C$. Also, $E_{\mathscr{L}} \subset In$ and $E_{\mathscr{R}} \subset Ou$ (or vice versa). The only thing remained is to show $V_{\mathscr{L}} \cap V_{\mathscr{R}} = \emptyset$. By contradiction, if $V_{\mathscr{L}} \cap V_{\mathscr{R}} \neq \emptyset$, then In and Ou have a vertex that is not on C in common and hence have an edge incident with the vertex, which is not on C, in common. This is a contradiction to $In \cap Ou = C$.

Conversely, from Lemma 18.14, we may assume that $G^*\backslash C^*$ is connected by contradiction. Then there exists a path P^* from v_1^* to v_2^* in $G^*\backslash C^*$ such that $V(f(v_1^*)) \cap V_{\mathscr{L}} \neq \emptyset$ and $V(f(v_2^*)) \cap V_{\mathscr{R}} \neq \emptyset$. Consider

$$H = \bigcup_{v^* \in P^*} f(v^*) \subseteq G.$$

Suppose $P = v_1 v_2 \cdots v_l$ is the shortest path in H from $V_{\mathscr{L}}$ to $V_{\mathscr{R}}$.

To show that P does not cross C. By contradiction, assume that v_{i+1}, which is the first vertex of P, crosses C. From the shortestness, v_i is not in $V_{\mathscr{R}}$. Suppose that subpath $v_{i+1} \cdots v_{j-1}$, $i + 2 \leq j < l$, lies on C and that v_j does not lie on C. By the definition of $E_{\mathscr{L}}$, $(v_{j-1}, v_j) \in E_{\mathscr{L}}$ and hence $v_j \in V_{\mathscr{L}}$. This is a contradiction to the shortestness. However, from that P does not cross C, $V_{\mathscr{L}} \cap V_{\mathscr{R}} \neq \emptyset$. This is a contradiction to the 3-JCP. □

Theorem C (Third Jordan curve theorem) Let $G = (V, E)$ be with an orientable polyhedron $\Sigma = \Sigma(G; F)$. Then, G has the 3-JCP if, and only if, G is planar.

Proof From Theorem B and Lemma 2, the sufficiency is obvious. Conversely, assume that G is not planar. By Lemma 4.2.6 in Liu [77], $\mathrm{Im}\partial_2 \subseteq \mathrm{Ker}\partial_1 = \mathscr{C}$, the cycle space of G. By Theorem 4.2.5 in Liu [77], $\mathrm{Im}\partial_2 \subset \mathrm{Ker}\partial_1$. Then, from Theorem B, there exists a circuit $C \in \mathscr{C} \backslash \mathrm{Im}\partial_2$ without the 2-JCP. Moreover, we also have that $V_{\mathscr{L}} \neq \emptyset$ and $V_{\mathscr{R}} \neq \emptyset$. If otherwise $V_{\mathscr{L}} = \emptyset$, let

$$D = \{f \mid \exists e \in E_{\mathscr{L}}, e \in f\} \subset F.$$

Because $V_{\mathscr{L}} = \emptyset$, and any $f \in D$ contains only edges and chords of C, we have

$$C = \sum_{f \in D} \partial_2 f,$$

which contradicts to $C \notin \mathrm{Im}\partial_2$. Therefore, from Lemma 18.15, C does not have the 3-JCP. The necessity holds. □

For S_g as a surface (orientable, or nonorientable) of genus g, if a graph H is not embedded on a surface S_g but what obtained by deleting an edge from H is embeddable on S_g, then H is said to be *reducible* for S_g. In a graph G, the subgraphs of G homeomorphic to H are called a type of *reducible configuration* of G, or in short a *reduction*. Robertson and Seymour [140] have shown that graphs have their types of reductions for a surface of genus given finite. However, even for projective plane the simplest nonorientable surface, the types of reductions are more than 100 in Archdeacon [2]) and Glover et al. [32].

For a surface S_g, $g \geq 1$, let $\mathscr{H}_{g-1}$ be the set of all reductions of surface S_{g-1}. For $H \in \mathscr{H}_{g-1}$, assume the embeddings of H on S_g have ϕ faces. If a graph G has a decomposition of ϕ subgraphs H_i, $1 \leq i \leq \phi$, such that

$$\bigcup_{i=1}^{\phi} H_i = G; \ \bigcup_{i\neq j}^{\phi} (H_i \bigcap H_j) = H; \tag{18.13}$$

all H_i, $1 \leq i \leq \phi$, are planar and the common vertices of each H_i with H in the boundary of a face, then G is said to be with the *reducibility* 1 for the surface S_g.

Let $\Sigma^* = (G^*; F^*)$ be a polyhedron that is the dual of the embedding $\Sigma = (G; F)$ of G on surface S_g. For surface S_{g-1}, a reduction $H \subseteq G$ is given. Denote $H^* = [e^* | \forall e \in E(H)]$. Naturally, $G^* - E(H^*)$ has at least $\phi = |F|$ connected components. If exact ϕ components and each component planar with all boundary vertices are successively on the boundary of a face, then Σ is said to be with the *reducibility* 2.

If a graph G has an embedding with reducibility 2, then G is said to be with *reducibility* 2 as well.

Given $\Sigma = (G; F)$ as a polyhedron with under graph $G = (V, E)$ and face set F. Let H be a reduction of surface S_{p-1} and $H \subseteq G$. Denote by C the set of edges on the boundary of H in G and E_C, the set of all edges of G incident to but not in H. Let us extend the relation $\sim_C$: $\forall a, b \in E_C$,

$$a \sim_C b \Leftrightarrow \exists f \in F_H,\ a, b \in \partial_2 f \tag{18.14}$$

by transitive law as a equivalence. Naturally, $|E_C / \sim_C| \leq \phi_H$. Denote by $\{E_i | 1 \leq i \leq \phi_C\}$ the set of equivalent classes on E_C. Note that $E_i = \emptyset$ can be missed without loss of generality. Let V_i, $1 \leq i \leq \phi_C$, be the set of vertices on a path between two edges of E_i in G avoiding boundary vertices. When $E_i = \emptyset$, $V_i = \emptyset$ is missed as well. By the connectedness of G, it is seen that

$$\bigcup_{i=1}^{\phi_C} V_i = V - V_H. \tag{18.15}$$

If for any $1 \leq i < j \leq \phi_C$, $V_i \cap V_j = \emptyset$, and all $[V_i]$ planar with all vertices incident to E_i on the boundary of a face, then H, G as well, is said to be with *reducibility* 3.

Theorem 18.15 A graph G can be embedded on a surface $S_g(g \geq 1)$ if, and only if, G is with the reducibility 1.

Proof Necessity. Let $\mu(G)$ be an embedding of G on surface $S_g(g \geq 1)$. If $H \in \mathscr{H}_{g-1}$, then $\mu(H)$ is an embedding on $S_g(g \geq 1)$ as well. Assume $\{f_i | 1 \leq i \leq \phi\}$ is the face set of $\mu(H)$, then $G_i = [\partial f_i + E([f_i]\text{in})]$, $1 \leq i \leq \phi$, provides a decomposition satisfied by (18.13). It is easy to show that all G_i, $1 \leq i \leq \phi$, are planar. And, all the common edges of G_i and H are successively in a face boundary. Thus, G is with reducibility 1.

Sufficiency. Because G has reducibility 1, let $H \in \mathscr{H}_{g-1}$, and assume the embedding $\mu(H)$ of H on surface S_g has ϕ faces. Let G have ϕ subgraphs H_i, $1 \leq i \leq \phi$, satisfied by (1), and all H_i planar with all common edges of H_i and H in a face boundary. Denote by $\mu_i(H_i)$ a planar embedding of H_i with one face whose boundary is in a face boundary of $\mu(H)$, $1 \leq i \leq \phi$. Put each $\mu_i(H_i)$ in the corresponding face of $\mu(H)$, an embedding of G on surface S_g $(g \geq 1)$ is obtained. □

Theorem 18.16 A graph G can be embedded on a surface $S_g(g \geq 1)$ if, and only if, G is with the reducibility 2.

Proof Necessity. Let $\mu(G) = \Sigma = (G; F)$ be an embedding of G on surface S_g $(g \geq 1)$ and $\mu^*(G) = \mu(G^*) = (G^*, F^*)(= \Sigma^*)$, its dual. Given $H \subseteq G$ as a reduction. From the duality between the two polyhedra $\mu(H)$ and $\mu^*(H)$, the interior domain of a face in $\mu(H)$ has at least a vertex of G^*, $G^* - E(H^*)$ has exactly $\phi = |F\mu(\mathrm{H})|$ connected components. Because each component on a planar disc with all boundary vertices is successively on the boundary of the disc, H is with the reducibility 2. Hence, G has the reducibility 2.

Sufficiency. By employing the embedding $\mu(H)$ of reduction H of G on surface $S_g(g \geq 1)$ with reducibility 2, put the planar embedding of the dual of each component of $G^* - E(H^*)$ in the corresponding face of $\mu(H)$ in agreement with common boundary, an embedding of $\mu(G)$ on surface S_g $(g \geq 1)$ is easily done. □

Theorem 18.17 A 3-connected graph G can be embedded on a surface S_g $(g \geq 1)$ if, and only if, G is with reducibility 3.

Proof Necessity. Assume $\mu(G) = (G, F)$ is an embedding of G on surface S_g $(g \geq 1)$. Given $H \subseteq G$ as a reduction of surface S_{p-1}. Because $H \subseteq G$, the restriction $\mu(H)$ of $\mu(G)$ on H is also an embedding of H on surface S_g $(g \geq 1)$. From the 3-connectedness of G, edges incident to a face of $\mu(H)$ are as an equivalent class in E_C. Moreover, the subgraph determined by a class is planar with boundary in coincidence, i.e., H has the reducibility 3. Hence, G has the reducibility 3.

Sufficiency. By employing the embedding $\mu(H)$ of the reduction H in G on surface $S_g(g \geq 1)$ with the reducibility 3, put each planar embedding of $[V_i]$ in the interior domain of the corresponding face of $\mu(H)$ in agreement with the boundary condition, an embedding $\mu(G)$ of G on S_g $(g \geq 1)$ is extended from $\mu(H)$. □

18.5 Notes

18.5.1 For the embeddability of a graph on the torus, double torus etc. or in general orientable surfaces of genus small, more efficient characterizations are still necessary to be further contemplated on the basis of Theorem 18.1.

18.5.2 For the embeddability of a graph on the projective plane (1-crosscap), Klein bottle (2-crosscap), 3-crosscap etc. or in general nonorientable surfaces of genus small, more efficient characterizations are also necessary to be further contemplated on the basis of Theorem 18.2.

18.5.3 Tree-travels can be extended to deal with all problems related to embedings of a graph on surfaces as joint trees in a constructive way.

18.5.4 For the embeddability of a graph on the torus, double torus etc. or in general orientable surfaces of genus small, more efficient characterizations are still necessary to be further contemplated on the basis of Theorems 18.3–18.4.

18.5.5 For planarity of a graph, it suffices to discuss nonseparable graphs instead of general ones and hence Corollary 18.1, Corollary 18.2 and Corollary 18.4 are, respectively, corresponding to Theorems D, C and B.

18.5.6 A number of quotient spaces looks useful for investigating global structural properties of a graph. Homological and cohomological spaces are only shown as an example. More others are still necessary to pay attention to.

18.5.7 Theorems 18.11 and 18.12 enable us to establish a procedure for finding all embeddings of a graph G in linear space of the size of G and in linear time of size of $\mathscr{H}(G)$. The implementation of this procedure on computers can be seen in Wang and Liu [162].

18.5.8 In Theorems 18.13 and 18.14, it is necessary to investigate a procedure to extract a sequence of transpositions considered for the corresponding purpose efficiently.

18.5.9 On the basis of the associate polyhegons, the recognition of operations from a polyhegon of genus P to that of genus $p + k$ for given $k \geq 0$ have not yet be investigated. However, for the case $k = 0$ the operations are just Operation 0–Operation 2 all topological that are shown in Liu[69, 72–74].

18.5.10 It is worth investigating the associate polyhegon graph of a graph further for accessing the determination of the maximum (orientable) and minimum (orientable or nonorientable) genus of a graph.

18.5.11 On the basis of Theorems 18.15–18.17, the surface embeddability of a graph on a surface (orientabl or nonorientable) of genus smaller can be easily found with better efficiency.

For an example, the sphere S_0 has its reductions in two classes described as $K_{3,3}$ and K_5. Based on these, the characterizations for the embeddability of a graph on the torus and the projective plane has been established in Liu and Liu [116]. Because of the number of distinct embeddings of K_5 and $K_{3,3}$ on torus and projective plane much smaller as shown in the Appendix of Liu [78], the characterizations can be realized by computers with an algorithm much more efficient compared with the existences e.g., in Glover et al. [32].

18.5.12 The three polyhedral forms of Jordan closed planar curve axiom as shown in Section 2 initiated from Chapter 4 of Liu [64, 65] are first used for surface embeddings of a graph in LiuYi-Liu [116]. However, characterizations in that chapter are with a mistake of missing the boundary conditions as shown in this chapter.

18.5.13 The condition of 3-connectedness in Theorem 18.17 is not essential. It is only for the simplicity in description.

18.5.14 In all of Theorems 18.15–18.17, the conditions on planarity can be replaced by the corresponding Jordan curve property as shown in Section 2 in LiuYi and LiuYP[116] with the attention of the boundary conditions.

Appendix 1: Concepts of Polyhedra, Surfaces, Embeddings and Maps

This appendix provides the basic concepts of polyhedra, surfaces, embeddings and maps from original to developed as a compensation for Chapters 1 and 2. Only those available in the usage from combinatorization to algebraication are particularly concentrated on.

A1.1 Polyhedra

A *polyhedron* P is a set $\{C_i|1 \leq i \leq k\}$, $k \geq 1$, of cycles of letters such that each letter occurs exactly twice with the same power (or index) or different powers: 1 (always omitted) and -1 and denoted by $P = (\{C_i|1 \leq i \leq k\})$. It is seen as a set of all the cycles in any cyclic order.

This is a general statement of Heffter (see [40]). More than half a century later, Edmonds' statement appeared (see Edmonds [30]) as dual case, which has the minimality of no proper subset as a polyhedron for the convenience of usages.

A polyhedron is *orientable* if there is an orientation of each cycle, clockwise or anticlockwise, such that the two occurrences of each letter are with different powers; *nonorientable*, otherwise.

The *support* of polyhedron $P = (\{C_i|1 \leq i \leq k\})$ is the graph $G = (V_P, E_P)$ with a weight w on E_P where $V_P = \{C_i|1 \leq i \leq k\}$, $(C_i, C_j) \in E_P$ if, and only if, C_i and C_j, $1 \leq i, j \leq k$, have a common letter, and

$$w(e) = \begin{cases} 0, \text{ when two powers are different;} \\ 1, \text{ otherwise} \end{cases} \tag{A1.1}$$

for $e \in E_P$.

The set of all the edges with weight 1 is called the 1-*set* of the polyhedron.

Theorem A1.1 A polyhedron $P = (\{C_i|1 \leq i \leq k\})$ is orientable if, and only if, one of the following statements is satisfied:

(1) What obtained by contracting all edges of weight 0 on the support is a bipartite graph;
(2) No odd weight fundamental circuit is on the support;
(3) No odd weight circuit is on the support;
(4) The 1-set forms a cocycle;

DOI 10.1515/9783110481846-019

(5) The equation system about $x_i = x_{C_i}$, $C_i \in V_P$, on GF(2)

$$x_i + x_j = w(C_i, C_j) \tag{A1.2}$$

for $(C_i, C_j) \in E_P$ has a solution.

Proof Because P is orientable, the two occurrences of each letter are with different powers. Since the weights of all edges are the constant 0, the equation system (A1.2) has a solution of $x_i = 0$ for all $C_i \in V_P$, $1 \leq i \leq k$. Further, by considering that the consistency of equation system (A1.2) is not changed on switching the orientation of a cycle between clockwise and anticlockwise while interchanging the weights between 0 and 1 of all the edges incident with the cycle on the support, statement (5) is satisfied for any orientable polyhedron.

On the basis of statement (5), from a solution of equation system (A1.2) the vertices of G_P are classified into two classes by $x_i = 1$ or 0: 1-class or 0-class, respectively. According to equation (A1.2), each edge with weight 1 has its two ends in different classes and hence the 1-set is a cocycle. This is statement (4).

On the basis of statement (4), since any circuit meets even number of edges with a cocycle, all circuits are with even weight. This means no odd weight circuit. Therefore, statement (3) is satisfied.

On the basis of statement (3), statement (2) is naturally deduced because a fundamental circuit is a circuit in its own right.

On the basis of statement (2), by contracting all edges of weight 0 in each fundamental circuit on the support, (1) is satisfied.

On the basis of statement (1), the vertices are partitioned into two classes by the equivalence that two vertices are joined by even weight path. By switching the orientation of all vertices in one of the two classes and keeping those in the other class unchanged, a polyhedron without weight 1 edge is found. This implies that P is orientable.

In summary, the theorem is proved. □

On the support $G_P = (V_P, E_P)$ of a polyhedron P, the operation of switching the orientations of all vertices in a subset of V_P between clockwise and anticlockwise and the weights of all edges incident with just one end in the subset interchanged between 0 and 1 is called a *switch* on P.

Theorem A1.2 The orientability of a polyhedron does not change under switches.

Proof From the definition of orientability, the theorem is true. □

Let T be a minimal set of edges having an edge in common with all cocycles in the support of a polyhedron. In fact, it can seen that T is a spanning tree.

All the polyhedra obtained by switching on a polyhedron P are seen to be the *same* as P; *different*, otherwise. From Theorem A1.2, in order to discuss all different polyhedra it enables us to consider all such polyhedra of the support with weight 0 on all tree edge for a spanning tree chosen independently in any convenient way. Such a polyhedron is said to be *classic*.

Theorem A1.3 A classic polyhedron is orientable if, and only if, all edges as letters have their two occurrences with different powers. A classic polyhedron is nonorientable if, and only if, the set of letters each of which has its two occurrences with same power does not contain a cocycle.

Proof The first statement is deduced from statement (3) of Theorem A1.1. The second statement is by contradiction derived from Theorems A1.1 and A1.2. □

Now, a polyhedron (always summed to be classic below) P is considered as a permutation formed by its cycles. Let δ be the permutation with which each cycle only consists of the two occurrences of each letter in P. Then, the *dual*, denoted by P^*, of P is defined as $P^* = P\delta$ such that their supports are with the same weight. The cycles in P are called *faces* and those in P^* are *vertices*. Cycles in δ are edges. Let $\nu(P)$, $\epsilon(P)$ and $\phi(P)$ be, respectively, the number of vertices, edges and faces on P, then $\nu(P) - \epsilon(P) + \phi(P)$ is the *Eulerian characteristic* of P. The graph that is formed by vertices and edges of P is called a *skeleton* of P. Of course, the skeleton of P is the support of P^*.

Theorem A1.4 P^* is a polyhedron and $P^{**} = P$. P^* is orientable if, and only if, so is P with the same Eulerian characteristic.

Proof It is easily checked that P^* is a polyhedron from P as a polyhedron. Since $\delta^2 = 1$, the identity, we have

$$P^{**} = P^*\delta = (P\delta)\delta = P(\delta^2) = P.$$

This is the first statement. From Theorem A1.3, the second statement is obtained. □

A1.2 Surfaces

Surfaces seen as polyhedral polygons can be topologically classified by a type of equivalence. Let **P** be the set of all such polygons.

For $P = (\{(A_i)|i \geq 1\}) \in \mathbf{P}$, the following three operations, including their inverses, are called *elementary transformation*:

Operation 0: For $(A_i) = (Xaa^{-1}Y)$, $(A_i) \Leftrightarrow (XY)$, where at least one of X and Y is not empty;

Operation 1: For $(A_i) = (XabYab)$(or $(XabYb^{-1}a^{-1})$),

$$(A_i) \Leftrightarrow (XaYa) \text{ or } (XaYa^{-1});$$

Operation 2: For $(A_i) = (Xa)$ and $(A_j) = (a^{-1}Y)$, $i \neq j$,

$$(\{(A_i), (A_j)\}) \Leftrightarrow (XY),$$

where at least one of X and Y is not empty. Particularly,

$$(\{(A_i), (A_j)\}) \Leftrightarrow (XaYa^{-1}),$$

when both (X) and (Y) are polyhedra.

If a polyhedron P can be obtained by elementary transformation into another polyhedron Q, then they are called *elementary equivalence*, denoted by $P \sim_{\text{el}} Q$. In topology, the elementary equivalence is topological in 2-dimensional sense.

Lemma A1.1 For $P \in \mathbf{P}$, there exists a polyhedron $Q = (X) \in \mathbf{P}$ where X is a linear order such that $P \sim_{\text{el}} Q$.

Proof Let $P = (\{(A_i) | 1 \leq i \leq k\})$. If $k = 1$, P is in the form as Q itself. If $k \geq 2$, by employing Operation 2 step by step to reduce the number of cycles by 1 if any, the form Q can be found. □

Lemma A1.2 For $P \in \mathbf{P}$, if $P = ((A)(B))$ with both (A) and (B) as polyhedra, then for any $x \notin A \cup B$, $P \sim_{\text{el}} ((A)x(B)x^{-1})$.

Proof It is seen that

$$\begin{aligned} P = (AB) &\sim_{\text{el}} (Axx^{-1}B) \text{ (by Operation 0)} \\ &\sim_{\text{el}} ((Ax)(x^{-1}B)) \text{ (by Operation 2)} \\ &= ((A)x(B)x^{-1}). \end{aligned}$$

□

From Lemmas A1.1 and A1.2, for classifying $\mathbf{P}$ it suffices to only discuss polygons as Q.

Lemma A1.3 Let $Q = (AxByCx^{-1}Dy^{-1})$, then

$$Q \sim_{\text{el}} (ADxyBx^{-1}Cy^{-1}). \tag{A1.3}$$

Proof It is seen that

$$\begin{aligned} Q &\sim_{\text{el}} ((Axz)(z^{-1}ByCx^{-1}Dy^{-1})) \text{ (by Operation 2)} \\ &\sim_{\text{el}} (zADy^{-1}z^{-1}ByC) \text{ (by Operation 2)} \\ &= (ADxyBx^{-1}Cy^{-1}). \end{aligned}$$

□

Lemma A1.4 Let $Q = (AxByCx^{-1}Dy^{-1})$, then

$$Q \sim_{el} (BAxyx^{-1}DCy^{-1}). \tag{A1.4}$$

Proof It is seen that

$$\begin{aligned} Q &\sim_{el} ((x^{-1}Dy^{-1}Axz)(ByCz^{-1})) \text{ (by Operation 2)} \\ &\sim_{el} (BAxzx^{-1}DCz^{-1}) \text{ (by Operation 2)} \\ &= (BAxyx^{-1}DCy^{-1}). \end{aligned}$$

□

Lemma A1.5 Let $Q = (AxByCx^{-1}Dy^{-1})$, then

$$Q \sim_{el} (ADCBxyx^{-1}y^{-1}). \tag{A1.5}$$

Proof From Lemma A1.4 and then Lemma A1.3, the lemma is easily done. □

According to Lemma A1.5, if A is replaced by EA in polyhedron $(ADCB)$, then the relation is soon derived as

Relation 1: $(AxByCx^{-1}Dy^{-1}E) \sim_{el} (ADCBExyx^{-1}y^{-1})$.

Lemma A1.6 Let $Q = (AxBx) \in \mathbf{P}$, then $Q \sim_{el} (AB^{-1}xx)$.

Proof It is seen that

$$\begin{aligned} Q &\sim_{el} ((Axz)(z^{-1}Bx)) = ((zAx)(x^{-1}B^{-1}z)) \text{ (by Operation 2)} \\ &\sim_{el} (zAB^{-1}z) = (AB^{-1}xx) \text{ (by Operation 2)}. \end{aligned}$$

□

According to Lemma A1.6, if A is replaced by CA in polyhedron (AB^{-1}), then the relation is soon derived as

Relation 2: $(AxBxC) \sim_{el} (AB^{-1}Cxx)$.

Lemma A1.7 Let $Q = (Axyx^{-1}y^{-1}zz) \in \mathbf{P}$, then $Q \sim_{el} (Axyzyxz)$.

Proof It is seen that

$$\begin{aligned} Q &\sim_{el} ((zAxyt)(t^{-1}x^{-1}y^{-1}z)) \text{ (by Operation 2)} \\ &\sim_{el} (Axytyxt) \text{ (by Operation 2)} \\ &= (Axyzyxz). \end{aligned}$$

□

According to Lemma A1.7, by Relation 2 twice for x and y, the relation is easily derived as

Relation 3: $(Axyx^{-1}y^{-1}zz) \sim_{\text{el}} (Axxyyzz)$.

Lemma A1.8 If $Q \in \mathbf{P}$ orientable not as $(AxByCx^{-1}Dy^{-1}E)$, then $Q \sim_{\text{el}} (xx^{-1})(= O_0)$.

Proof Because Q is not in the above form, Q has to be in form as $(Axx^{-1}B)$. If both A and B are empty, then $Q \sim_{\text{el}} (xx^{-1})$; otherwise, $Q \sim_{\text{el}} (AB)$. Because (AB) still satisfies the given condition, by the finite recursion principle, (xx^{-1}) can be found. □

Theorem A1.5 For any $Q \in \mathbf{P}$ orientable, if $Q \nsim_{\text{el}} (xx^{-1})$, then there exists an integer $p \geq 1$ such that

$$Q \sim_{\text{el}} \left(\prod_{i=1}^{p} x_i y_i x_i^{-1} y_i^{-1} \right) (= O_p). \tag{A1.6}$$

Proof Because $Q \sim_{\text{el}} (Ax_1By_1Cx_1^{-1}Dy_1^{-1}E)$, by Relation 1 we have

$$Q \sim_{\text{el}} (ADCBEx_1y_1x_1^{-1}y_1^{-1}).$$

If $(ADCBE) \sim_{\text{el}} (xx^{-1})$, the $Q \sim_{\text{el}} (x_1y_1x_1^{-1}y_1^{-1})$. That is, the case $p = 1$. Otherwise $(ADCBE) = (A_1x_2B_1y_2C_1x_2^{-1}D_1y_2^{-1}E_1)$. Because

$$(ADCBEx_1y_1x_1^{-1}y_1^{-1}) = (A_1x_2B_1y_2C_1x_2^{-1}D_1y_2^{-1}E_1)x_1y_1x_1^{-1}y_1^{-1}$$

is still in the given condition. By the finite recursion principle, (A1.6) is found. □

Theorem A1.6 For any $Q \in \mathbf{P}$ nonorientable, there exists an integer $q \geq 1$ such that

$$Q \sim_{\text{el}} \left(\prod_{i=1}^{q} x_i x_i \right) (= Q_q). \tag{A1.7}$$

Proof Because Q is nonorientable, there is a letter x_1 in Q such that $Q = (Ax_1Bx_1C)$. By Relation 2, $D \sim_{\text{el}} (AB^{-1}Cx_1x_1)$. If $(AB^{-1}C) \sim_{\text{el}} (xx^{-1})$, then by Operation 0 we have $Q \sim_{\text{el}} (x_1x_1)$. This is the case of $q = 1$. Otherwise, there exits an integer $k \geq 1$ such that

$$Q \sim_{\text{el}} \left(A \prod_{i=1}^{k} x_i x_i \right) (= Q_q)$$

and $(A) \nsim_{\text{el}} (xx^{-1})$ is orientable. By Theorem 1.5, there exists an integer $s \geq 1$ such that

$$(A) \sim_{\text{el}} \left(\prod_{i=1}^{s} x_i y_i x_i^{-1} y_i^{-1} \right) (= O_p).$$

Thus, by Relation 3 for s times, we have

$$Q \sim_{\mathrm{el}} \left(\prod_{i=1}^{2s+k} x_i x_i \right) (= Q_q).$$

That is, $q = 2s + k \geq 1$. □

On the basis of Lemma A1.8 and Theorems A1.5 and A1.6, surfaces in topology are in fact the classes of polyhedra under the elementary equivalence. Surfaces O_0, O_p, $p \geq 1$, are, respectively, orientable *standard surfaces* of genus 0, p, $p \geq 1$. Surfaces Q_q, $q \geq 1$, are nonorientable standard surfaces of genus q.

A1.3 Embeddings

An *embedding* (i.e., *cellular embedding* in early references particularly in topology and geometry) of a graph is such a polyhedron whose skeleton is the graph.

The distinction of embeddings are the same as polyhedra. Precisely speaking, two distinct embeddings on a 2-dimensional manifold are not equivalent topologically in 1-dimensional sense.

According to A1.1, all embeddings always imply to be classic.

For a graph $G = (V, E)$, Heffter-Edmonds' model of an embedding of G by rotation system at vertices, in fact, only for orientable case Heffter [40] and Edmonds [30].

Let $\sigma = \{\sigma_v | v \in V\}$ be the rotation system on G where σ_v is the cyclic order of semiedges at $v \in V$. Then, by the following procedure find an embedding of G.

Procedure A1.1 First, put different vertices in different positions marked by hole circles or bold points on the plane. Draw lines for edges such that no interior point passes through a vertex and σ_v is clockwise when v is a hole circle; anticlockwise, otherwise.

Then, by travelling along an edge in the rule: passing through on the same side when the two ends of the edges are in same type; crossing to the other side, otherwise. Find all cycles such that each edge occurs just twice. The set of cycles is denoted by P_G.

Lemma A1.9 P_G is a polyhedron.

Proof It can be easily checked from the definition of a polyhedron. □

Lemma A1.10 P_G is orientable.

Proof Because the dual is orientable, from Theorem A1.4, the lemma is true. □

Theorem A1.7 The dual of P_G is an orientable embedding of G.

Proof Because the support of the dual of P_G is G itself, the theorem is deduced. □

However, P_G in general is not classic except for all vertices are of same type.

Theorem A1.8 For a given rotation system σ of a graph G, let $P_G(\sigma; 0)$ be the polyhedron obtained by the procedure above for all vertices of same type, then $P_G(\sigma; 0)$ is unique.

Proof From the uniqueness of classic polyhedron in this case, the theorem is done. □

On the basis of Theorem A1.8, it suffices only to make all vertices with the same type, e.g., in clockwise. Further, in order to extend to nonorientable case, on account of Theorem A1.3, edges in a set not containing cocycle are marked for crossing one side to the other in the Heffter-Edmonds' model. The marked edges are called *twist*. This model as well as the Procedure A1.1 here are called an *expansion*.

Theorem A1.9 The dual of what is obtained in an expansion is a unique nonorientable embedding of G for twist edges fixed.

Proof Because one obtained in an expansion is a classic polyhedron, from the uniqueness of the dual of a polyhedron, the theorem is deduced. □

Theorem A1.10 All embeddings of a graph G obtained by expansions for all possible rotation system and twist edges in a subset of the cotree $\bar{T}$ of a given spanning tree T on G are distinct.

Proof As a result of Theorem A1.9. □

This theorem enables us to choose a spanning tree T on a graph G for discussing all embeddings of G on surfaces.

Let T_1 and T_2 be two spanning trees of a graph G. The sets of all embeddings of G as shown in Theorem A1.10 for T_1 and T_2 are, respectively, denoted by ϵ_1 and ϵ_2.

Theorem A1.11 Let ϵ_1^g and ϵ_2^g be, respectively, the subsets of ϵ_1 and ϵ_2 on surfaces of genus g (orientable $g = p \geq 0$, or nonorientable $g = q \geq 1$). Then $\epsilon_1^g = \epsilon_2^g$.

Proof Because of Theorem A1.10, it suffices only to discuss expansions for T_1 and T_2. Since $|\bar{T}_1| = |\bar{T}_2|$, Theorems A1.8 and A1.9 imply the theorem. □

For an embedding $P \in \epsilon_1^g$, if $P \notin \epsilon_2^g$, then there exits a twist edge e in T_2. By doing a switch with the fundamental cocircuit containing e for T_2, an embedding P' in the same distinct class with P is found. If no twist edge is in $\bar{T}_2$, then P' is the classic embedding in ϵ_2^g corresponding to P. Otherwise, by the finite recursion, a classic embedding $Q \in \mathcal{E}_2^g$ in the same distinct class with P is finally found. In this way, the one-to-one correspondence between $\mathcal{E}_1^g$ and $\mathcal{E}_2^g$ is established.

The last two theorems form the foundation of the joint tree model shown in Liu [72, 73]. Related topics are referred to Stahl [147, 148].

A1.4 Maps

Maps as polyhedra or embeddings of its underlying graph had no specific meaning until 1979 when Tutte (William T., 1917–2002) clarified that a map is a particular type of permutation on a set formed as a union of quadricells in Tutte [150–152]. All quadricells are with similar construction that four elements have the symmetry as a straight line segment with two ends and two sides.

This idea would go back to Klein (Felix, 1849–1925), who considered a triangulation of an embedding on a surface by inserting a vertex in the interior of each face and each edge and then connecting all line segments from a vertex in the interior of a face to all vertices on the boundary of the face. It is seen that each edge is adjacent to four triangles called *flags* as a quadricell. So, such a pattern of map used in this course can be named as Klein-Tutte's model. Related topics are referred to in Vince [153, 154].

Now, we have seen that a surface is determined by an elementary class of polyhedra, an embedding is by a distinct class of polyhedra and a map is by an isomorphic class of embeddings. The distinction of embeddings is based on edges labelled by letters, or numbers. This is also a kind of asymmetrization. But edges on a map are without labelling. Isomorphic maps are combinatorially considered with symmetry. So, a map is an isomorphic class of embeddings of its underlying graph.

Let $G = (V, E)$ be a graph. As shown in Section 1.1, $V = \mathrm{Par}(X)$ and $E = \{Bx|x \in X\}$, where $\mathrm{Par}(X)$ is a partition on $B(X) = \cup_{x\in X}Bx$, $Bx = \{x(0), x(1)\}$ for a set X. Two graphs $G_1 = (V_1, E_1)$ and $G_2 = (V_2, E_2)$ are *isomorphic* if, and only if, there exists a bijection ι: $X_1 \to X_2$ such that the diagrams

$$\begin{array}{ccc} X_1 & \xrightarrow{\iota} & X_2 \\ \downarrow{\scriptstyle\sigma_1} & & \downarrow{\scriptstyle\sigma_2} \\ X_1 & \xrightarrow[\iota]{} & X_2 \end{array} \tag{A1.8}$$

for $\sigma_i = B_i, \mathrm{Par}_i$, $i = 1, 2$, are commutative. Let $\mathrm{Aut}(G)$ be the automorphism group of G.

On the other hand, a *semi-arc isomorphism* between two graphs $G_1 = (V_1, E_1)$ and $G_2 = (V_2, E_2)$ is defined as a bijection τ: $B_1(X_1) \to B_2(X_2)$ that

$$\begin{array}{ccc} B_1(X_1) & \xrightarrow{\tau} & B_2(X_2) \\ \downarrow{\scriptstyle\sigma_1} & & \downarrow{\scriptstyle\sigma_2} \\ B_1(X_1) & \xrightarrow[\tau]{} & B_2(X_2) \end{array} \tag{A1.9}$$

for $\sigma_i = B_i, \mathrm{Par}_i$, $i = 1, 2$, are commutative. Let $\mathrm{Aut}_{1/2}(G)$ be the semi-arc automorphism group of G.

Theorem A1.12 If $\mathrm{Aut}(G)$ and $\mathrm{Aut}_{1/2}(G)$ are, respectively, the automorphism and semi-arc automorphism groups of graph G, then

$$\mathrm{Aut}_{1/2}(G) = \mathrm{Aut}(G) \times S_2^l, \tag{A1.10}$$

where l is the number of self-loops on G and S_2 is the symmetric group of degree 2.

Proof Because each automorphism of G just induces two semi-arc isomorphisms of G for a self-loop, the theorem is true. □

For map $M = (\mathscr{X}_{\alpha,\beta}, \mathscr{P})$, its automorphisms are discussed with asymmetrization in Chapter 8. Let $\mathscr{M}(G)$ be the set of all nonisomorphic maps with underlying graph G.

Lemma A1.11 For an automorphism ζ on map $M = (\mathscr{X}_{\alpha,\beta}(X), \mathscr{P})$, we have exhaustively $\zeta|_{B(X)} \in \mathrm{Aut}_{1/2}(G)$ and $\zeta\alpha|_{B(X)} \in \mathrm{Aut}_{1/2}(G)$, where $G = G(M)$, the underlying graph of M, and $B(X) = X + \beta X$.

Proof Because $\mathscr{X}_{\alpha,\beta}(X) = (X+\beta X)+(\alpha X+\alpha\beta X) = B(X)+\alpha B(X)$, by Conjugate Axiom each $\zeta \in \mathrm{Aut}(M)$ has exhaustively two possibilities: $\zeta|_{B(X)} \in \mathrm{Aut}_{1/2}(G)$ and $\zeta\alpha|_{B(X)} \in \mathrm{Aut}_{1/2}(G)$. □

Theorem A1.13 Let $\mathscr{E}_g(G)$ be the set of all embeddings of a graph G on a surface of genus g (orientable or nonorientable), then the number of nonisomorphic maps in $\mathscr{E}_g(G)$ is

$$m_g(G) = \frac{1}{2 \times \mathrm{aut}_{1/2}(G)} \sum_{\tau \in \mathrm{Aut}_{1/2}(G)} \left|\Phi(\tau)\right| \tag{A1.11}$$

where $\Phi(\tau) = \{M \in \mathscr{E}_g(G) | \tau(M) = M \text{ or } \tau\alpha(M) = M\}$.

Proof Suppose $X_1, X_2, \ldots, X_m$ are all the equivalent classes of $X = \mathscr{E}_g(G)$ under the group $\mathrm{Aut}_{1/2}(G) \times \langle\alpha\rangle$, then $m = m_g(G)$. Let $S(x) = \{\tau \in \mathrm{Aut}_{1/2}(G) \times \langle\alpha\rangle | \tau(x) = x\}$ be the stabilizer at x, a subgroup of $\mathrm{Aut}_{1/2}(G) \times \langle\alpha\rangle$. Because $|\mathrm{Aut}_{1/2}(G) \times \langle\alpha\rangle| = |S(x_i)||X_i|$, $x_i \in X_i$, $i = 1, 2, \ldots, m$, we have

$$m|\mathrm{Aut}_{1/2}(G) \times \langle\alpha\rangle| = \sum_{i=1}^{m} |S(x_i)||X_i|. \tag{1}$$

By observing $|S(x_i)|$ independent of the choice of x_i in the class X_i, the right side of (1) is

$$\begin{aligned}\sum_{x\in X}|S(x)| &= \sum_{x\in X}\sum_{\tau\in S(x)} 1\\ &= \sum_{\tau\in \mathrm{Aut}_{1/2}(G)\times\langle\alpha\rangle}\ \sum_{x=\tau(x)} 1 \\ &= \sum_{\tau\in \mathrm{Aut}_{1/2}(G)\times\langle\alpha\rangle}\left|\Phi(\tau)\right|.\end{aligned} \tag{2}$$

From (1) and (2), the theorem can be soon derived. □

The theorem above shows how to find nonisomorphic upper maps of a graph when the automorphism group of the graph is known.

Theorem A1.14 For a graph G, let $\mathscr{R}_g(G)$ and $\mathscr{E}_g(G)$ be, respectively, the sets of all non-isomorphic rooted upper maps and all distinct embeddings of G with size $\epsilon(G)$ on a surface of genus g (orientable or nonorientable). Then,

$$|\mathscr{R}_g(G)| = \frac{2\epsilon(G)}{\mathrm{aut}_{1/2}(G)}|\mathscr{E}_g(G)|. \tag{A1.12}$$

Proof Let $\mathscr{M}_g(G)$ be the set of all non-isomorphic upper maps of G. By (11.3),

$$|\mathscr{R}_g(G)| = \frac{4\epsilon(G)}{2\times \mathrm{aut}_{1/2}(G)}\sum_{M\in\mathscr{M}_g(G)}\frac{2\times \mathrm{aut}_{1/2}(G)}{\mathrm{aut}(M)}.$$

By considering that

$$\begin{aligned}2\times \mathrm{aut}_{1/2}(G) &= |\mathrm{Aut}_{1/2}(G)\times\langle\alpha\rangle|\\ &= \left|(\mathrm{Aut}_{1/2}(G)\times\langle\alpha\rangle)|_M\right|\times\left|\mathrm{Aut}_{1/2}(G)\times\langle\alpha\rangle(M)\right|\end{aligned}$$

and $(\mathrm{Aut}_{1/2}(G)\times\langle\alpha\rangle)|_M = \mathrm{Aut}(M)$, we have

$$\begin{aligned}|\mathscr{R}_g(G)| &= \frac{4\epsilon(G)}{|\mathrm{Aut}_{1/2}(G)\times\langle\alpha\rangle|}\sum_{M\in\mathscr{M}_g(G)}\left|\mathrm{Aut}_{1/2}(G)\times\langle\alpha\rangle(M)\right|\\ &= \frac{2\epsilon(G)}{\mathrm{aut}_{1/2}(G)}|\mathscr{E}_g(G)|.\end{aligned}$$

This is (A1.12). □

This theorem enables us to determine all the upper rooted maps of a graph when the automorphism group of the graph is known. However, the problem of finding an automorphism of a graph is much more difficult than that of finding an automorphism of a map on the basis of Chapter 8 in general. For asymmetric graphs, Theorems A1.13 and A1.14 provide much simpler results. More results are referred to in Mao et al. [126].

Appendix 2: Table of Genus Polynomials for Embeddings and Maps of Small Size

For a graph G, let $p_G(x)$, $\mu_G(x)$ and $\mu^r_G(x)$ be, respectively, the orientable genus distributions of embeddings, upper maps and rooted upper maps of G, or called *orientable genus polynomials*. Similarly, let $q_G(x^{-1})$, $\nu_G(x^{-1})$ and $\nu^r_G(x^{-1})$ be, respectively, the nonorientable genus distributions of embeddings, upper maps and rooted upper maps of G, or called *nonorientable genus polynomials*.

A2.1 Triconnected cubic graphs

First, list all non-isomorphic 3-connected cubic graphs from size 6 through 15.

Size 6

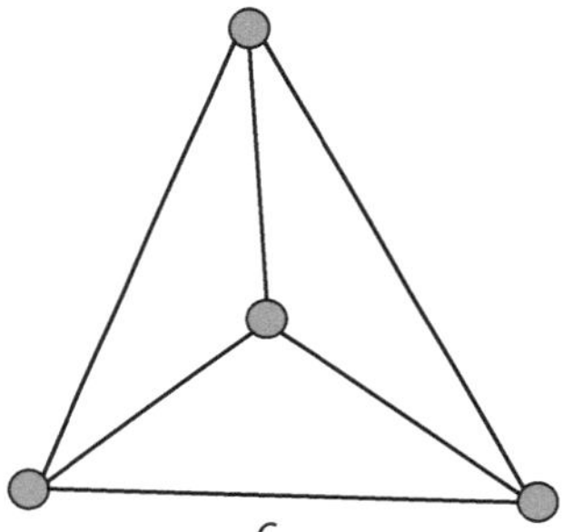

$C_{6,1}$

Size 9

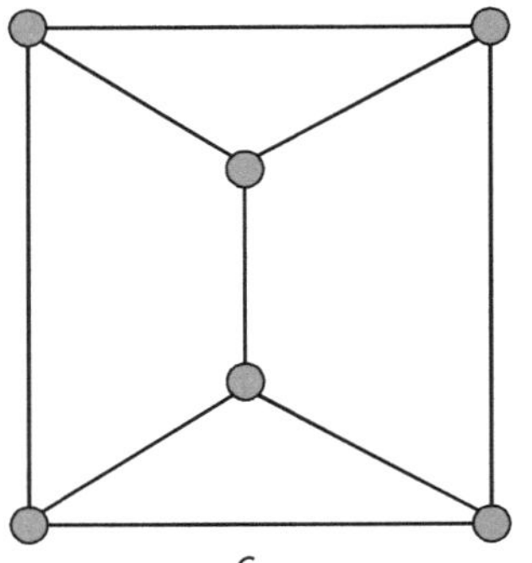

$C_{9,1}$

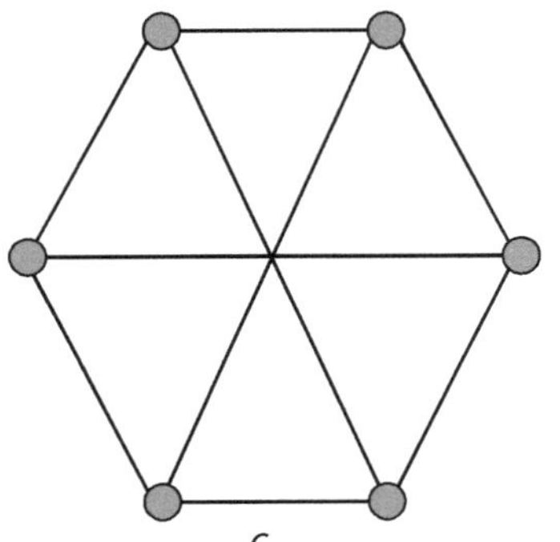

$C_{9,2}$

DOI 10.1515/9783110481846-020

Size 12

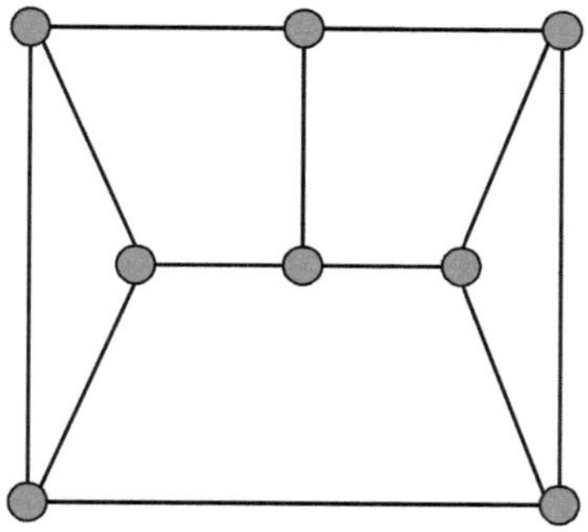

$C_{12,1}$

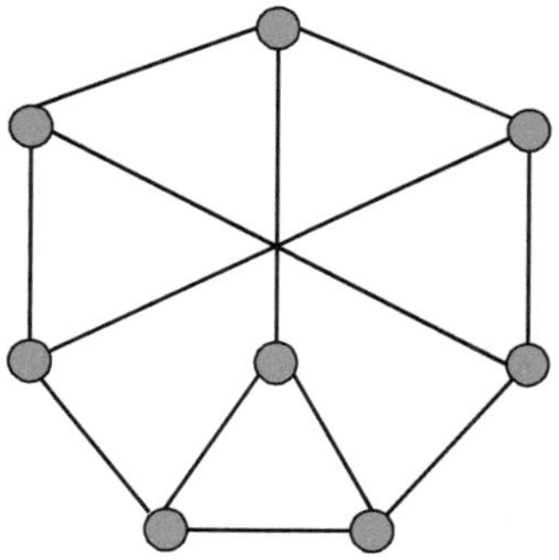

$C_{12,2}$

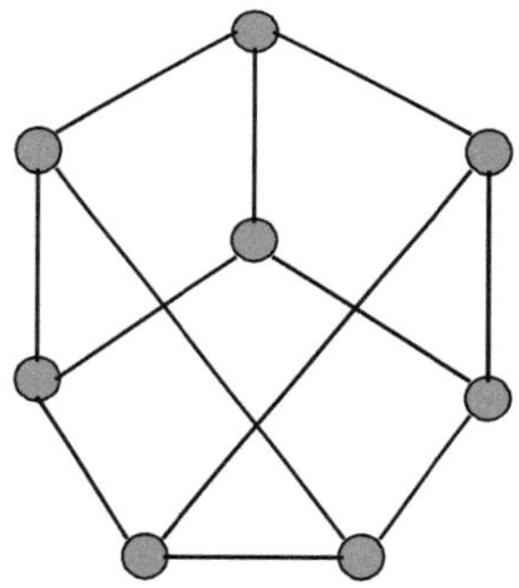

$C_{12,3}$

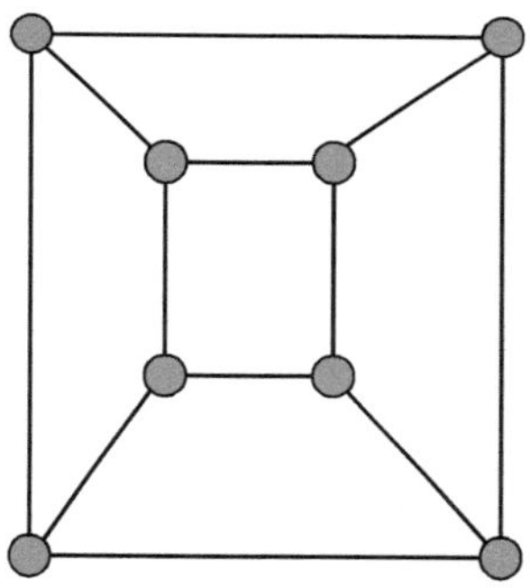

$C_{12,4}$

Size 15

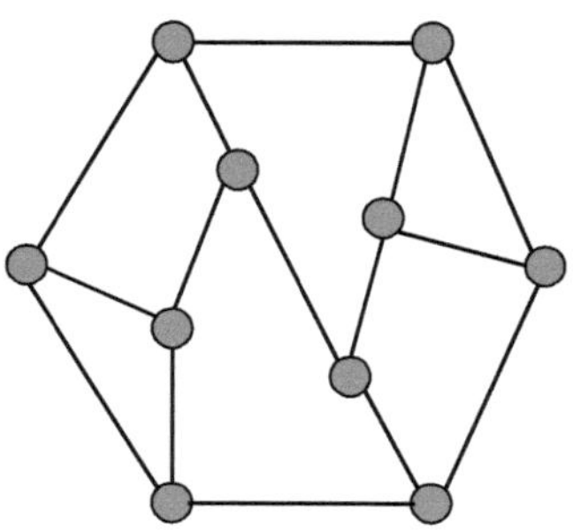

$C_{15,1}$

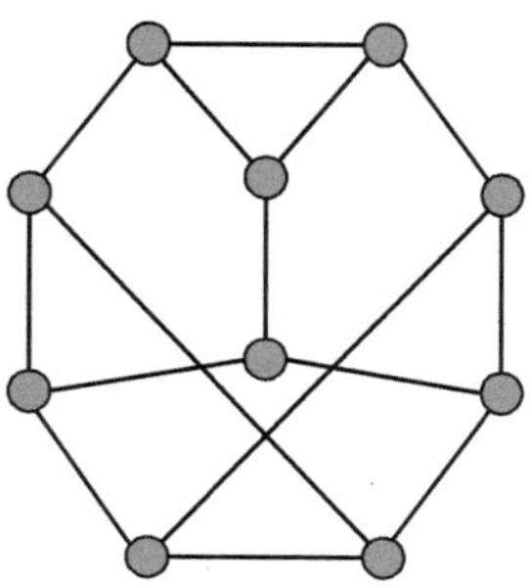

$C_{15,2}$

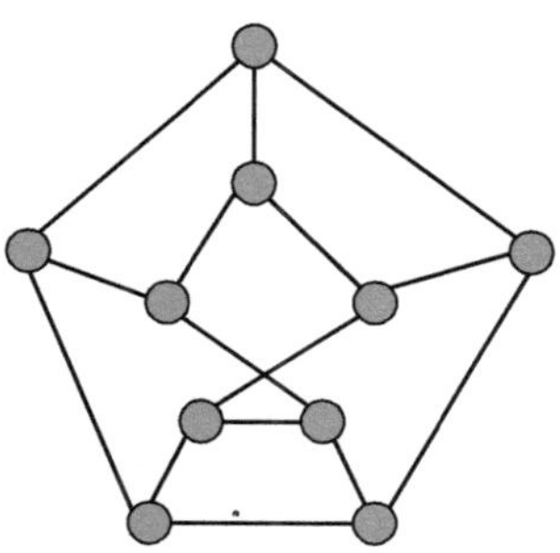

$C_{15,3}$

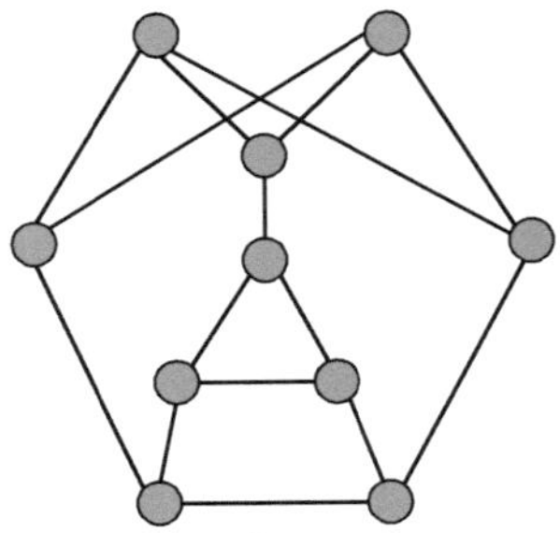

$C_{15,4}$

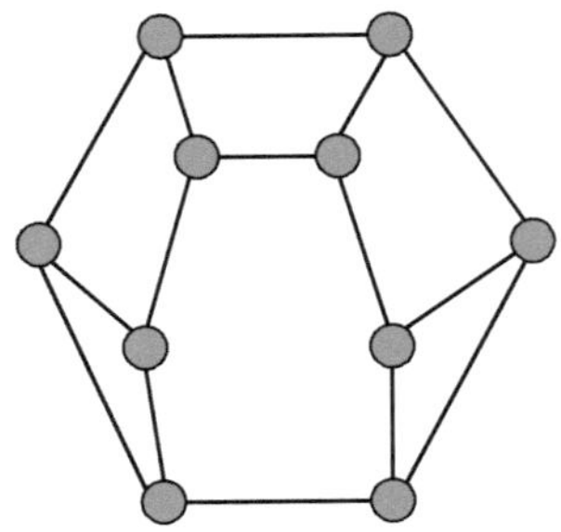

$C_{15,5}$

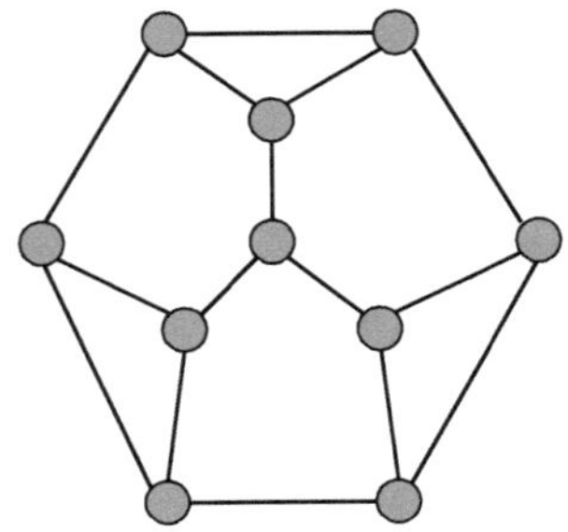

$C_{15,6}$

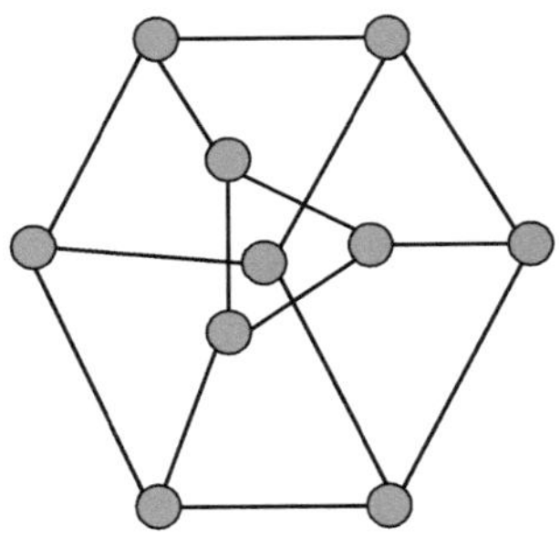

$C_{15,7}$

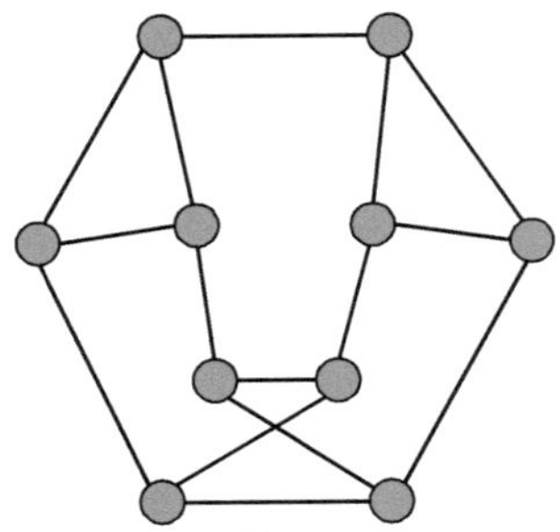

$C_{15,8}$

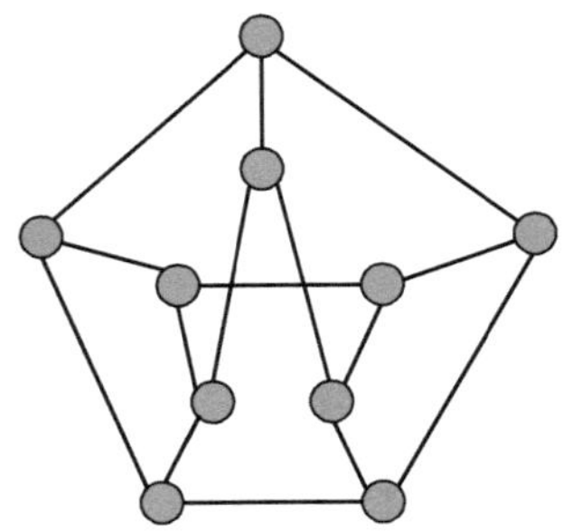

$C_{15,9}$

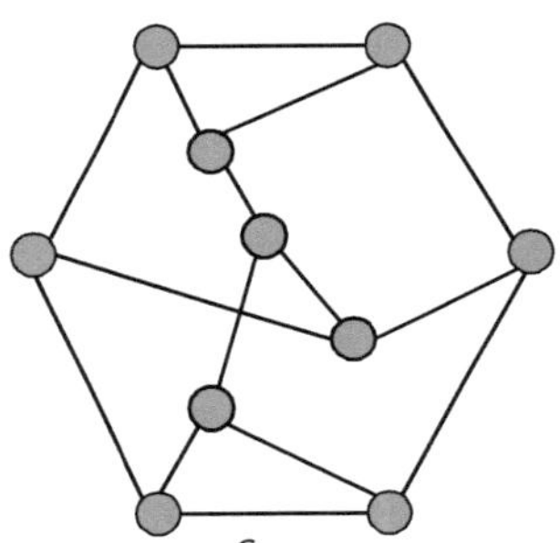

$C_{15,10}$

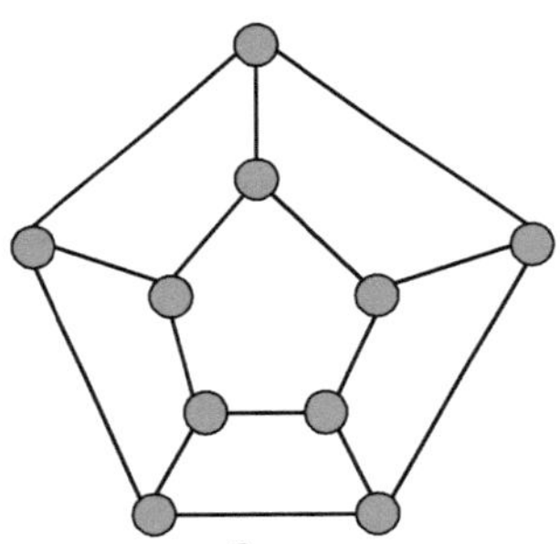

$C_{15,11}$

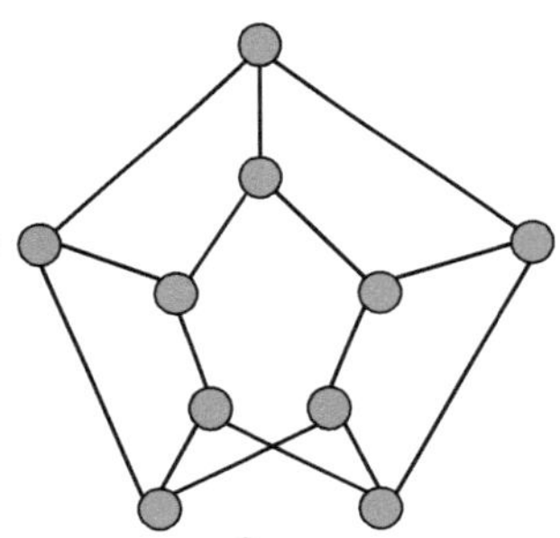

$C_{15,12}$

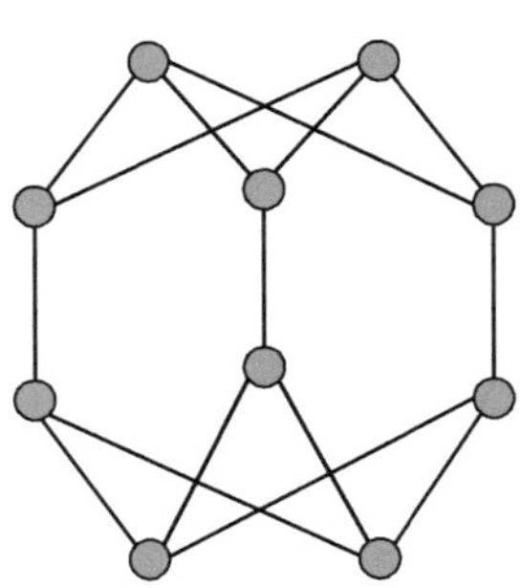

$C_{15,13}$

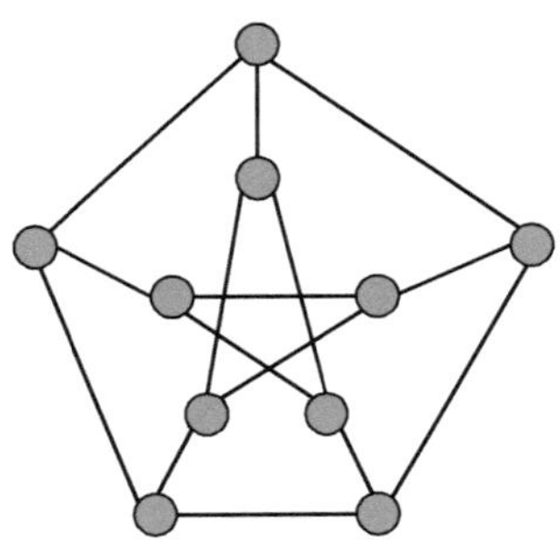

$C_{15,14}$

In what follows, the orientable and nonorientable genus polynomials of embeddings, upper maps and rooted upper maps of 3-connected cubic graphs shown above are provided.

Case of size 6:

Orientable

$$p_{C_{6,1}}(x) = 2 + 14x,$$
$$\mu_{C_{6,1}}(x) = 1 + 2x,$$
$$\mu^r_{C_{6,1}}(x) = 1 + 7x.$$

Nonorientable

$$q_{C_{6,1}}(x^{-1}) = 14x + 42x^2 + 56x^3,$$
$$\nu_{C_{6,1}}(x^{-1}) = 2x + 3x^2 + 3x^3,$$
$$\nu^r_{C_{6,1}}(x^{-1}) = 7x + 21x^2 + 28x^3.$$

Case of size 9:

Orientable

$$p_{C_{9,1}}(x) = 2 + 38x + 24x^2,$$
$$\mu_{C_{9,1}}(x) = 1 + 5x + 2x^2,$$
$$\mu^r_{C_{9,1}}(x) = 3 + 57x + 36x^2;$$

$$p_{C_{9,2}}(x) = 40x + 24x^2,$$
$$\mu_{C_{9,2}}(x) = 2x + x^2,$$
$$\mu^r_{C_{9,2}}(x) = 10x + 6x^2.$$

Nonorientable

$$q_{C_{9,1}}(x^{-1}) = 22x + 122x^2 + 424x^3 + 392x^4,$$
$$\nu_{C_{9,1}}(x^{-1}) = 3x + 12x^2 + 28x^3 + 23x^4,$$
$$\nu^r_{C_{9,1}}(x^{-1}) = 33x + 183x^2 + 636x^3 + 588x^4;$$

$$q_{C_{9,2}}(x^{-1}) = 12x + 108x^2 + 432x^3 + 408x^4,$$
$$\nu_{C_{9,2}}(x^{-1}) = x + 2x^2 + 6x^3 + 6x^4,$$
$$\nu^r_{C_{9,2}}(x^{-1}) = 3x + 27x^2 + 108x^3 + 102x^4.$$

Case of size 12:

Orientable

$$p_{C_{12,1}}(x) = 2 + 70x + 184x^2,$$
$$\mu_{C_{12,1}}(x) = 1 + 15x + 28x^2,$$
$$\mu^r_{C_{12,1}}(x) = 12 + 420x + 1104x^2;$$

$$p_{C_{12,2}}(x) = 64x + 192x^2,$$
$$\mu_{C_{12,2}}(x) = 4x + 12x^2,$$
$$\mu^r_{C_{12,2}}(x) = 128x + 384x^2;$$

$$p_{C_{12,3}}(x) = 56x + 200x^2,$$
$$\mu_{C_{12,3}}(x) = 5x + 13x^2,$$
$$\mu^r_{C_{12,3}}(x) = 84x + 300x^2;$$

$$p_{C_{12,4}}(x) = 2 + 54x + 200x^2,$$
$$\mu_{C_{12,4}}(x) = 1 + 5x + 8x^2,$$
$$\mu^r_{C_{12,4}}(x) = 1 + 27x + 100x^2.$$

Nonorientable

$$q_{C_{12,1}}(x^{-1}) = 30x + 242x^2 + 1448x^3 + 3272x^4 + 2944x^5,$$
$$\nu_{C_{12,1}}(x^{-1}) = 7x + 44x^2 + 217x^3 + 452x^4 + 38x^5,$$
$$\nu^r_{C_{12,1}}(x^{-1}) = 180x + 1452x^2 + 8688x^3 + 19632x^4 + 17664x^5;$$

$$q_{C_{12,2}}(x^{-1}) = 12x + 180x^2 + 1360x^3 + 3312x^4 + 3072x^5,$$
$$\nu_{C_{12,2}}(x^{-1}) = x + 9x^2 + 64x^3 + 149x^4 + 137x^5,$$
$$\nu^r_{C_{12,2}}(x^{-1}) = 24x + 360x^2 + 2720x^3 + 6624x^4 + 6144x^5;$$

$$q_{C_{12,3}}(x^{-1}) = 10x + 158x^2 + 1272x^3 + 3296x^4 + 3200x^5,$$
$$\nu_{C_{12,3}}(x^{-1}) = 2x + 11x^2 + 57x^3 + 133x^4 + 118x^5,$$
$$\nu^r_{C_{12,3}}(x^{-1}) = 15x + 237x^2 + 1908x^3 + 2944x^4 + 4800x^5;$$

$$q_{C_{12,4}}(x^{-1}) = 24x + 192x^2 + 1288x^3 + 3264x^4 + 3168x^5,$$
$$\nu_{C_{12,4}}(x^{-1}) = x + 7x^2 + 24x^3 + 58x^4 + 40x^5,$$
$$\nu^r_{C_{12,4}}(x^{-1}) = 12x + 96x^2 + 644x^3 + 1632x^4 + 1584x^5.$$

Case of size 15:

Orientable

$$p_{C_{15,1}}(x) = 2 + 102x + 664x^2 + 256x^3,$$
$$\mu_{C_{15,1}}(x) = 1 + 27x + 176x^2 + 68x^3,$$
$$\mu^r_{C_{15,1}}(x) = 30 + 1530x + 9960x^2 + 3840x^3;$$

$$p_{C_{15,2}}(x) = 72x + 664x^2 + 288x^3,$$
$$\mu_{C_{15,2}}(x) = 20x + 180x^2 + 672x^3,$$
$$\mu^r_{C_{15,2}}(x) = 1080x + 9960x^2 + 4320x^3;$$

$$p_{C_{15,3}}(x) = 56x + 648x^2 + 320x^3,$$
$$\mu_{C_{15,3}}(x) = 12x + 96x^2 + 44x^3,$$
$$\mu^r_{C_{15,3}}(x) = 420x + 4860x^2 + 2400x^3;$$

$$p_{C_{15,4}}(x) = 80x + 688x^2 + 256x^3,$$
$$\mu_{C_{15,4}}(x) = 11x + 93x^2 + 32x^3,$$
$$\mu^r_{C_{15,4}}(x) = 600x + 5160x^2 + 1920x^3;$$

$$p_{C_{15,5}}(x) = 2 + 118x + 648x^2 + 256x^3,$$
$$\mu_{C_{15,5}}(x) = 1 + 27x + 88x^2 + 36x^3,$$
$$\mu^r_{C_{15,5}}(x) = 15 + 885x + 4860x^2 + 1920x^3;$$

$$p_{C_{15,6}}(x) = 2 + 110x + 688x^2 + 224x^3,$$
$$\mu_{C_{15,6}}(x) = 1 + 14x + 69x^2 + 20x^3,$$
$$\mu^r_{C_{15,6}}(x) = 10 + 550x + 3440x^2 + 1120x^3;$$

$$p_{C_{15,7}}(x) = 2 + 78x + 656x^2 + 288x^3,$$
$$\mu_{C_{15,7}}(x) = 1 + 14x + 81x^2 + 24x^3,$$
$$\mu^r_{C_{15,7}}(x) = 10 + 390x + 3280x^2 + 1440x^3;$$

$$p_{C_{15,8}}(x) = 96x + 672x^2 + 256x^3,$$
$$\mu_{C_{15,8}}(x) = 9x + 49x^2 + 18x^3,$$
$$\mu^r_{C_{15,8}}(x) = 360x + 2520x^2 + 960x^3;$$

$$p_{C_{15,9}}(x) = 48x + 656x^2 + 320x^3,$$
$$\mu_{C_{15,9}}(x) = 8x + 59x^2 + 25x^3,$$
$$\mu^r_{C_{15,9}}(x) = 180x + 2460x^2 + 1200x^3;$$

$$p_{C_{15,10}}(x) = 88x + 648x^2 + 288x^3,$$
$$\mu_{C_{15,10}}(x) = 5x + 31x^2 + 16x^3,$$
$$\mu^r_{C_{15,10}}(x) = 220x + 1620x^2 + 720x^3;$$

$$p_{C_{15,11}}(x) = 2 + 70x + 632x^2 + 320x^3,$$
$$\mu_{C_{15,11}}(x) = 1 + 5x + 28x^2 + 10x^3,$$
$$\mu^r_{C_{15,11}}(x) = 3 + 105x + 948x^2 + 480x^3;$$

$$p_{C_{15,12}}(x) = 72x + 632x^2 + 320x^3,$$
$$\mu_{C_{15,12}}(x) = 6x + 24x^2 + 14x^3,$$
$$\mu^r_{C_{15,12}}(x) = 108x + 948x^2 + 480x^3;$$

$$p_{C_{15,13}}(x) = 48x + 720x^2 + 256x^3,$$
$$\mu_{C_{15,13}}(x) = 2x + 15x^2 + 6x^3,$$
$$\mu^r_{C_{15,13}}(x) = 30x + 450x^2 + 160x^3;$$

$$p_{C_{15,14}}(x) = 40x + 664x^2 + 320x^3,$$
$$\mu_{C_{15,14}}(x) = x + 7x^2 + 2x^3,$$
$$\mu^r_{C_{15,14}}(x) = 10x + 166x^2 + 80x^3.$$

Nonorientable

$$q_{C_{15,1}}(x^{-1}) = 38x + 394x^2 + 3336x^3 + 12744x^4 + 27008x^5 + 20992x^6,$$
$$\nu_{C_{15,1}}(x^{-1}) = 10x + 104x^2 + 838x^3 + 3220x^4 + 6768x^5 + 5300x^6,$$
$$\nu^r_{C_{15,1}}(x^{-1}) = 570x + 5910x^2 + 50040x^3 + 191160x^4 + 405120x^5 + 314880x^6;$$

$$q_{C_{15,2}}(x^{-1}) = 10x + 214x^2 + 2576x^3 + 11664x^4 + 27424x^5 + 22624x^6,$$
$$\nu_{C_{15,2}}(x^{-1}) = 4x + 60x^2 + 676x^3 + 2988x^4 + 6952x^5 + 5688x^6,$$
$$\nu^r_{C_{15,2}}(x^{-1}) = 150x + 3210x^2 + 38640x^3 + 174960x^4 + 411360x^5 + 339360x^6;$$

$$q_{C_{15,3}}(x^{-1}) = 6x + 158x^2 + 2188x^3 + 10912x^4 + 27504x^5 + 23744x^6,$$
$$v_{C_{15,3}}(x^{-1}) = 2x + 27x^2 + 313x^3 + 1466x^4 + 3572x^5 + 3044x^6,$$
$$v^r_{C_{15,3}}(x^{-1}) = 45x + 1185x^2 + 16410x^3 + 81840x^4 + 206280x^5 + 178080x^6;$$

$$q_{C_{15,4}}(x^{-1}) = 12x + 244x^2 + 2816x^3 + 12224x^4 + 27456x^5 + 21760x^6,$$
$$v_{C_{15,4}}(x^{-1}) = 2x + 33x^2 + 368x^3 + 1565x^4 + 3480x^5 + 2736x^6,$$
$$v^r_{C_{15,4}}(x^{-1}) = 90x + 1830x^2 + 21120x^3 + 91680x^4 + 205920x^5 + 163200x^6;$$

$$q_{C_{15,5}}(x^{-1}) = 38x + 410x^2 + 3496x^3 + 12952x^4 + 26880x^5 + 20736x^6,$$
$$v_{C_{15,5}}(x^{-1}) = 8x + 76x^2 + 524x^3 + 1768x^4 + 3460x^5 + 2652x^6,$$
$$v^r_{C_{15,5}}(x^{-1}) = 385x + 3075x^2 + 26220x^3 + 97140x^4 + 201600x^5 + 155520x^6;$$

$$q_{C_{15,6}}(x^{-1}) = 38x + 402x^2 + 3448x^3 + 13040x^4 + 27072x^5 + 20512x^6,$$
$$v_{C_{15,6}}(x^{-1}) = 6x + 44x^2 + 319x^3 + 1157x^4 + 2354x^5 + 1744x^6,$$
$$v^r_{C_{15,6}}(x^{-1}) = 190x + 2010x^2 + 17240x^3 + 65200x^4 + 135360x^5 + 102560x^6;$$

$$q_{C_{15,7}}(x^{-1}) = 32x + 312x^2 + 2800x^3 + 11800x^4 + 27200x^5 + 22368x^6,$$
$$v_{C_{15,7}}(x^{-1}) = 5x + 35x^2 + 267x^3 + 1077x^4 + 2358x^5 + 1866x^6,$$
$$v^r_{C_{15,7}}(x^{-1}) = 160x + 1560x^2 + 14000x^3 + 59000x^4 + 136000x^5 + 111840x^6;$$

$$q_{C_{15,8}}(x^{-1}) = 12x + 260x^2 + 2976x^3 + 12432x^4 + 27328x^5 + 21504x^6,$$
$$v_{C_{15,8}}(x^{-1}) = 2x + 21x^2 + 207x^3 + 828x^4 + 1772x^5 + 1382x^6,$$
$$v^r_{C_{15,8}}(x^{-1}) = 45x + 975x^2 + 11160x^3 + 26620x^4 + 192480x^5 + 80640x^6;$$

$$q_{C_{15,9}}(x^{-1}) = 4x + 132x^2 + 2049x^3 + 10720x^4 + 27616x^5 + 24000x^6,$$
$$v_{C_{15,9}}(x^{-1}) = x + 16x^2 + 152x^3 + 753x^4 + 1811x^5 + 1559x^6,$$
$$v^r_{C_{15,9}}(x^{-1}) = 15x + 495x^2 + 7650x^3 + 40200x^4 + 103560x^5 + 90000x^6;$$

$$q_{C_{15,10}}(x^{-1}) = 12x + 252x^2 + 2864x^3 + 12136x^4 + 27264x^5 + 21984x^6,$$
$$v_{C_{15,10}}(x^{-1}) = x + 11x^2 + 124x^3 + 517x^4 + 1154x^5 + 941x^6,$$
$$v^r_{C_{15,10}}(x^{-1}) = 30x + 630x^2 + 7160x^3 + 30340x^4 + 681560x^5 + 54960x^6;$$

$$q_{C_{15,11}}(x^{-1}) = 30x + 282x^2 + 2560x^3 + 11240x^4 + 27168x^5 + 23232x^6,$$
$$v_{C_{15,11}}(x^{-1}) = 2x + 17x^2 + 92x^3 + 351x^4 + 754x^5 + 624x^6,$$
$$v^r_{C_{15,11}}(x^{-1}) = 45x + 423x^2 + 3840x^3 + 16860x^4 + 40752x^5 + 34848x^6;$$

$$q_{C_{15,12}}(x^{-1}) = 12x + 220x^2 + 2480x^3 + 11240x^4 + 27264x^5 + 23296x^6,$$
$$v_{C_{15,12}}(x^{-1}) = 2x + 11x^2 + 90x^3 + 343x^4 + 756x^5 + 638x^6,$$
$$v^r_{C_{15,12}}(x^{-1}) = 18x + 330x^2 + 3720x^3 + 16860x^4 + 40896x^5 + 34944x^6;$$

$$q_{C_{15,13}}(x^{-1}) = 120x^2 + 2232x^3 + 11568x^4 + 27936x^5 + 22656x^6,$$
$$v_{C_{15,13}}(x^{-1}) = 4x^2 + 28x^3 + 144x^4 + 307x^5 + 259x^6,$$
$$v^r_{C_{15,13}}(x^{-1}) = 75x^2 + 1395x^3 + 7230x^4 + 17460x^5 + 14160x^6;$$

$$q_{C_{15,14}}(x^{-1}) = 4x + 120x^2 + 1900x^3 + 10440x^4 + 27664x^5 + 24384x^6,$$
$$v_{C_{15,14}}(x^{-1}) = x + 2x^2 + 16x^3 + 62x^4 + 142x^5 + 111x^6,$$
$$v^r_{C_{15,14}}(x^{-1}) = x + 30x^2 + 474x^3 + 2610x^4 + 6916x^5 + 6039x^6.$$

A2.2 Bouquets

Let B_m be the bouquet of size m, $m \geq 1$.

Case of $m = 1$**:**

Orientable

$$p_{B_1}(x) = 1,$$
$$\mu_{B_1}(x) = 1,$$
$$\mu^r_{B_1}(x) = 1.$$

Nonorientable

$$q_{B_1}(x^{-1}) = x,$$
$$v_{B_1}(x^{-1}) = x,$$
$$v^r_{B_1}(x^{-1}) = x.$$

Case of $m = 2$**:**

Orientable

$$p_{B_2}(x) = 4 + 2x,$$
$$\mu_{B_2}(x) = 1 + x,$$
$$\mu^r_{B_2}(x) = 2 + x.$$

Nonorientable

$$q_{B_2}(x^{-1}) = 10x + 8x^2,$$
$$v_{B_2}(x^{-1}) = 2x + 2x^2,$$
$$v^r_{B_2}(x^{-1}) = 5x + 4x^2.$$

Case of $m = 3$**:**

Orientable

$$p_{B_3}(x) = 40 + 80x,$$
$$\mu_{B_3}(x) = 2 + 3x,$$
$$\mu^r_{B_3}(x) = 5 + 10x.$$

Nonorientable

$$q_{B_3}(x^{-1}) = 176x + 336x^2 + 328x^3,$$
$$v_{B_3}(x^{-1}) = 5x + 8x^2 + 8x^3,$$
$$v^r_{B_3}(x^{-1}) = 22x + 42x^2 + 41x^3.$$

Case of $m = 4$**:**

Orientable

$$p_{B_4}(x) = 672 + 3360x + 1008x^2,$$
$$\mu_{B_4}(x) = 3 + 10x + 4x^2,$$
$$\mu^r_{B_4}(x) = 14 + 70x + 21x^2.$$

Nonorientable

$$q_{B_4}(x^{-1}) = 4464x + 14592x^2 + 33120x^3 + 23424x^4,$$
$$v_{B_4}(x^{-1}) = 12x + 33x^2 + 64x^3 + 47x^4,$$
$$v^r_{B_4}(x^{-1}) = 93x + 304x^2 + 690x^3 + 488x^4.$$

Case of $m = 5$**:**

Orientable

$$p_{B_5}(x) = 16128 + 161280x + 185472x^2,$$
$$\mu_{B_5}(x) = 6 + 35x + 38x^2,$$
$$\mu^r_{B_5}(x) = 42 + 420x + 483x^2.$$

Nonorientable

$$q_{B_5}(x^{-1}) = 148224x + 718080x^2 + 2745600x^3 + 4477440x^4 + 3159936x^5,$$
$$\nu_{B_5}(x^{-1}) = 33x + 131x^2 + 442x^3 + 686x^4 + 473x^5,$$
$$\nu^r_{B_5}(x^{-1}) = 386x + 1870x^2 + 7150x^3 + 11660x^4 + 8229x^5.$$

A2.3 Wheels

Let W_n be the wheel of order n, $n \geq 4$, i.e., all vertices are of valency (or degree) 3 but one and all 3-valent vertices form a circuit.

Case of $n = 4$**:**

Orientable

$$p_{W_4}(x) = 2 + 14x,$$
$$\mu_{W_4}(x) = 1 + 2x,$$
$$\mu^r_{W_4}(x) = 1 + 7x.$$

Nonorientable

$$q_{W_4}(x^{-1}) = 14x + 42x^2 + 56x^3,$$
$$\nu_{W_4}(x^{-1}) = 2x + 3x^2 + 3x^3,$$
$$\nu^r_{W_4}(x^{-1}) = 7x + 21x^2 + 28x^3.$$

Case of $n = 5$**:**

Orientable

$$p_{W_5}(x) = 2 + 58x + 36x^2,$$
$$\mu_{W_5}(x) = 1 + 8x + 4x^2,$$
$$\mu^r_{W_5}(x) = 4 + 116x + 72x^2.$$

Nonorientable

$$q_{W_5}(x^{-1}) = 28x + 176x^2 + 640x^3 + 596x^4,$$
$$\nu_{W_5}(x^{-1}) = 4x + 18x^2 + 52x^3 + 48x^4,$$
$$\nu^r_{W_5}(x^{-1}) = 56x + 352x^2 + 1280x^3 + 1192x^4.$$

Case of $n = 6$:

Orientable

$$p_{W_6}(x) = 2 + 190x + 576x^2,$$
$$\mu_{W_6}(x) = 1 + 14x + 41x^2,$$
$$\mu^r_{W_6}(x) = 4 + 380x + 1152x^2.$$

Nonorientable

$$q_{W_6}(x^{-1}) = 52x + 580x^2 + 4080x^3 + 9880x^4 + 9216x^5,$$
$$\nu_{W_6}(x^{-1}) = 6x + 38x^2 + 227x^3 + 539x^4 + 494x^5,$$
$$\nu^r_{W_6}(x^{-1}) = 104x + 1160x^2 + 8160x^3 + 19760x^4 + 18432x^5.$$

Case of $n = 7$:

Orientable

$$p_{W_7}(x) = 2 + 550x + 4968x^2 + 2160x^3,$$
$$\mu_{W_7}(x) = 1 + 34x + 240x^2 + 106x^3,$$
$$\mu^r_{W_7}(x) = 4 + 1100x + 9936x^2 + 4320x^3.$$

Nonorientable

$$q_{W_7}(x^{-1}) = 94x + 1680x^2 + 19482x^3 + 87536x^4 + 205496x^5 + 169552x^6,$$
$$\nu_{W_7}(x^{-1}) = 8x + 89x^2 + 878x^3 + 3829x^4 + 8788x^5 + 7241x^6,$$
$$\nu^r_{W_7}(x^{-1}) = 188x + 3360x^2 + 38964x^3 + 175072x^4 + 410992x^5 + 339104x^6.$$

Case of order $n = 8$:

Orientable

$$p_{W_8}(x) = 2 + 1484x + 31178x^2 + 59496x^3,$$
$$\mu_{W_8}(x) = 1 + 63x + 1176x^2 + 2246x^3,$$
$$\mu^r_{W_8}(x) = 4 + 2968x + 62356x^2 + 118992x^3.$$

A2.4 Link bundles

Let P_m be the link bundle of size m, $m \geq 3$. A *link bundle* is a graph of order 2 without loop.

Case of size $m = 3$**:**

Orientable

$$p_{P_3}(x) = 2 + 2x,$$
$$\mu_{P_3}(x) = 1 + x,$$
$$\mu^r_{P_3}(x) = 1 + 1x.$$

Nonorientable

$$q_{P_3}(x^{-1}) = 6x + 6x^2,$$
$$\nu_{P_3}(x^{-1}) = 1x + 2x^2,$$
$$\nu^r_{P_3}(x^{-1}) = 3x + 3x^2.$$

Case of size $m = 4$**:**

Orientable

$$p_{P_4}(x) = 6 + 30x,$$
$$\mu_{P_4}(x) = 1 + 2x,$$
$$\mu^r_{P_4}(x) = 1 + 5x.$$

Nonorientable

$$q_{P_4}(x^{-1}) = 36x + 96x^2 + 120x^3,$$
$$\nu_{P_4}(x^{-1}) = 2x + 4x^2 + 3x^3,$$
$$\nu^r_{P_4}(x^{-1}) = 6x + 16x^2 + 20x^3.$$

Case of size $m = 5$**:**

Orientable

$$p_{P_5}(x) = 24 + 360x + 192x^2,$$
$$\mu_{P_5}(x) = 1 + 3x + 3x^2,$$
$$\mu^r_{P_5}(x) = 1 + 15x + 8x^2.$$

Nonorientable

$$q_{P_5}(x^{-1}) = 240x + 1200x^2 + 3840x^3 + 3360x^4,$$
$$\nu_{P_5}(x^{-1}) = 2x + 7x^2 + 14x^3 + 14x^4,$$
$$\nu^r_{P_5}(x^{-1}) = 10x + 50x^2 + 160x^3 + 140x^4.$$

Case of size $m = 6$**:**

Orientable

$$p_{P_6}(x) = 120 + 4200x + 10080x^2,$$
$$\mu_{P_6}(x) = 1 + 6x + 10x^2,$$
$$\mu^r_{P_6}(x) = 1 + 35x + 84x^2.$$

Nonorientable

$$q_{P_6}(x^{-1}) = 1800x + 14400x^2 + 84120x^3 + 184320x^4 + 161760x^5,$$
$$\nu_{P_6}(x^{-1}) = 3x + 14x^2 + 48x^3 + 96x^4 + 72x^5,$$
$$\nu^r_{P_6}(x^{-1}) = 15x + 120x^2 + 701x^3 + 1536x^4 + 1348x^5.$$

A2.5 Complete bipartite graphs

Let $K_{m,n}$ be the complete bipartite graph of order $m + n$, $m, n \geq 3$.

Case of order $m + n = 6$**:**

Orientable

$$p_{K3,3}(x) = 40x + 24x^2,$$
$$\mu_{K3,3}(x) = 2x + x^2,$$
$$\mu^r_{K3,3}(x) = 10x + 6x^2.$$

Nonorientable

$$q_{K3,3}(x^{-1}) = 12x + 108x^2 + 432x^3 + 408x^4,$$
$$\nu_{K3,3}(x^{-1}) = x + 2x^2 + 6x^3 + 6x^4,$$
$$\nu^r_{K3,3}(x^{-1}) = 3x + 27x^2 + 108x^3 + 102x^4.$$

Case of order $m + n = 7$**:**

Orientable

$$p_{K3,4}(x) = 156x + 2244x^2 + 1056x^3,$$
$$\mu_{K3,4}(x) = 3x + 16x^2 + 10x^3,$$
$$\mu^r_{K3,4}(x) = 26x + 374x^2 + 176x^3.$$

Nonorientable

$$q_{K3,4}(x^{-1}) = 12x + 432x^2 + 6852x^3 + 36288x^4 + 93360x^5 + 80784x^6,$$
$$\nu_{K3,4}(x^{-1}) = x + 4x^2 + 33x^3 + 156x^4 + 358x^5 + 317x^6,$$
$$\nu^r_{K3,4}(x^{-1}) = 2x + 72x^2 + 1142x^3 + 6048x^4 + 15560x^5 + 13464x^6.$$

Case of order $m + n = 8$**:**

Orientable

$$p_{K4,4}(x) = 108x + 24984x^2 + 565020x^3 + 1089504x^4,$$
$$\mu_{K4,4}(x) = 2x + 25x^2 + 318x^3 + 530x^4,$$
$$\mu^r_{K4,4}(x) = 3x + 694x^2 + 15695x^3 + 30264x^4.$$

$$p_{K3,5}(x) = 240x + 37584x^2 + 290880x^3 + 113664x^4,$$
$$\mu_{K3,5}(x) = x + 33x^2 + 225x^3 + 105x^4,$$
$$\mu^r_{K3,5}(x) = 10x + 1566x^2 + 12120x^3 + 4736x^4.$$

Appendix 3: Atlas of Rooted and Unrooted Maps for Small Graphs

In the symbol $X : a, b, c$ for a map appearing under a figure below, X is the under graph of the map, $a = oy$ or qy are, respectively, orientable or nonorientable genus y, b is the series number with two digits and c is the number of ways to assign a root. And, $\bar{x}$ on a surface is for x^{-1}, or $-x$, $x = 1, 2, \ldots$.

A3.1 Bouquets B_m of size $4 \geq m \geq 1$

Case $m = 1$**:**

Orientable genus 0

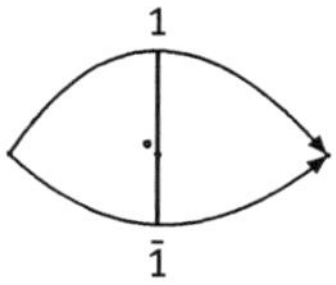

$B_1 : o0 - 01 - 01$

Nonorientable genus 1

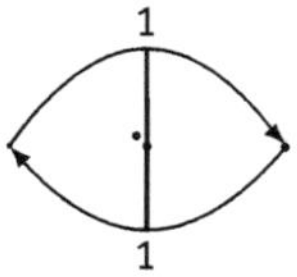

$B_1 : q0 - 01 - 01$

Case $m = 2$**:**

Orientable genus 0

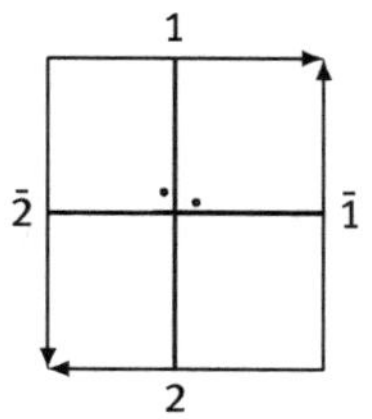

$B_2 : o0 - 01 - 02$

DOI 10.1515/9783110481846-021

Orientable genus 1

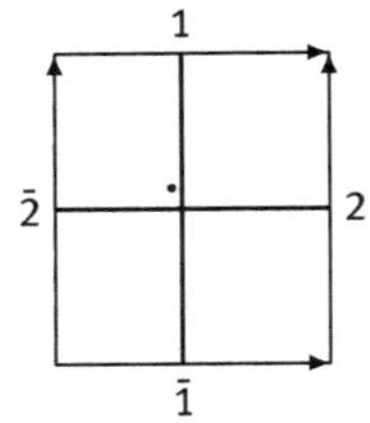

$B_2 : o1-01-01$

Nonorientable genus 1

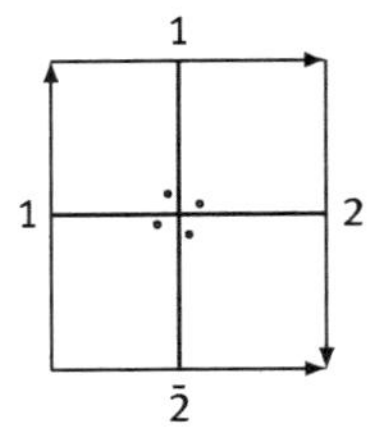

$B_2 : q1-01-04$

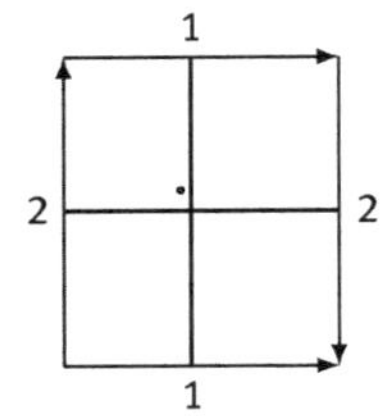

$B_2 : q1-02-01$

Nonorientable genus 2

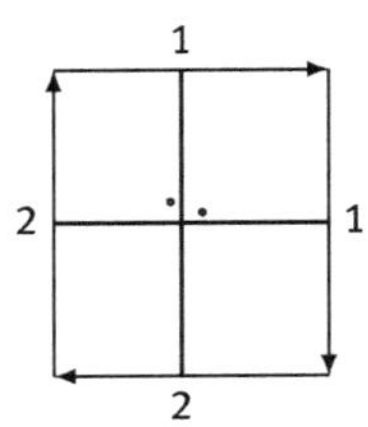

$B_2 : q2-01-02$

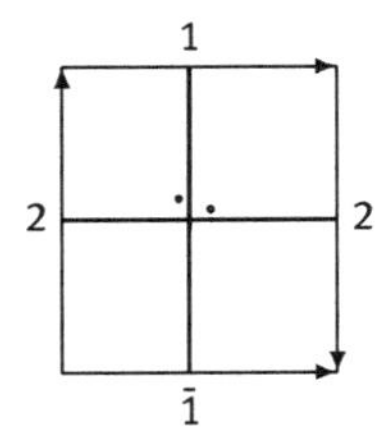

$B_2 : q2-02-02$

Case $m = 3$**:**

Orientable genus 0

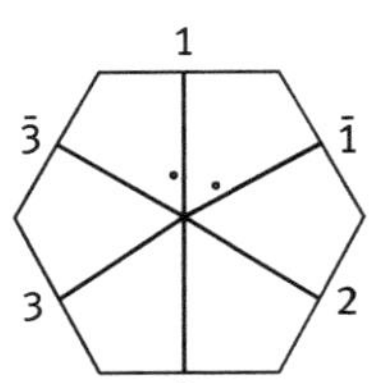

$B_3 : o0-01-02$

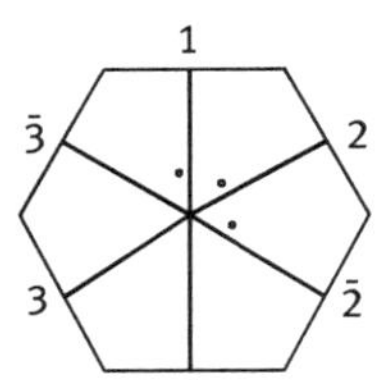

$B_3 : o0-02-03$

Orientable genus 1

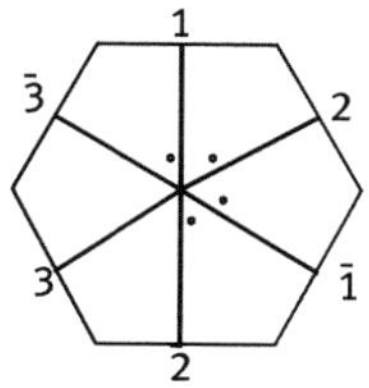

$B_3 : o1-01-04$

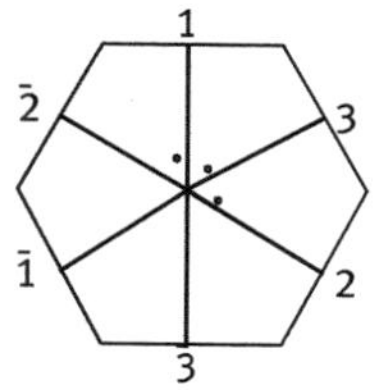

$B_3 : o1-02-03$

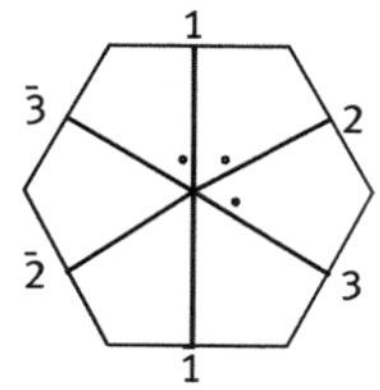

$B_3 : o1-03-03$

Nonorientable genus 1

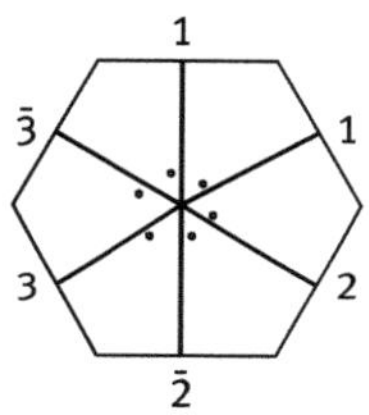

$B_3 : q1-01-06$

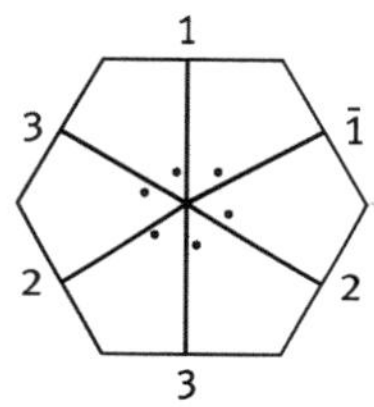

$B_3 : q1-02-06$

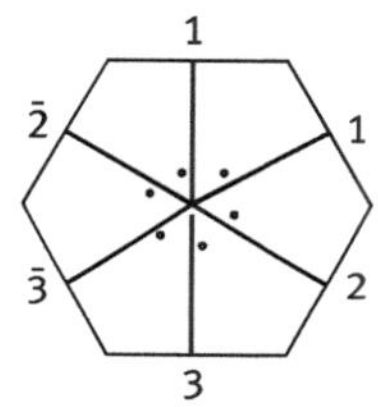

$B_3 : q1-03-06$

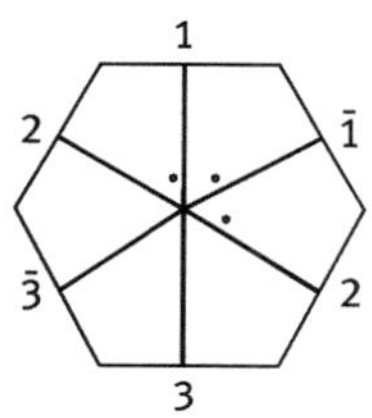

$B_3 : q1-04-03$

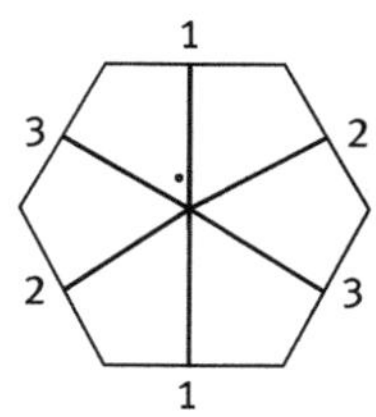

$B_3 : q1-05-01$

Nonorientable genus 2

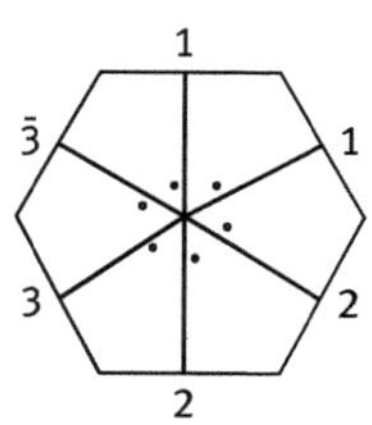

$B_3 : q2-01-06$

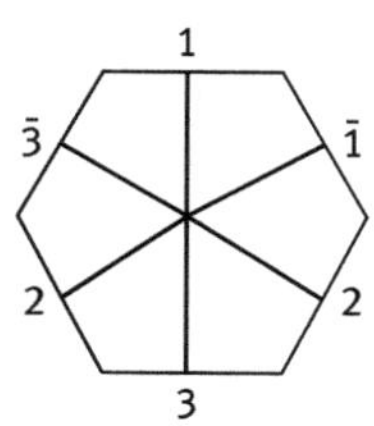

$B_3 : q2-01-12$

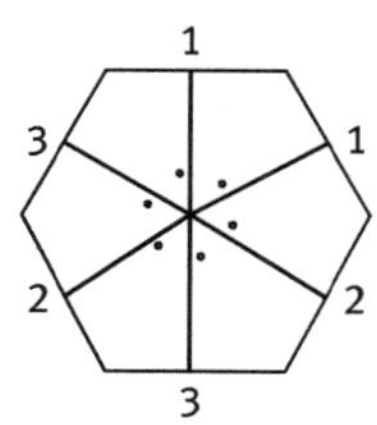

$B_3 : q2-03-06$

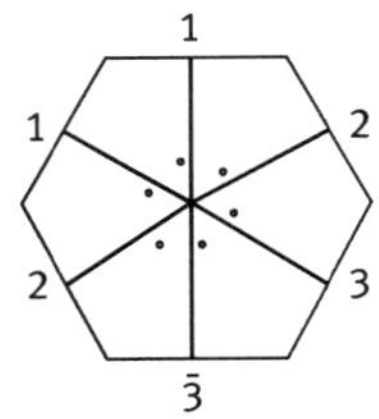

$B_3 : q2-04-06$

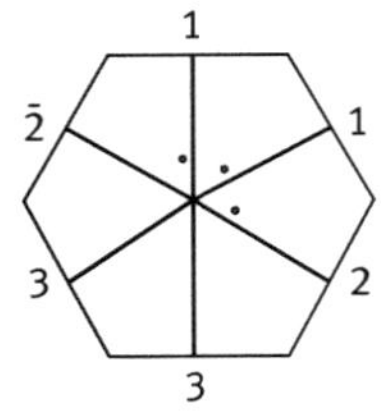

$B_3 : q2-05-03$

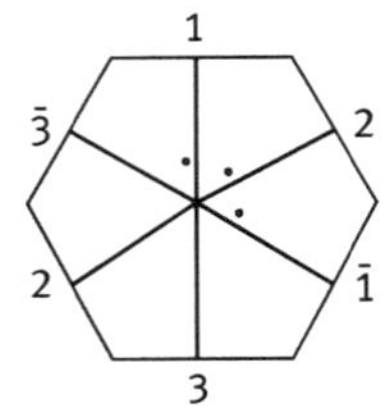

$B_3 : q2-06-03$

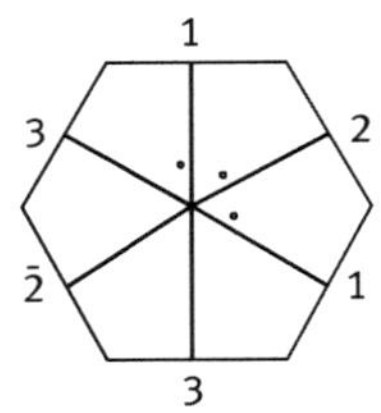

$B_3 : q2-07-03$

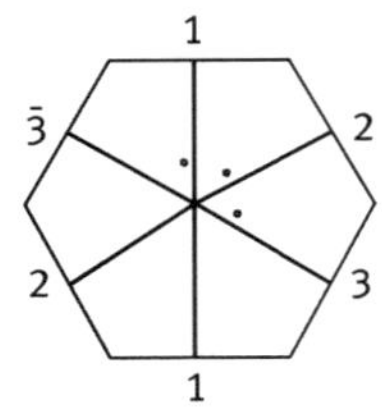

$B_3 : q2-08-03$

Nonorientable genus 3

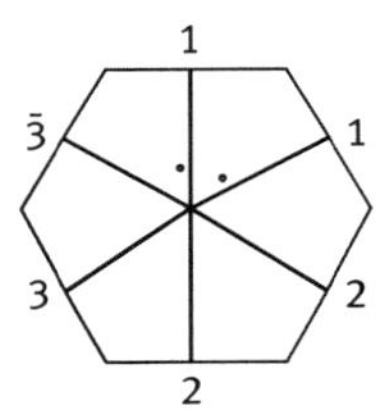

$B_3 : q3-01-02$

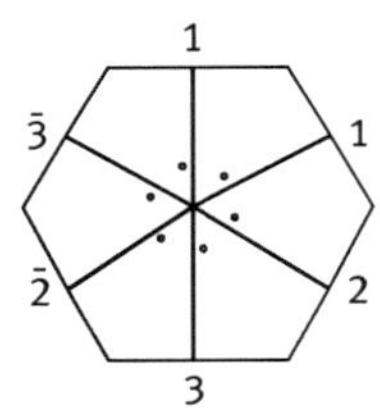

$B_3 : q3-02-06$

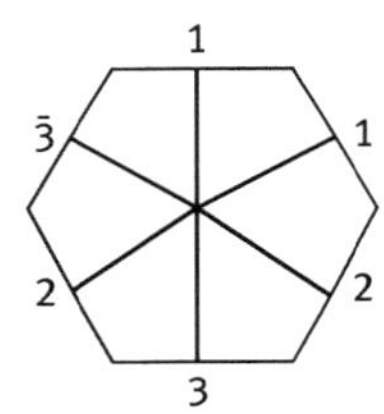

$B_3 : q3-03-12$

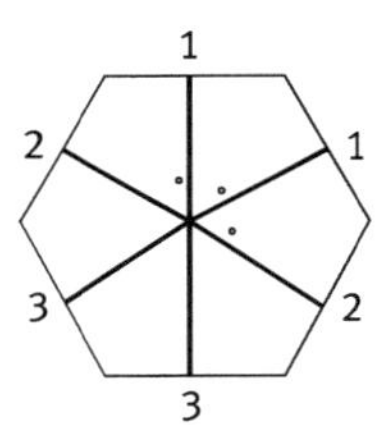

$B_3 : q3-04-03$

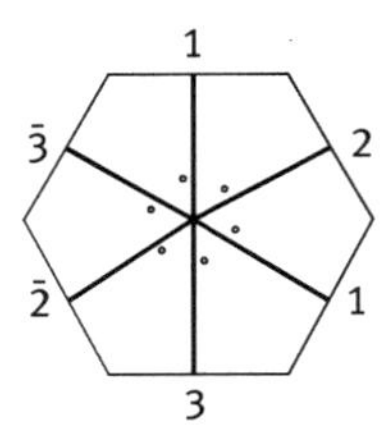

$B_3 : q3-05-06$

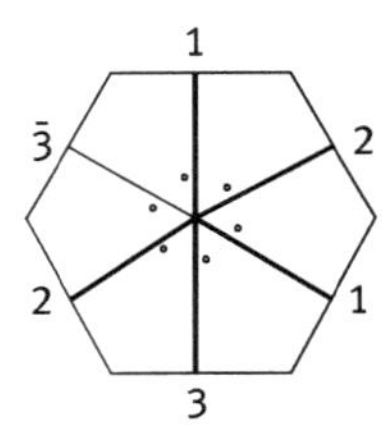

$B_3 : q3-06-06$

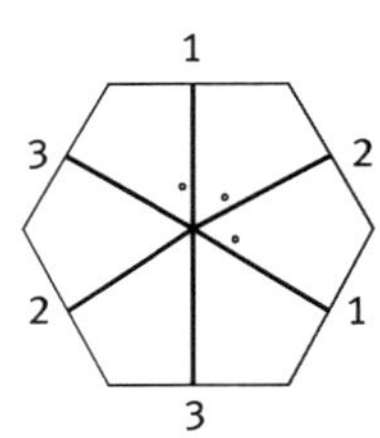

$B_3 : q3-07-03$

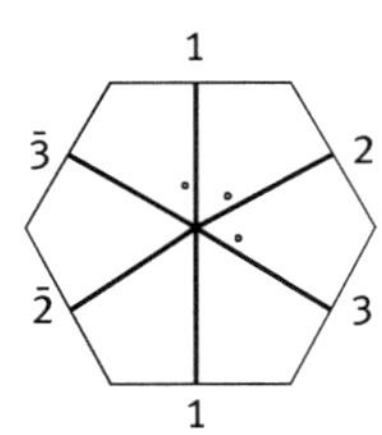

$B_3 : q3-08-03$

Case $m = 4$**:**

Orientable genus 0

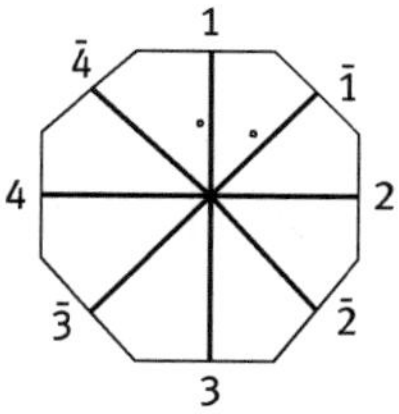

$B_4 : o0-01-02$

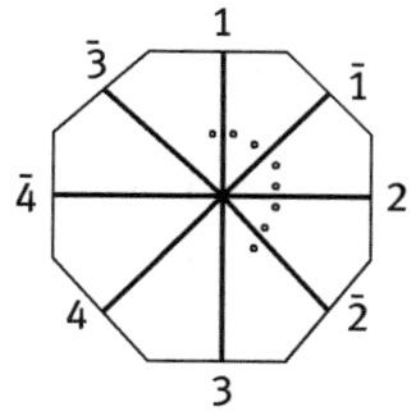

$B_4 : o3-02-08$

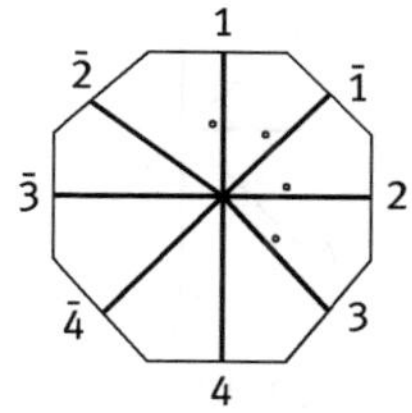

$B_4 : o0-03-04$

Orientable genus 1

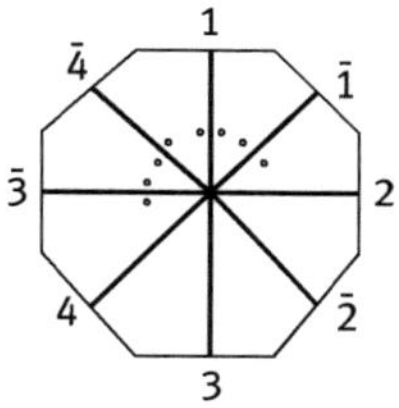

$B_4 : o0-01-08$

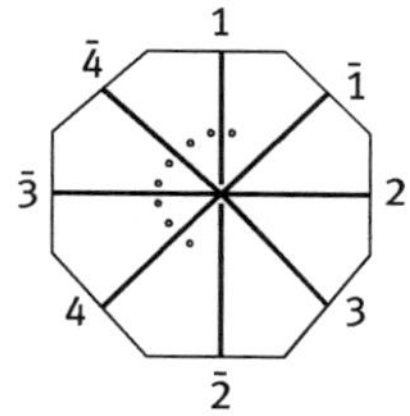

$B_4 : o1-02-08$

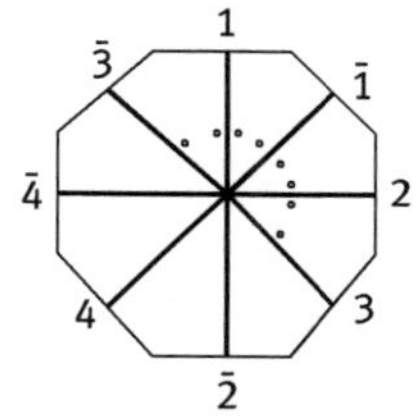

$B_4 : o1-03-08$

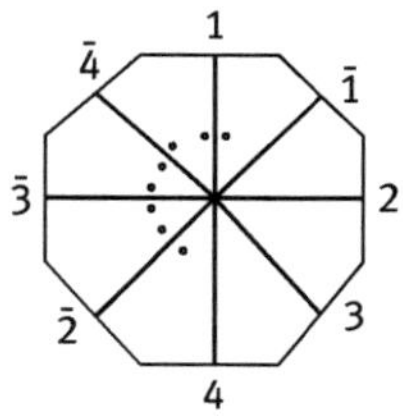

$B_4 : o1-04-08$

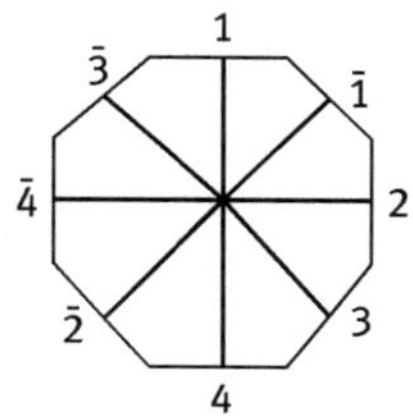

$B_4 : o1-05-16$

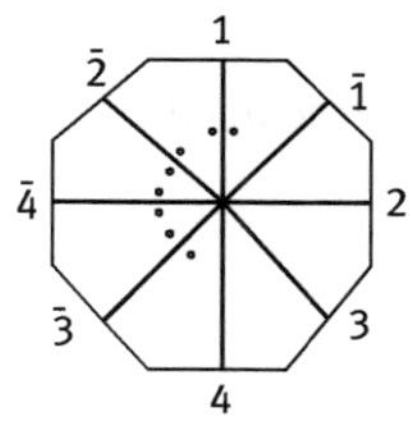

$B_4 : o1-06-08$

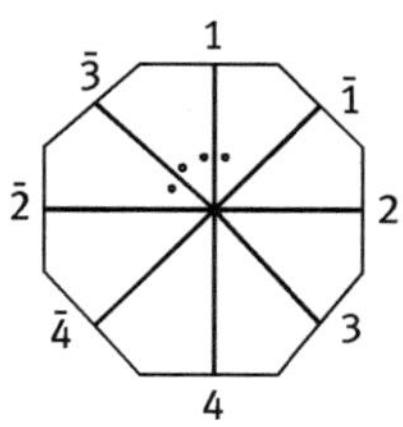

$B_4 : o1-07-04$

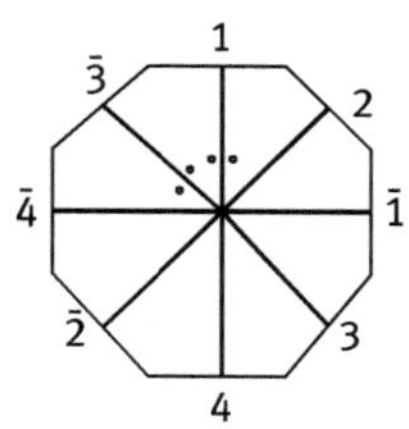

$B_4 : o1-08-04$

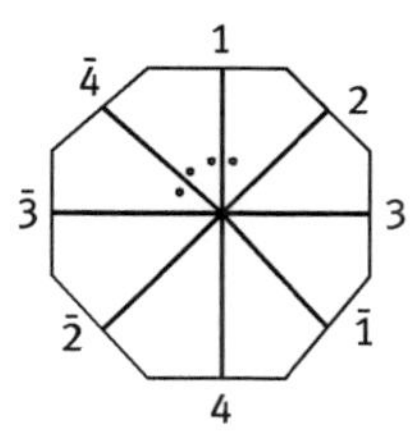

$B_4 : o1-09-04$

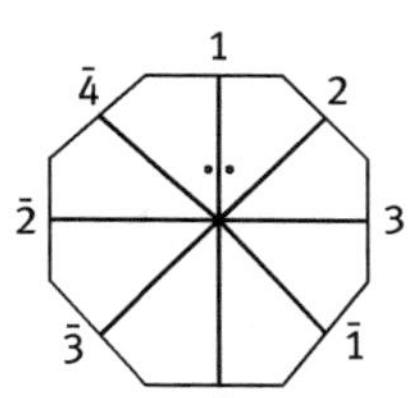

$B_4 : o1-10-02$

Orientable genus 2

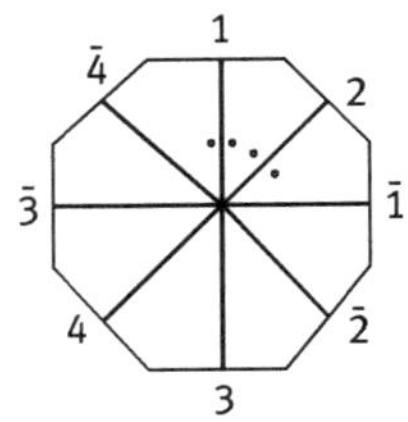

$B_4 : o2-01-04$

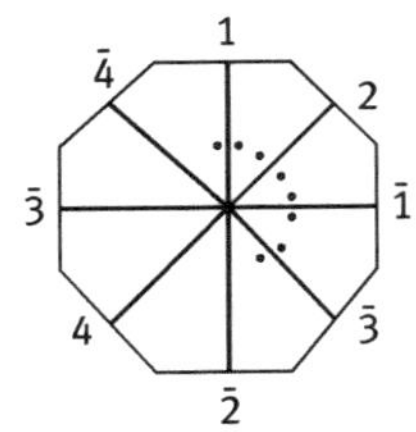

$B_4 : o2-02-08$

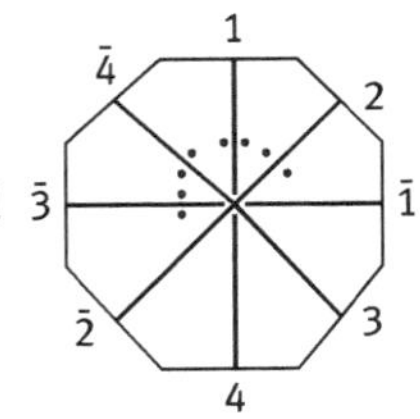

$B_4 : o2-03-08$

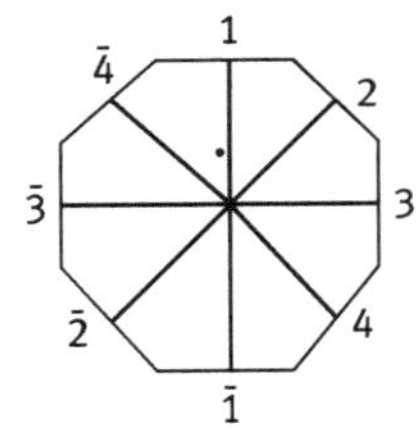

$B_4 : o2-04-01$

A3.2 Link bundles L_m, $6 \geq m \geq 3$

Case $m = 3$**:**

Orientable genus 0

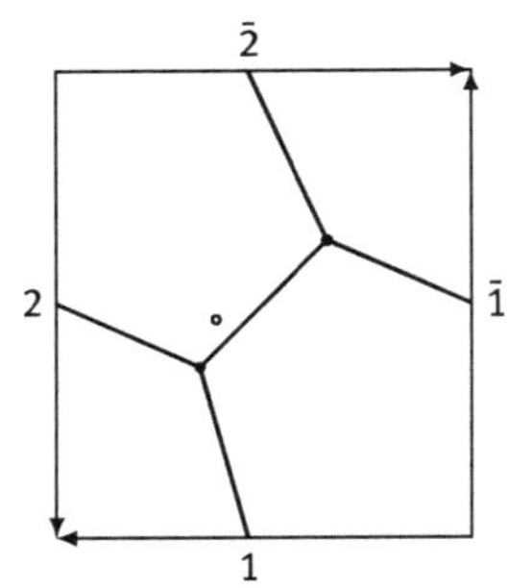

$L_3 : o0-01-01$

Orientable genus 1

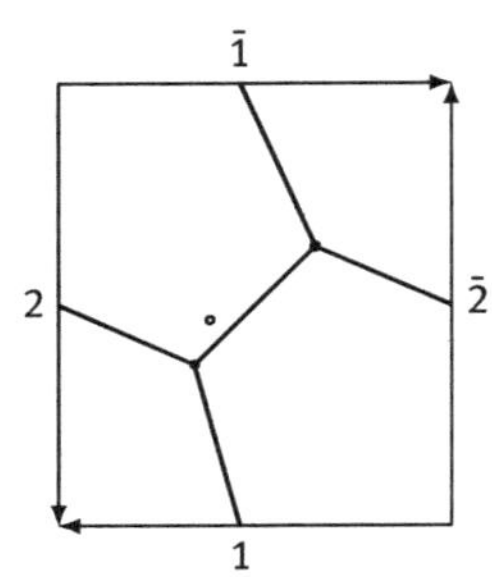

$L_3 : o1-01-01$

Nonorientable genus 1

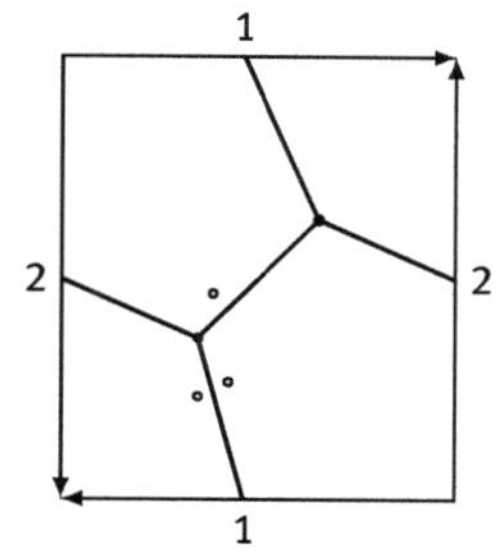

$L_3 : q1-01-03$

Nonorientable genus 2

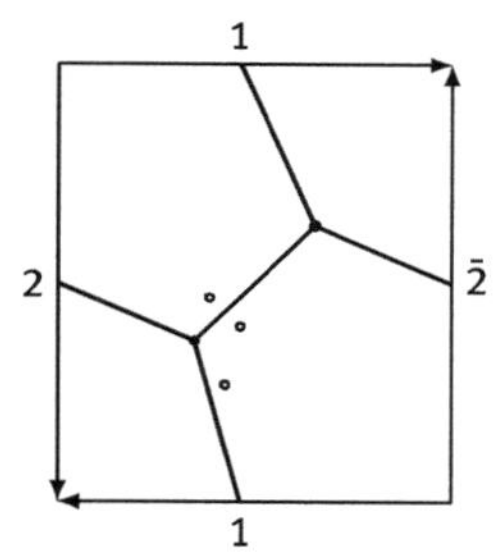

$L_3 : q2-01-03$

Case $m = 4$**:**

Orientable genus 0

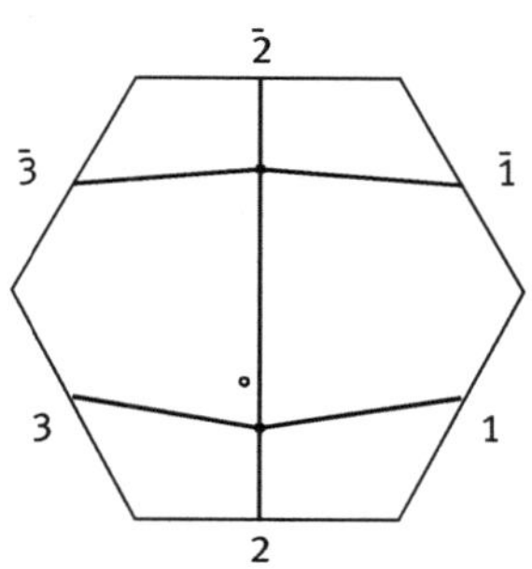

$L_4 : o0-01-01$

Orientable genus 1

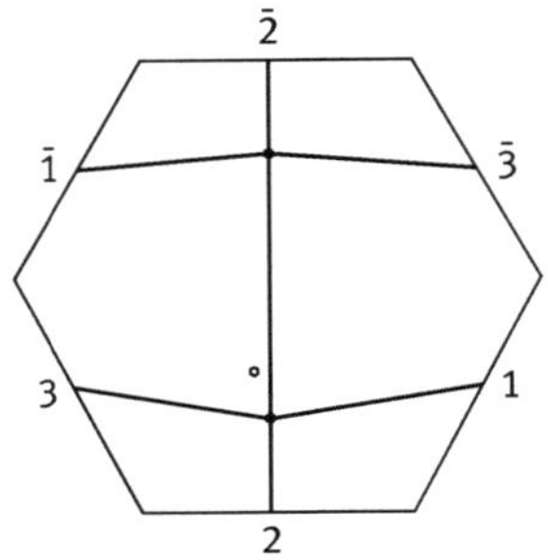

$L_4: o1-01-01$

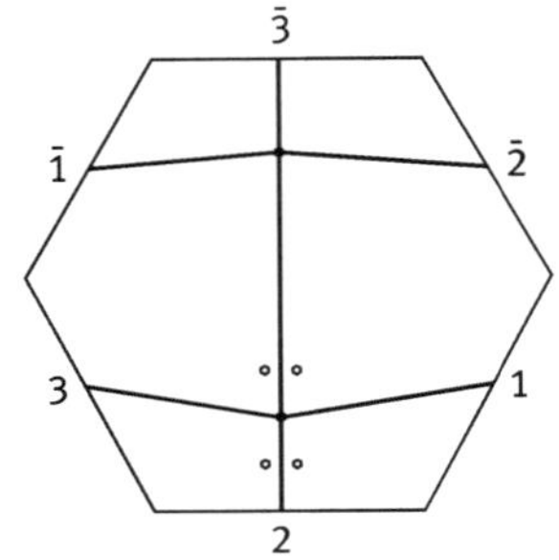

$L_4: o1-02-04$

Nonorientable genus 1

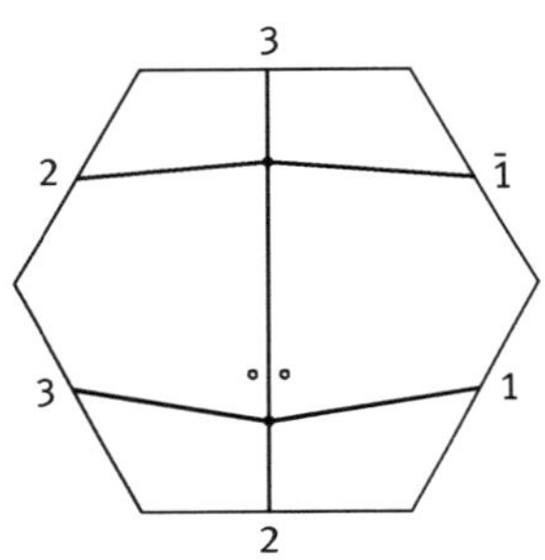

$L_4: q1-01-02$

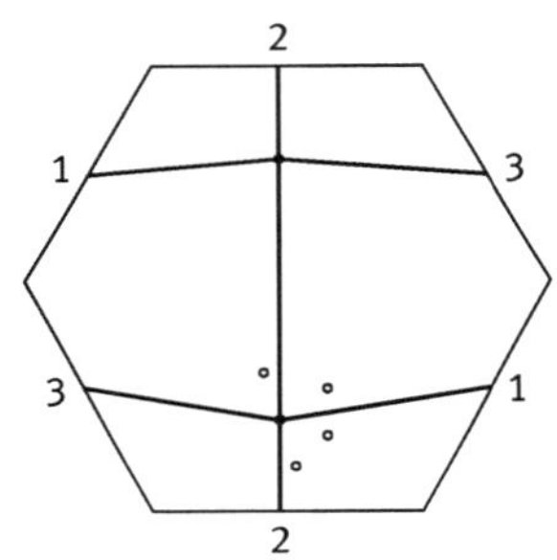

$L_4: q1-02-04$

Nonorientable genus 2

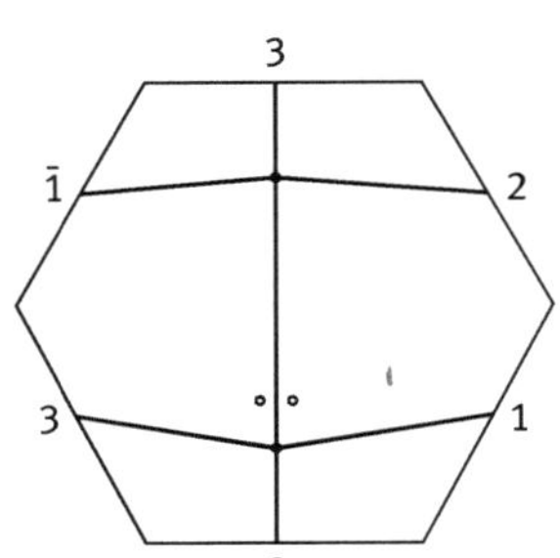

$L_4: q2-01-02$

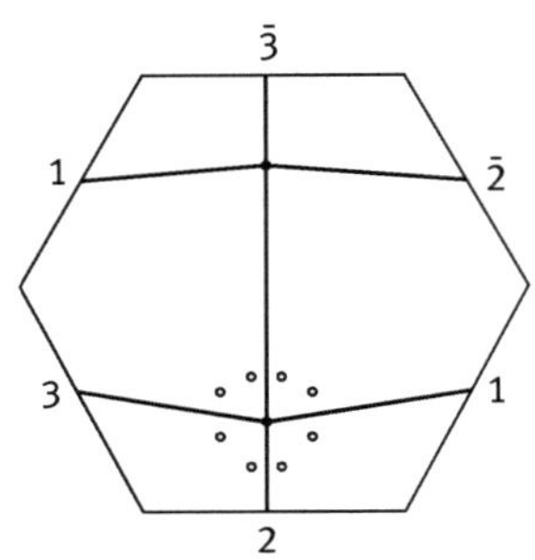

$L_4: q2-02-08$

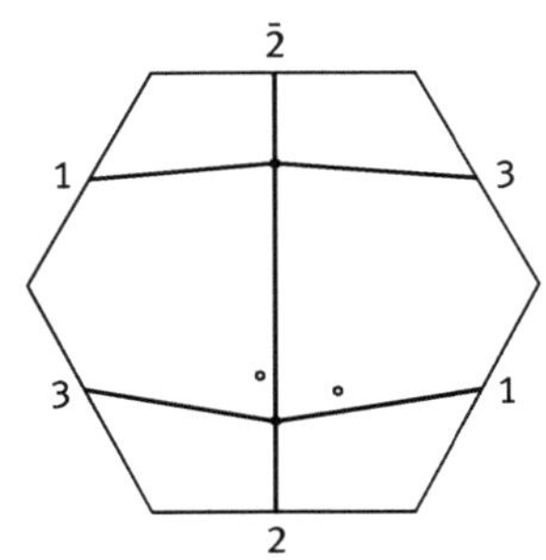

$L_4: q2-03-02$

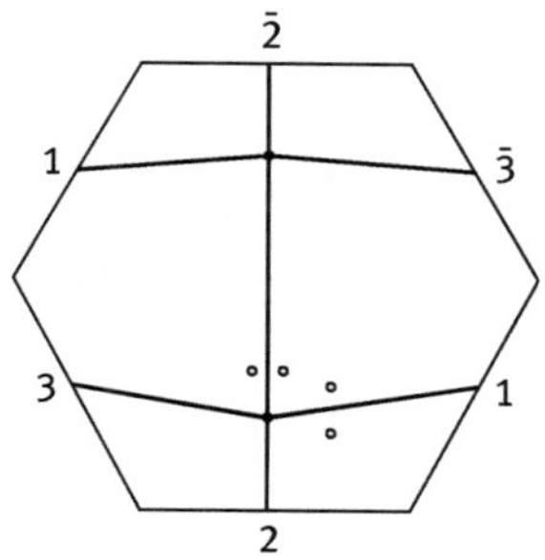

$L_4 : q2-04-04$

Nonorientable genus 3

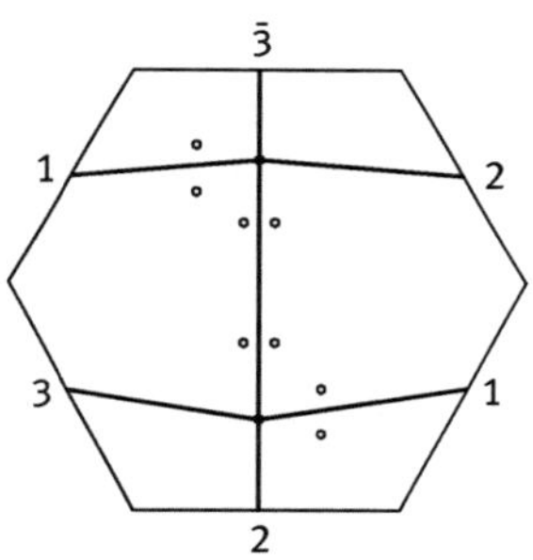

$L_4 : q3-01-08$

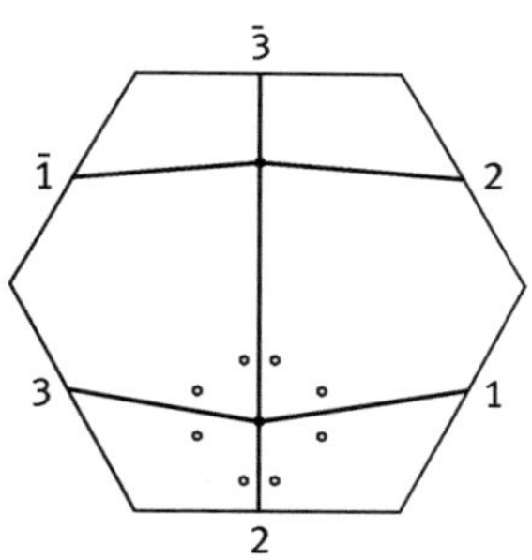

$L_4 : q3-02-08$

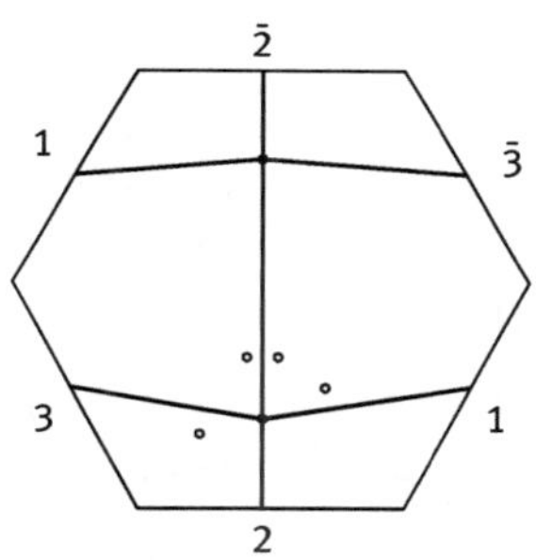

$L_4 : q3-03-04$

Case $m = 5$**:**

Orientable genus 0

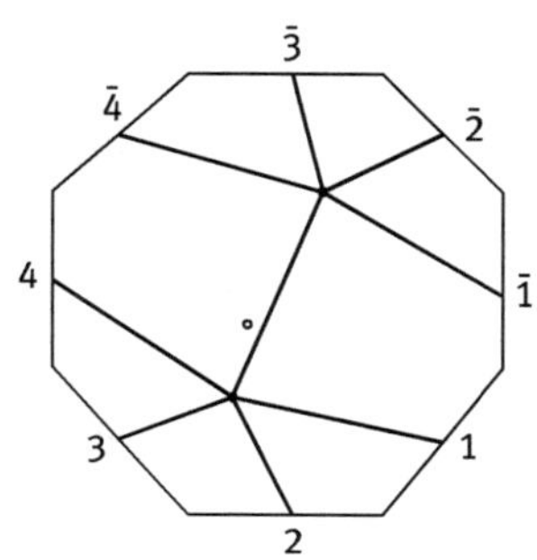

$L_5 : o0-01-01$

Orientable genus 1

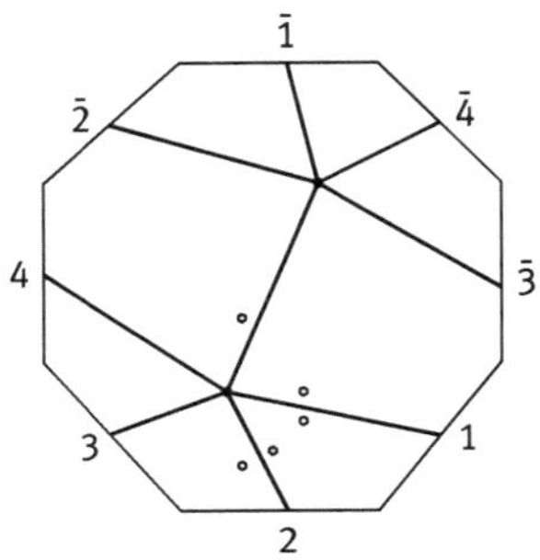

$L_5 : o1-01-05$

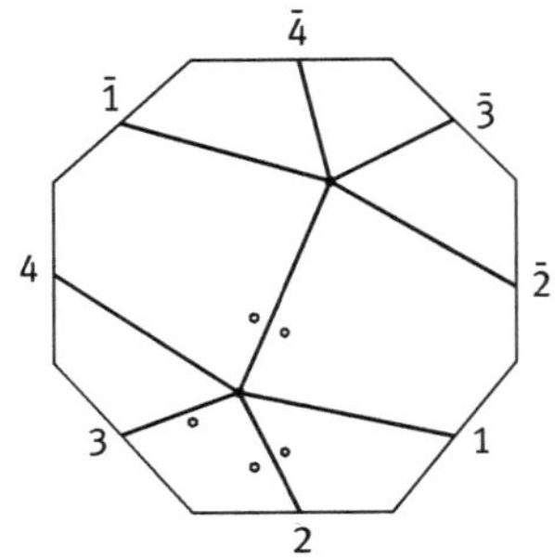

$L_5 : o1-02-05$

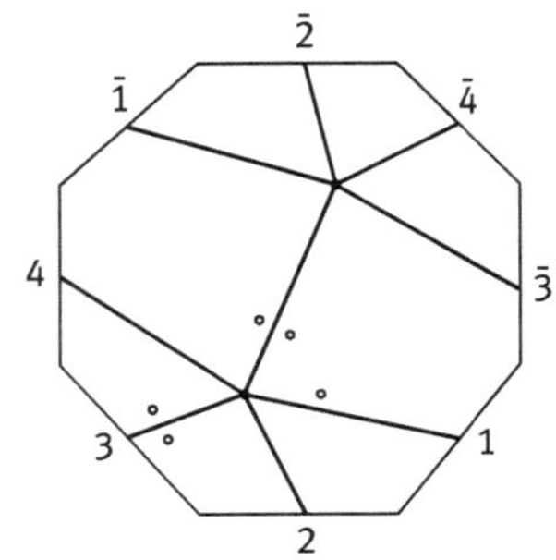

$L_5 : o1-03-05$

Orientable genus 2

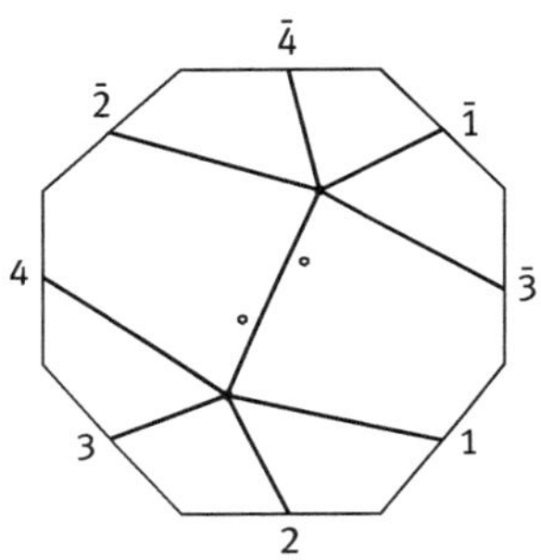

$L_5 : o2-01-02$

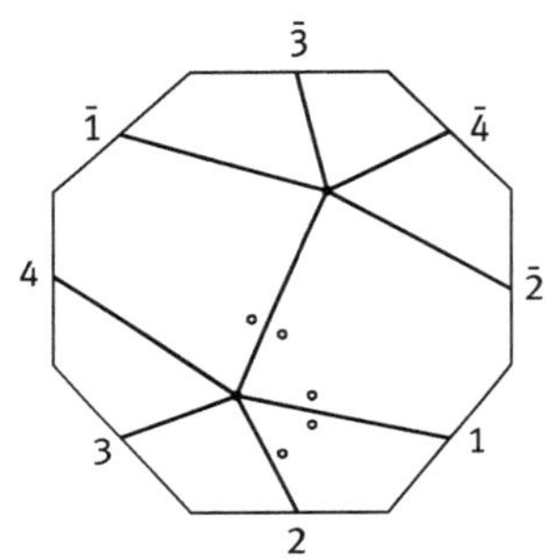

$L_5 : o2-02-05$

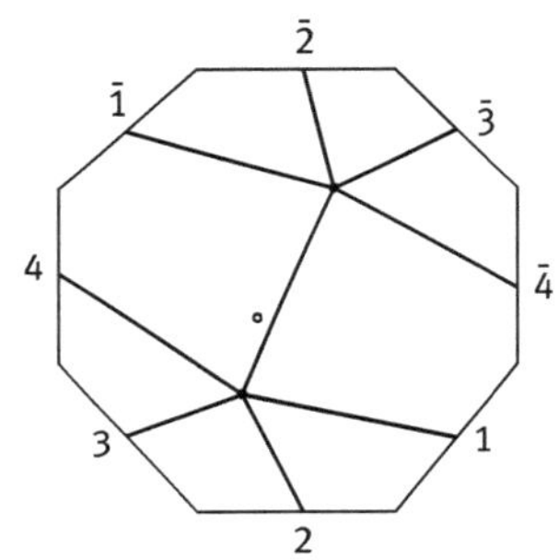

$L_5 : o2-03-01$

Nonorientable genus 1

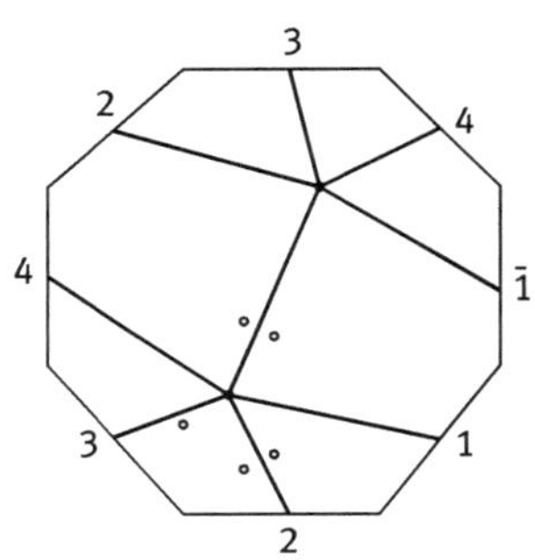

$L_5 : q1-01-05$

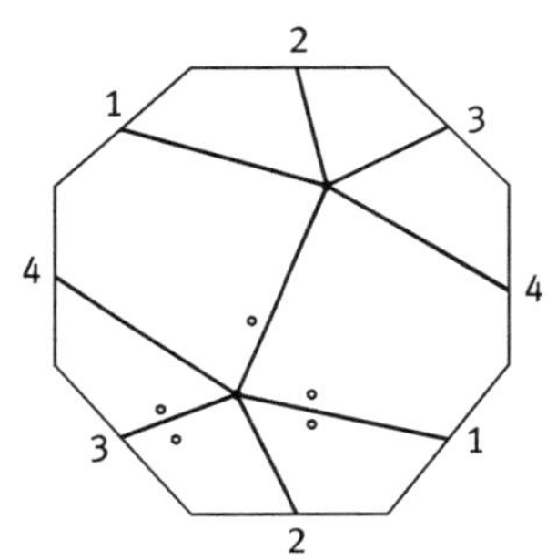

$L_5 : q1-02-05$

Nonorientable genus 2

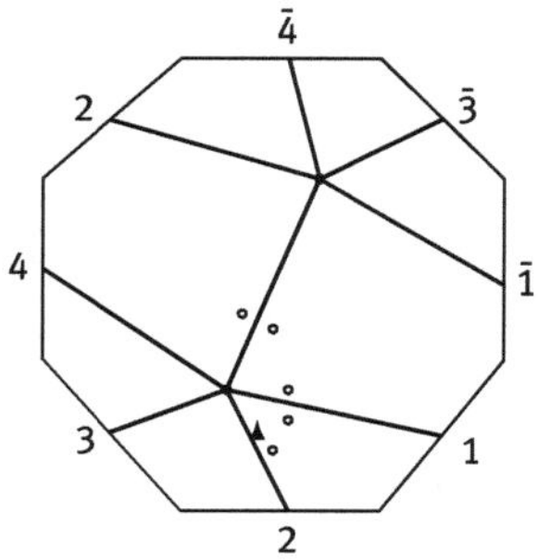

$L_5: q2-01-05$

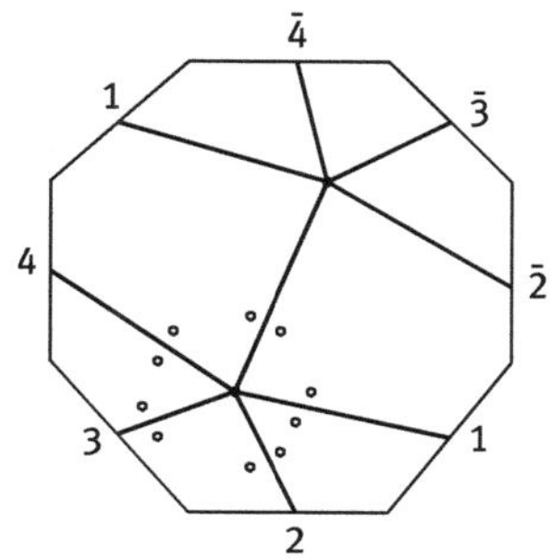

$L_5: q2-02-10$

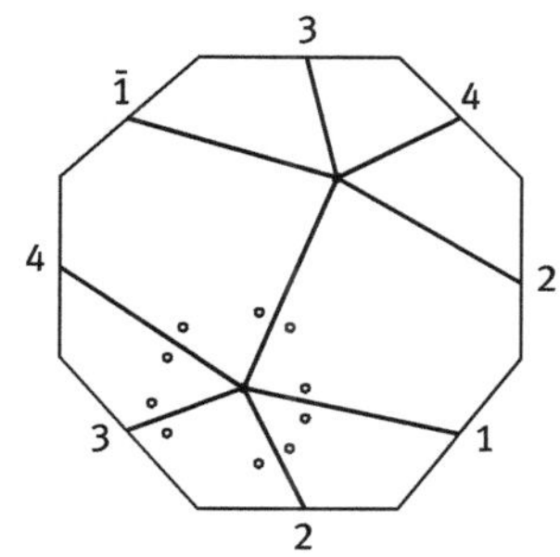

$L_5: q2-03-10$

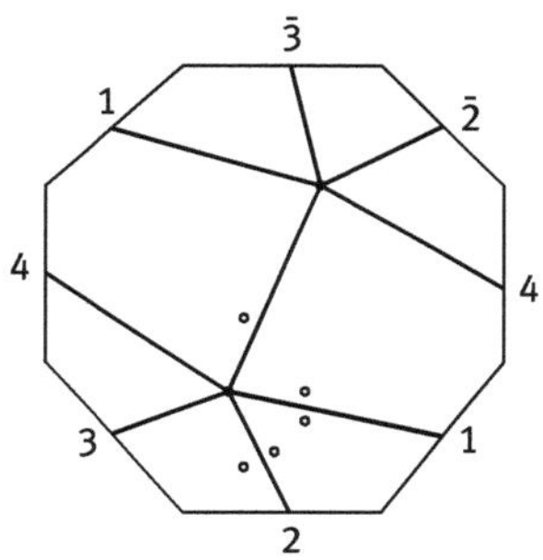

$L_5: q2-04-05$

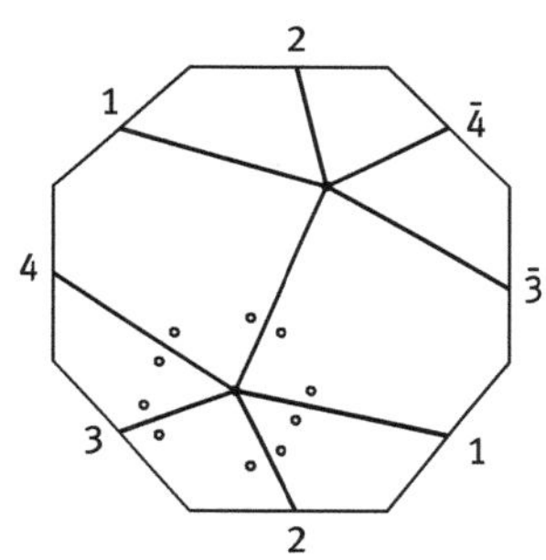

$L_5: q2-05-10$

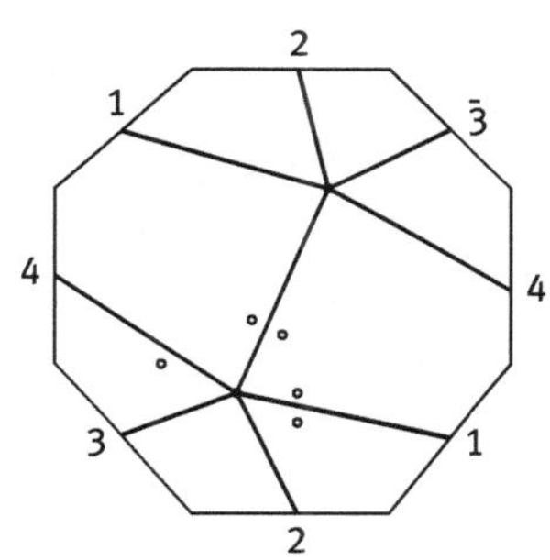

$L_5: q2-06-05$

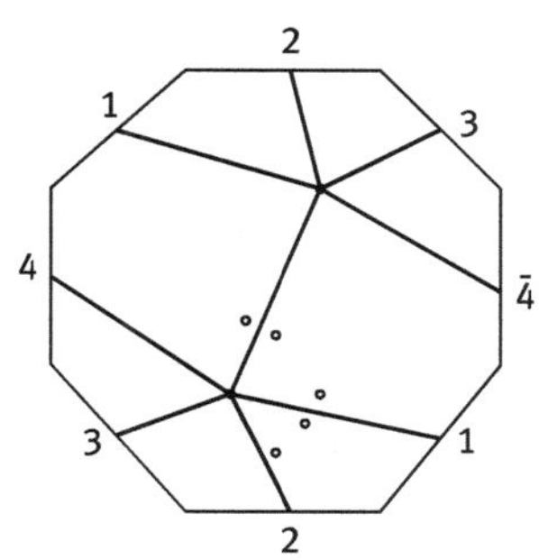

$L_5: q2-07-05$

Nonorientable genus 3

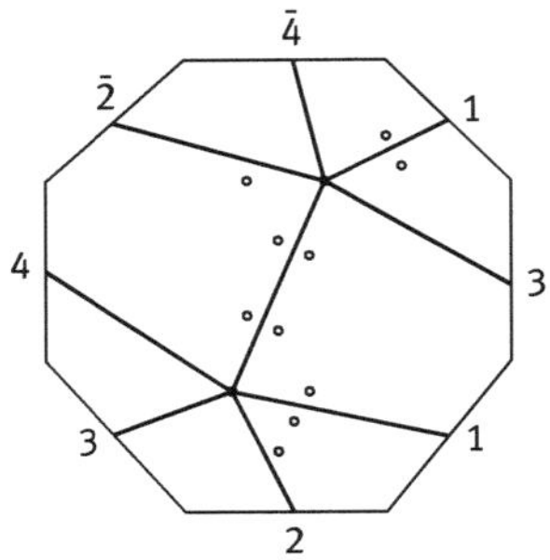

$L_5 : q3-01-10$

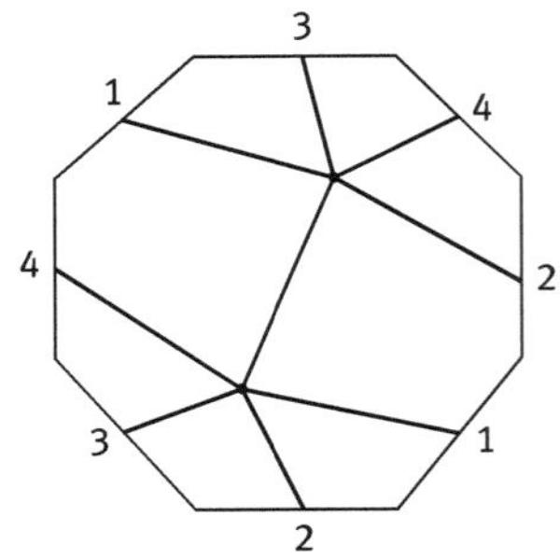

$L_5 : q3-02-20$

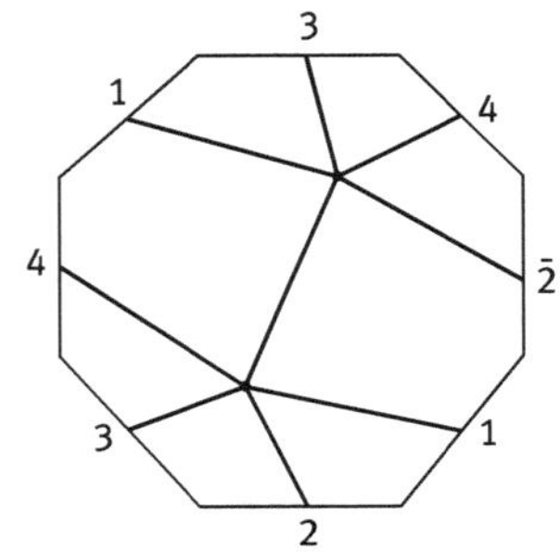

$L_5 : q3-03-20$

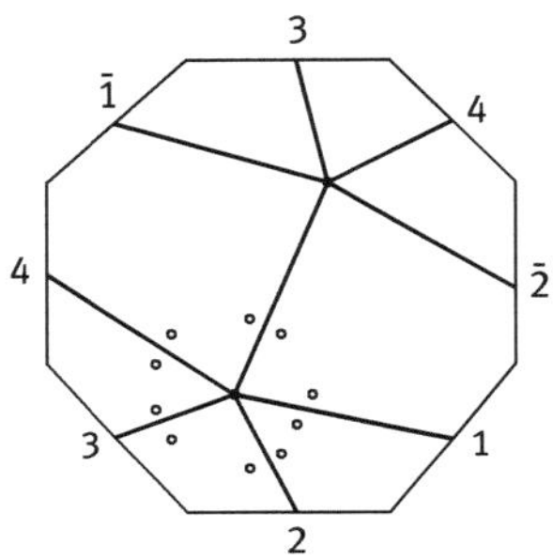

$L_5 : q3-04-10$

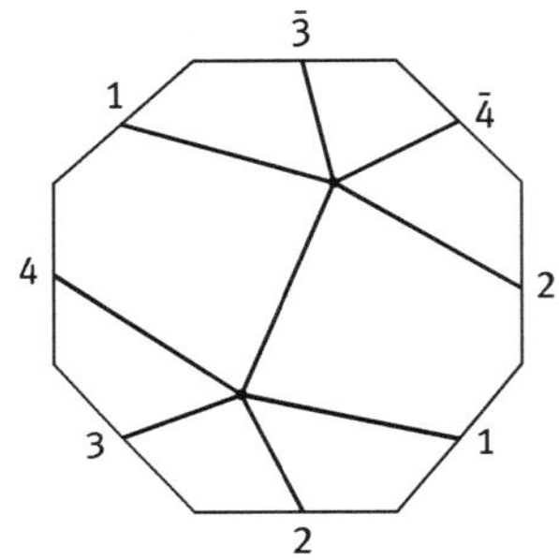

$L_5 : q3-05-20$

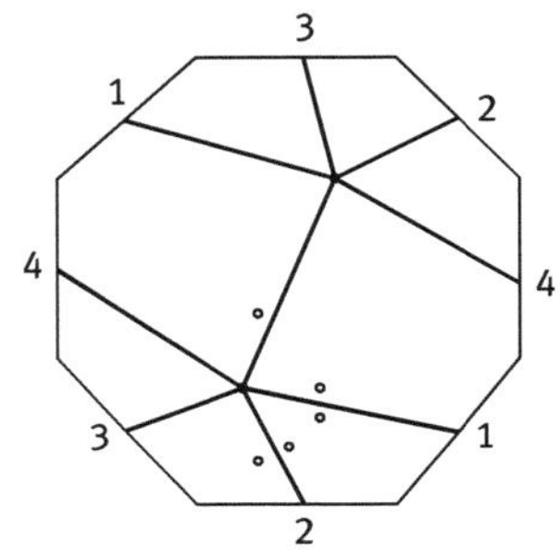

$L_5 : q3-06-05$

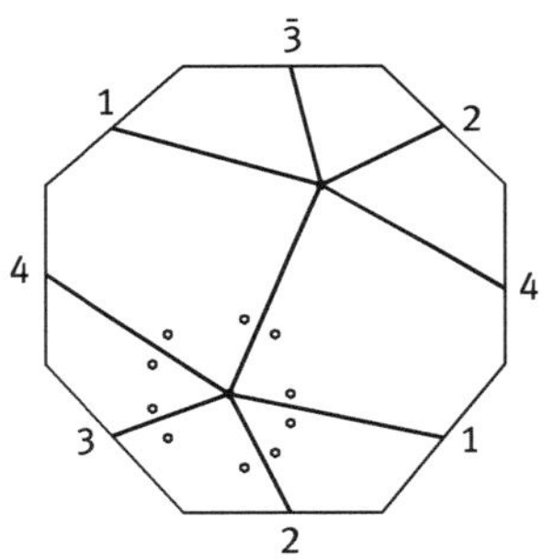

$L_5 : q3-07-10$

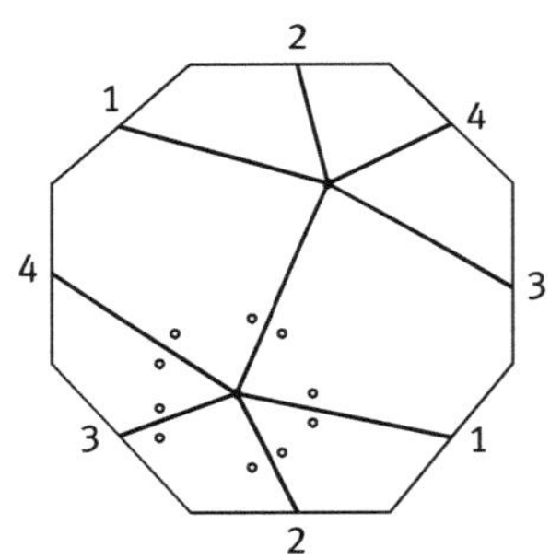

$L_5 : q3-08-10$

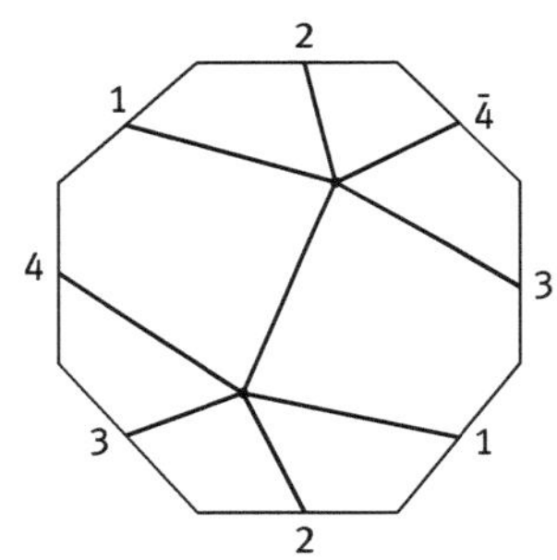

$L_5 : q3-09-20$

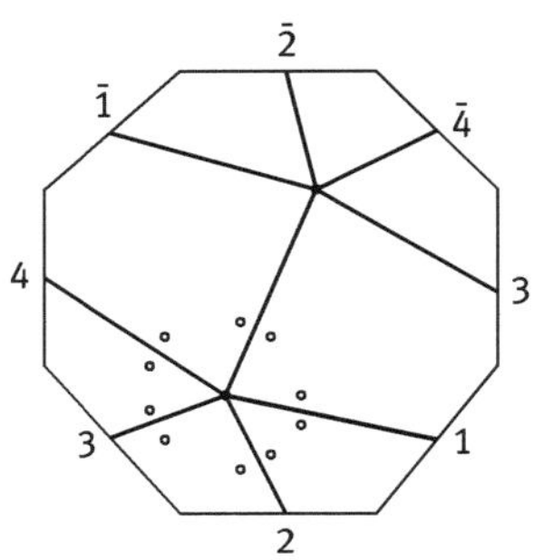

$L_5 : q3-10-10$

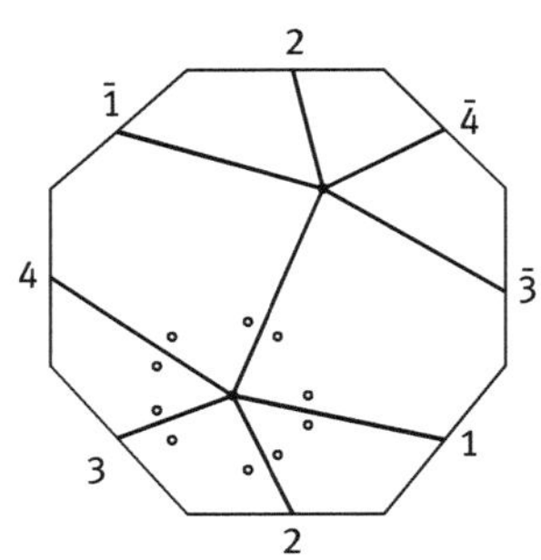

$L_5 : q3-11-10$

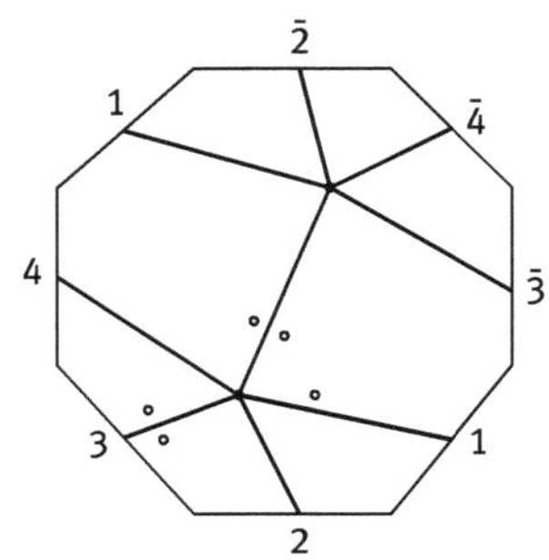

$L_5 : q3-12-05$

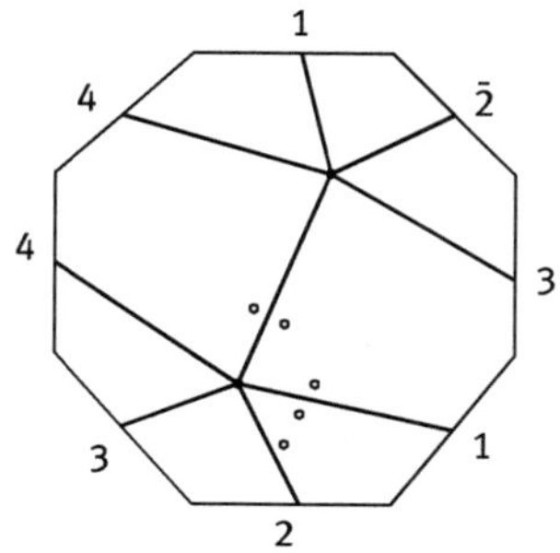

L_5:$q3$ – 13 – 05

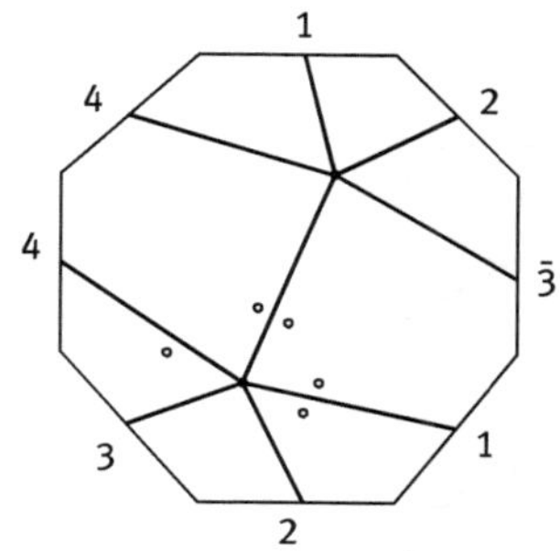

L_5:$q3$ – 14 – 05

Nonorientable genus 4

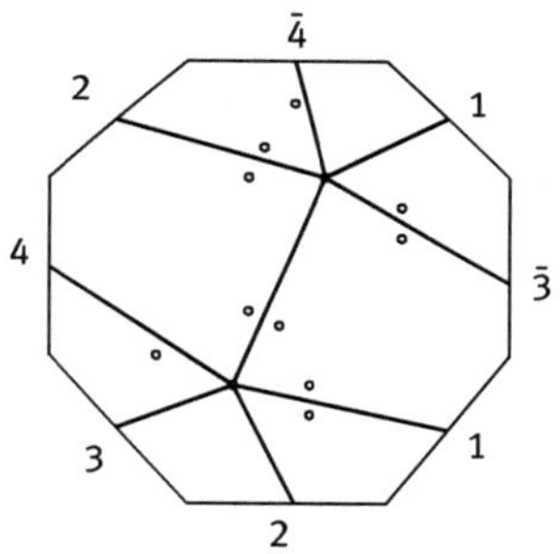

L_5:$q4$ – 01 – 10

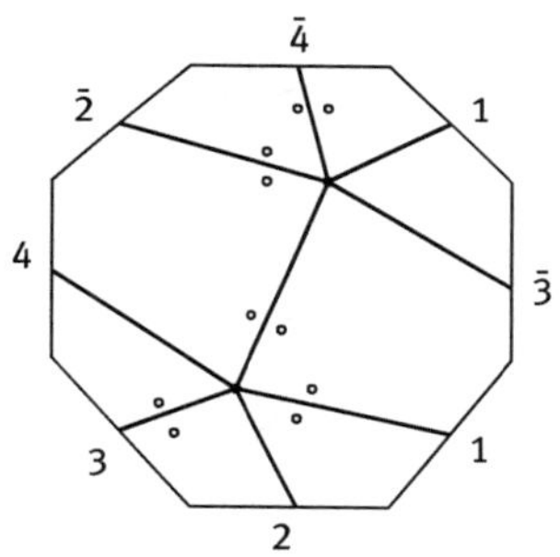

L_5:$q4$ – 02 – 10

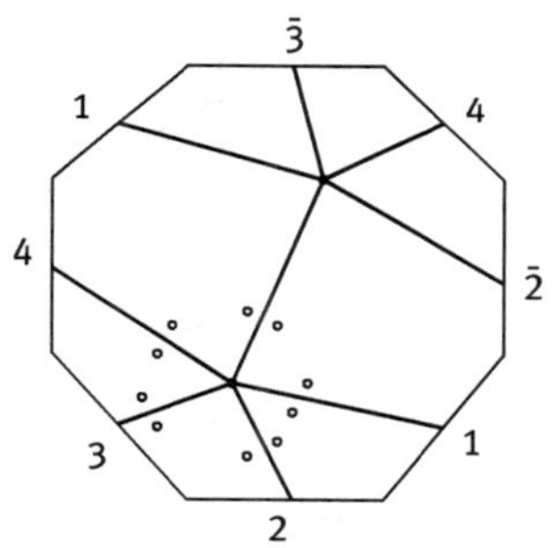

L_5:$q4$ – 03 – 10

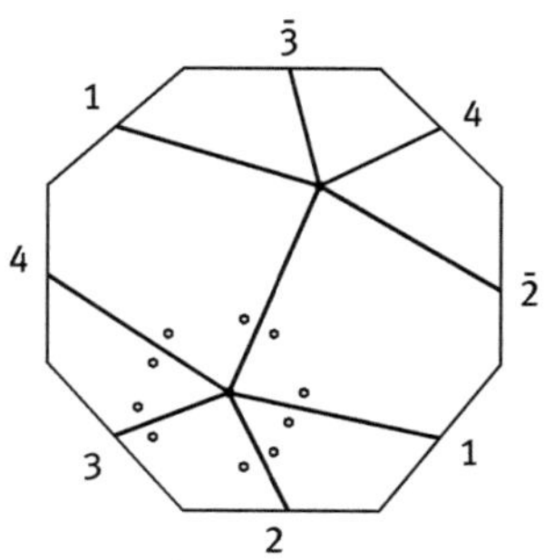

L_5:$q4$ – 04 – 10

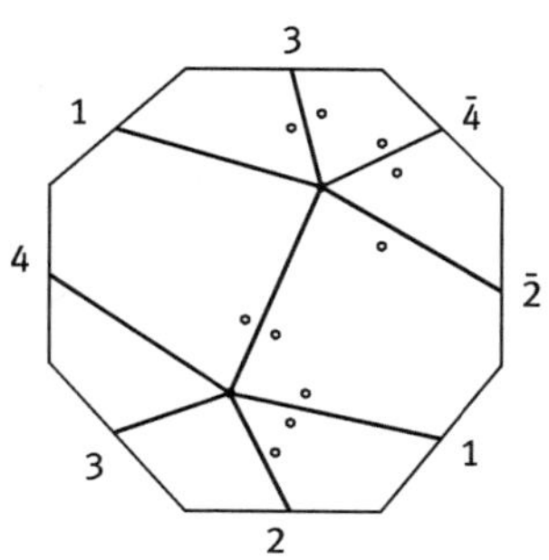

L_5:$q4$ – 05 – 10

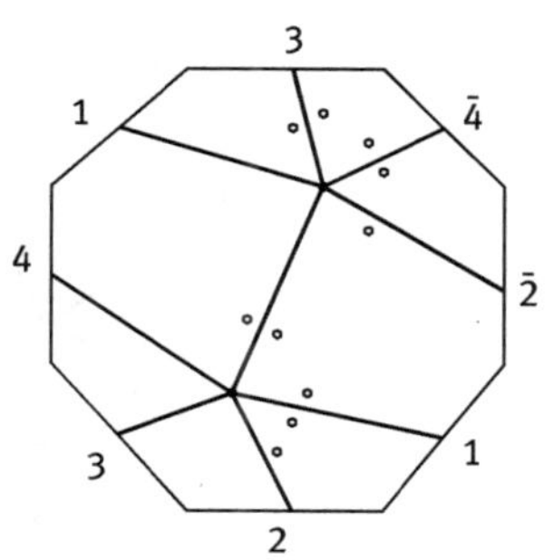

L_5:$q4$ – 06 – 10

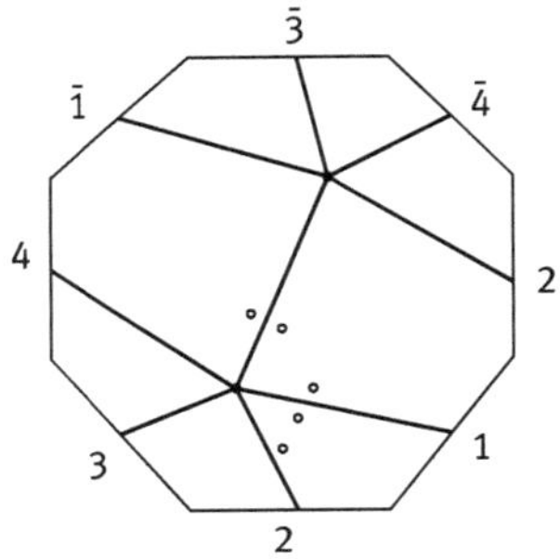

L_5:$q4 - 07 - 05$

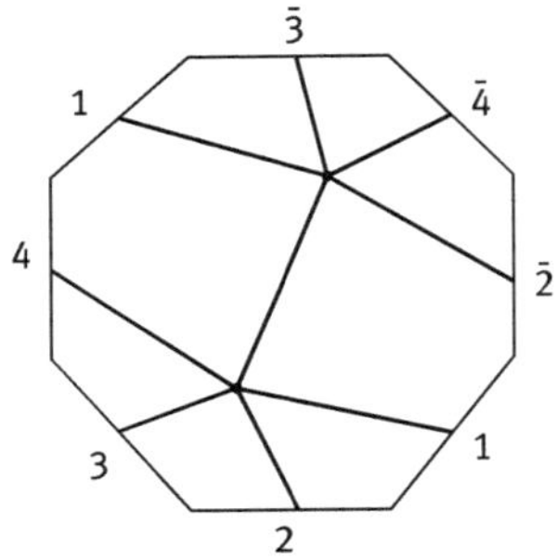

L_5:$q4 - 08 - 20$

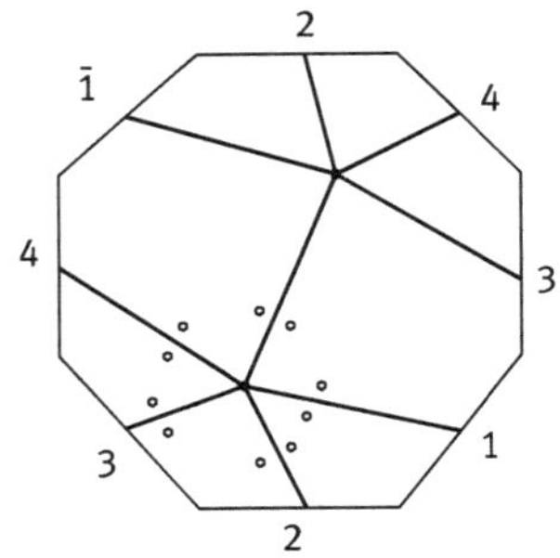

L_5:$q4 - 09 - 10$

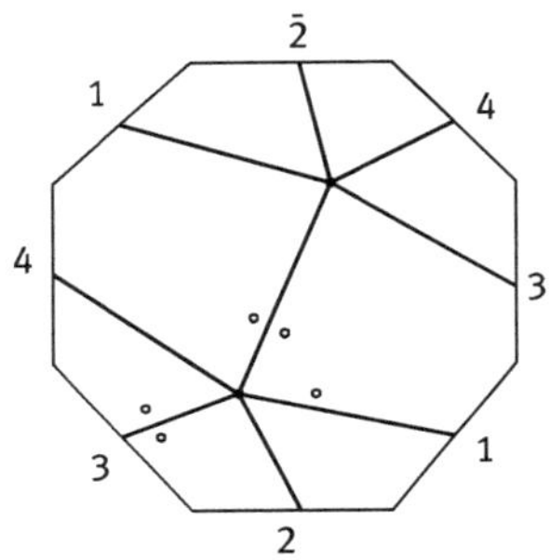

L_5:$q4 - 10 - 05$

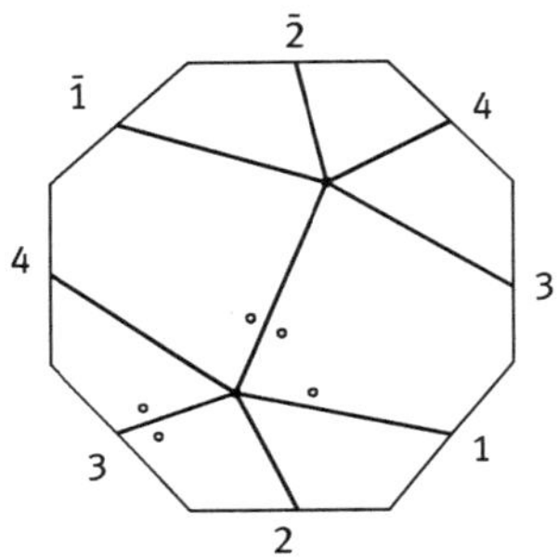

L_5:$q4 - 11 - 05$

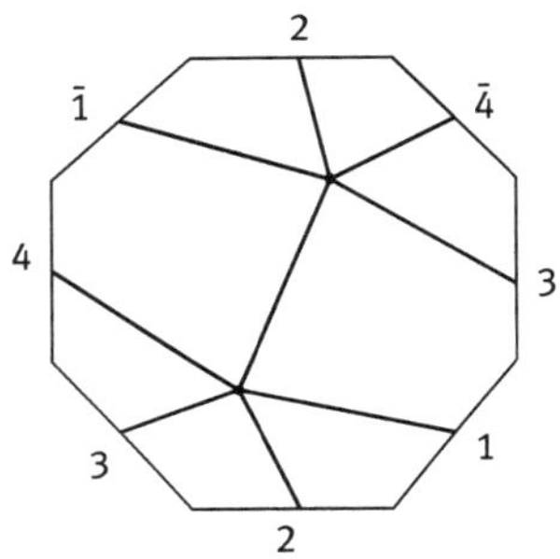

L_5:$q4 - 12 - 20$

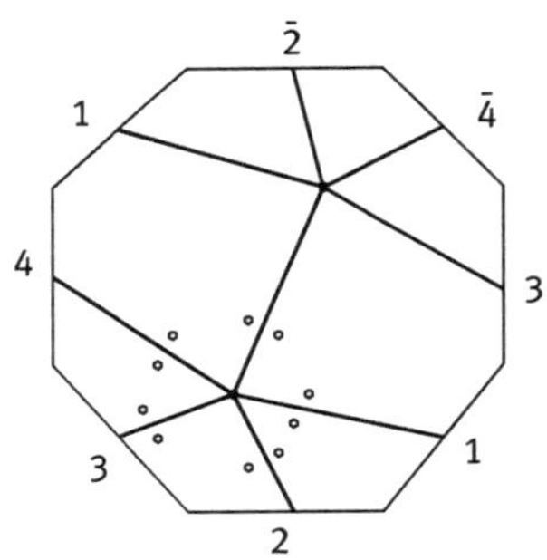

L_5:$q4 - 13 - 10$

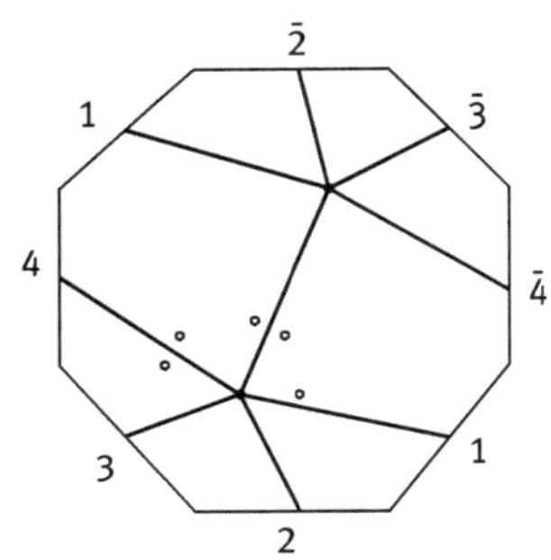

L_5:$q4 - 14 - 05$

Case $m = 6$**:**

Orientable genus 0

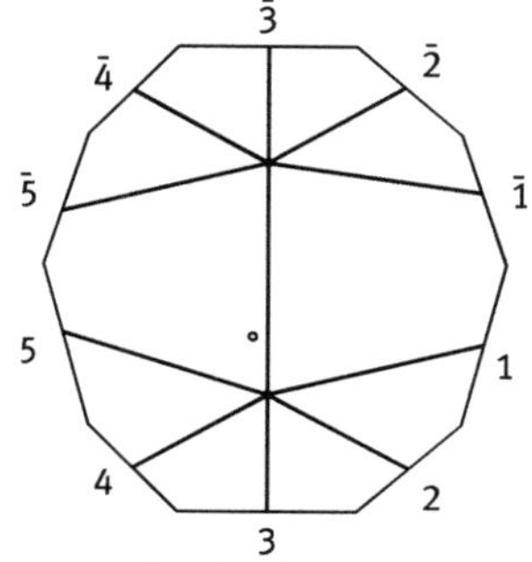

L_6:o0 – 01 – 01

Orientable genus 1

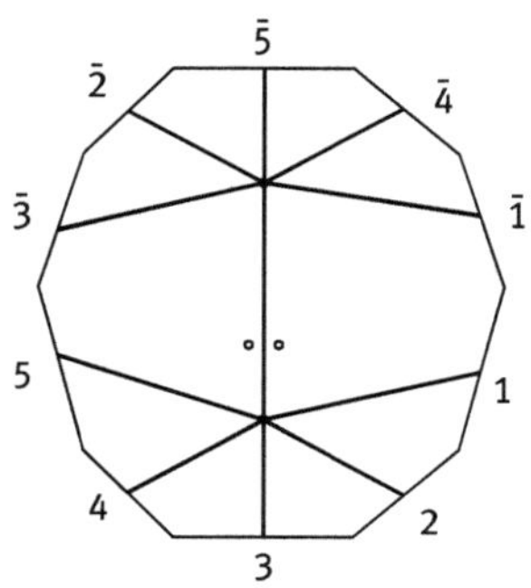

L_6:o1 – 01 – 02

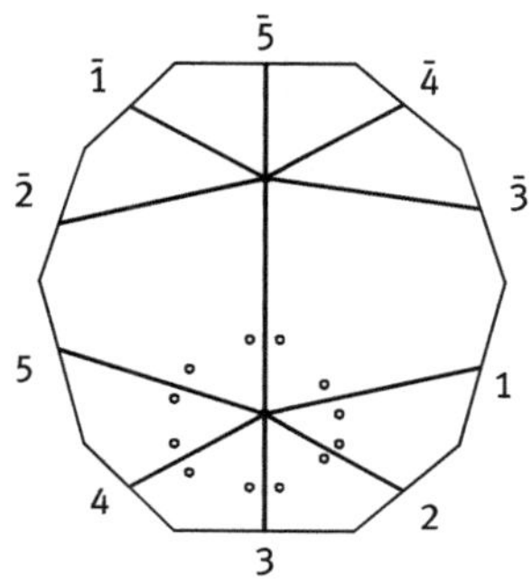

L_6:o1 – 02 – 12

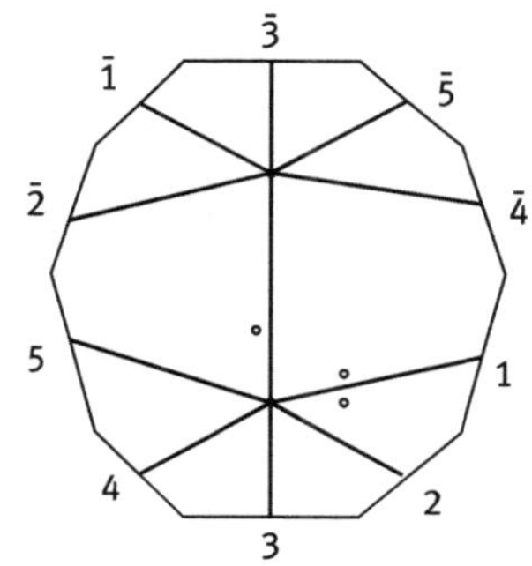

L_6:o1 – 03 – 03

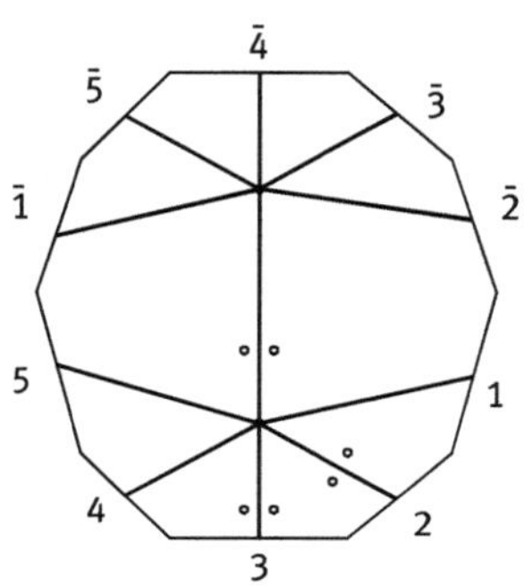

L_6:o1 – 04 – 06

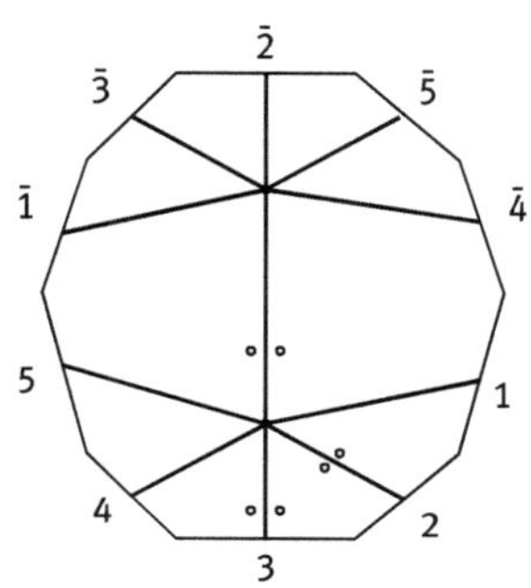

L_6:o1 – 05 – 06

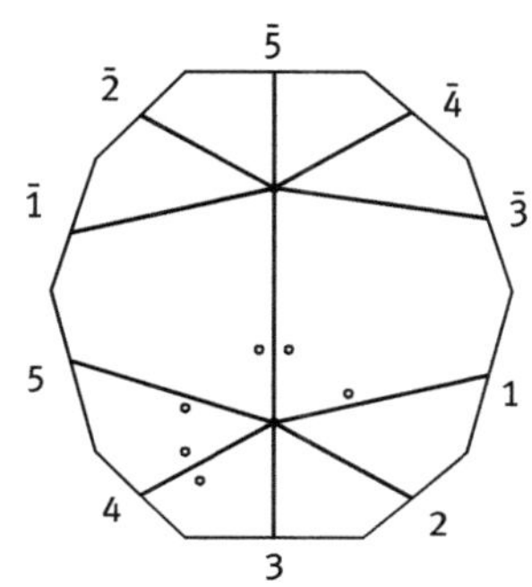

L_6:o1 – 06 – 06

Orientable genus 2

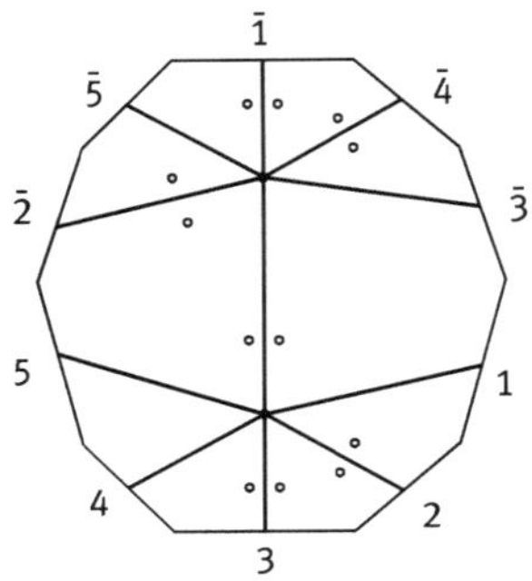

L_6:o2 – 01 – 12

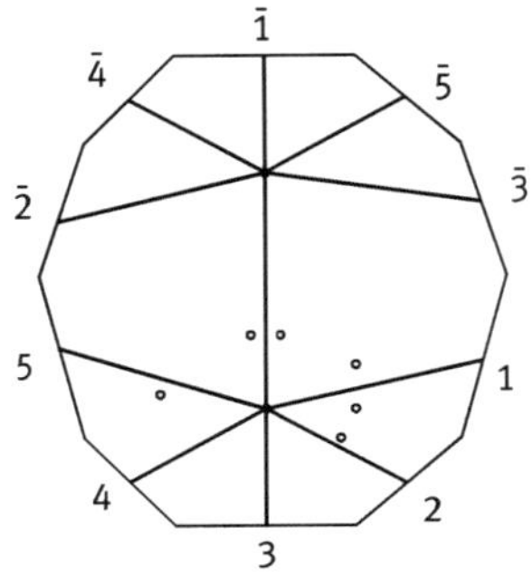

L_6:o2 – 02 – 06

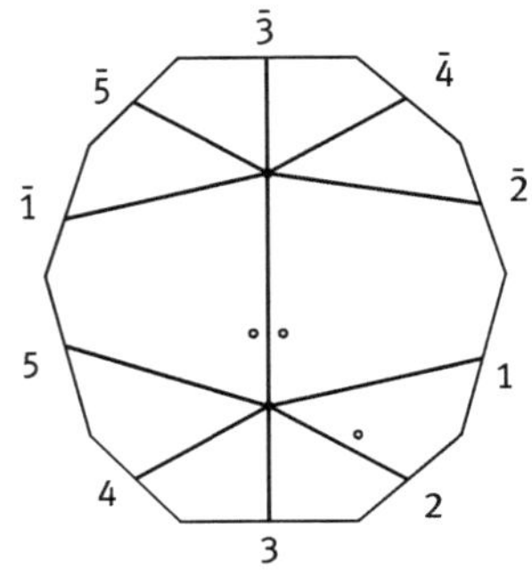

L_6:o2 – 03 – 03

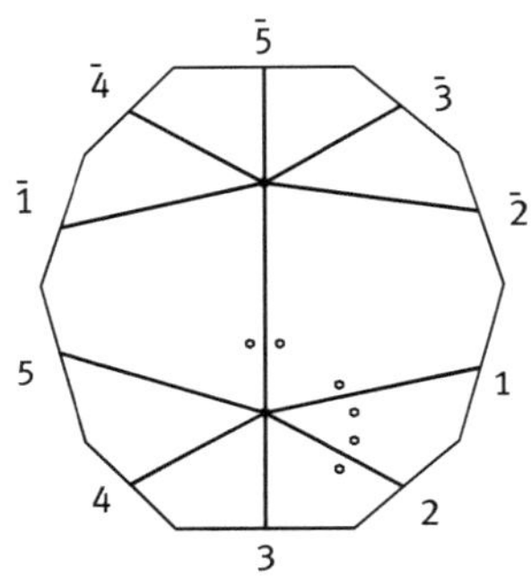

L_6:o2–04–06

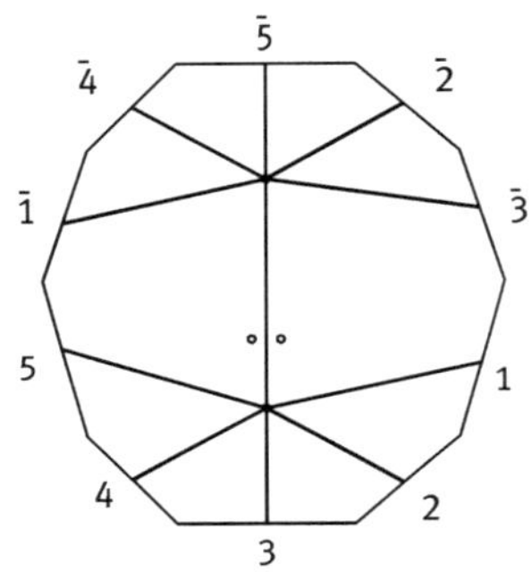

L_6:o2–05–02

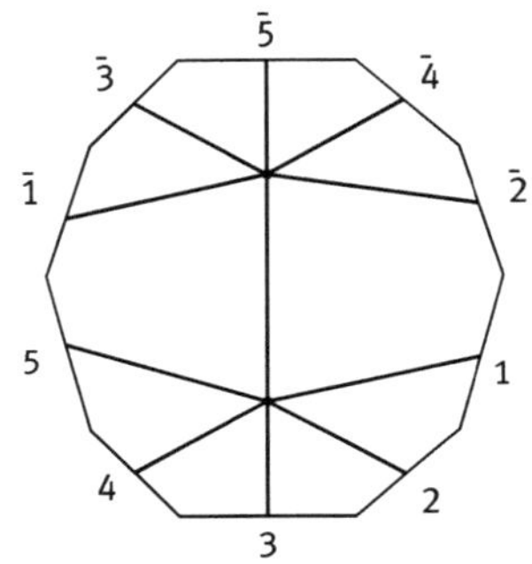

L_6:o6–06–24

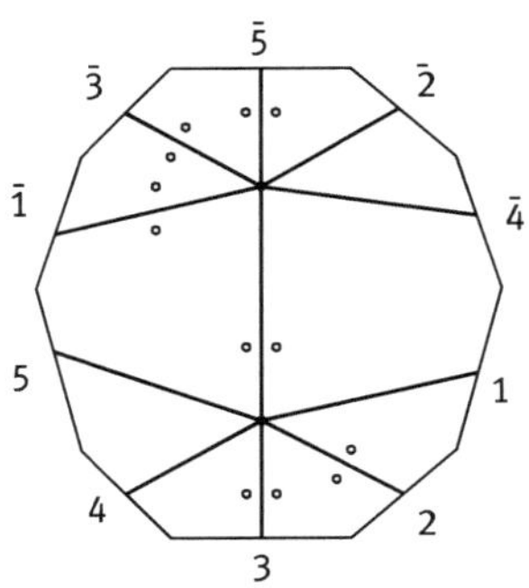

L_6:o2 – 07 – 12

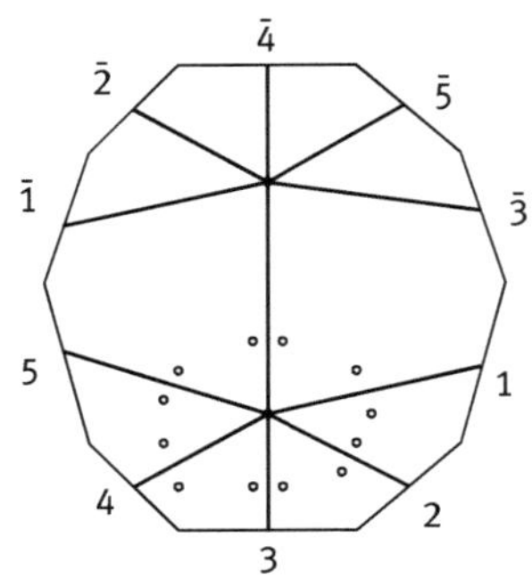

L_6:o2 – 08 – 12

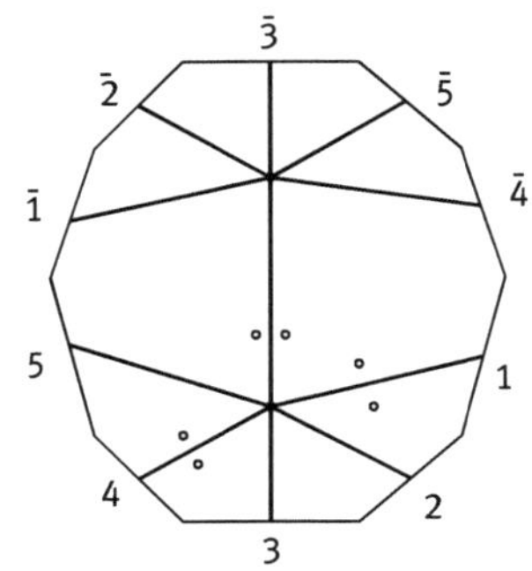

L_6:o2 – 09 – 06

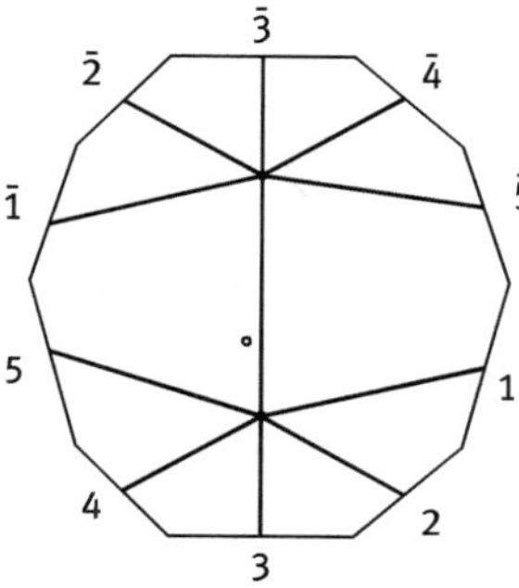

L_6:o2 – 10 – 01

A3.3 Complete bipartite graphs $K_{m,n}$, $4 \geq m, n \geq 3$

Case $m + n = 6$**:**

Orientable genus 1

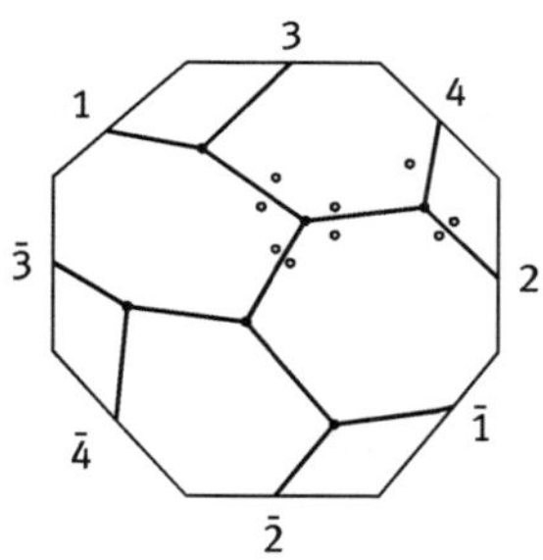

$K_{3,3}$:o1 – 01 – 09

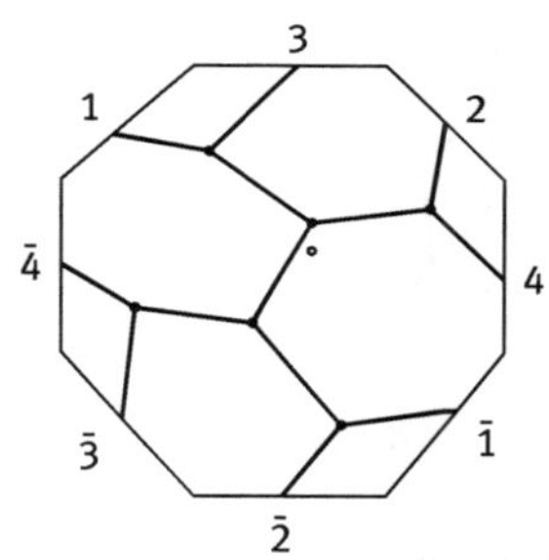

$K_{3,3}$:o1 – 02 – 01

Orientable genus 2

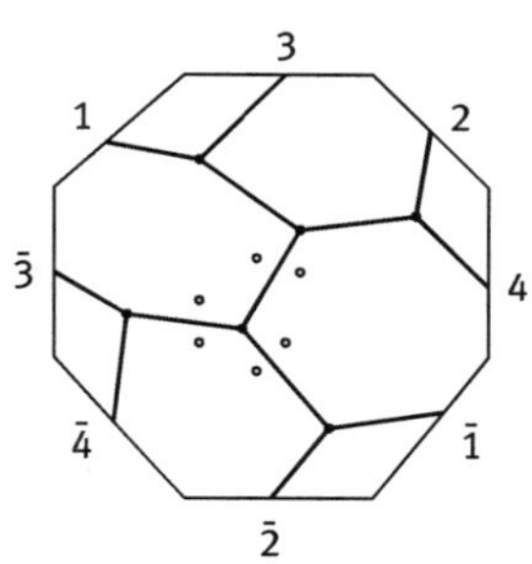

$K_{3,3}$:o2 –01–06

Nonorientable genus 1

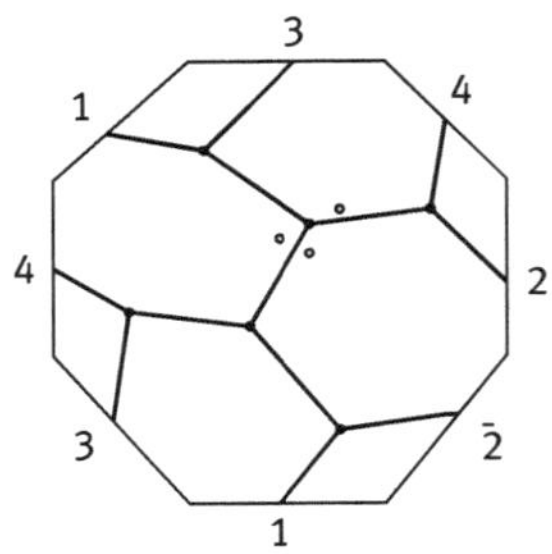

$K_{3,3}$:*q*1 – 01 – 03

Nonorientable genus 2

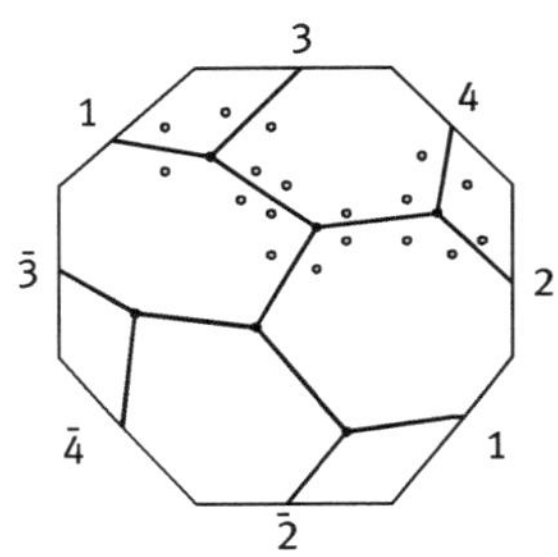

$K_{3,3}$:*q*2 – 01 – 18

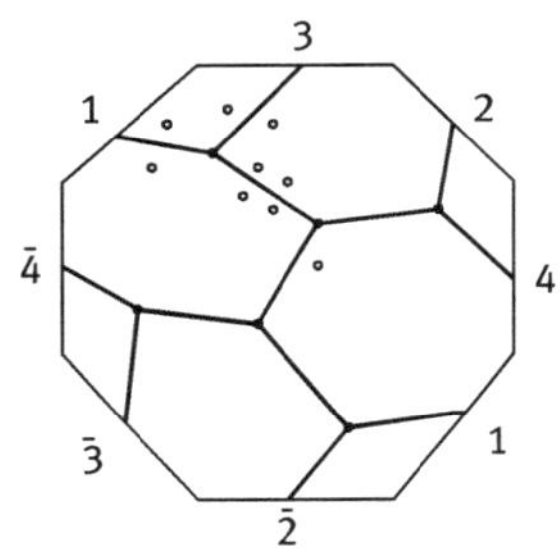

$K_{3,3}$:*q*2 – 02 – 09

Nonorientable genus 3

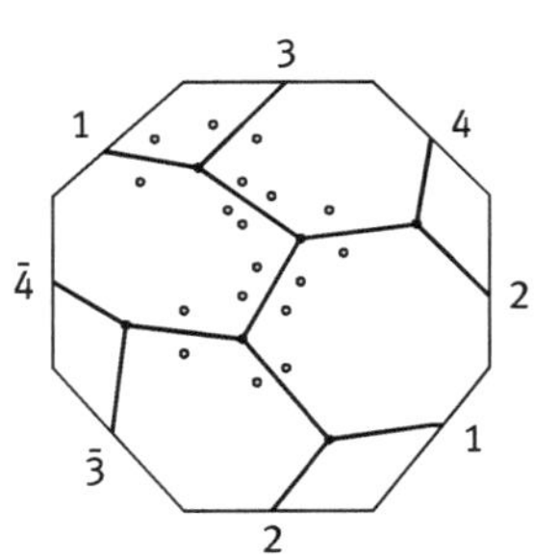

$K_{3,3}$:*q*3 – 01 – 18

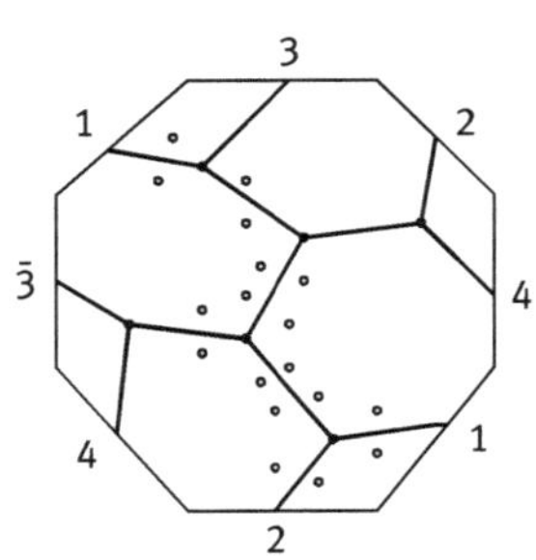

$K_{3,3}$:*q*3 – 02 – 18

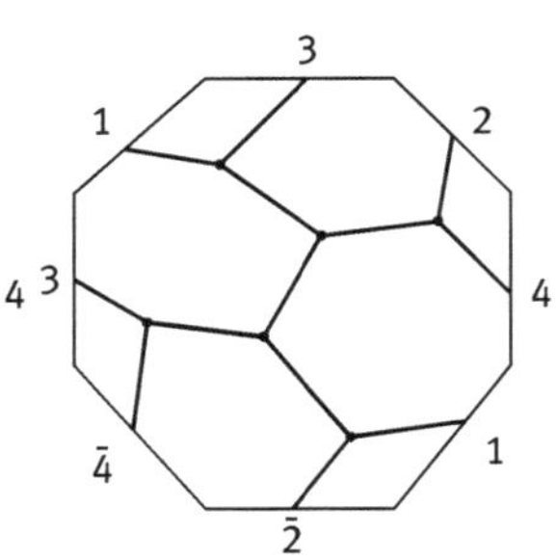

$K_{3,3}$:*q*3 – 03 – 36

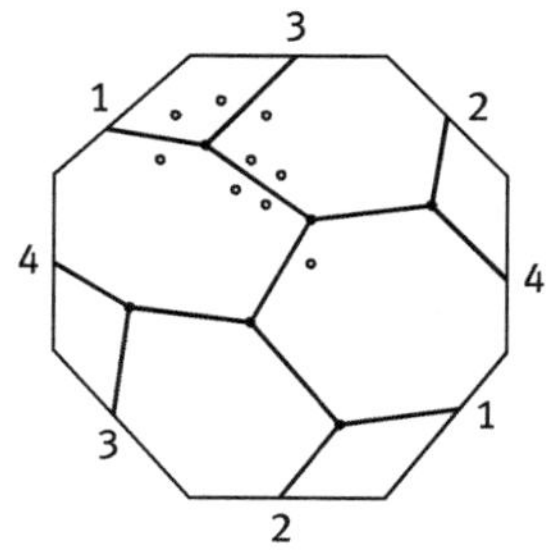

$K_{3,3}: q3-04-09$

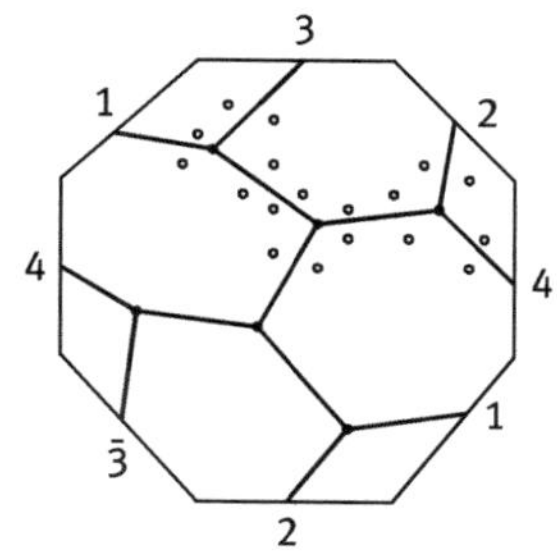

$K_{3,3}: q3-05-18$

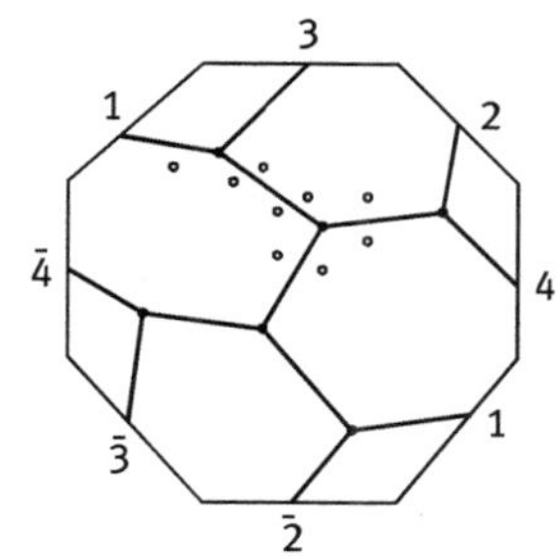

$K_{3,3}: q3-06-09$

Nonorientable genus 4

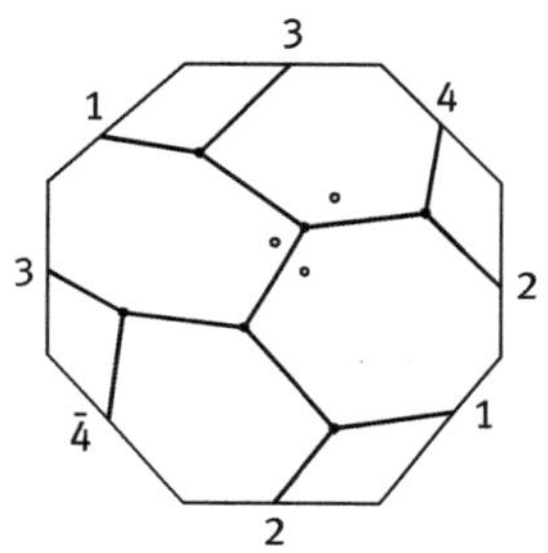

$K_{3,3}: q4-01-03$

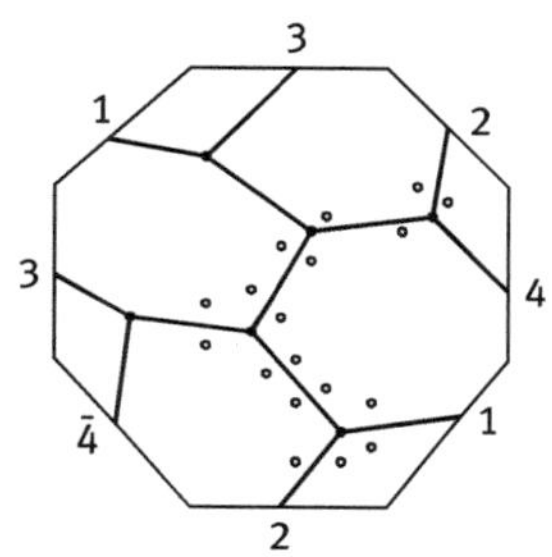

$K_{3,3}: q4-02-18$

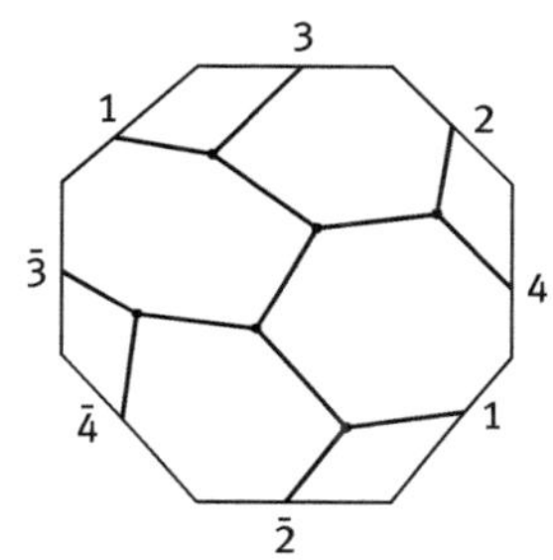

$K_{3,3}: q4-03-36$

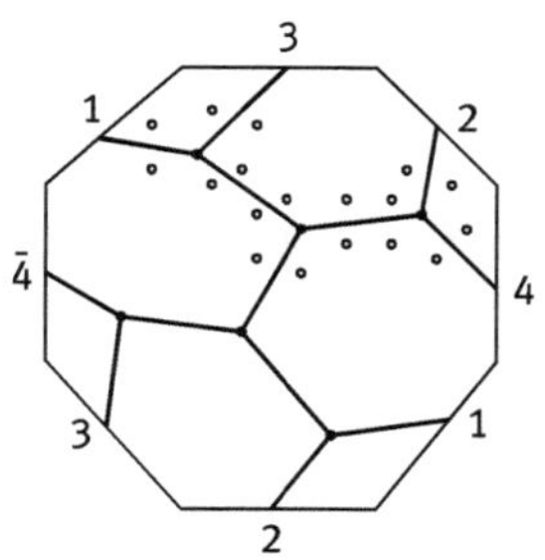

$K_{3,3}: q4-04-18$

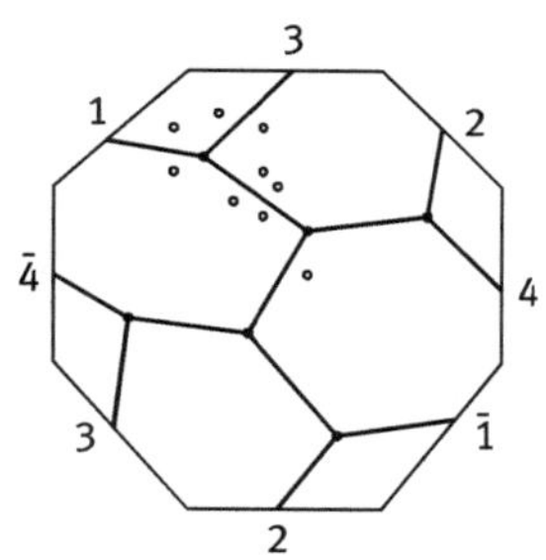

$K_{3,3}: q4-05-09$

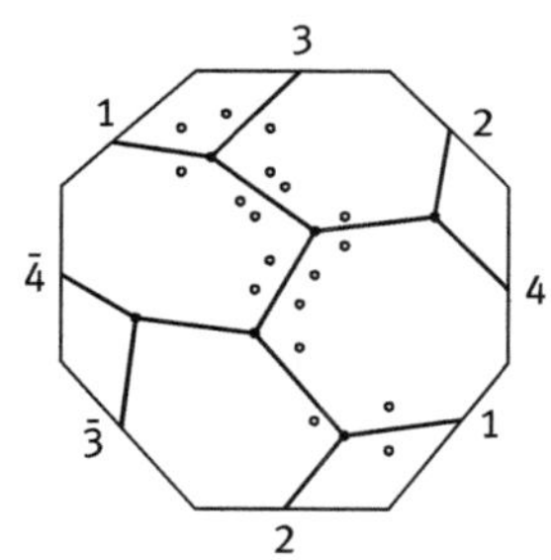

$K_{3,3}: q4-06-18$

Case $m + n = 7$**:**

Orientable genus 1

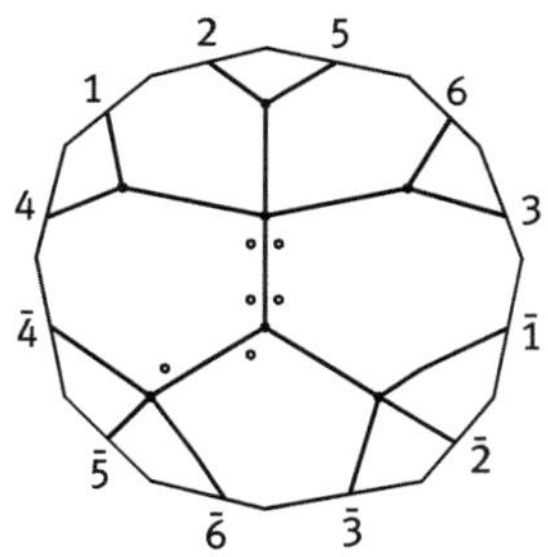

$K_{4,3}: o1-01-06$

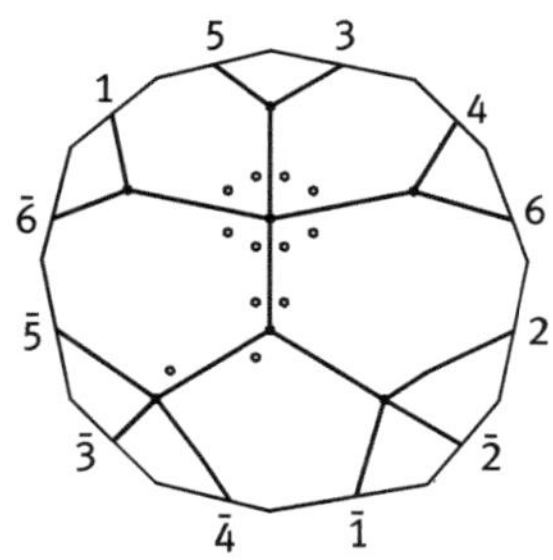

$K_{4,3}: o1-02-12$

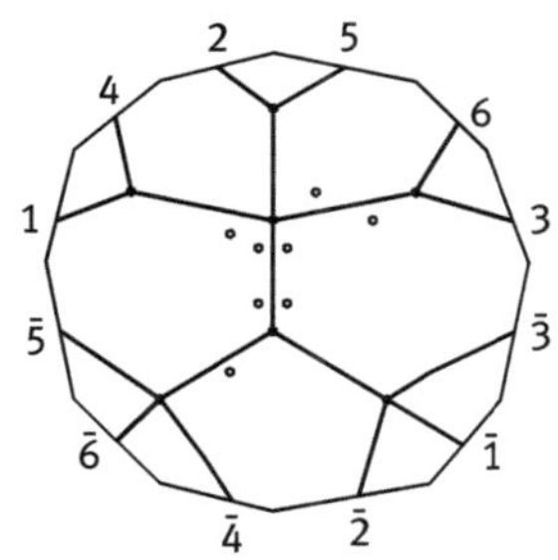

$K_{4,3}: o1-03-08$

Orientable genus 2

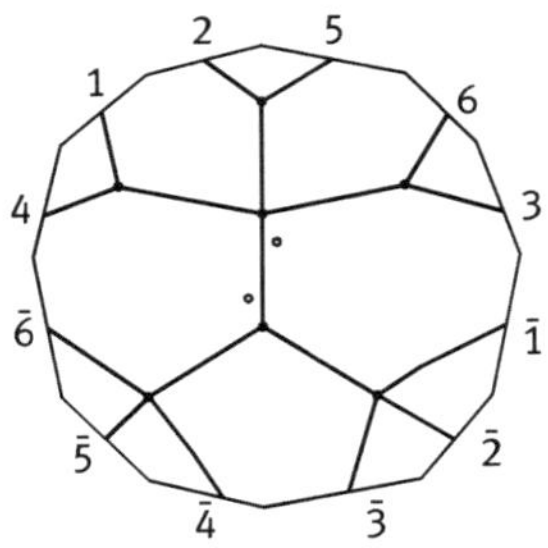

$K_{4,3}$: $o2-01-02$

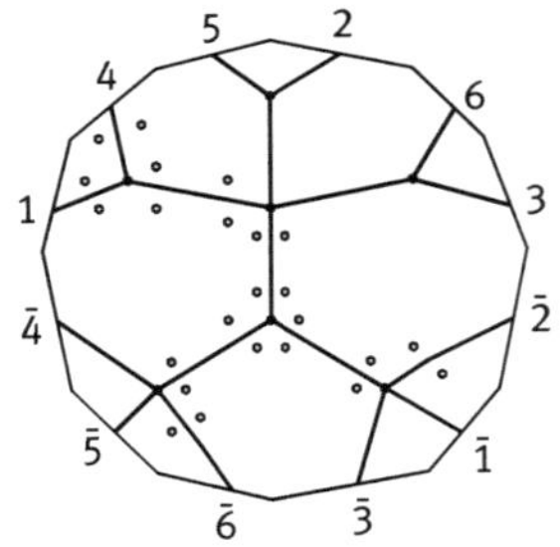

$K_{4,3}$: $o2-02-24$

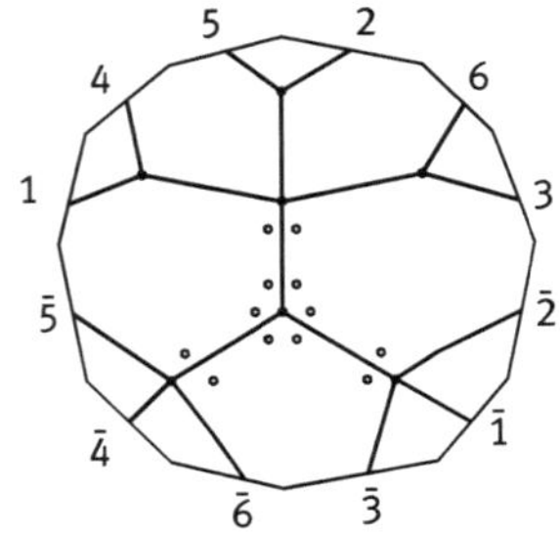

$K_{4,3}$: $o2-03-12$

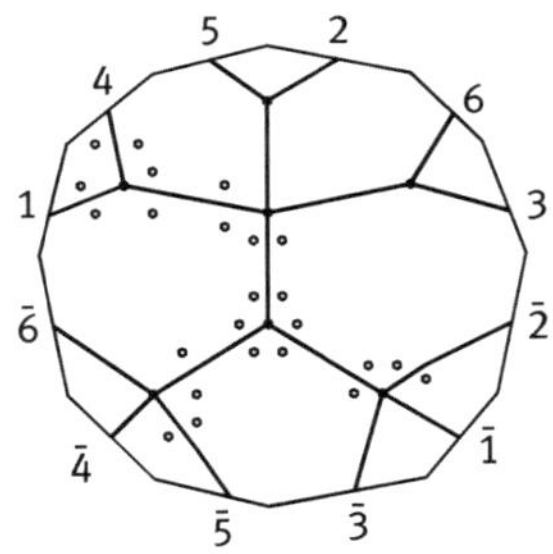

$K_{4,3}$: $o2-04-24$

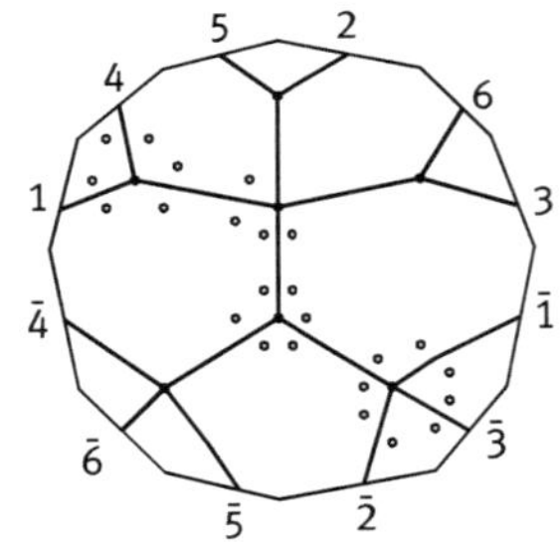

$K_{4,3}$: $o2-05-24$

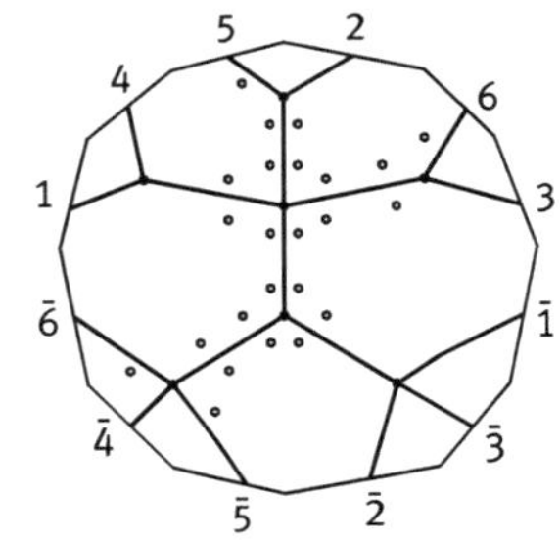

$K_{4,3}$: $o2-06-24$

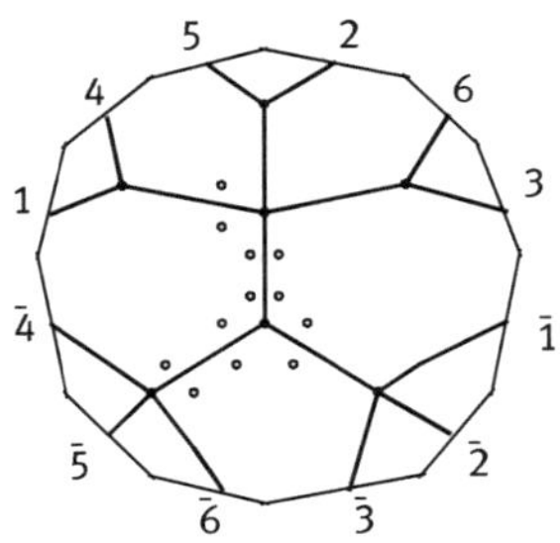

$K_{4,3}$: $o2-07-12$

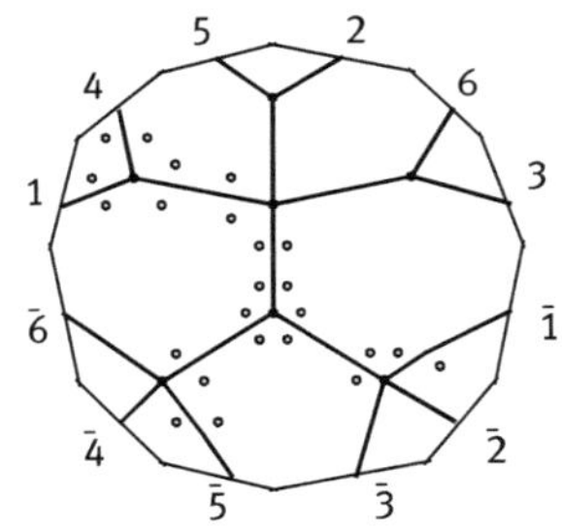

$K_{4,3}$: $o2-08-24$

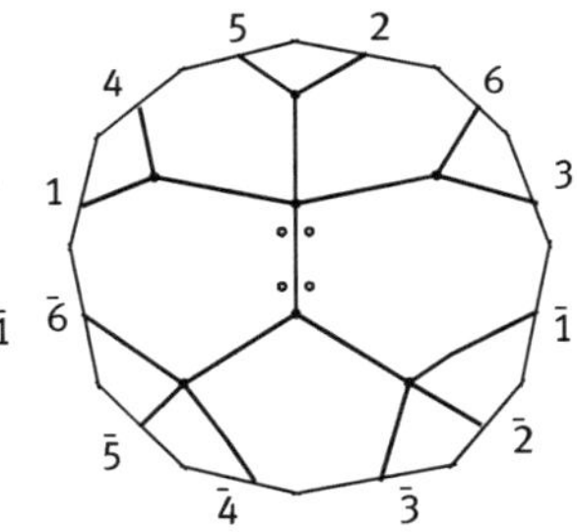

$K_{4,3}$: $o2-09-04$

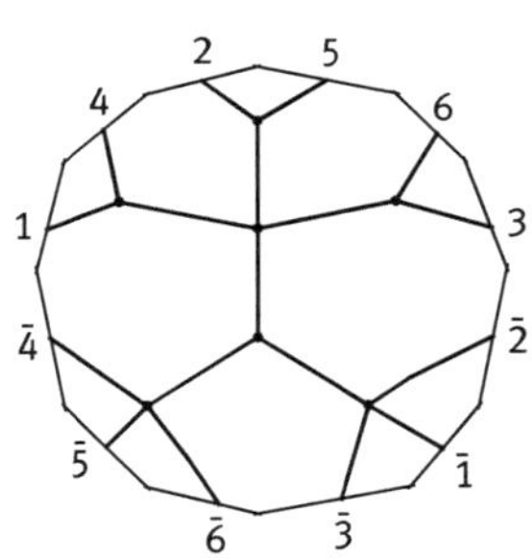

$K_{4,3}$: $o2-10-48$

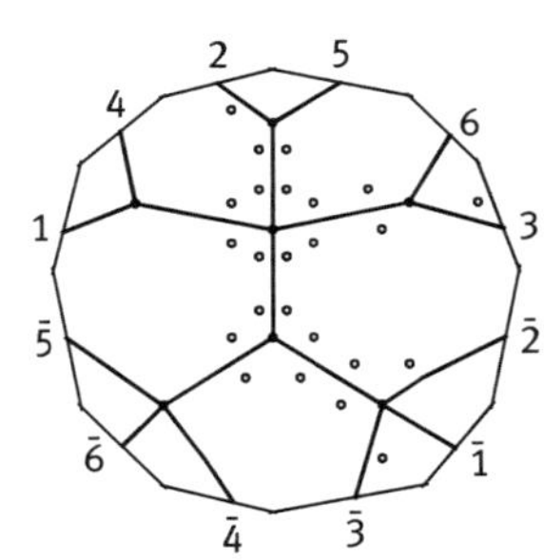

$K_{4,3}$: $o2-11-24$

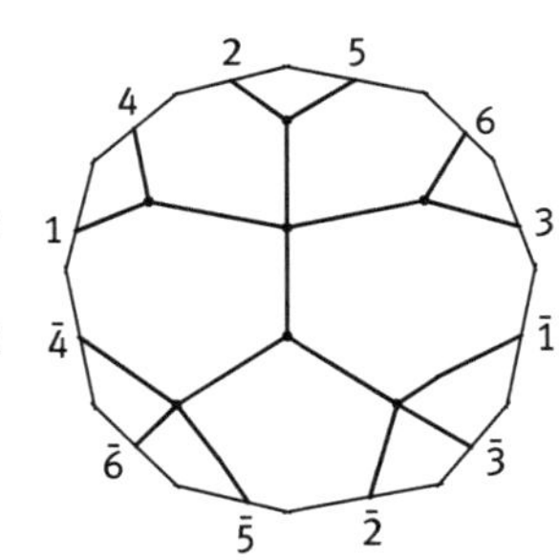

$K_{4,3}$: $o2-12-48$

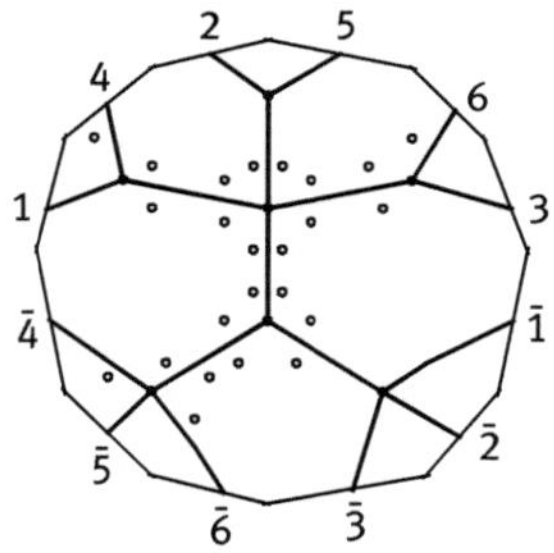

$K_{4,3}: o2-13-24$

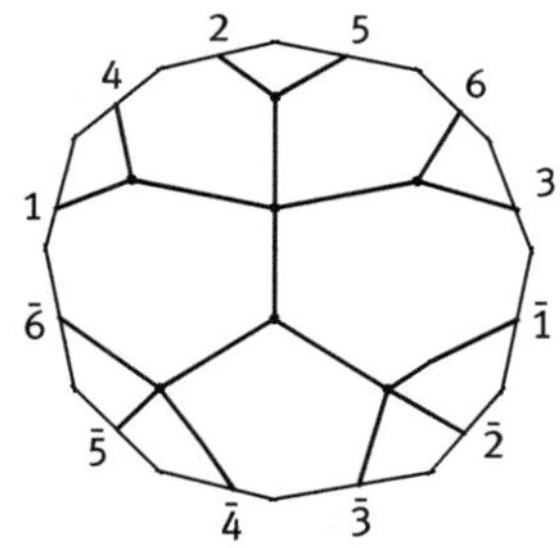

$K_{4,3}: o2-14-48$

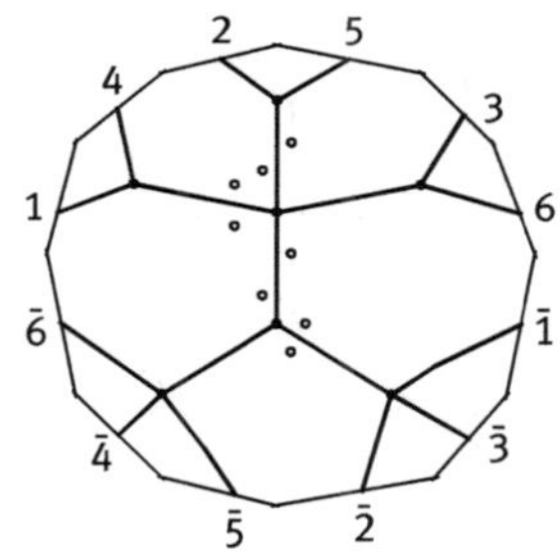

$K_{4,3}: o2-15-08$

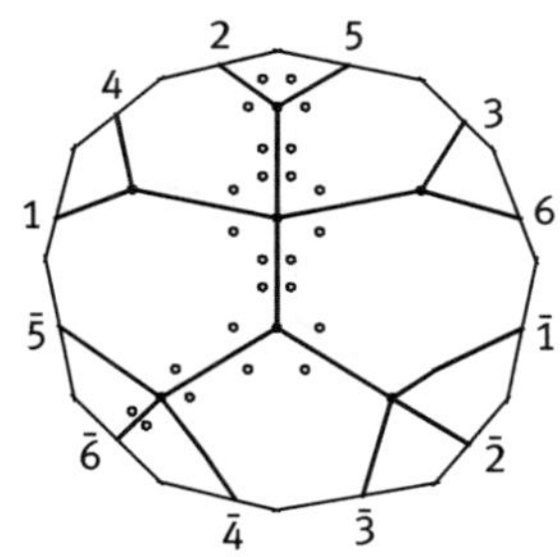

$K_{4,3}: o2-16-24$

Orientable genus 3

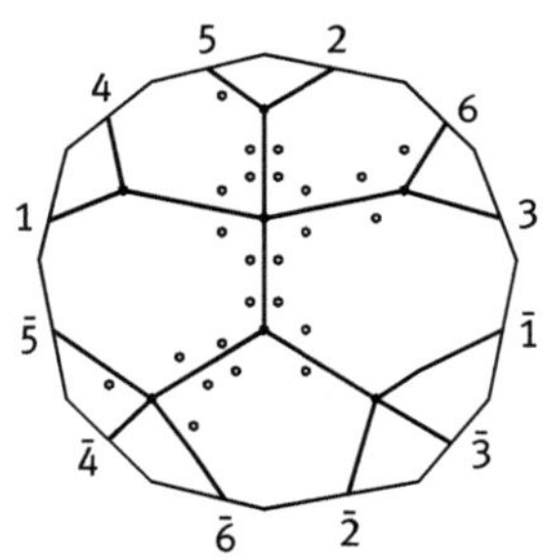

$K_{4,3}: o3-01-24$

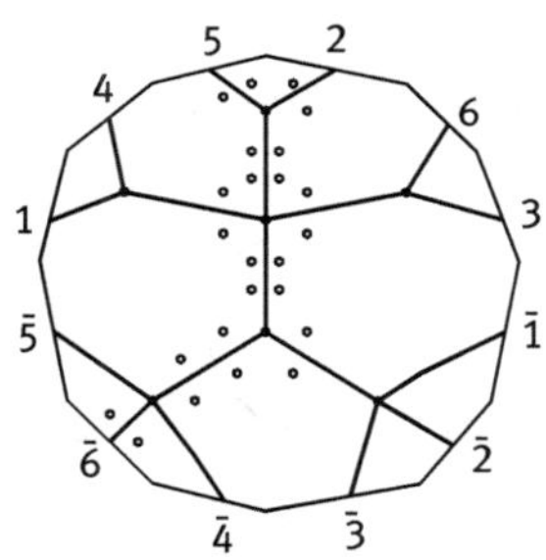

$K_{4,3}: o3-02-24$

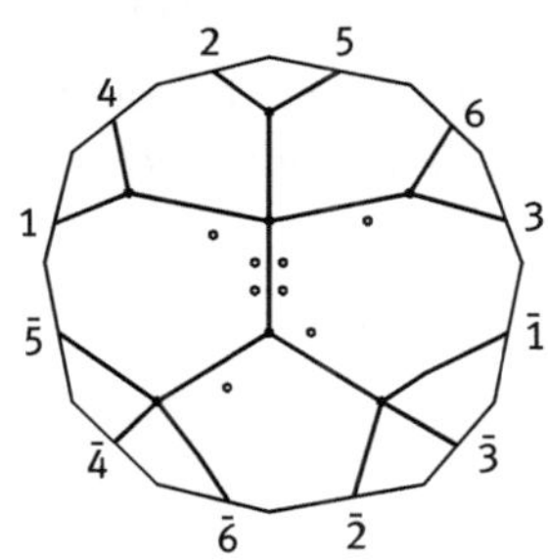

$K_{4,3}: o3-03-08$

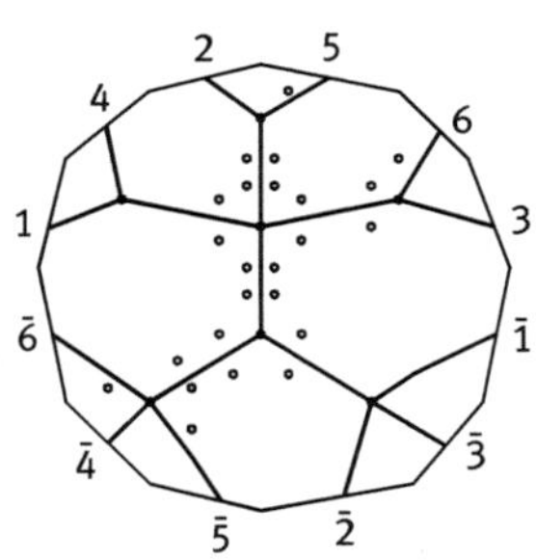

$K_{4,3}: o3-04-24$

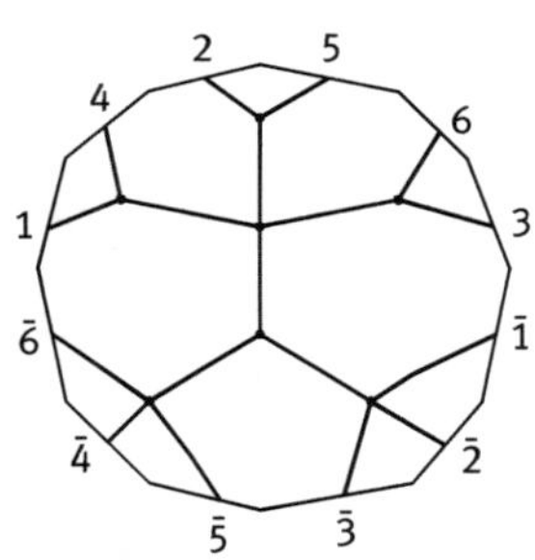

$K_{4,3}: o3-05-48$

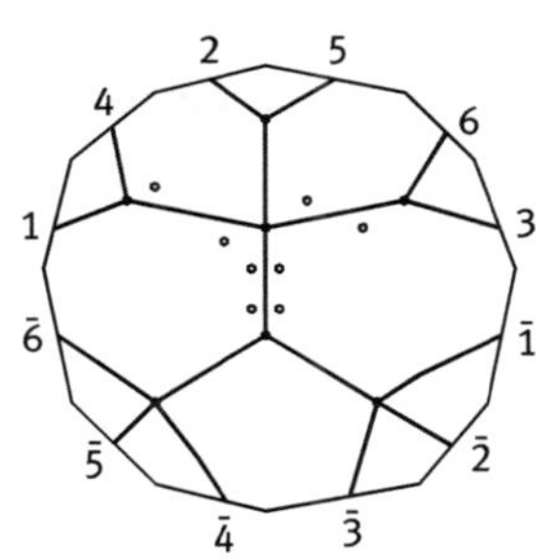

$K_{4,3}: o3-06-08$

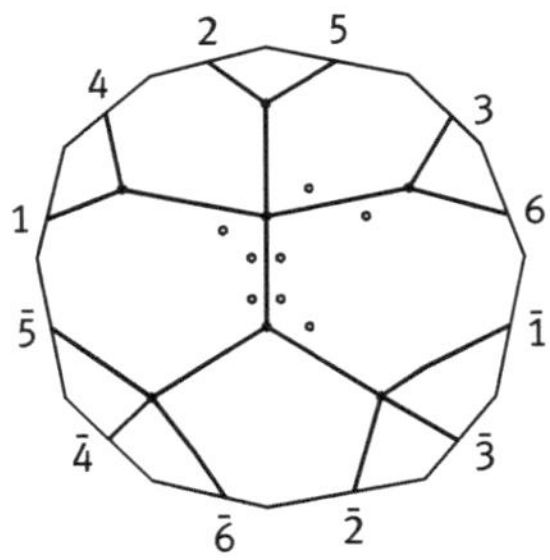

$K_{4,3}$: o3 – 07 – 08

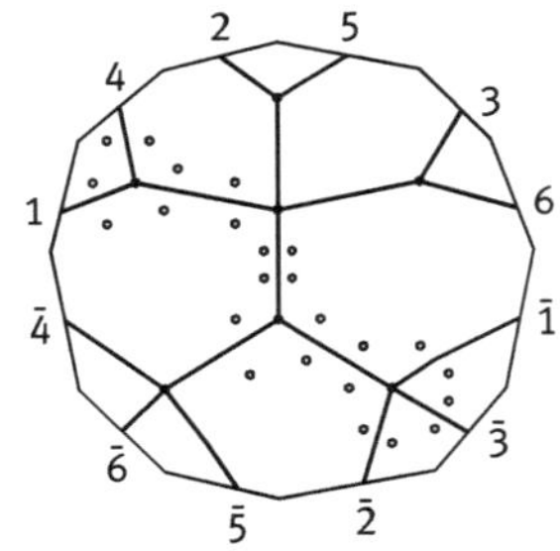

$K_{4,3}$: o3 – 08 – 24

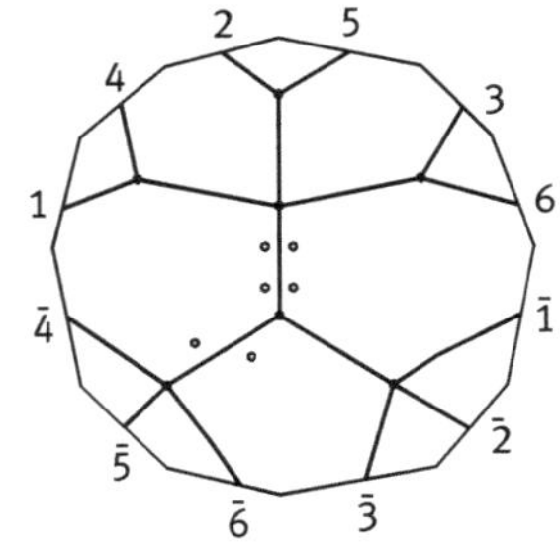

$K_{4,3}$: o3 – 09 – 06

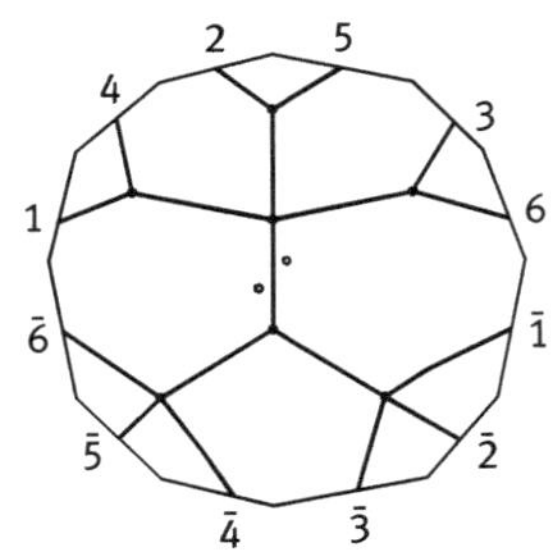

$K_{4,3}$: o3 – 10 – 02

A3.4 Wheels W_n, $5 \geq n \geq 4$

Case $n = 4$ (i.e., the complete graph K_4 of order 4):

Orientable genus 0

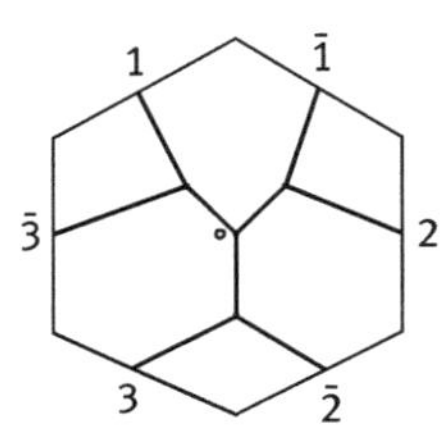

W_4: o0 – 01 – 01

Orientable genus 1

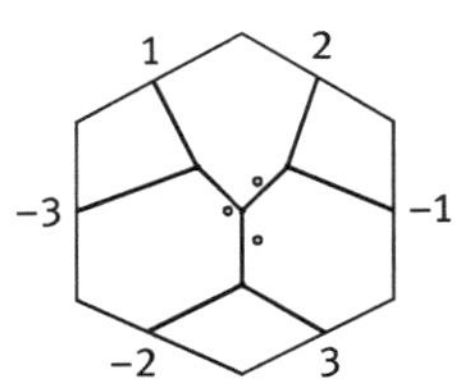

W_4: o1 – 01 – 03

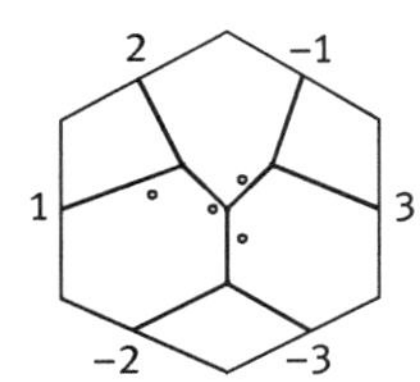

W_4: o1 – 02 – 04

Nonorientable genus 1

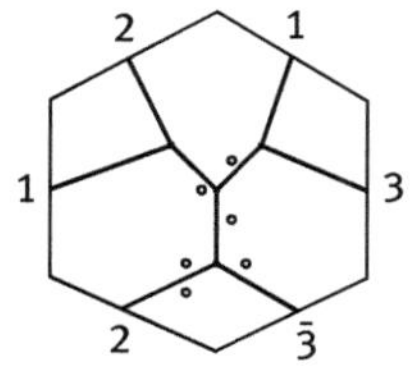

$W_4 : q1 - 01 - 06$

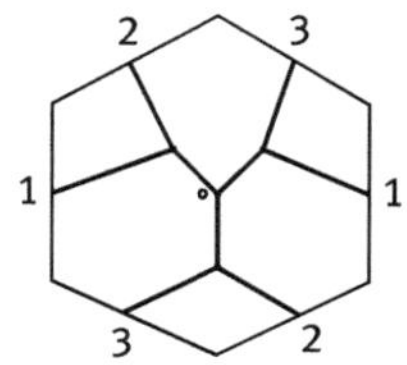

$W_4 : q1 - 02 - 01$

Nonorientable genus 2

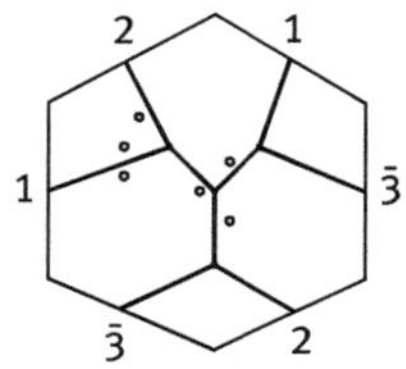

$W_4 : q2 - 01 - 06$

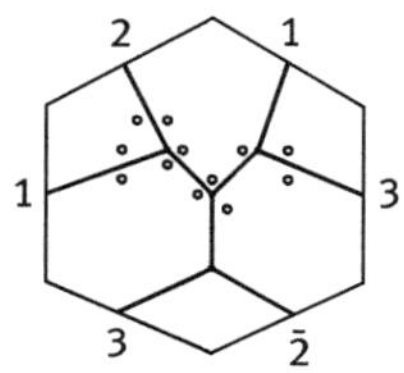

$W_4 : q2 - 02 - 12$

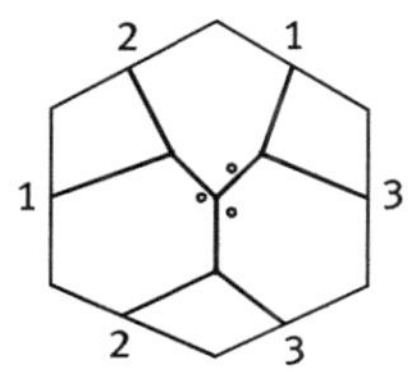

$W_4 : q2 - 03 - 03$

Nonorientable genus 3

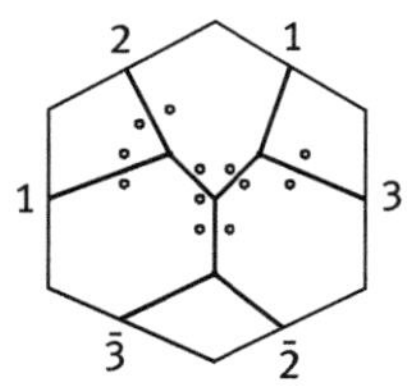

$W_4 : q3 - 01 - 12$

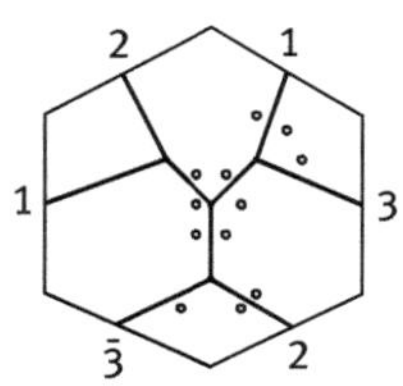

$W_4 : q3 - 02 - 12$

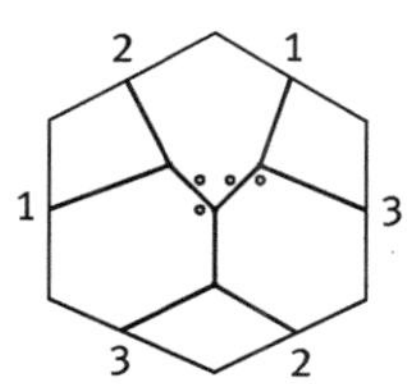

$W_4 : q3 - 03 - 04$

Case $n = 5$**:**

Orientable genus 0

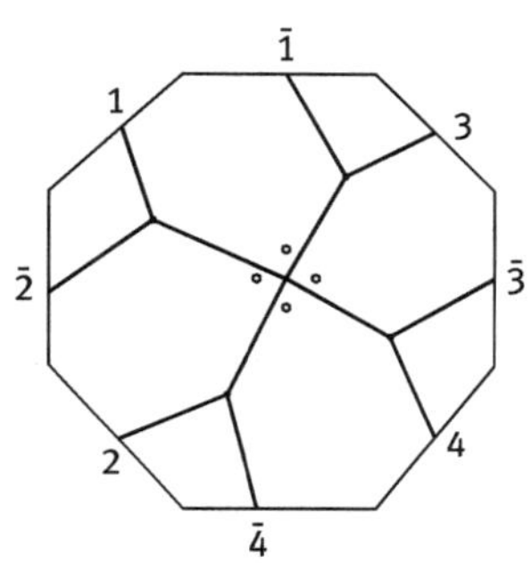

$W_5 : o0 - 01 - 04$

Orientable genus 1

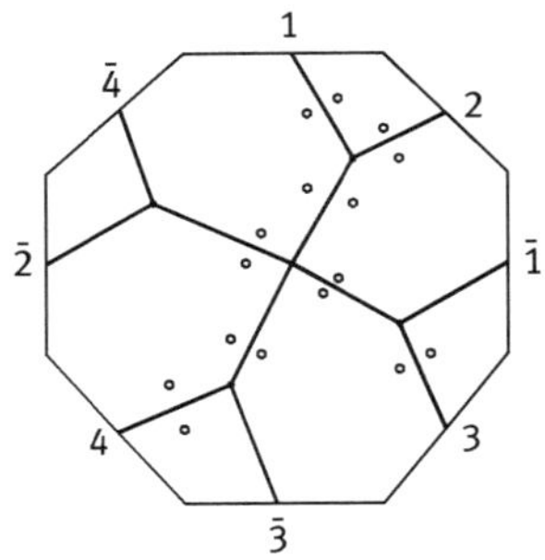

$W_5: o1-01-16$

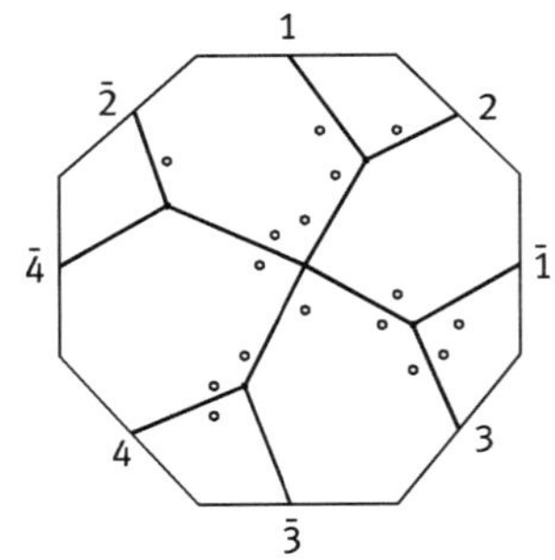

$W_5: o1-02-16$

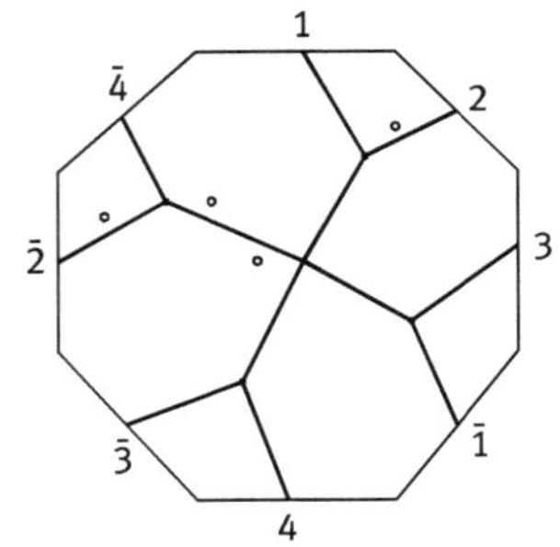

$W_5: o1-03-04$

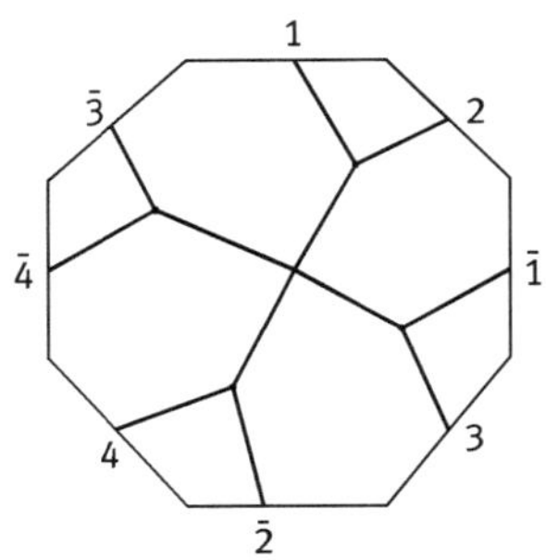

$W_5: o1-04-32$

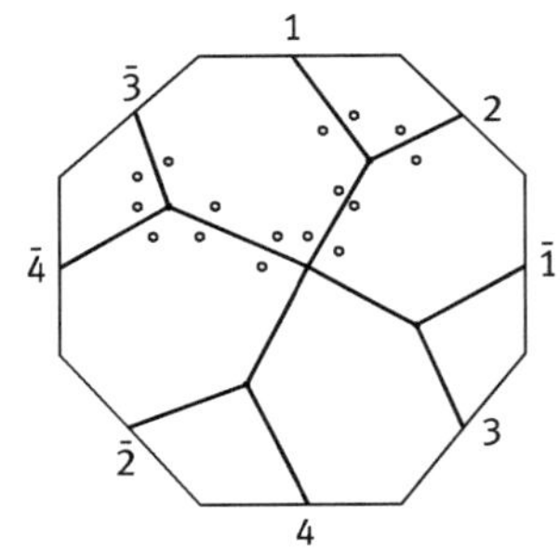

$W_5: o1-05-16$

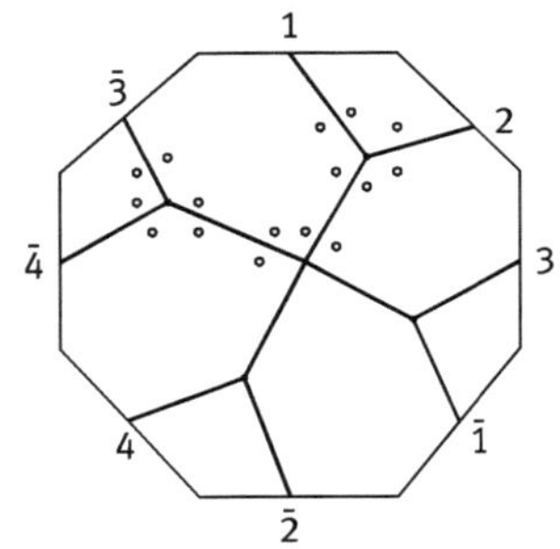

$W_5: o1-06-16$

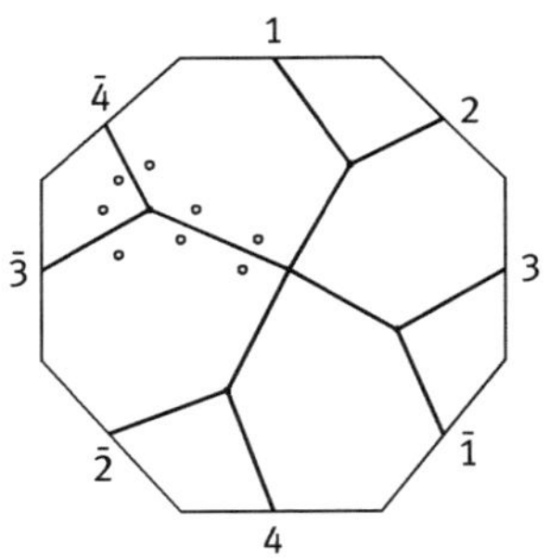

$W_5: o1-07-08$

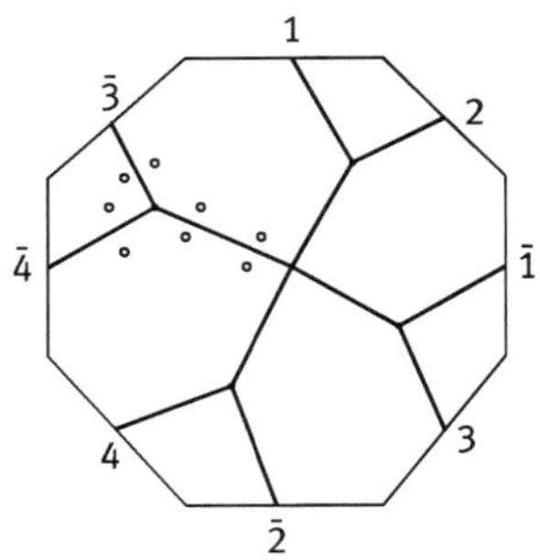

$W_5: o1-08-08$

Orientable genus 2

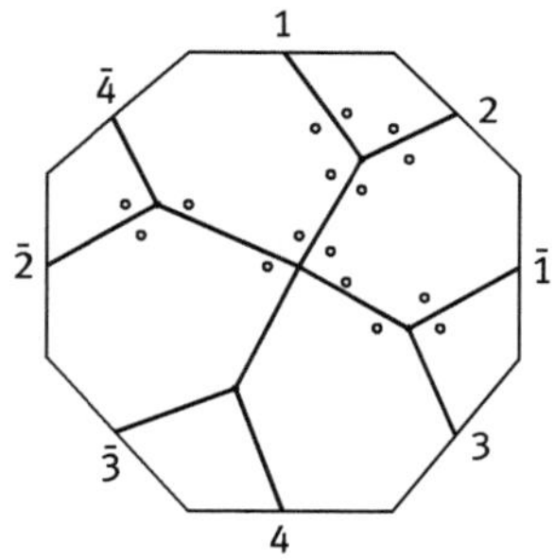

$W_5: o2-01-16$

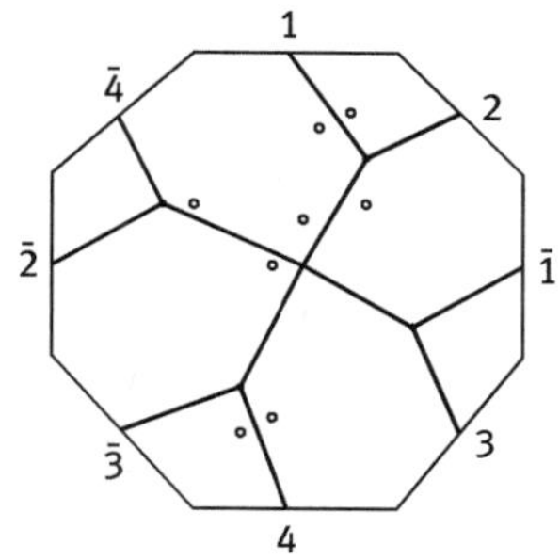

$W_5: o2-02-08$

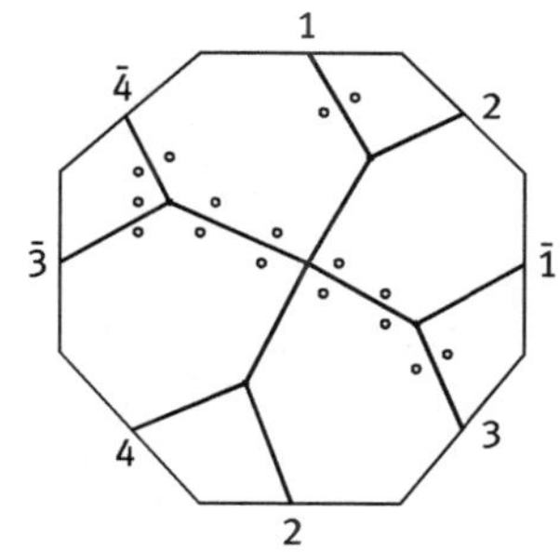

$W_5: o2-03-16$

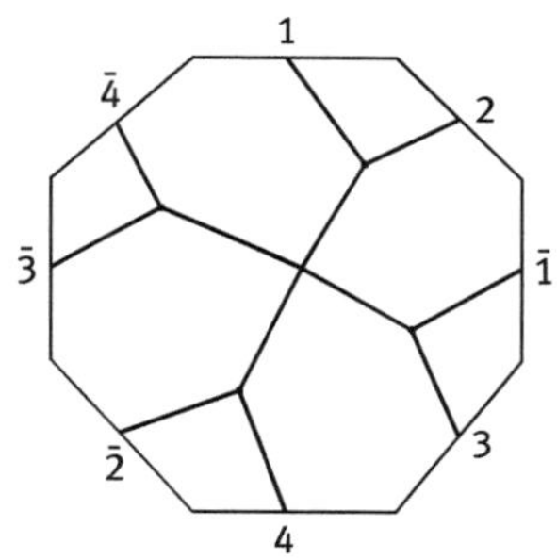

$W_5: o2-04-32$

A3.5 Triconnected cubic graphs of size in [6, 15]

Size 6: ($C_{6,1} = K_4 = W_4$, W_4 is known above).

Size 9: ($C_{9,2} = K_{3,3}$ in A3.3)

$C_{9,1}$:

Orientable genus 0

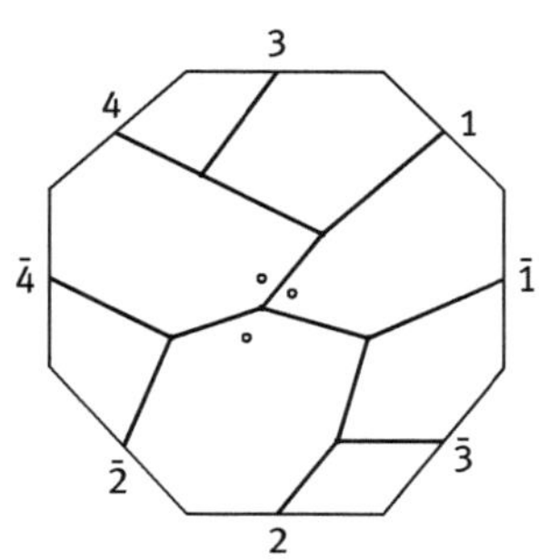

$W_{9,1}: o0-01-03$

Orientable genus 1

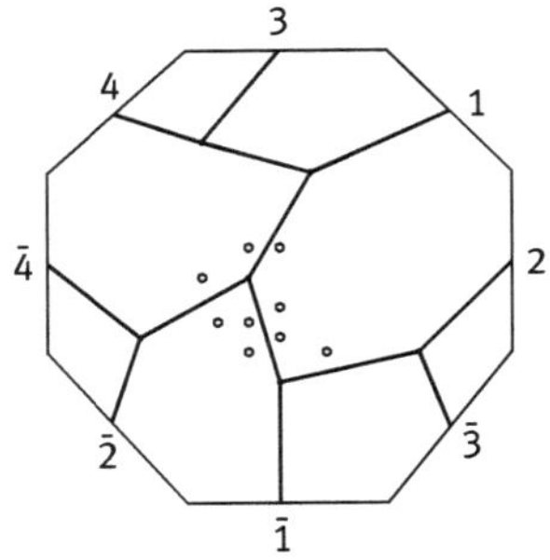

$C_{9,1}: o1-01-09$

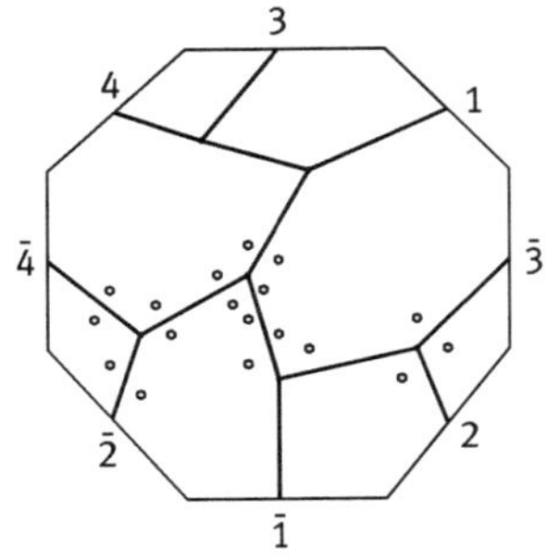

$C_{9,1}: o1-02-18$

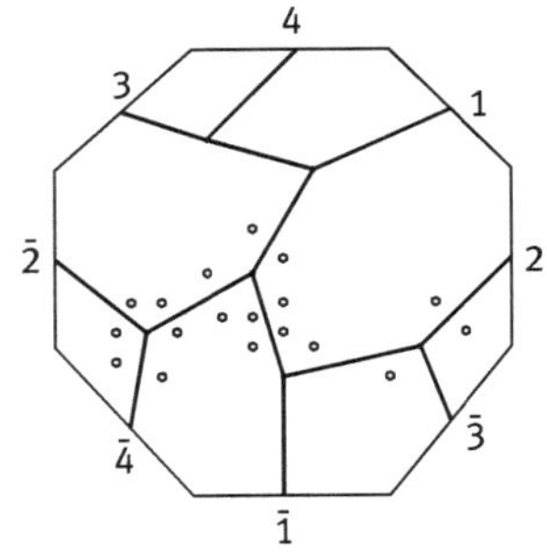

$C_{9,1}: o1-03-18$

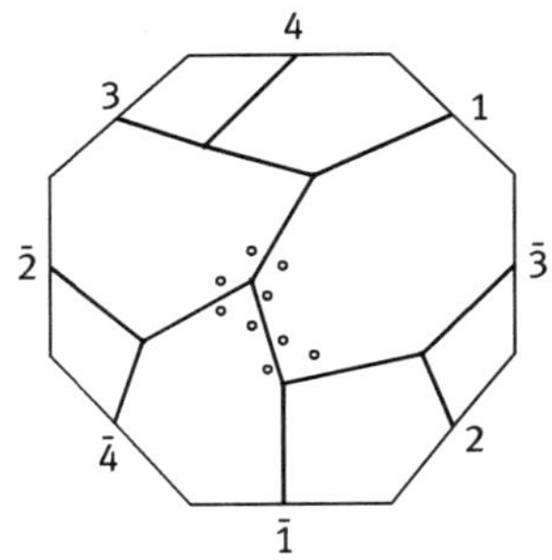

$C_{9,1}: o1-04-09$

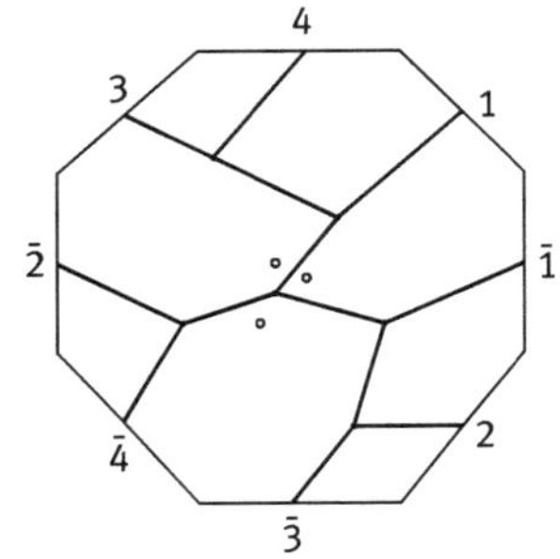

$C_{9,1}: o1-05-03$

Orientable genus 2

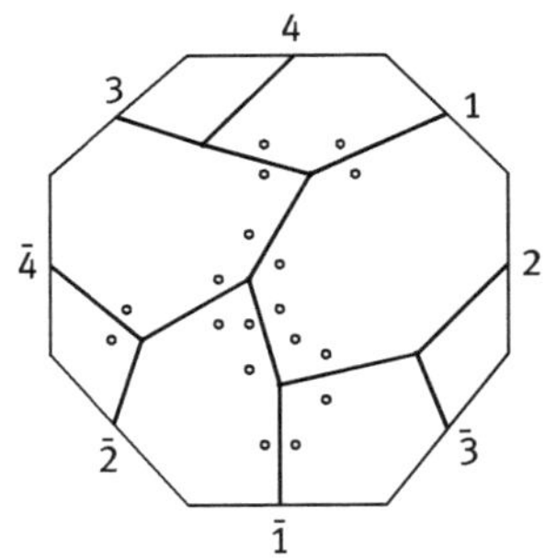

$C_{9,1}: o2-01-18$

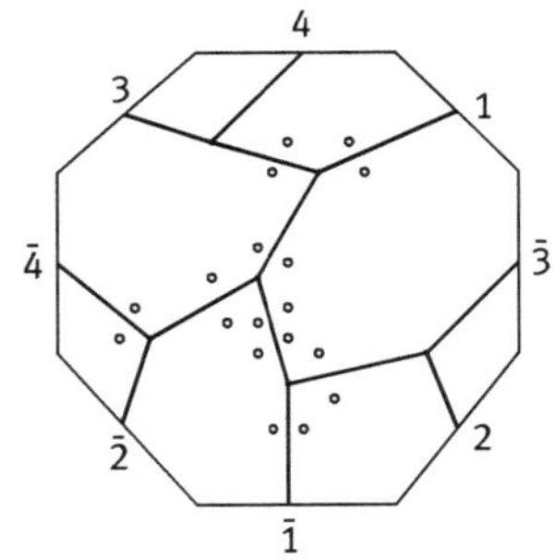

$C_{9,1}: o2-02-18$

Size 12: ($C_{12,4}$ is the cube)

$C_{12,1}$:

Orientable genus 0

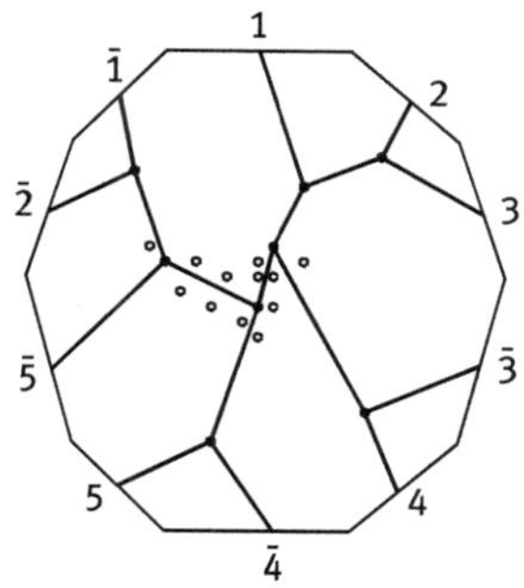

$C_{12,1}$: $o0-01-12$

Orientable genus 1

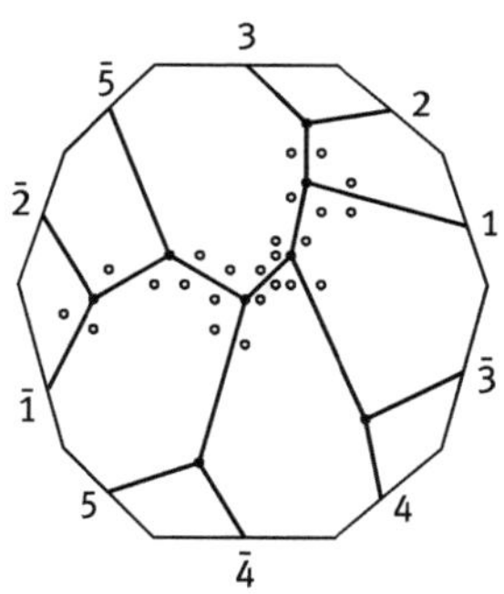

$C_{12,1}$: $o1-01-24$

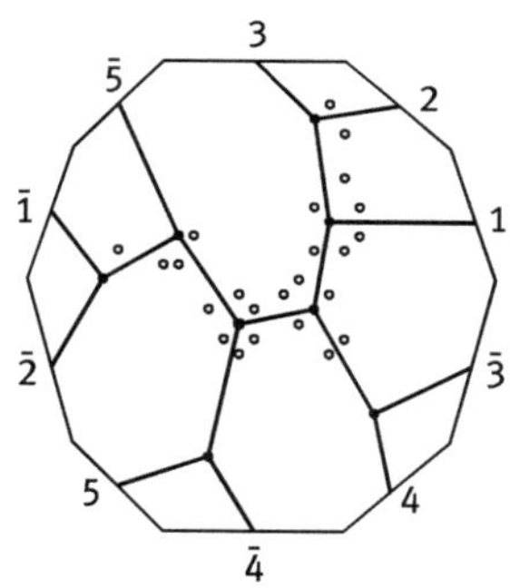

$C_{12,1}$: $o1-02-24$

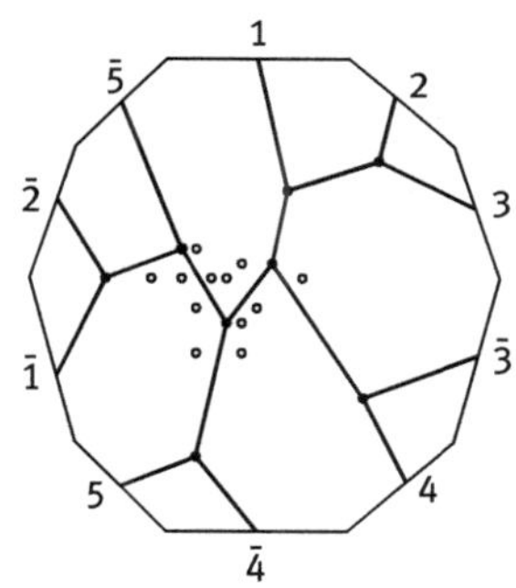

$C_{12,1}$: $o1-03-12$

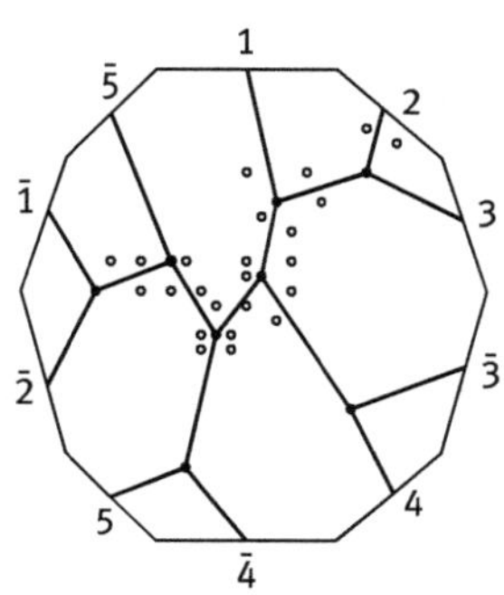

$C_{12,1}$: $o1-04-24$

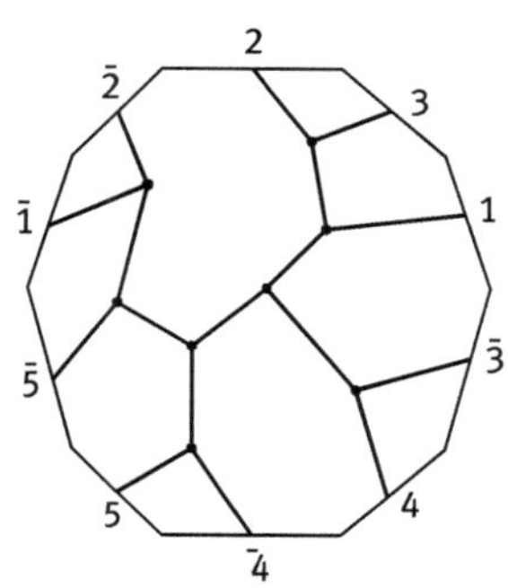

$C_{12,1}$: $o1-05-48$

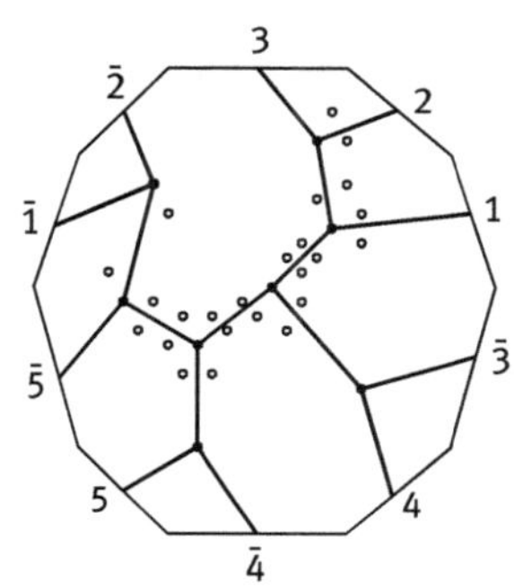

$C_{12,1}$: $o1-06-24$

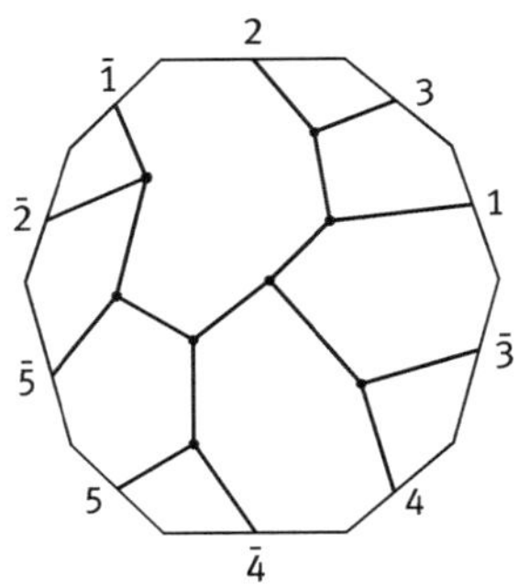

$C_{12,1}$: o1 – 01 – 48

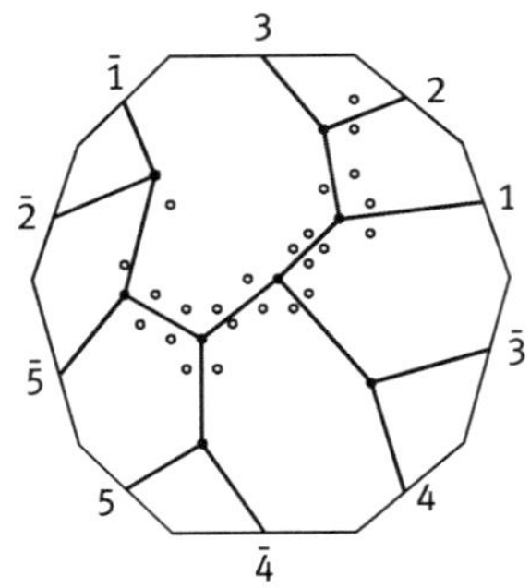

$C_{12,1}$: o1 – 08 – 24

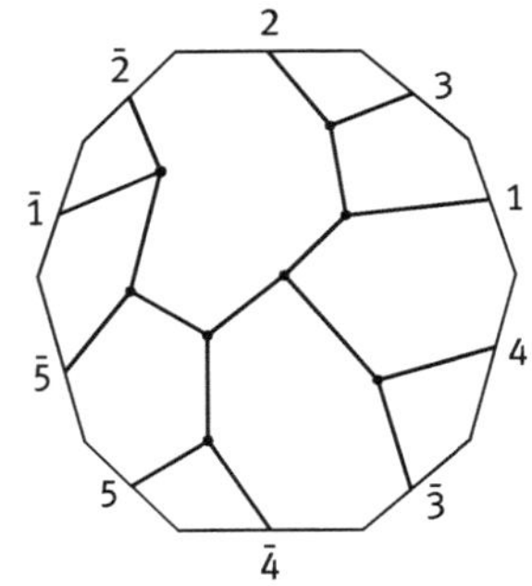

$C_{12,1}$: o1 – 09 – 48

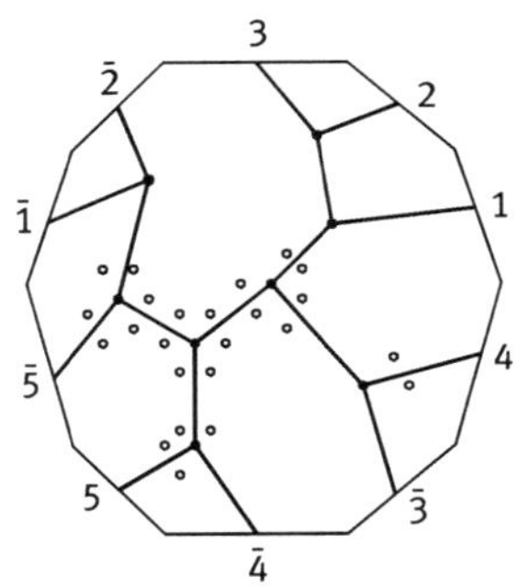

$C_{12,1}$: o1 – 10 – 24

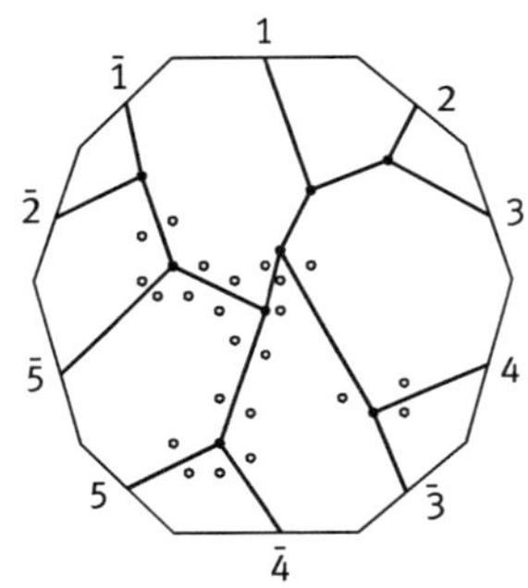

$C_{12,1}$: o1 – 11 – 24

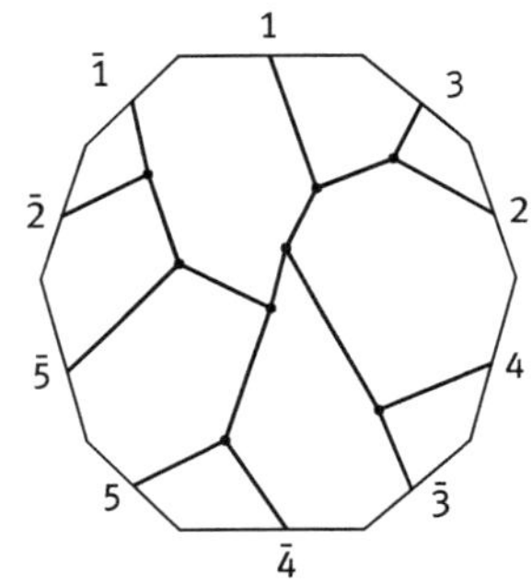

$C_{12,1}$: o1 – 12 – 48

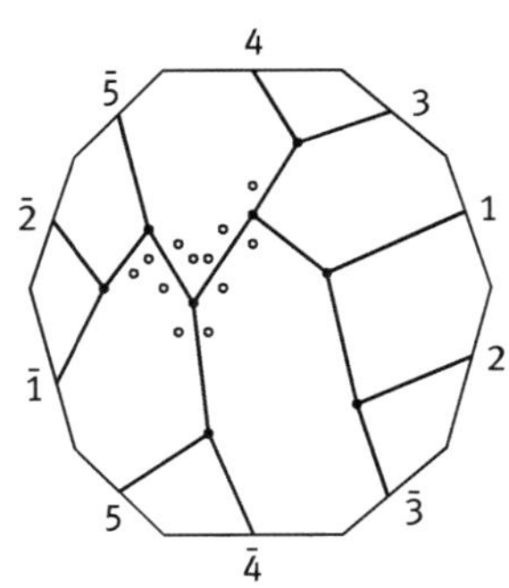

$C_{12,1}$: o1 – 13 – 12

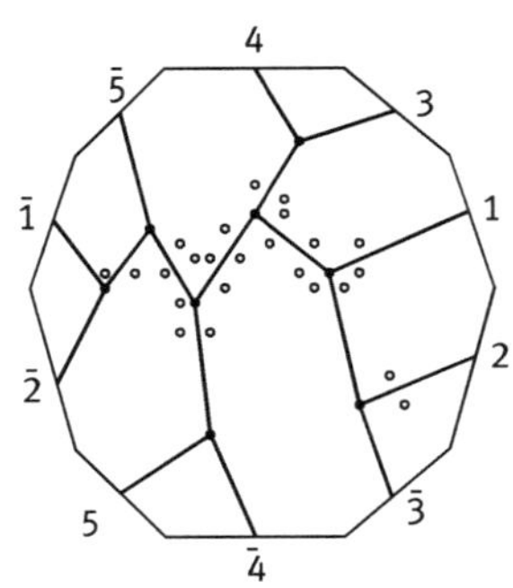

$C_{12,1}$: o1 – 14 – 24

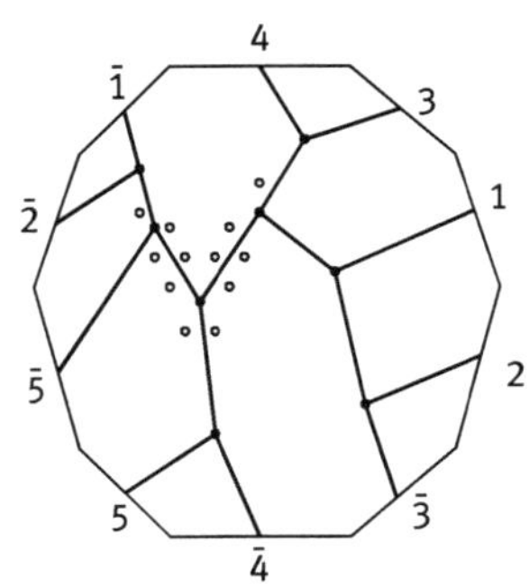

$C_{12,1}$: o1 – 15 – 12

Orientable genus 2

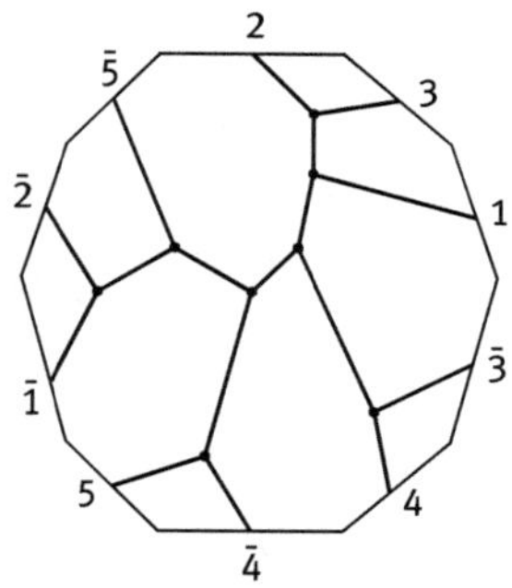

$C_{12,1}$: $o2 - 01 - 48$

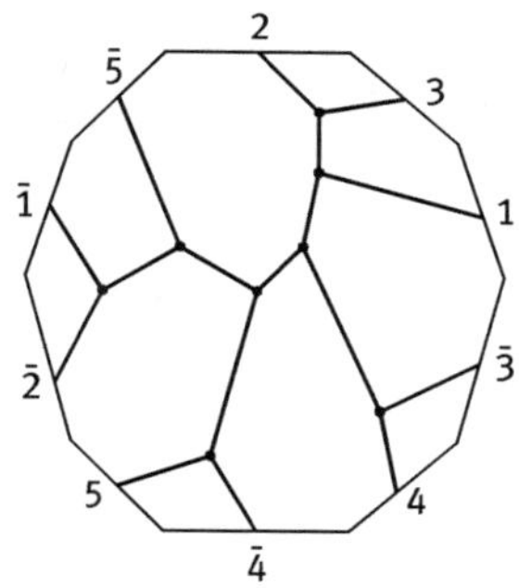

$C_{12,1}$: $o2 - 02 - 48$

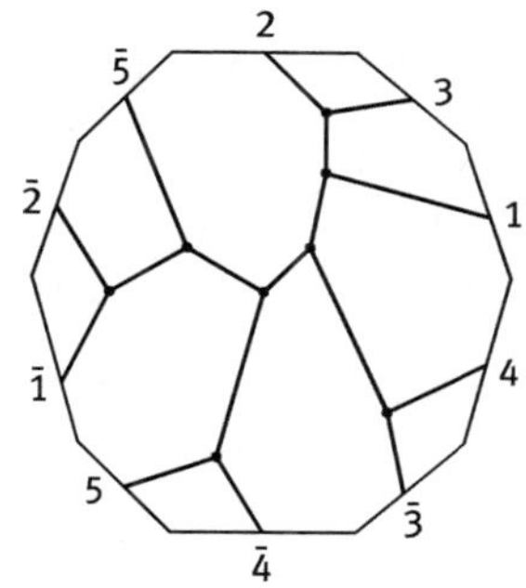

$C_{12,1}$: $o2 - 03 - 48$

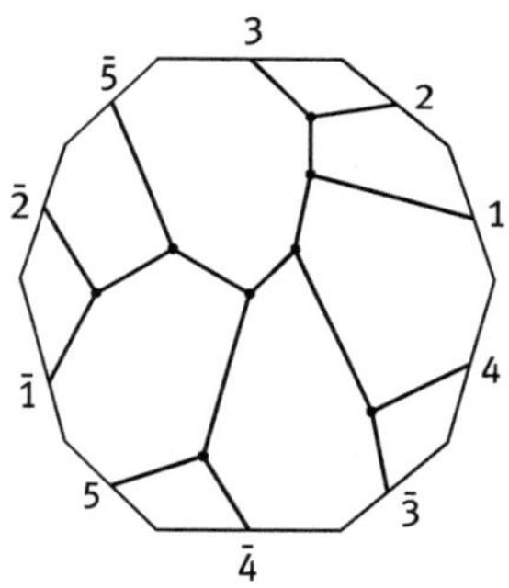

$C_{12,1}$: $o2 - 04 - 48$

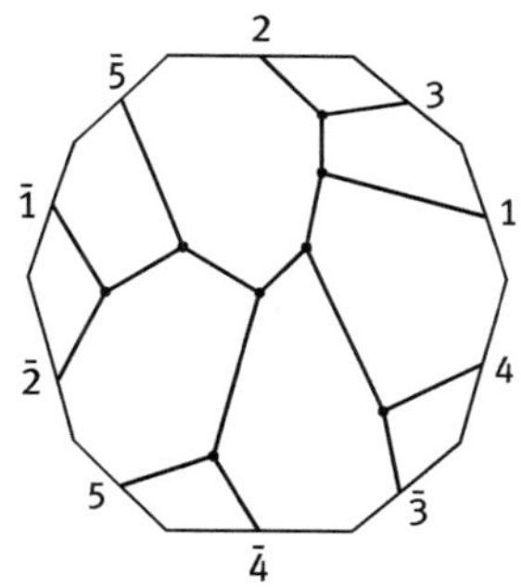

$C_{12,1}$: $o2 - 05 - 48$

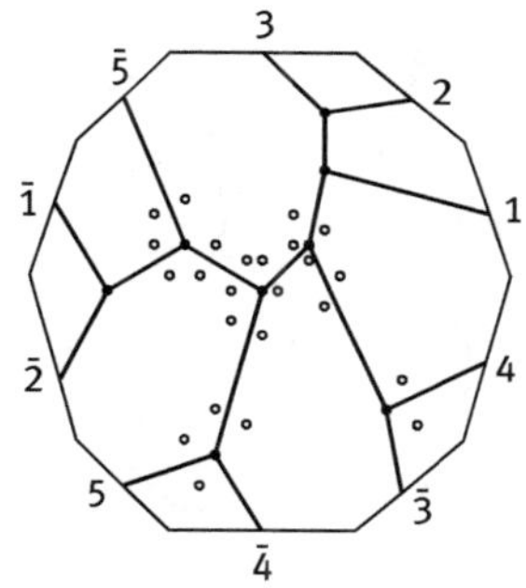

$C_{12,1}$: $o2 - 06 - 24$

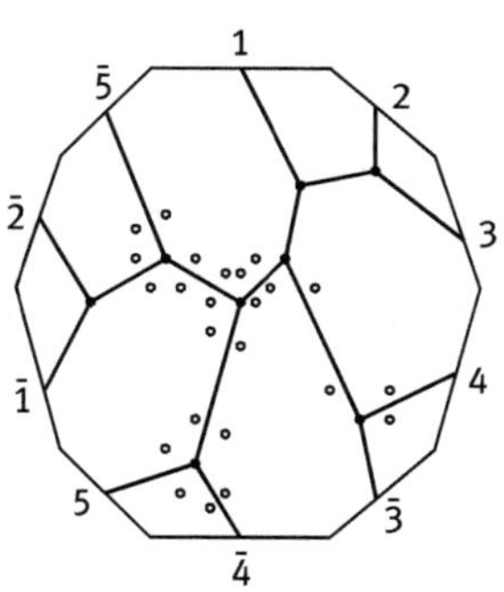

$C_{12,1}$: $o2 - 07 - 24$

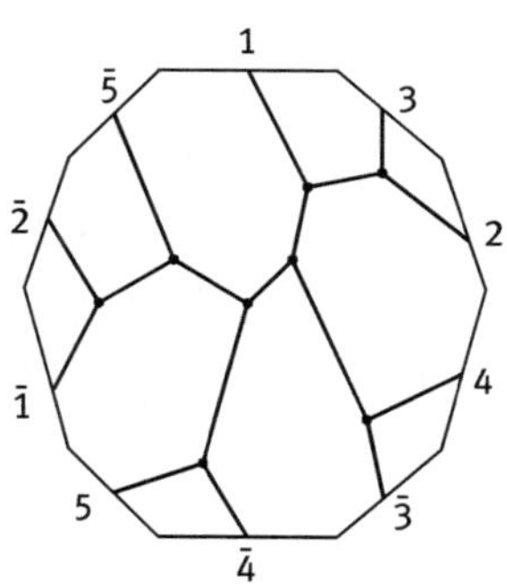

$C_{12,1}$: $o2 - 08 - 48$

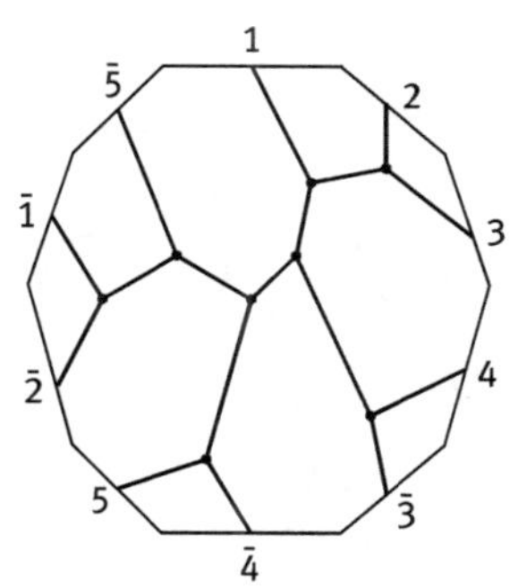

$C_{12,1}$: $o2 - 09 - 48$

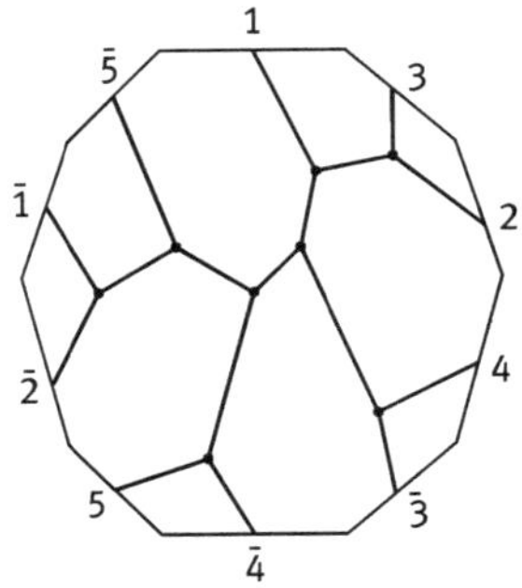

$C_{12,1}$: $o2-10-48$

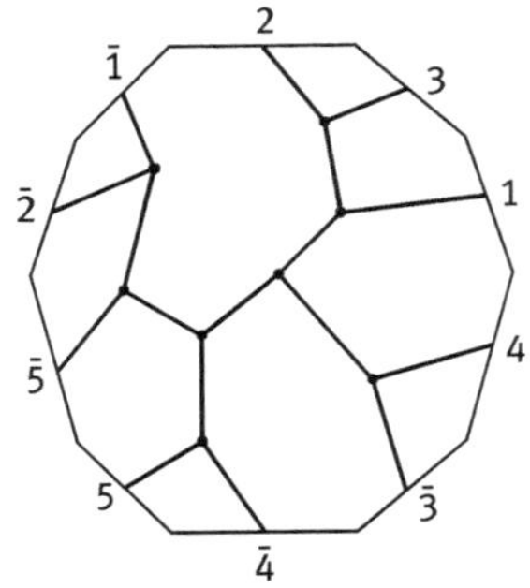

$C_{12,1}$: $o2-11-48$

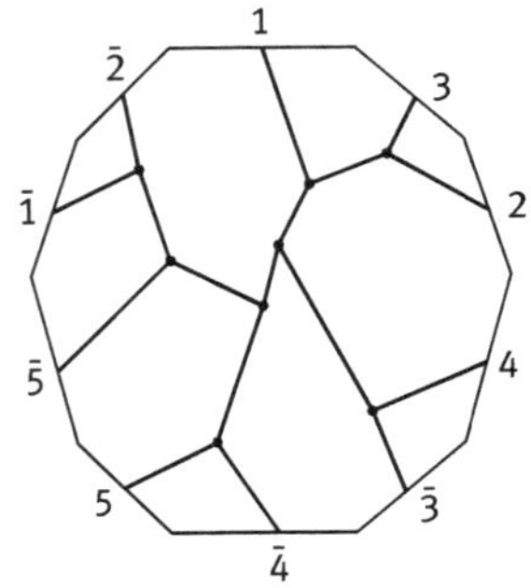

$C_{12,1}$: $o2-12-48$

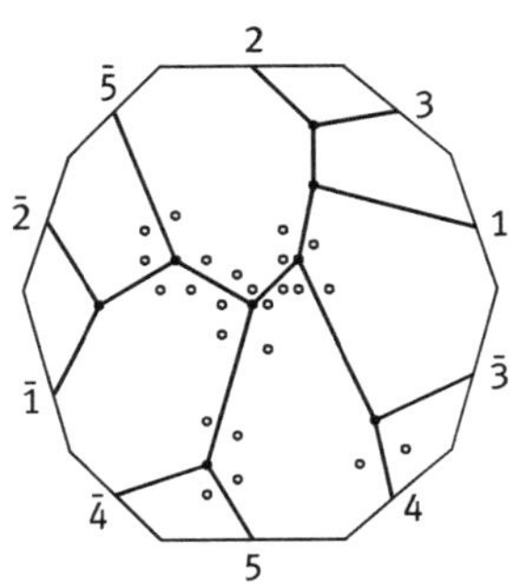

$C_{12,1}$: $o2-13-24$

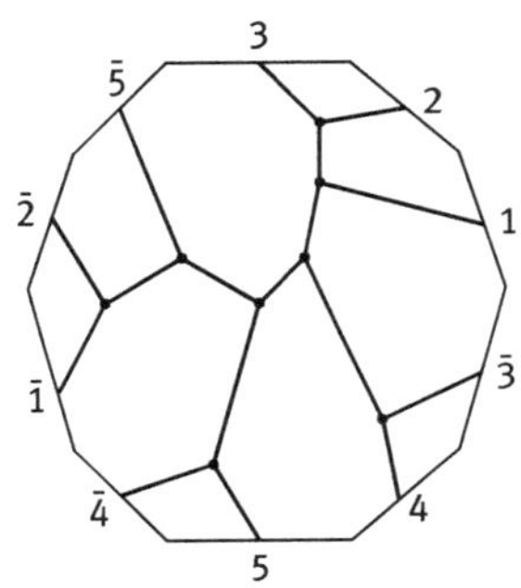

$C_{12,1}$: $o2-14-48$

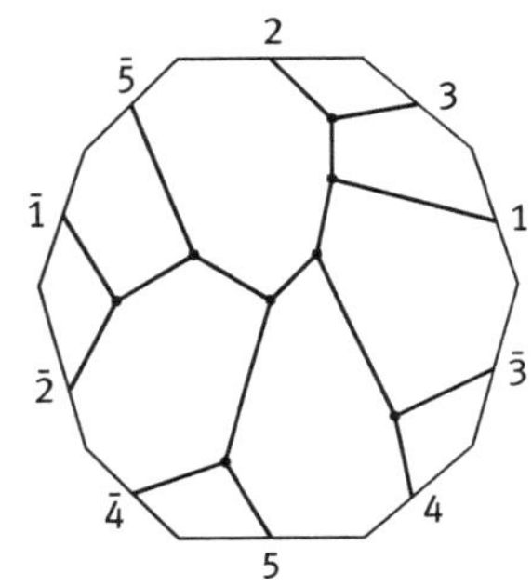

$C_{12,1}$: $o2-15-48$

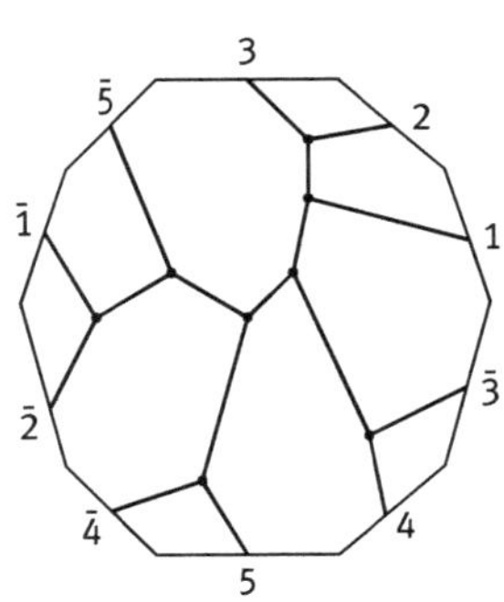

$C_{12,1}$: $o2-16-48$

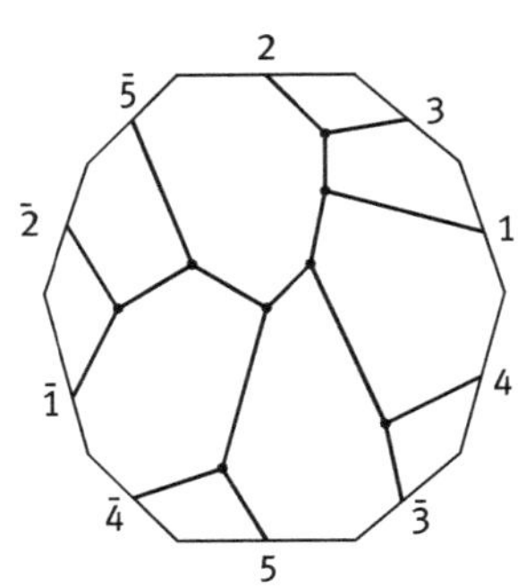

$C_{12,1}$: $o2-17-48$

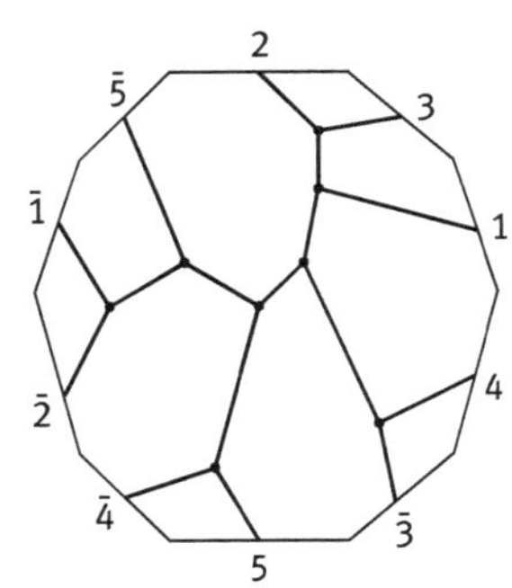

$C_{12,1}$: $o2-18-48$

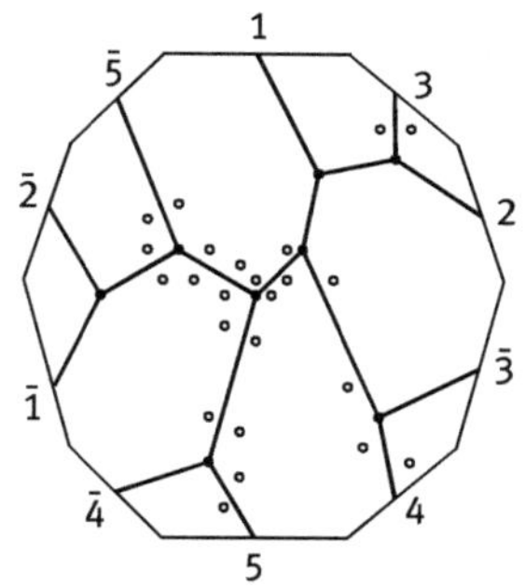

$C_{12,1}: o2 - 19 - 24$

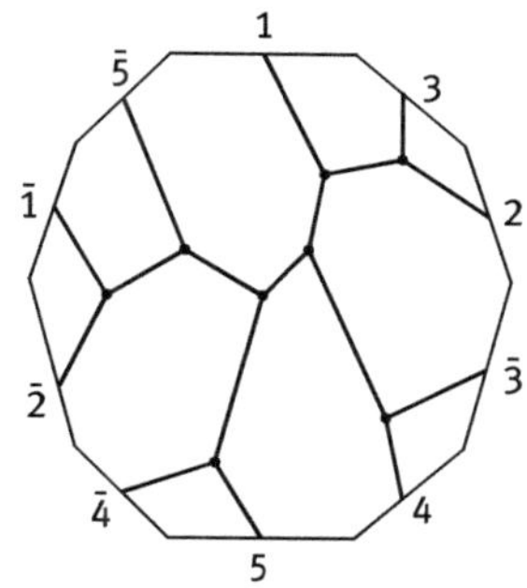

$C_{12,1}: o2 - 20 - 48$

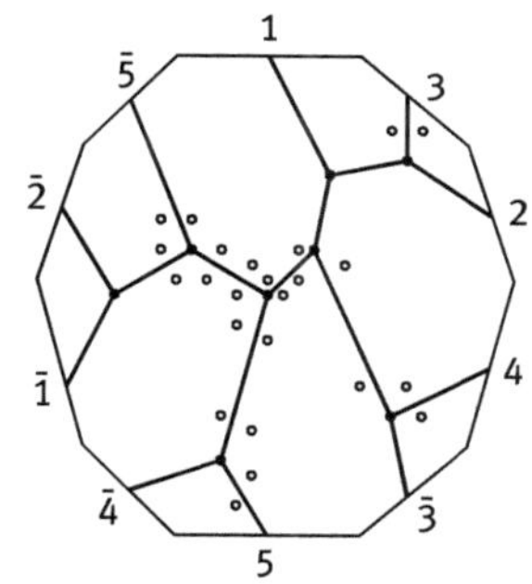

$C_{12,1}: o2 - 21 - 24$

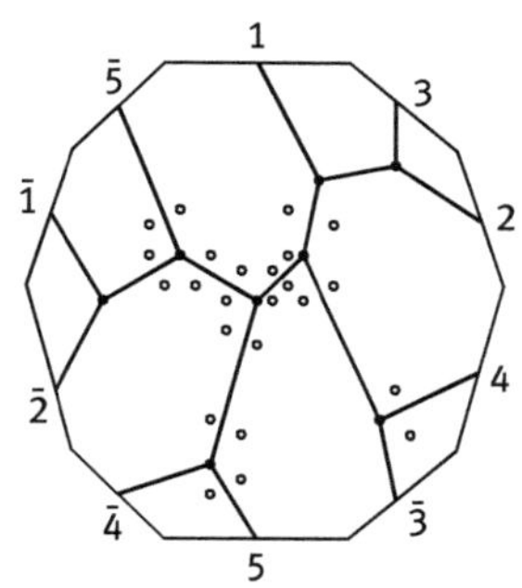

$C_{12,1}: o2 - 22 - 24$

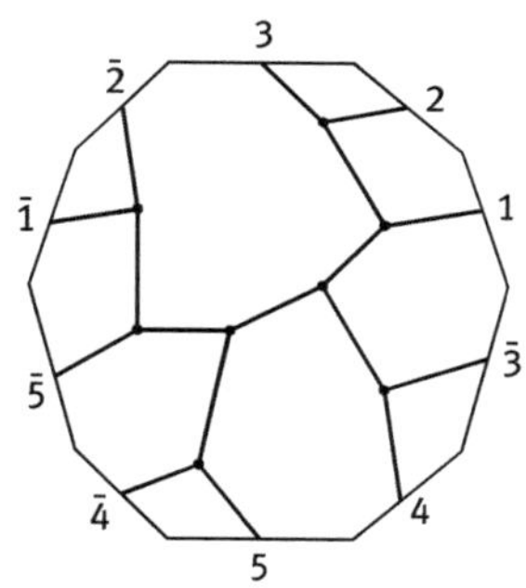

$C_{12,1}: o2 - 23 - 48$

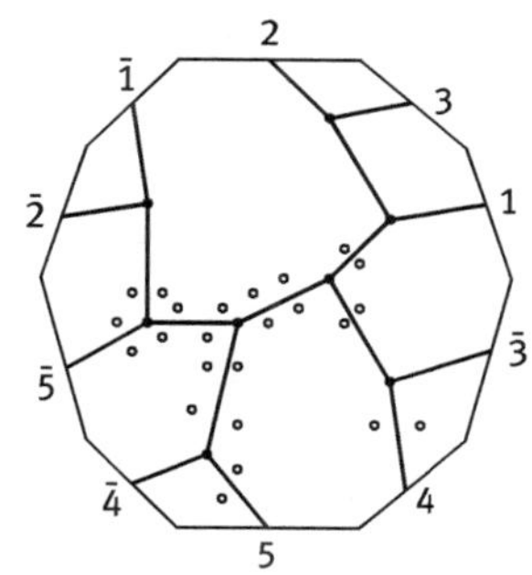

$C_{12,1}: o2 - 24 - 24$

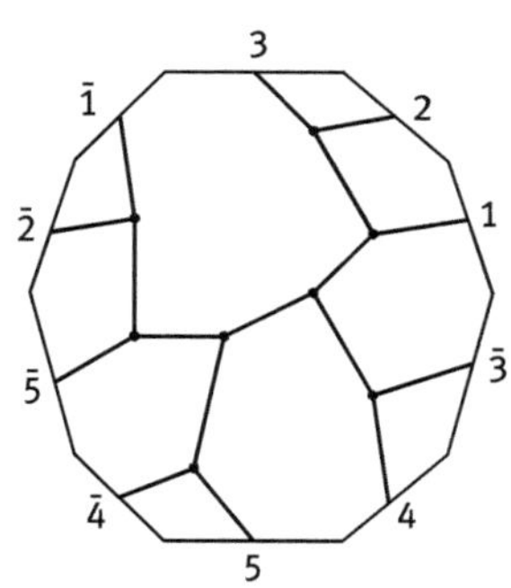

$C_{12,1}: o2 - 25 - 48$

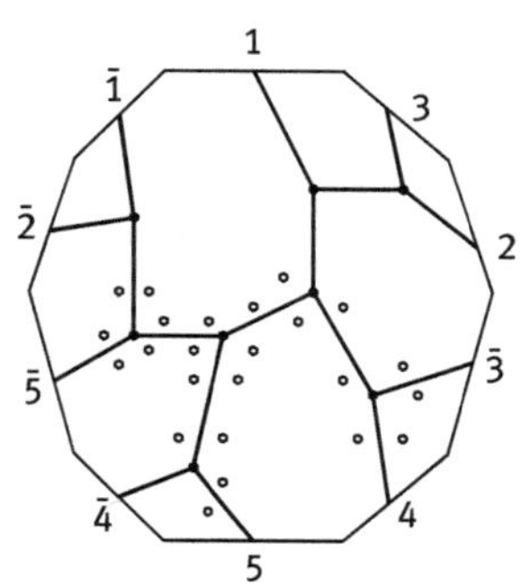

$C_{12,1}: o2 - 26 - 24$

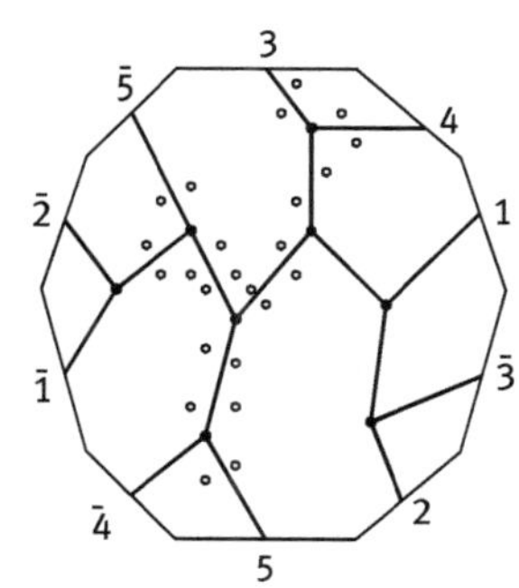

$C_{12,1}: o2 - 27 - 24$

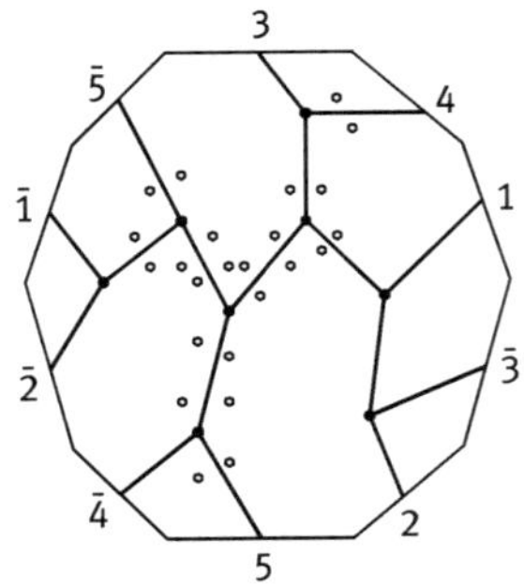

$C_{12,1}$: o2 – 28 – 24

$C_{12,2}$:

Orientable genus 1

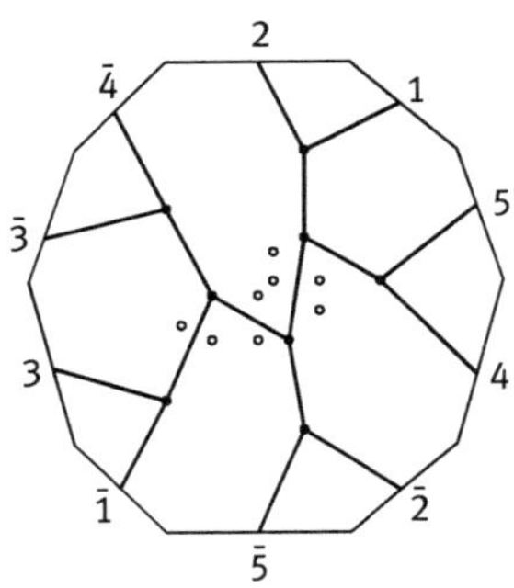

$C_{12,2}$: o2 – 01 – 08

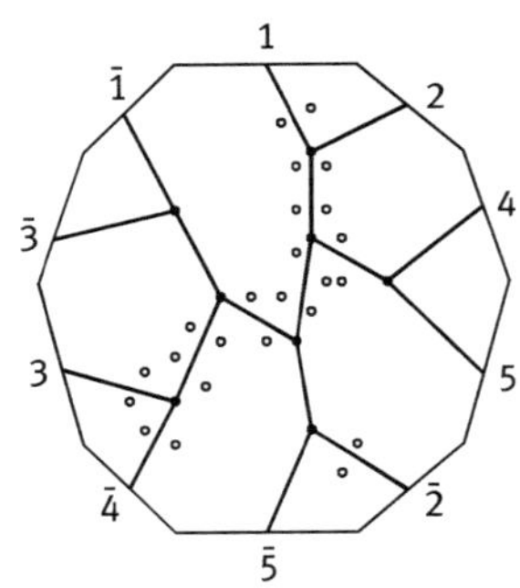

$C_{12,2}$: o1 – 02 – 24

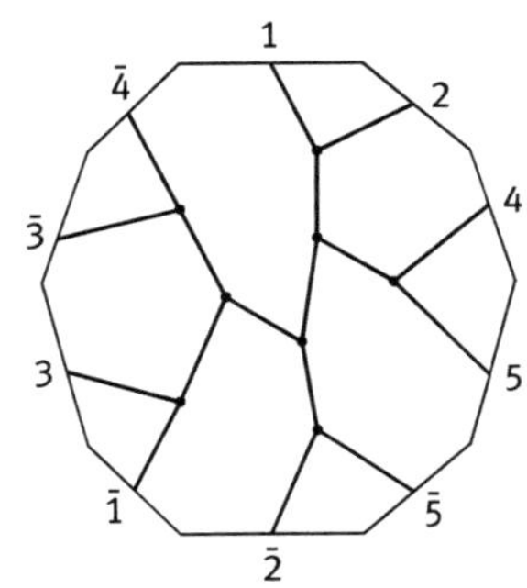

$C_{12,2}$: o1 – 03 – 48

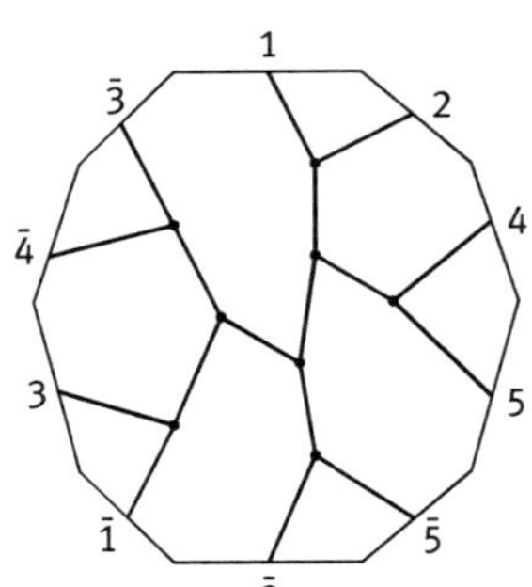

$C_{12,2}$: o1 – 04 – 48

Orientable genus 2

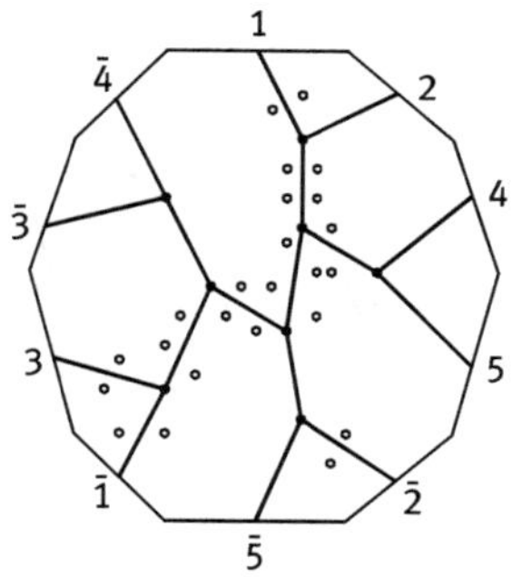

$C_{12,2}$: o2 – 01 – 24

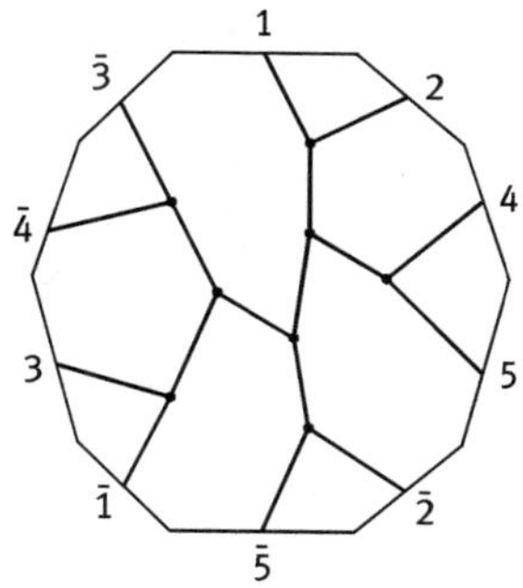

$C_{12,2}$: o2 – 02 – 48

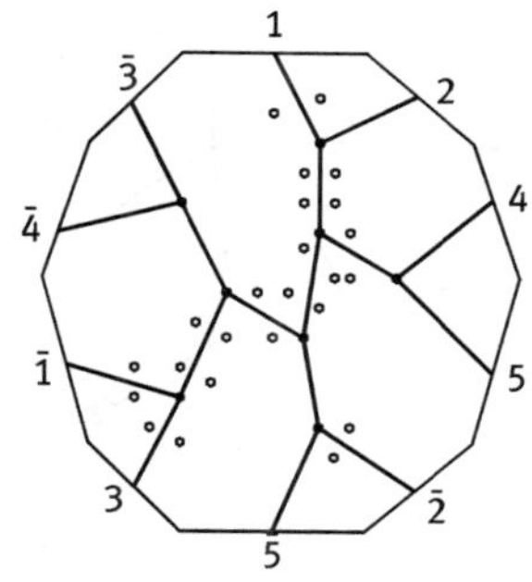

$C_{12,2}$: o2 – 03 – 24

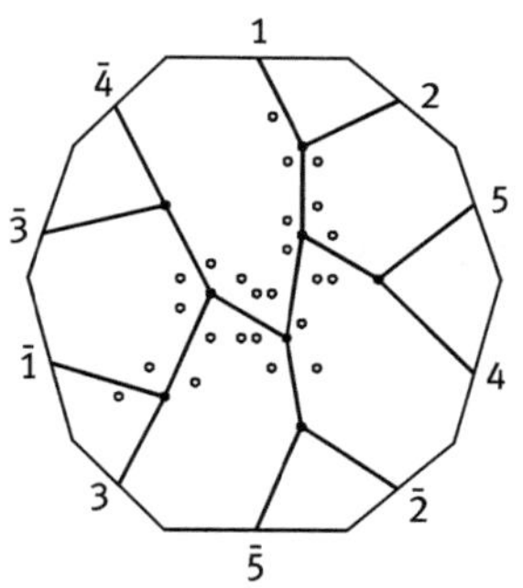

$C_{12,2}$: o2 – 02 – 24

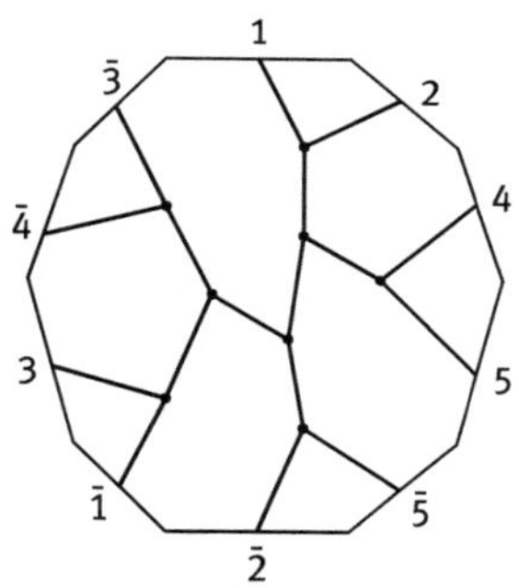

$C_{12,2}$: o2 – 05 – 48

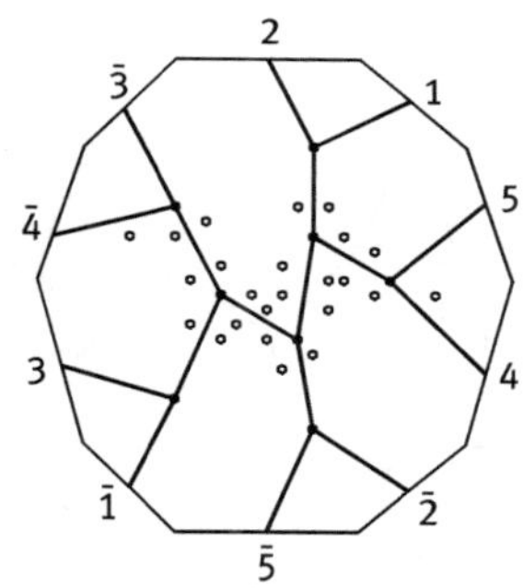

$C_{12,2}$: o2 – 06 – 24

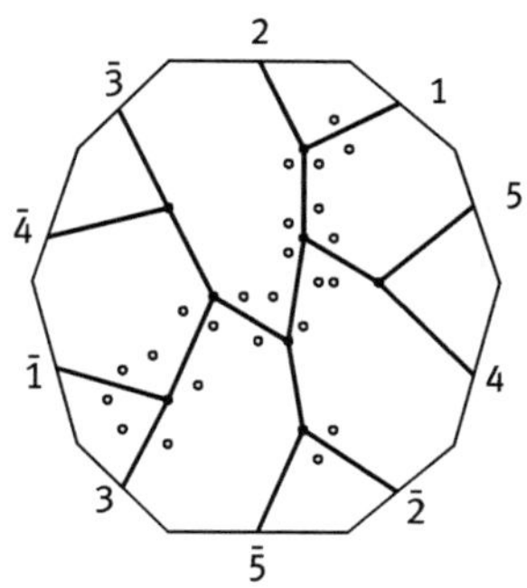

$C_{12,2}$: o2 – 07 – 24

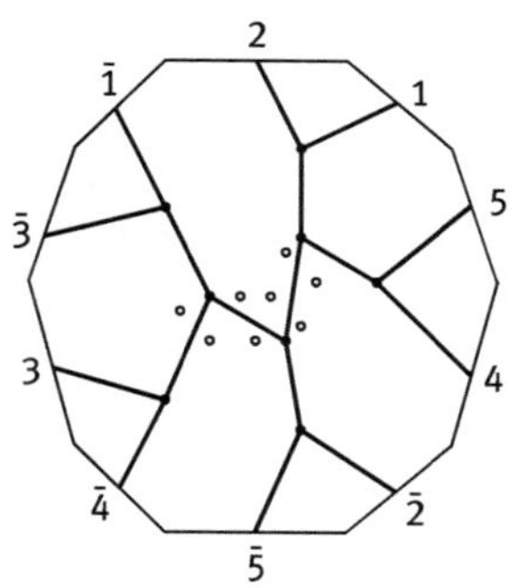

$C_{12,2}$: o2 – 08 – 08

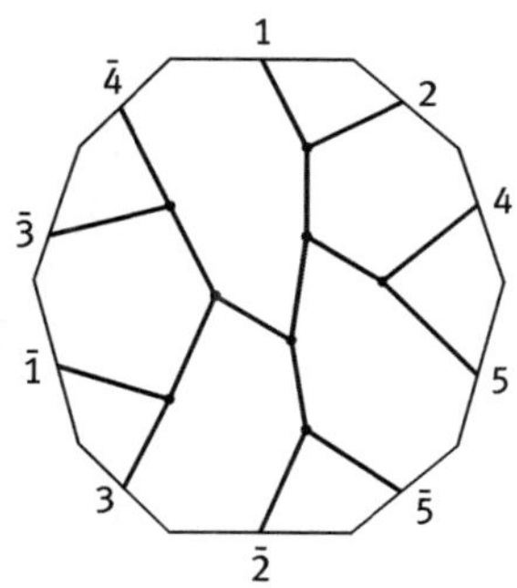

$C_{12,2}$: o2 – 09 – 48

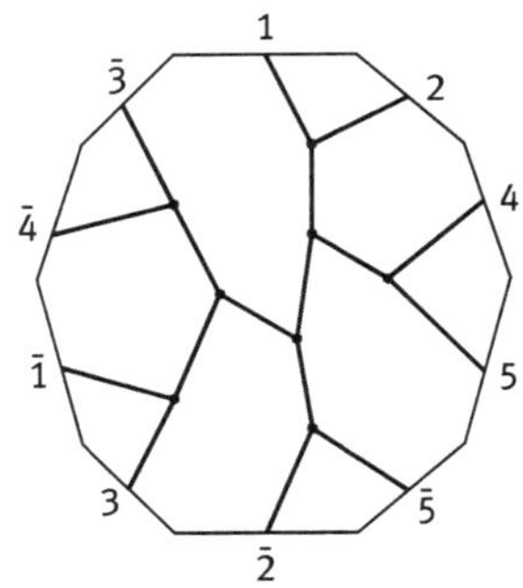

$C_{12,2}$: $o2-10-48$

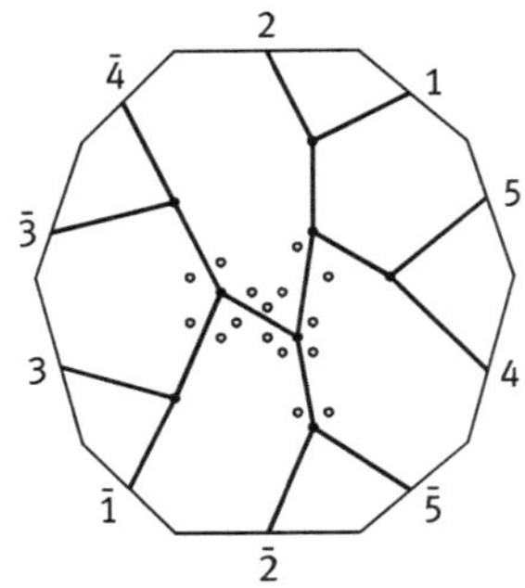

$C_{12,2}$: $o2-11-16$

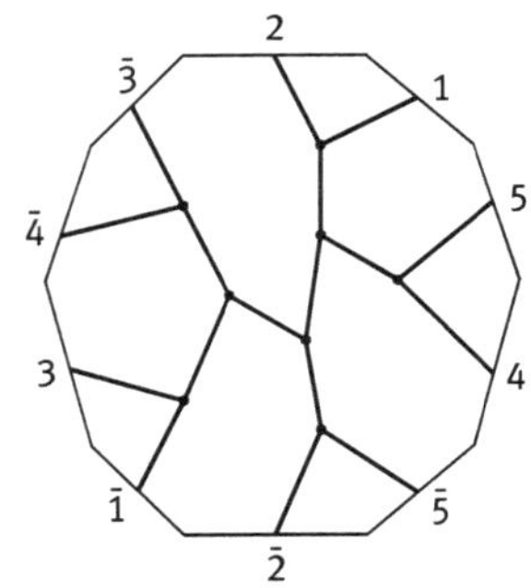

$C_{12,2}$: $o2-12-48$

$C_{12,3}$:

Orientable genus 1

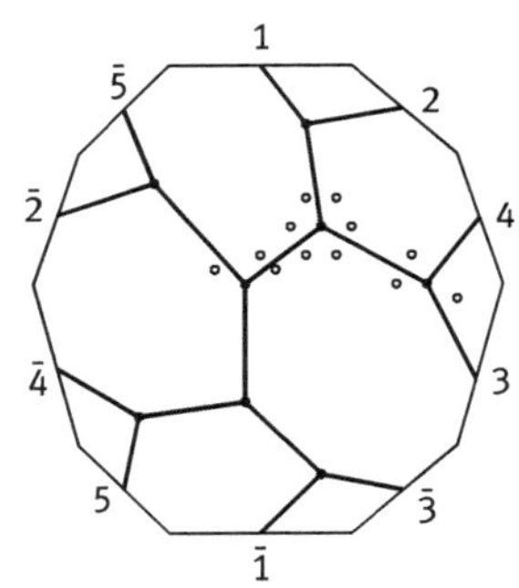

$C_{12,3}$: $o1-01-12$

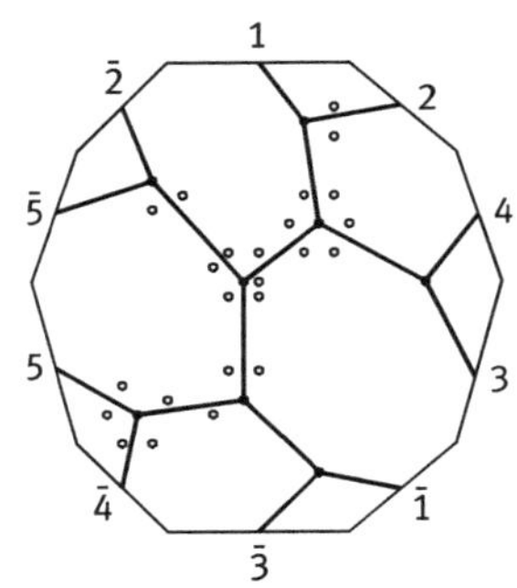

$C_{12,3}$: $o1-02-24$

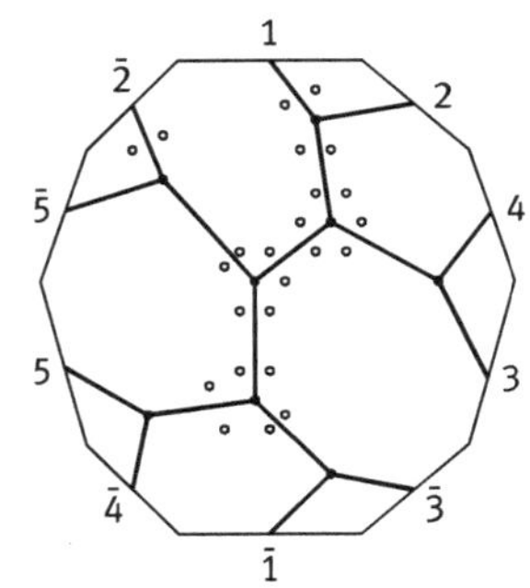

$C_{12,3}$: $o1-03-24$

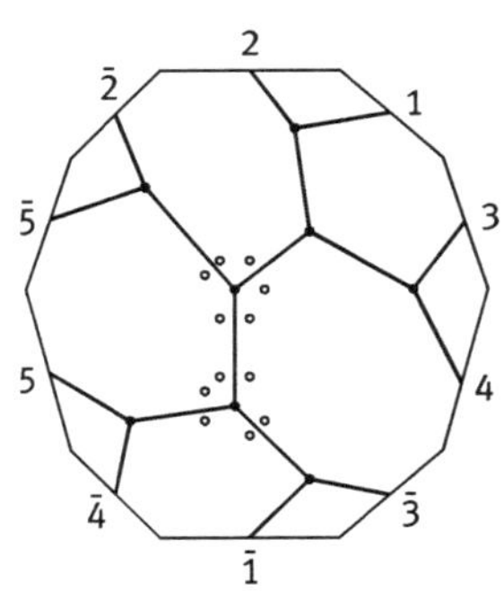

$C_{12,3}$: $o1-04-12$

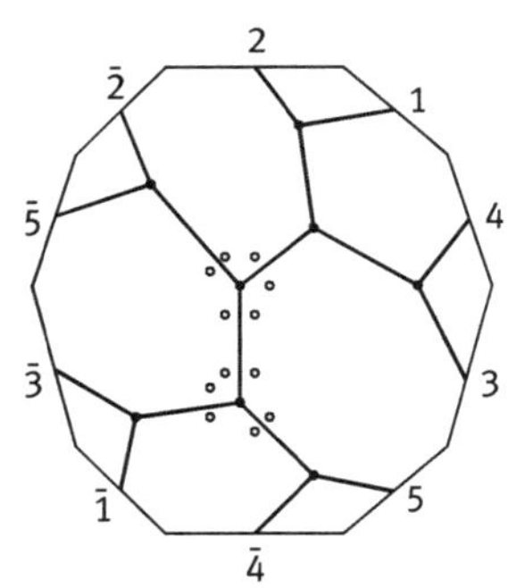

$C_{12,3}$: $o1-05-12$

Orientable genus 2

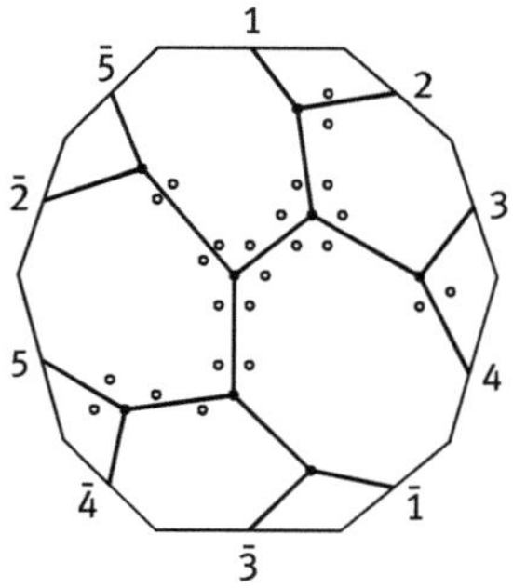

$C_{12,3}$: o2 – 01 – 24

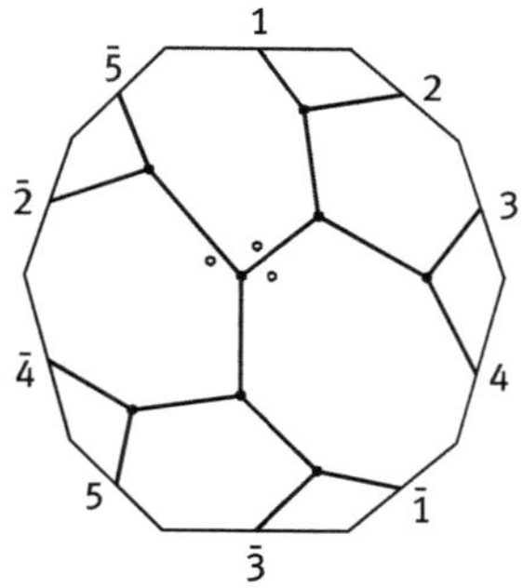

$C_{12,3}$: o2 – 02 – 03

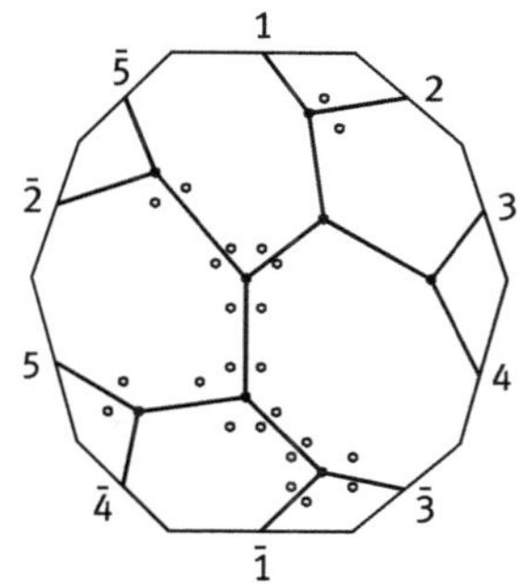

$C_{12,3}$: o2 – 03 – 24

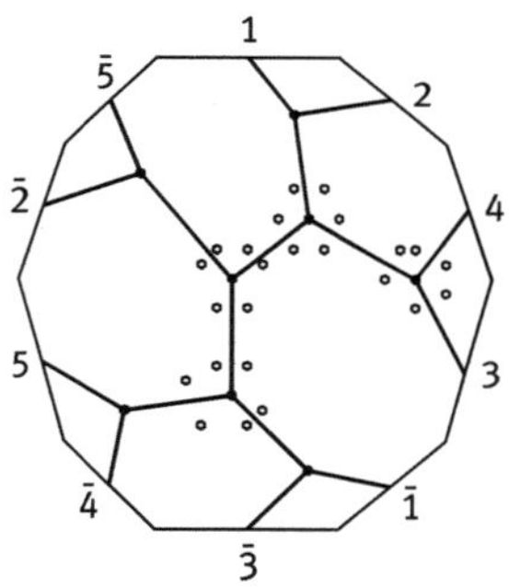

$C_{12,3}$: o2 – 04 – 24

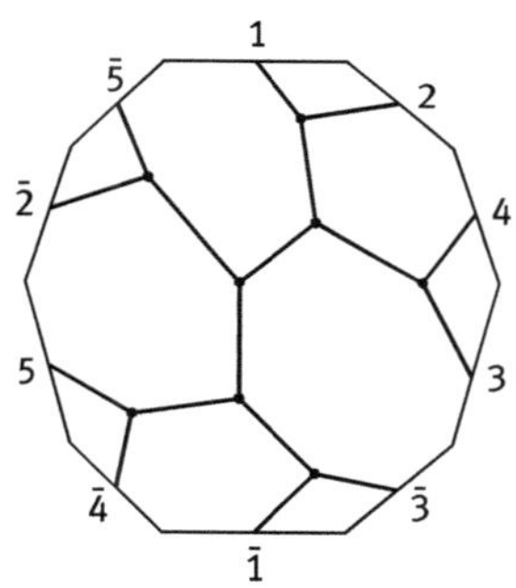

$C_{12,3}$: o2 – 05 – 48

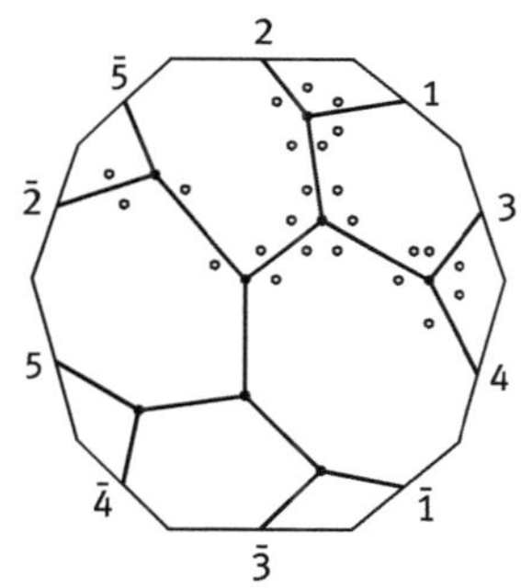

$C_{12,3}$: o2 – 06 – 24

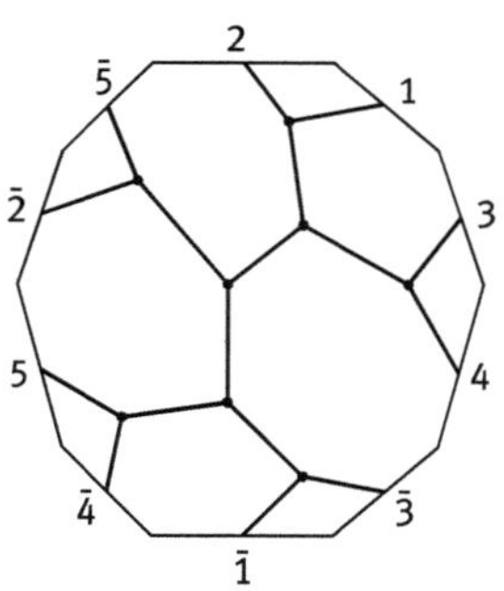

$C_{12,3}$: o2 – 07 – 48

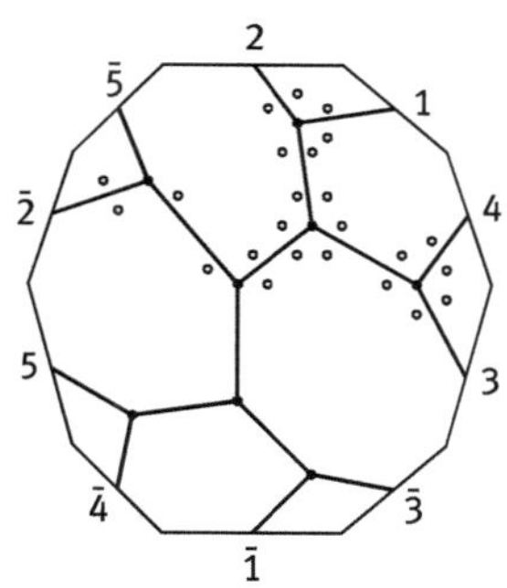

$C_{12,3}$: o2 – 08 – 24

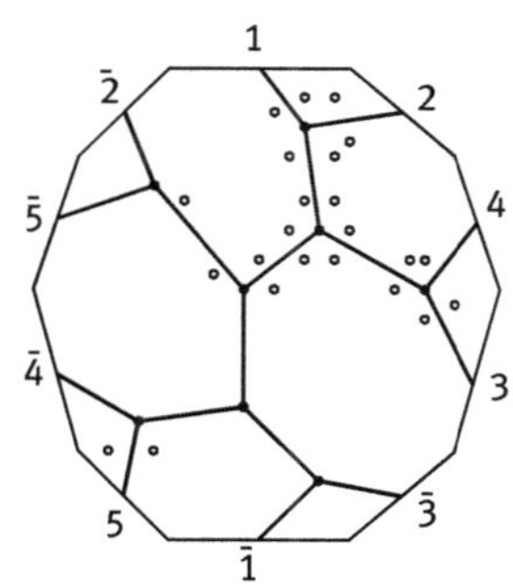

$C_{12,3}$: o2 – 09 – 24

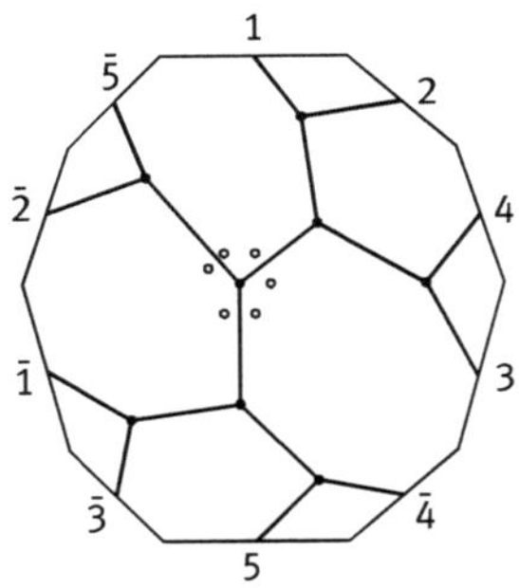

$C_{12,3}: o2-10-06$

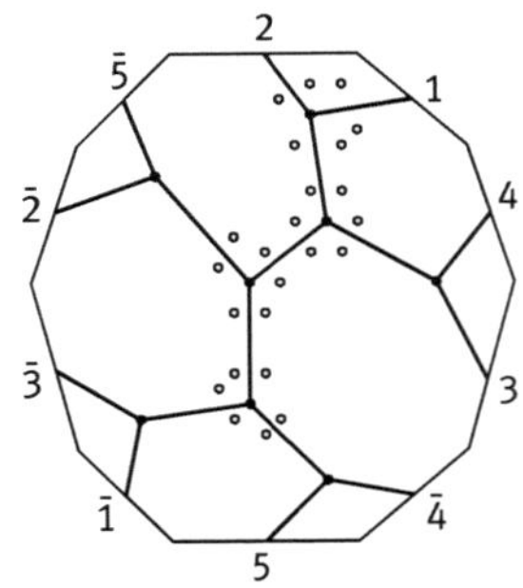

$C_{12,3}: o2-11-24$

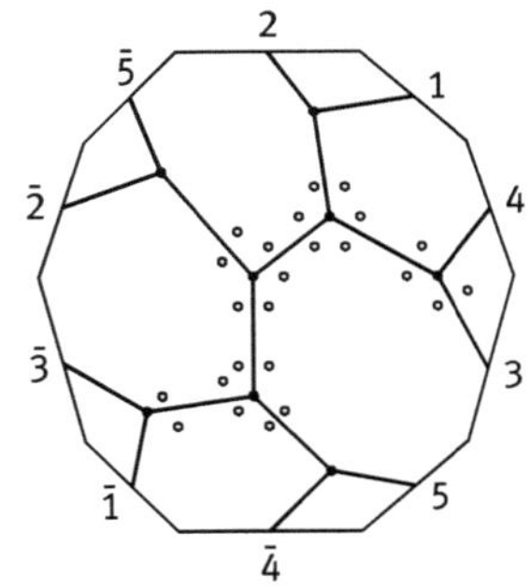

$C_{12,3}: o2-12-24$

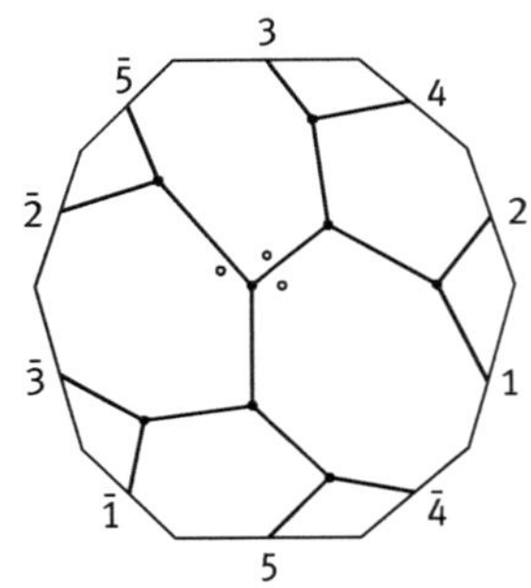

$C_{12,3}: o2-13-03$

$C_{12,4}$:

Orientable genus 0

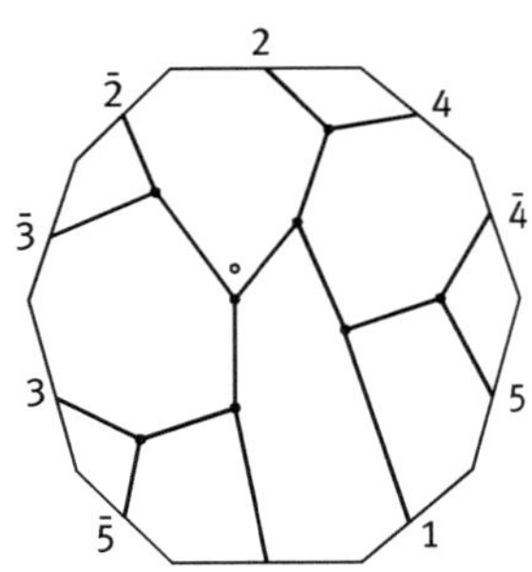

$C_{12,4}: o0-01-01$

Orientable genus 1

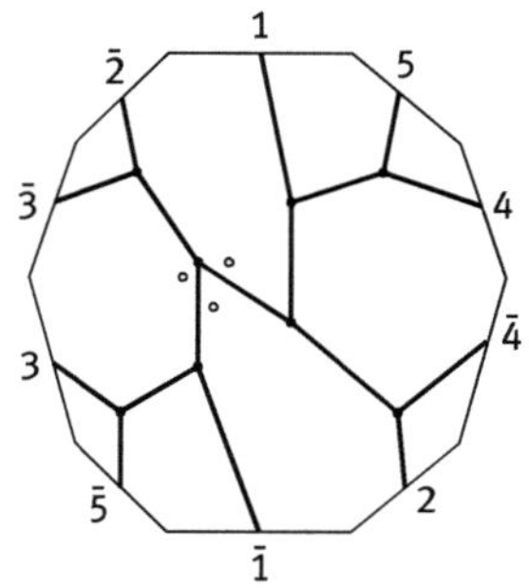

$C_{12,4}: o1-01-03$

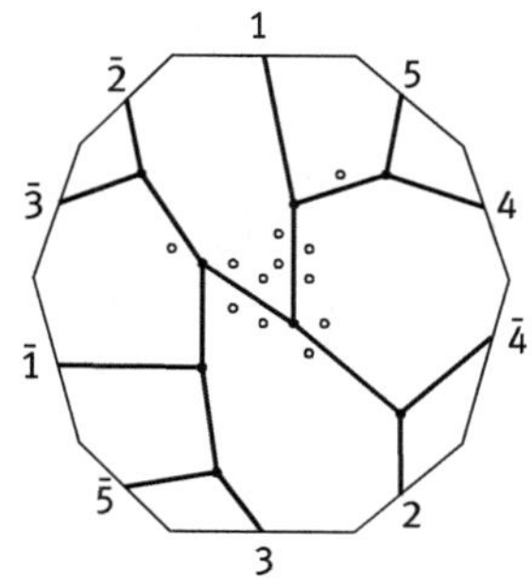

$C_{12,4}: o1-02-12$

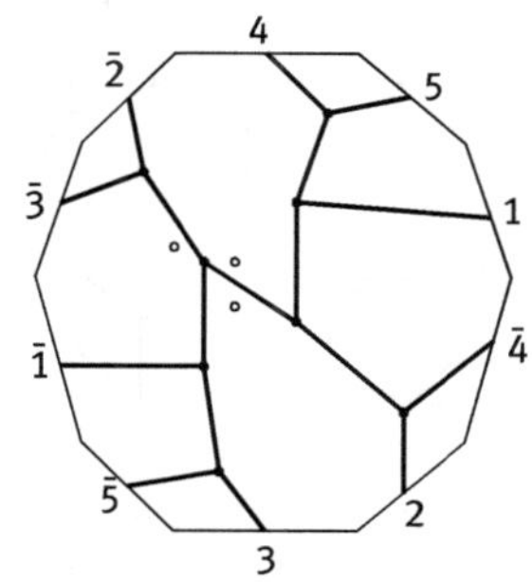

$C_{12,4}: o1-03-03$

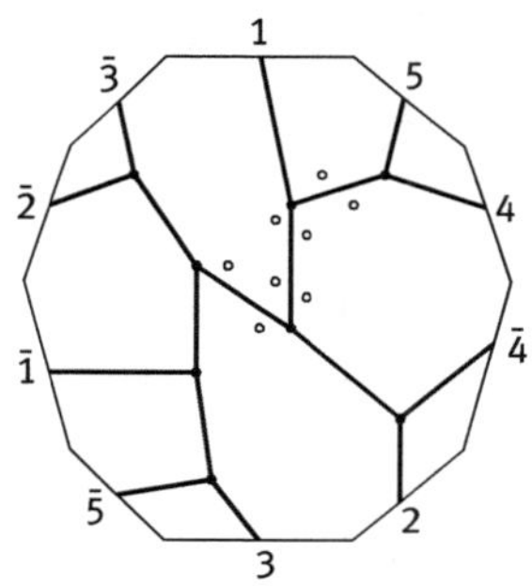

$C_{12,4}: o1-04-08$

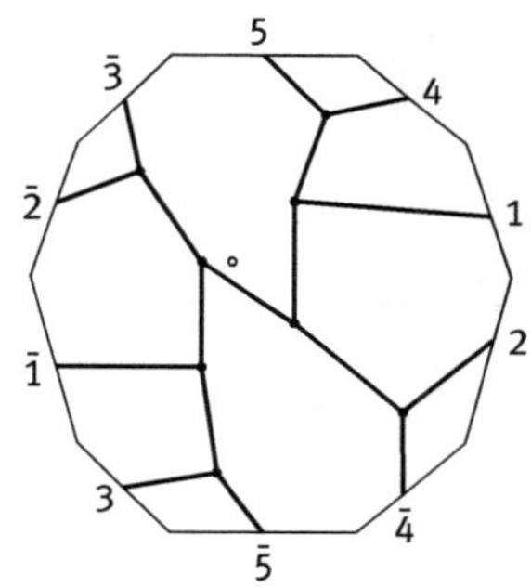

$C_{12,4}: o1-05-01$

Orientable genus 2

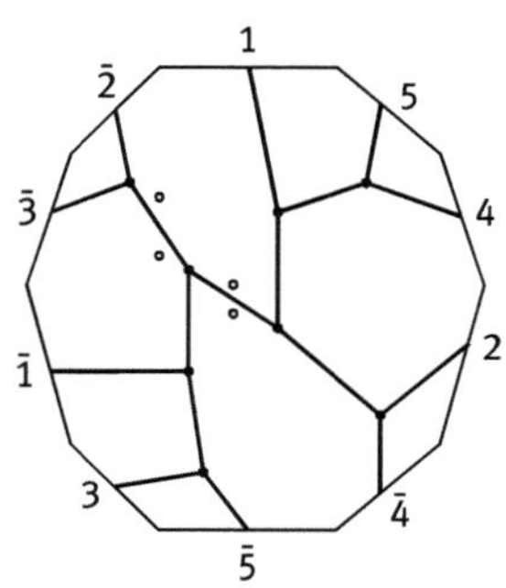

$C_{12,4}: o2-01-04$

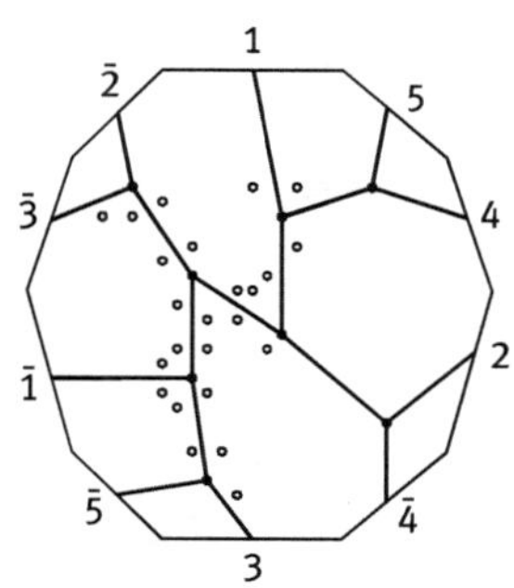

$C_{12,4}: o2-02-24$

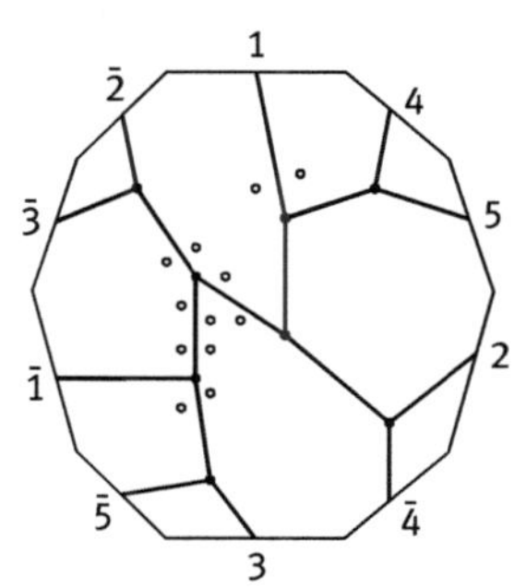

$C_{12,4}: o2-03-12$

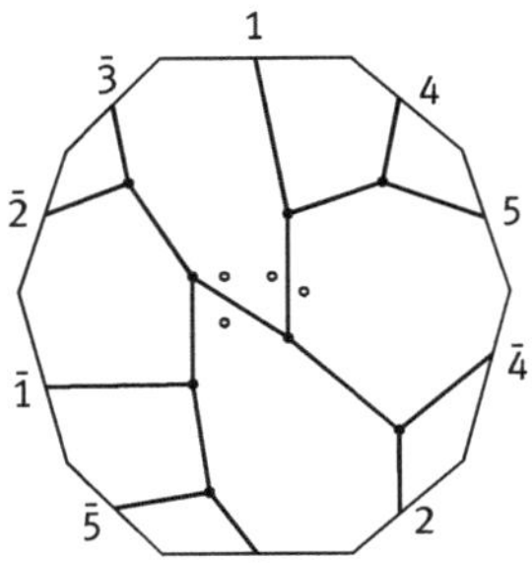

$C_{12,4}: o2-04-04$

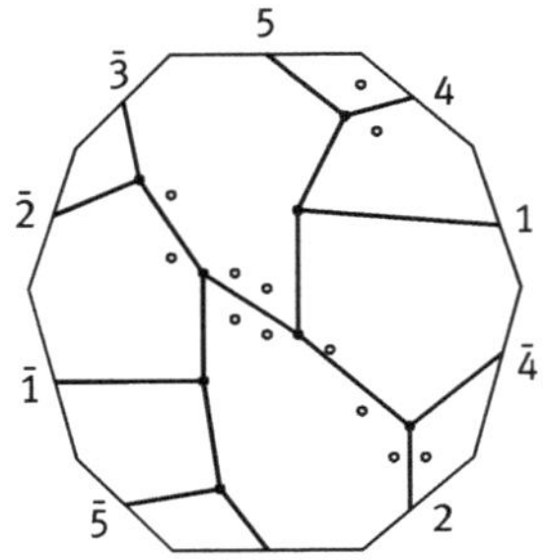

$C_{12,4}: o2-05-12$

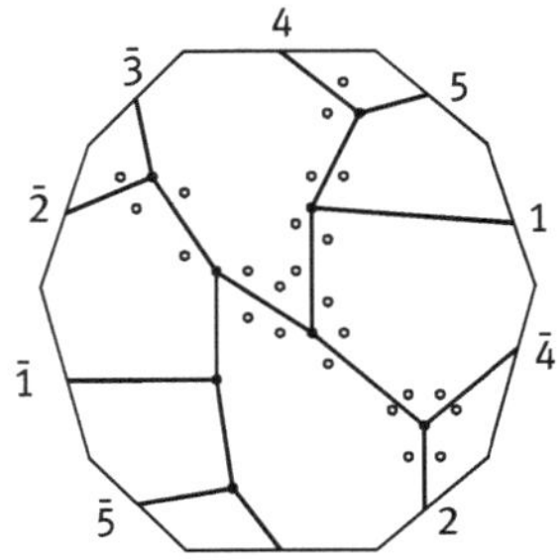

$C_{12,4}: o2-06-24$

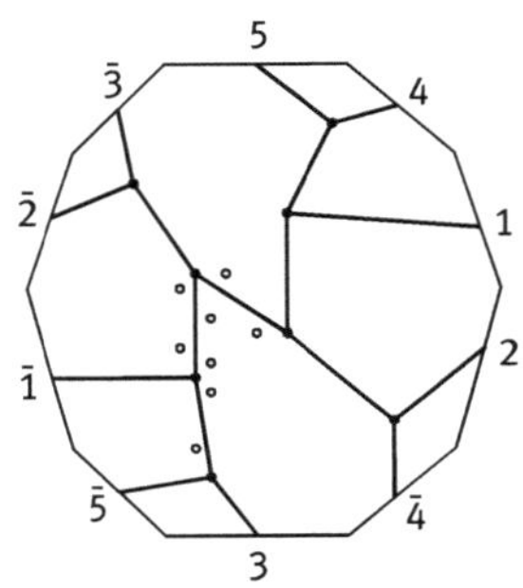

C12,4:o2–07–08

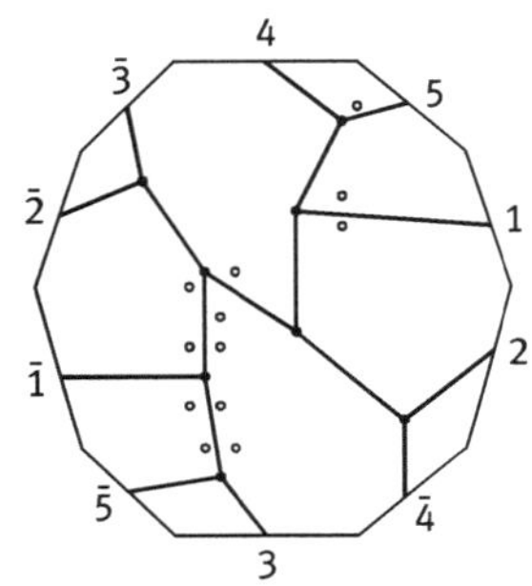

$C_{12,4}: o2-08-12$

Size 15: Two chosen where $C_{15,14}$ is the Petersen graph.

$C_{15,11}$:

Orientable genus 0

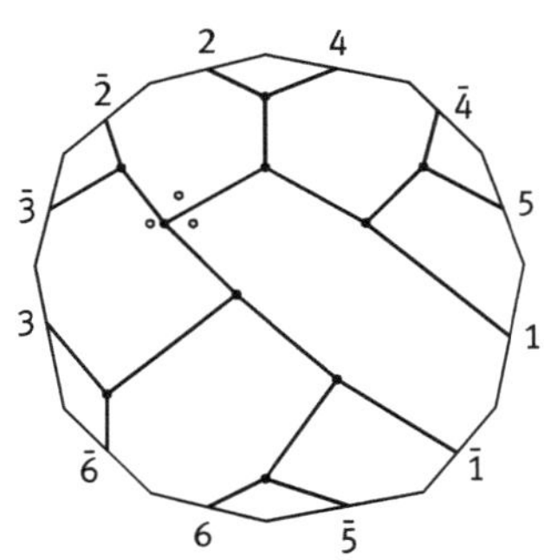

$C_{15,11}: o0-01-03$

Orientable genus 1

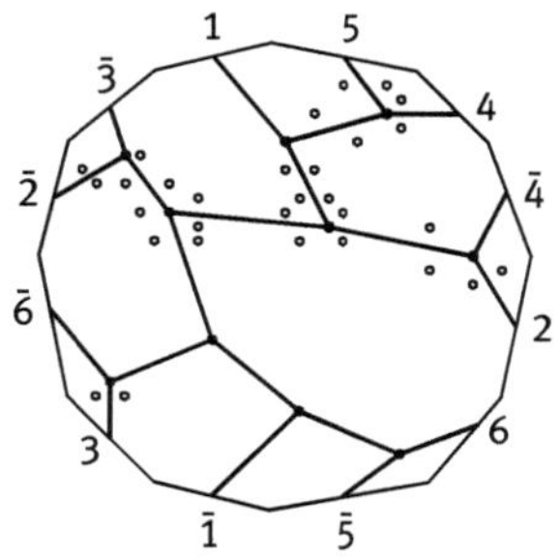

$C_{15,11}$: o1 – 01 – 30

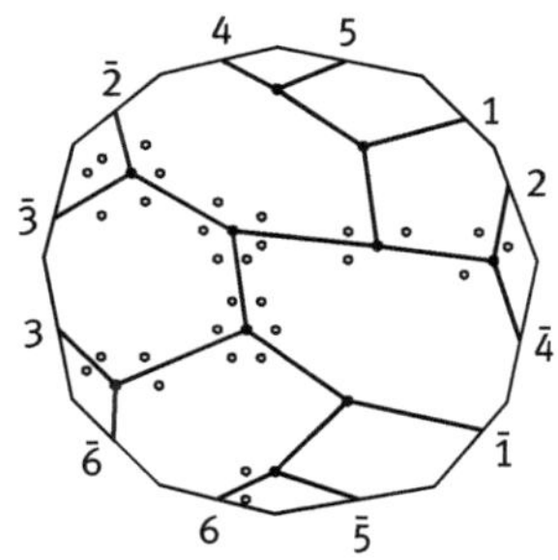

$C_{15,11}$: o1 – 02 – 30

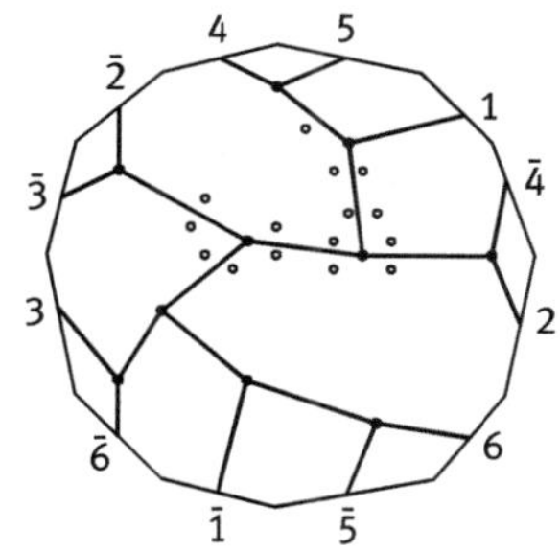

$C_{15,11}$: o2 – 03 – 15

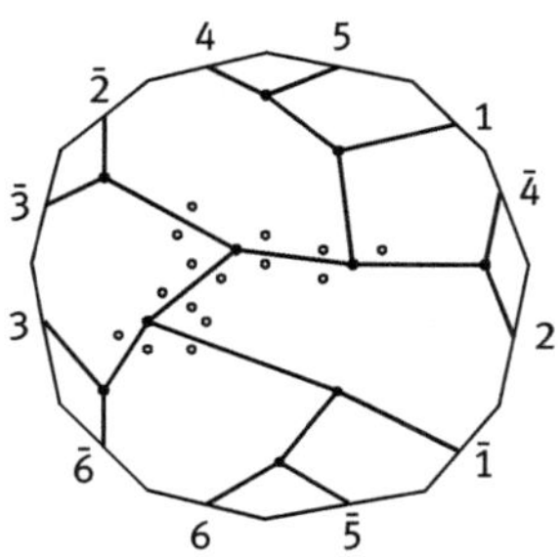

$C_{15,11}$: o1 – 04 – 15

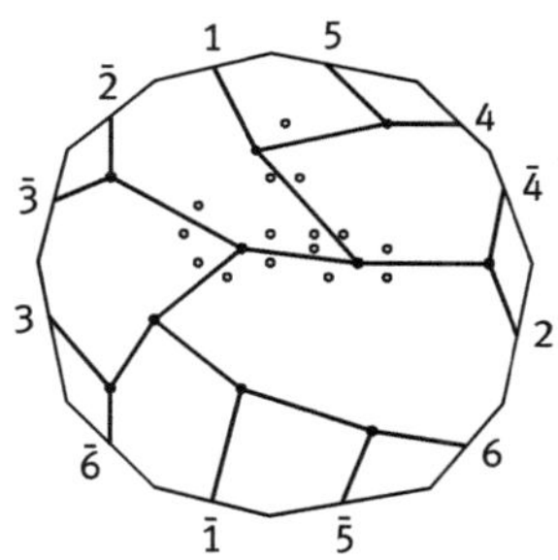

$C_{15,11}$: o1 – 05 – 15

Orientable genus 2

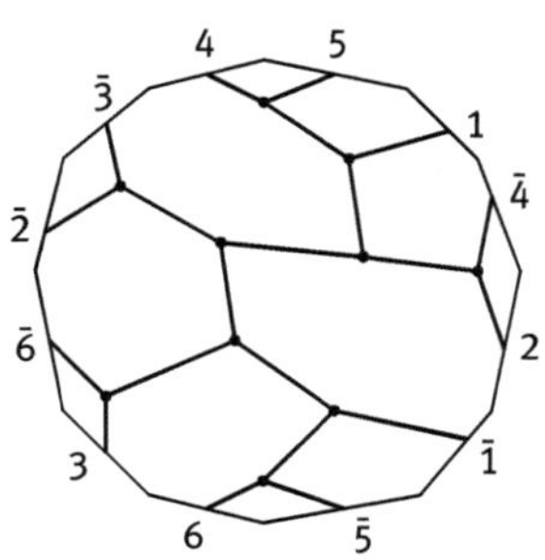

$C_{15,11}$: o2 – 01 – 60

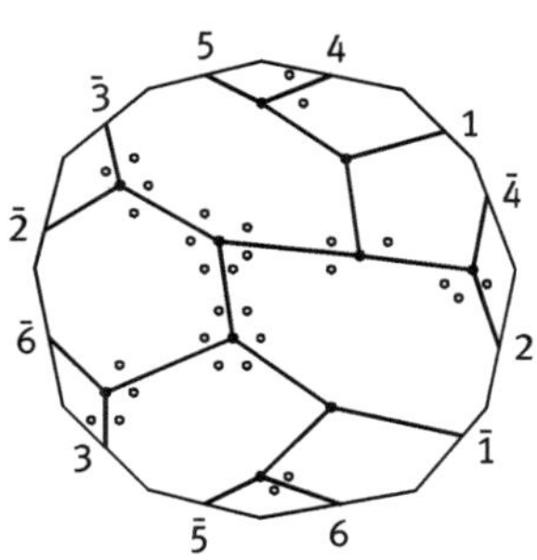

$C_{15,11}$: o2 – 02 – 30

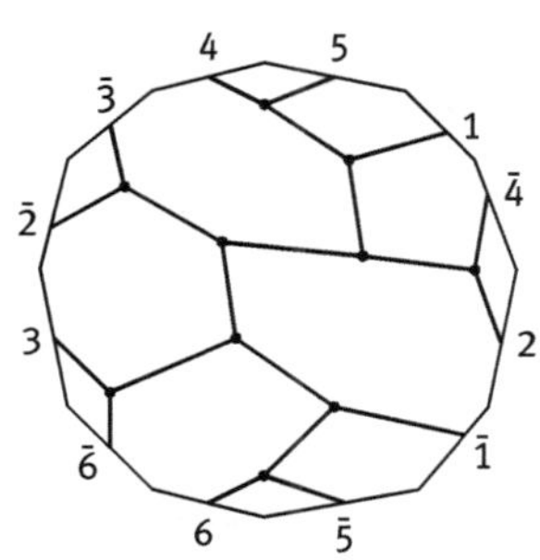

$C_{15,11}$: o2 – 03 – 60

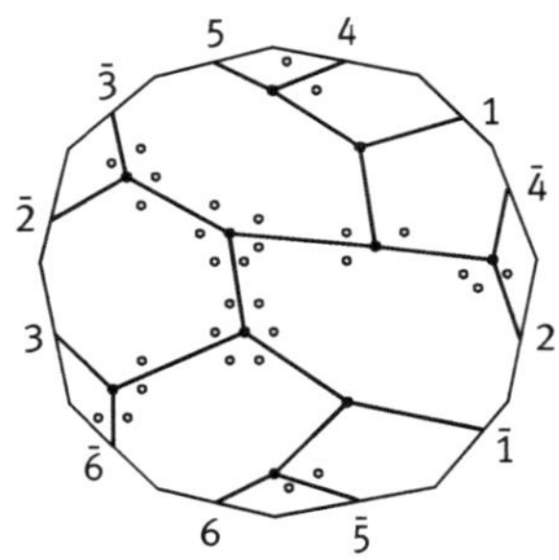

$C_{15,11}$: o2 – 04 – 30

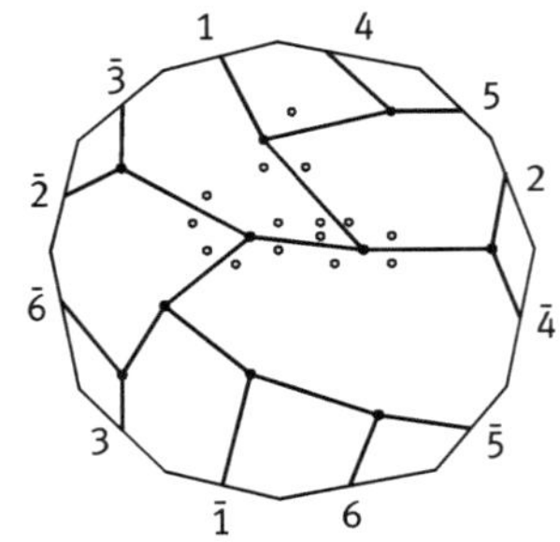

$C_{15,11}$: o2 – 05 – 15

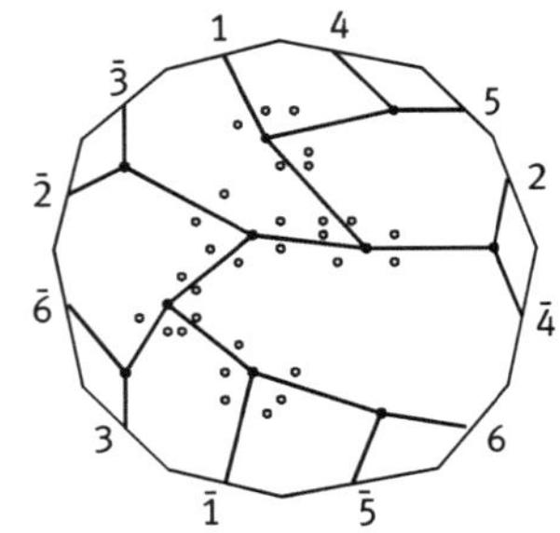

$C_{15,11}$: o2 – 06 – 30

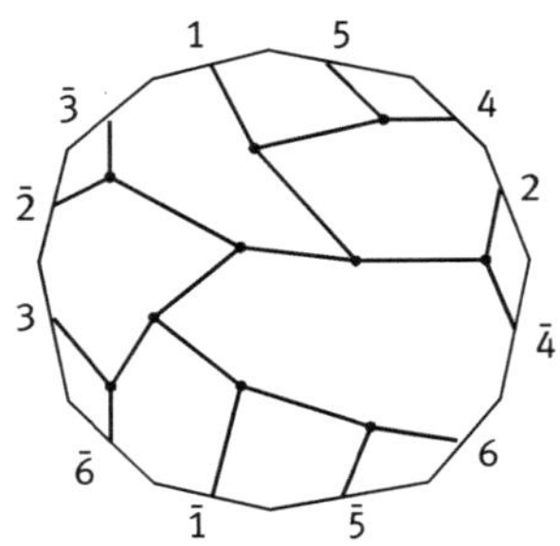

$C_{15,11}$: o2 – 07 – 60

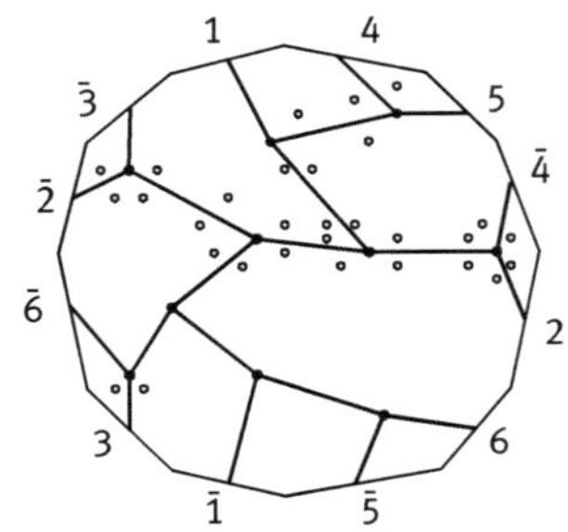

$C_{15,11}$: o2 – 08 – 30

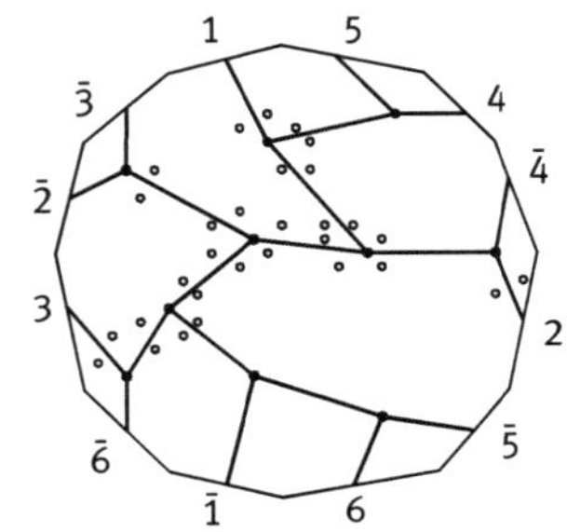

$C_{15,11}$: o2 – 09 – 30

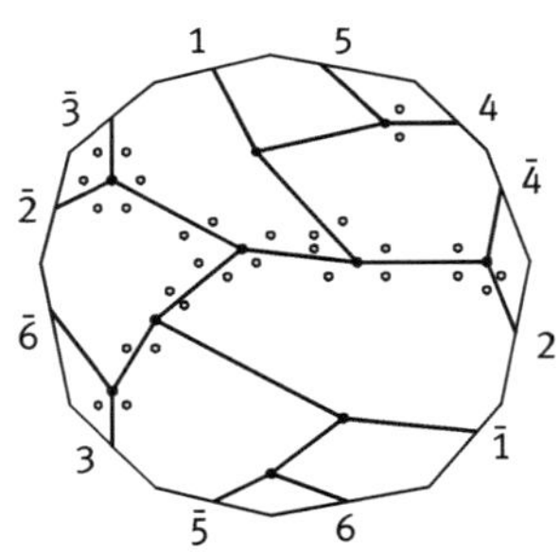

$C1_{5,11}$: o2 – 10 – 30

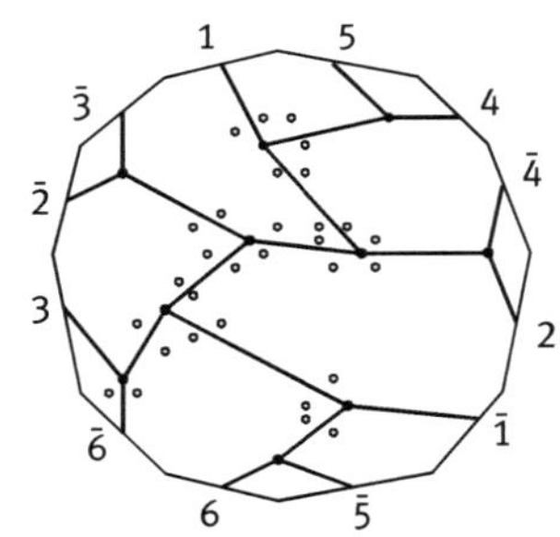

$C_{15,11}$: o2 – 11 – 30

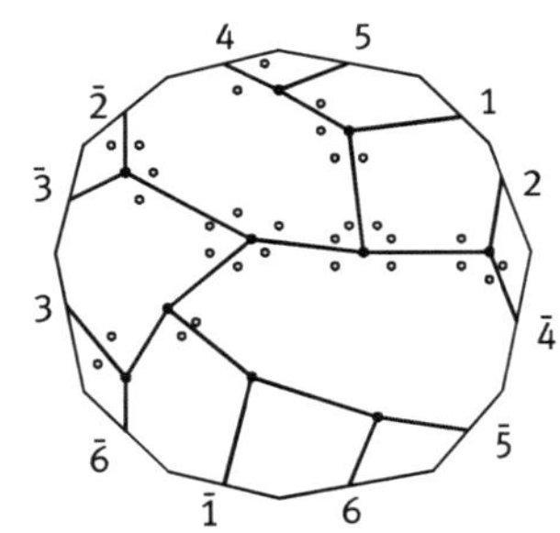

$C_{15,11}$: o2 – 12 – 30

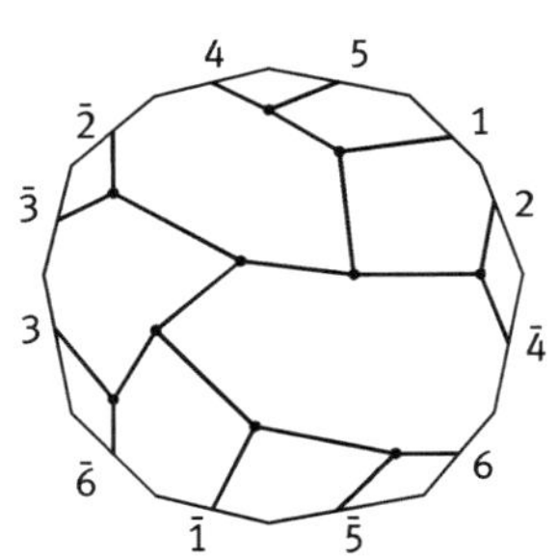

$C_{15,11}$: o2 – 13 – 60

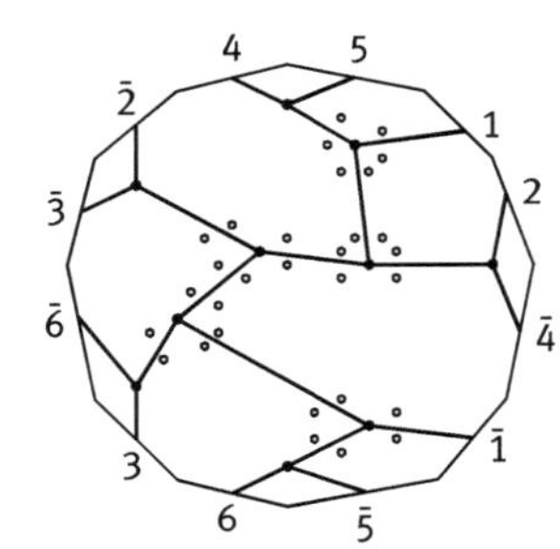

$C_{15,11}$: o2 – 14 – 30

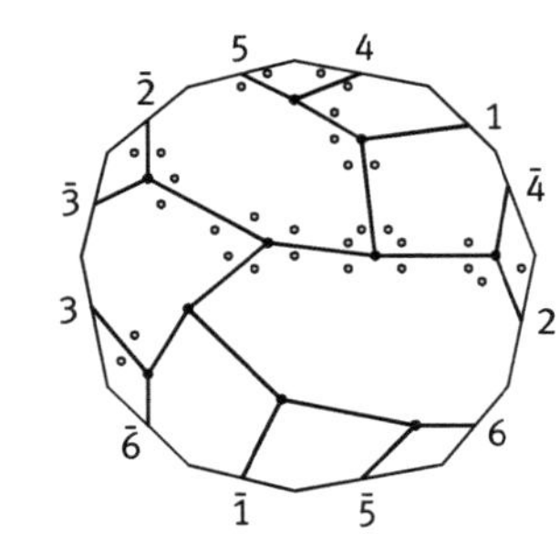

$C_{15,11}$: o2 – 15 – 30

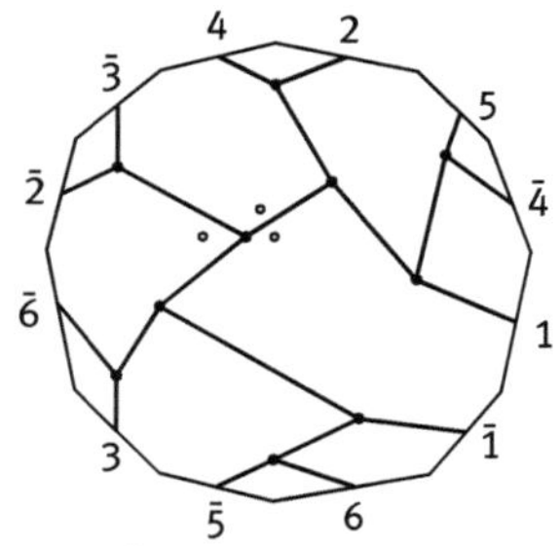

$C_{15,11}: o2-16-03$

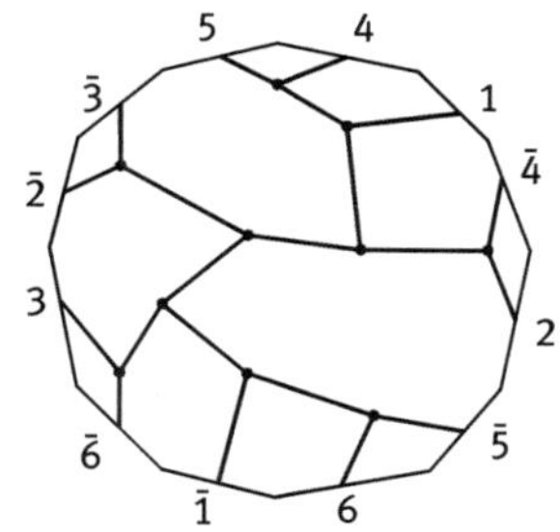

$C_{15,11}: o2-17-60$

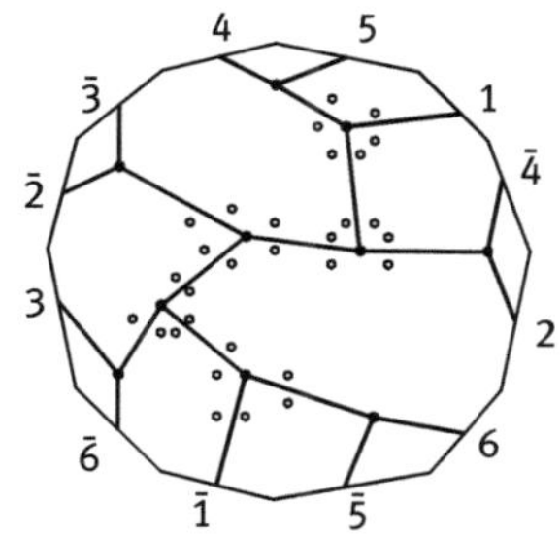

$C_{15,11}: o2-18-30$

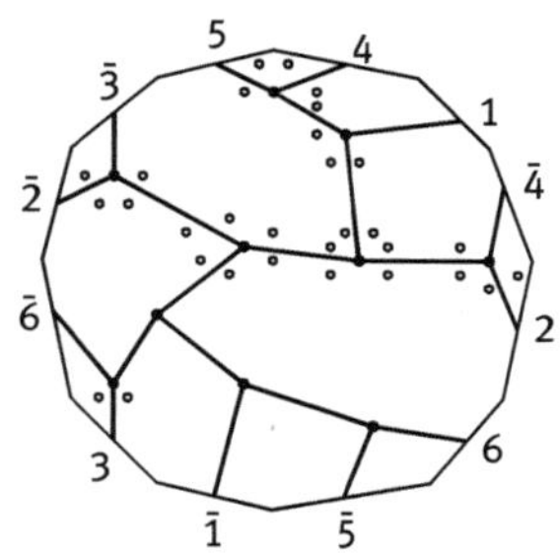

$C_{15,11}: o2-19-30$

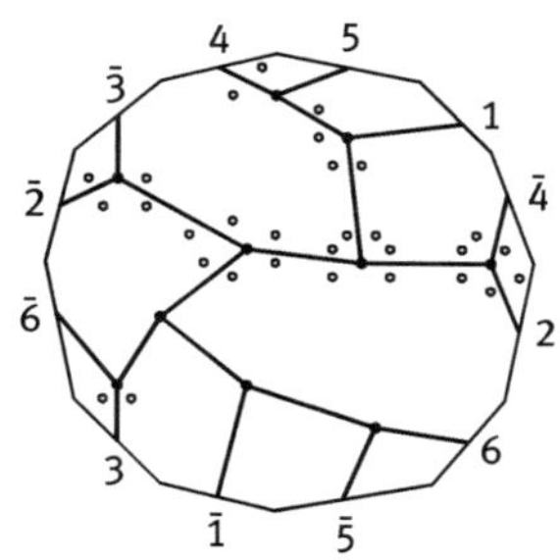

$C_{15,11}: o2-20-30$

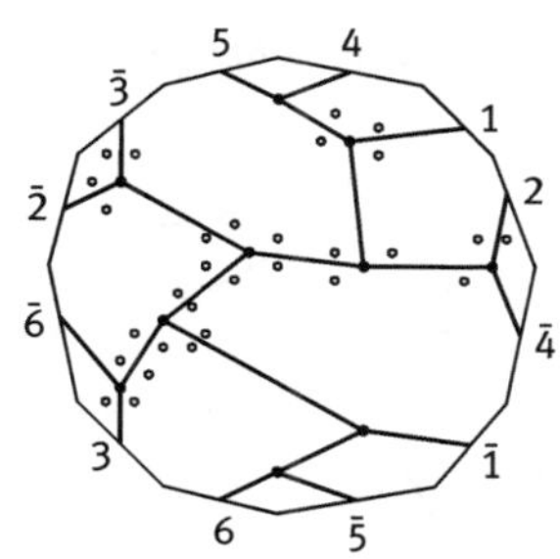

$C_{15,11}: o2-21-30$

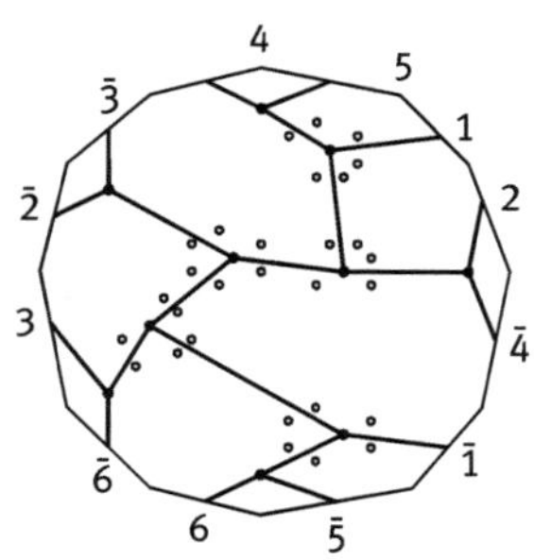

$C_{15,11}: o2-22-30$

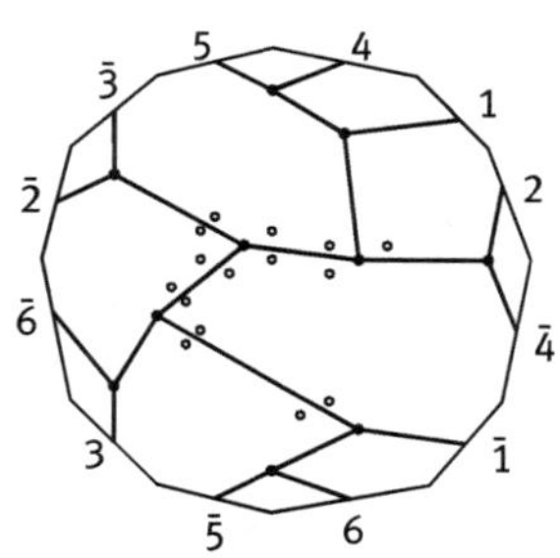

$C_{15,11}: o2-23-15$

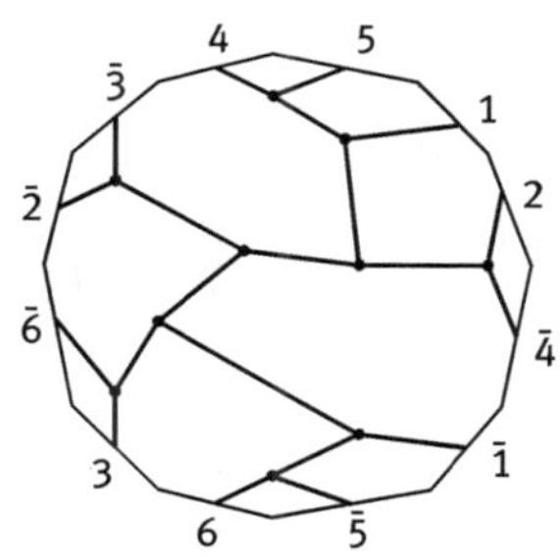

$C_{15,11}: o2-24-60$

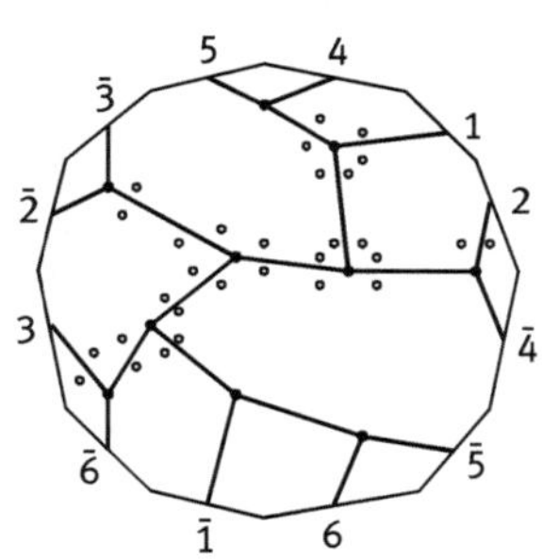

$C_{15,11}: o2-25-30$

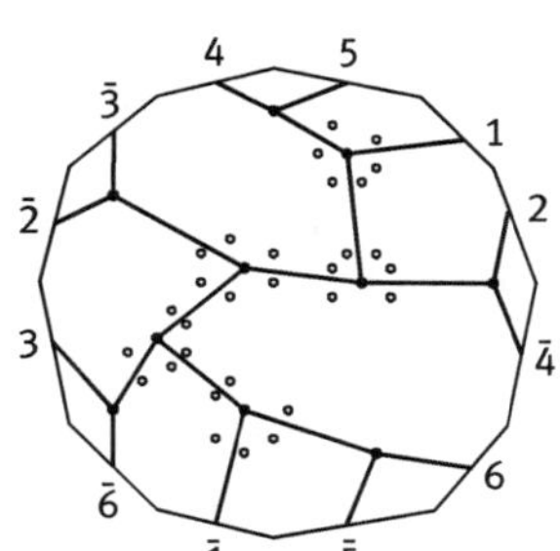

$C_{15,11}: o2-26-30$

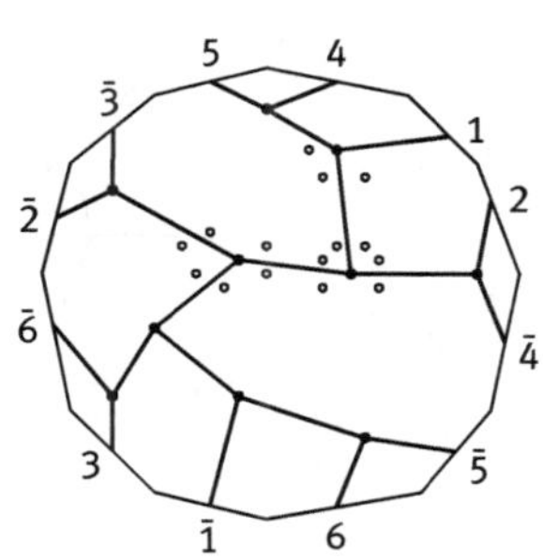

$C_{15,11}: o2-27-15$

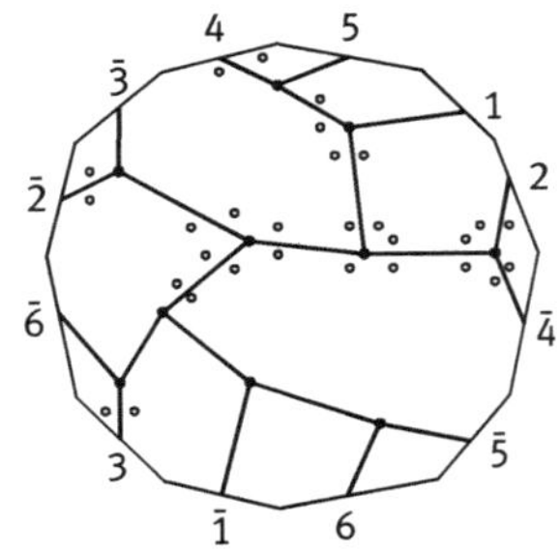

$C_{15,11}$: $o2-28-30$

Orientable genus 3

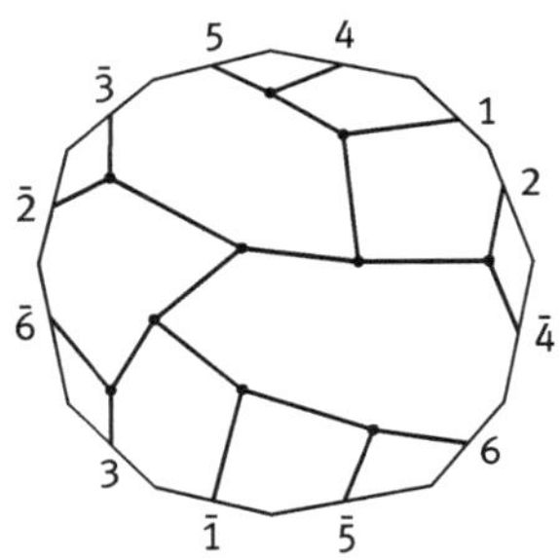

$C_{15,11}$: $o3-01-60$

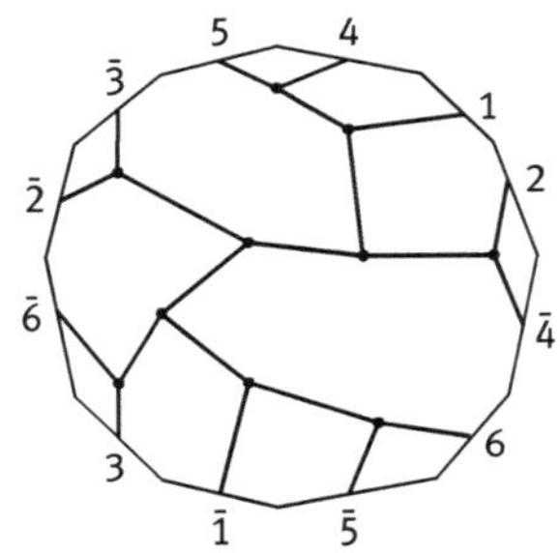

$C_{15,11}$: $o3-02-60$

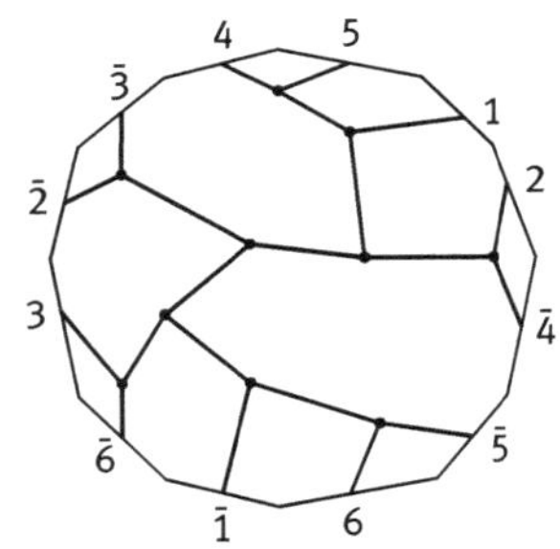

$C_{15,11}$: $o3-03-60$

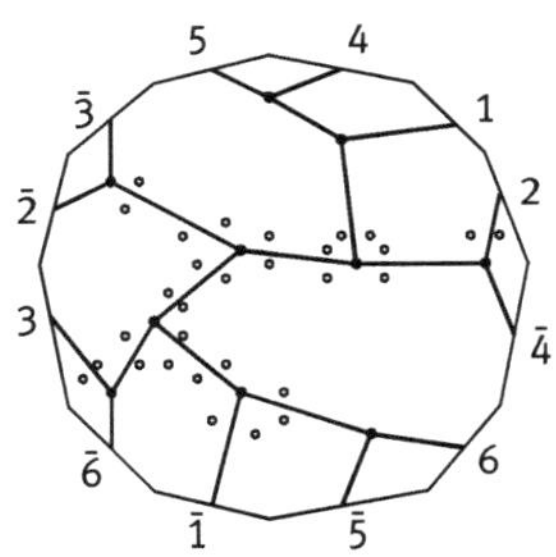

$C_{15,11}$: $o3-04-30$

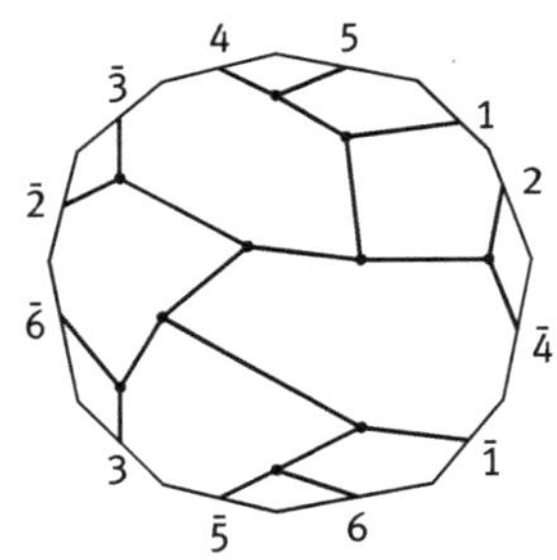

$C_{15,11}$: $o3-05-60$

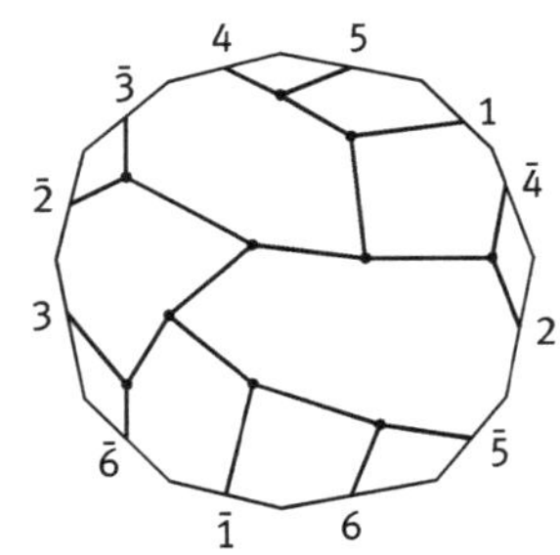

$C_{15,11}$: $o3-06-60$

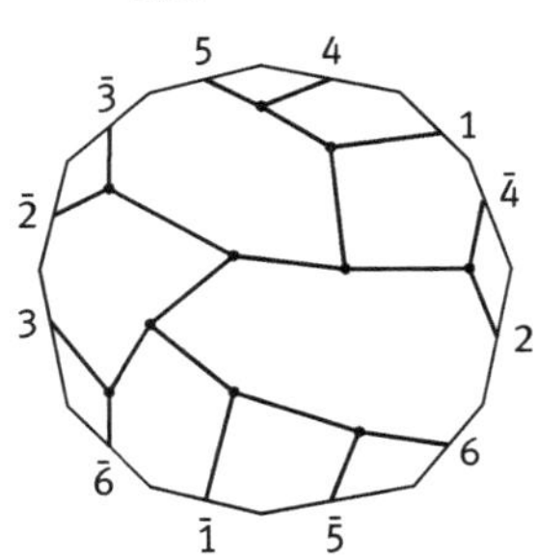

$C_{15,11}$: $o3-07-60$

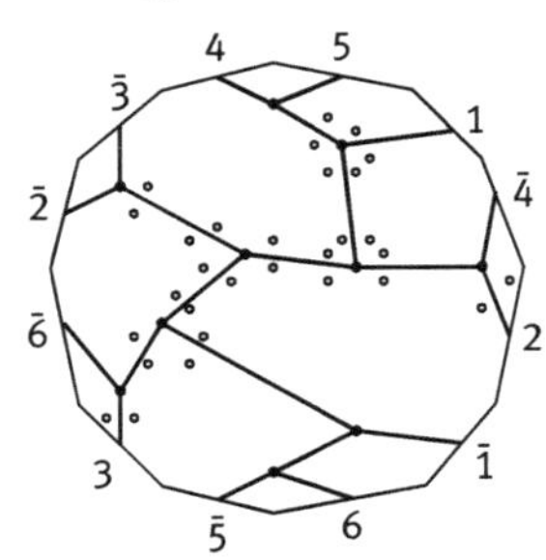

$C_{15,11}$: $o3-08-30$

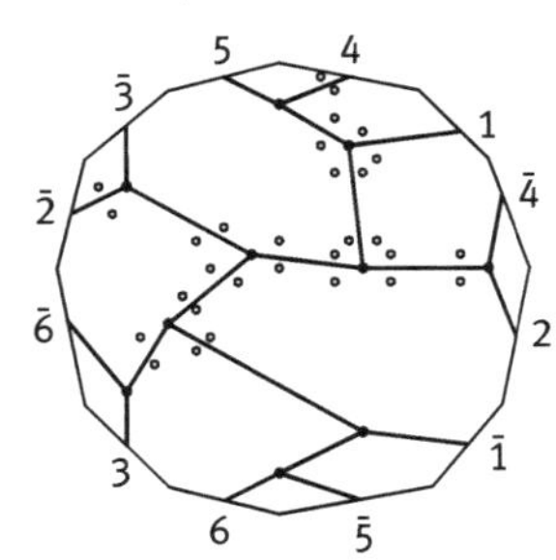

$C_{15,11}$: $o3-09-30$

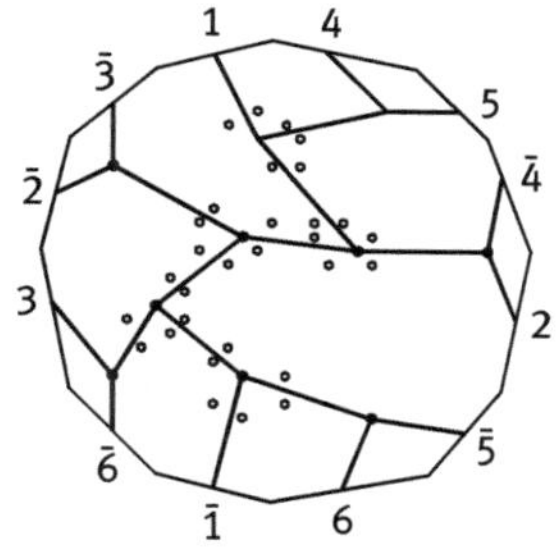

$C_{15,11}$: o3 – 10 – 30

$C_{15,14}$:

Orientable genus 1

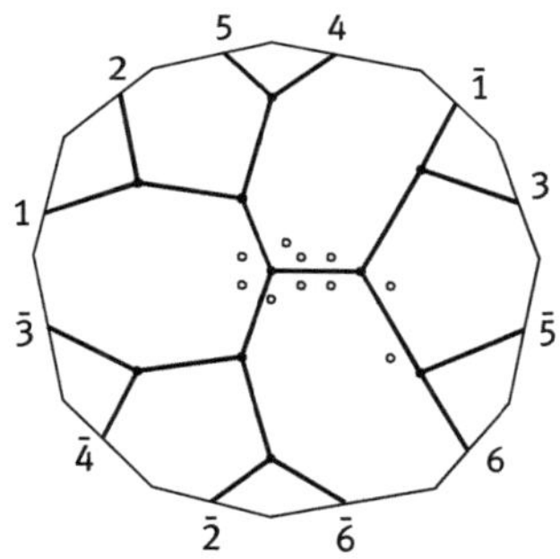

$C_{15,14}$: o1 – 01 – 10

Orientable genus 2

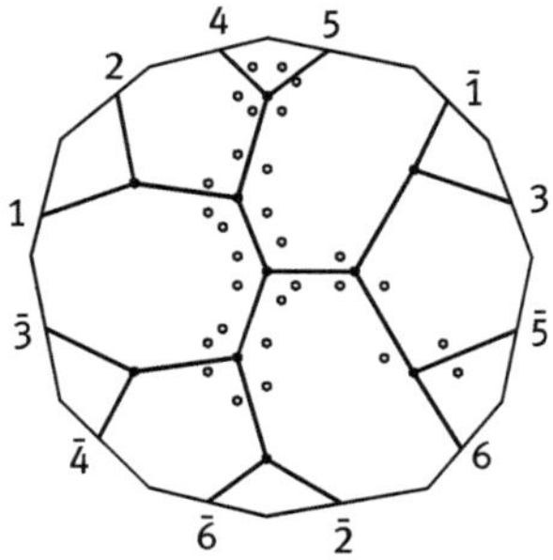

$C_{15,14}$: o2 – 01 – 30

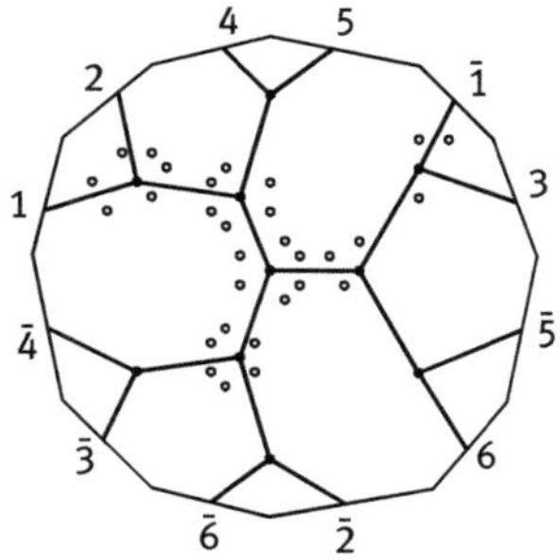

$C_{15,14}$: o2 – 02 – 30

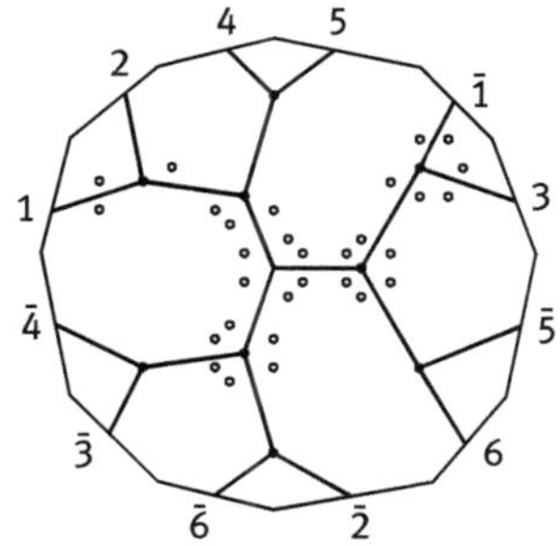

$C_{15,14}$: o2 – 03 – 30

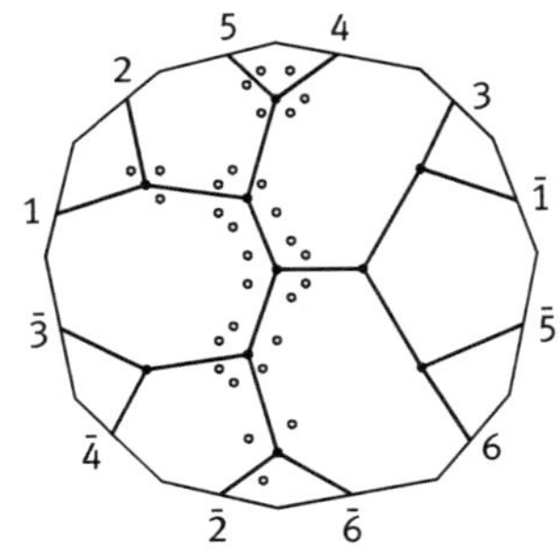

$C_{15,14}$: o2 – 04 – 30

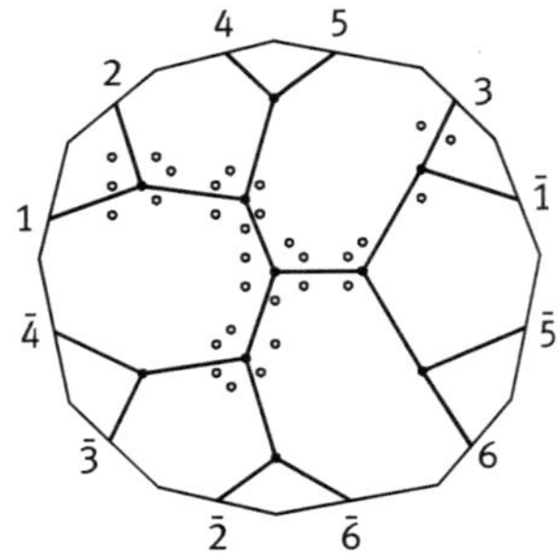

$C_{15,14}$: o2 – 05 – 30

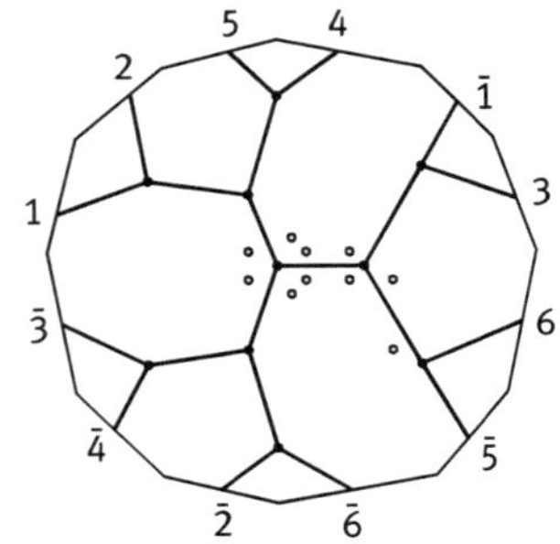

C15,14 : o2 – 06 – 10

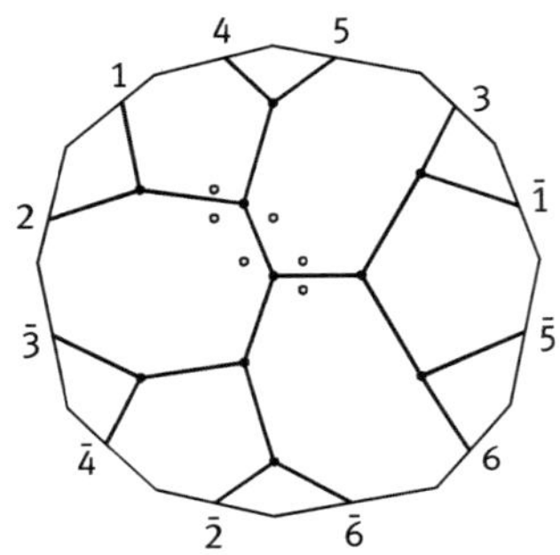

$C_{15,14}$: o2 – 07 – 06

Orientable genus 3

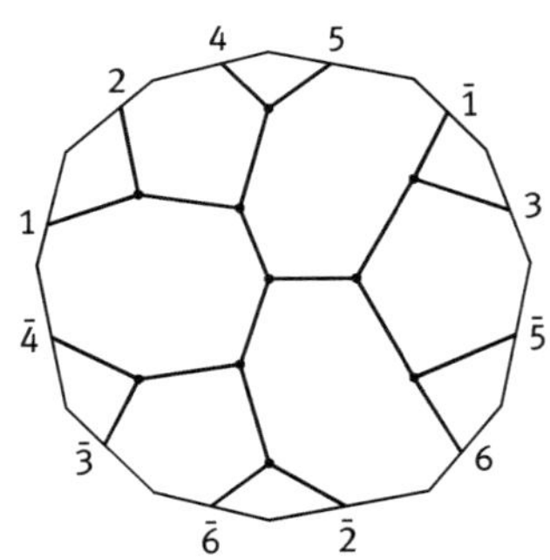

$C_{15,14}$: o3 – 01– 60

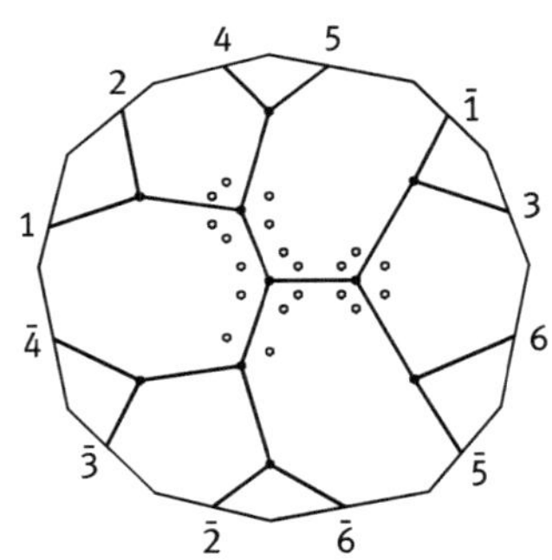

$C_{15,14}$: o3 – 02 – 20

Bibliography

[1] Abrams, L. and D.C. Slilaty, An algebraic characterization of projective-planar graphs, *J. Graph Theory*, **42**(2003), 320–331.

[2] Archdeacon, D., A Kuratowski theorem for the projective plane, *J. Comb. Theory*, **5**(1981), 243–246.

[3] Archdeacon, D. and J. Siran, Characterizing planarity using theta graphs, *J. Graph Theory*, **27**(1998), 17–20.

[4] Baxter, R.J., Dichromatic polynomials and Potts models summed over rooted maps, *Annals Comb.* **9**(2001), 17–36.

[5] Bender, E.A, Z.C. Gao, L.B. Richmond and N.C. Wormald, Asymptotic properties of rooted 3-connected maps on surfaces, *J. Austral. Math. Soc.*, **A60**(1996), 31–41.

[6] Biggs, N., *Algebraic Graph theory*, Cambridge University Press, 1974.

[7] Birkhoff, G.D., The reducibility of maps, *Amer, J. Math.* **35**(1913), 115–128.

[8] Cai, J.L., G.H. Dong and Y.P. Liu, A sufficient condition on upper embeddability of graphs, *Sci. China, Math.* **53**(2010), 5: 1377–1384.

[9] Cai, J.L., R.X. Hao and Y.P. Liu, Enumeratimg rooted Eulerian planar maps, *Acta Math. Sci.*, English Series **21B**(2001), 3: 280–294.

[10] Cai, J.L. and Y.P. Liu, The number of rooted nearly cubic c-nets, *Acta Math. Appl. Sinica*, **17**(2001), 29–37.

[11] Cai, J.L. and Y.P. Liu, The enumeration of rooted nonseparable nearly cubic maps, *Discrete Math.*, **207**(1999), 9–24.

[12] Cai, J.L. and Y.P. Liu, Counting rooted Eulerian planar maps, *International J. Math. Comb.* or *Math. Comb. Book Series*, **1**(2010), 65–73.

[13] Chai, Z. and Y.P. Liu, Up-embeddability of a graph with maximum face degree 6 (in Chinese with English abstract), *J. Beijing Jiaotong Univ.*, **26**(2005), 6: 94–97.

[14] Chen, Y.C. and Y.P. Liu, Maximum genus, girth and maximum non-adjacent edge set, *Ars Comb.*, **79**(2006), 145–159.

[15] Chen, Y.C. and Y.P. Liu, On the average crosscap number II: bounds for a graph, *Sci. China*, **A50**(2007), 292–304.

[16] Chen, Y.C. and Y.P. Liu, Up-embeddability of a graph by order and girth, *Graphs Comb.*, **23**(2007), 521–527.

[17] Chen, Y.C. and Y.P. Liu, On a conjecture of S. Stahl, *Canad. J. Math.*, **62**(2010), 5: 1058–1059.

[18] Chen, Y.C., Y.P. Liu and R.X. Hao, Structure for a graph with average genus, *J. Appl. Math. & Comput.*, **21**(2006), 45–55.

[19] Dehn, M., Uber Kombinatoriche Topologie, *Acta Math.*, **67**(1936), 123–168.

[20] Dixon, J.M. and B. Mortimer, *Permutation Groups*, Springer, 1996.

[21] Dong, F.M. and Y.P. Liu, Counting rooted near triangulations on the sphere, *Discrete Math.*, **123**(1993), 35–45.

[22] Dong, F.M. and J.Y. Yan, On enumeration of outerplanar rooted maps (in Chinese), *Acta Math. Sinica*, **32**(1989), 4: 501–511.

[23] Dong, G.H. and Y.P. Liu, Results of the maximum genus of graphs, *Sci. China*, **A50**(2007), 1563–1570.

[24] Dong, G.H. and Y.P. Liu, Lower bound on the maximum genus of triangle-free graphs, *Adv. Math.*, (China) **38**(2009), 6: 692–698.

[25] Dong, G.H., Y.P. Liu and N. Wang, Up-embeddable graphs via the degree-sum of nonadjacent vertices: non-simple graphs, *Ars Comb.*, **95**(2010), 129–141.

[26] Dong, G.H., N. Wang, Y.Q. Huang and Y.P. Liu, Joint-tree model and the maximum genus of graphs, *Int. J. Math. Comb.*, **4**(2012), 80–91.

DOI 10.1515/9783110481846-022

[27] Dong, G.H., N. Wang, Y.Q. Huang, H. Ren and Y.P. Liu, Vertex splitting and upper embeddable graphs, *CORR* (Computing Reseach Repository), March (2012) [abs/1203.0840], 1–14, *Adv. Math*, **43**(2014), 1–14.

[28] Dong, J.Q. and Y.P. Liu, Partition of a travel into circuits, *J. Math. Res. & Exposition*, **21**(2001), 505–512.

[29] Duke, R.A., The genus, regional number, and Betti number of a graph, *Canad. J. Math.*, **18**(1966), 817–822

[30] Edmonds, J.R., A combinatorial representation for polyhedral surfaces, *Notices Amer. Math. Soc.*, **7**(1960), 646.

[31] Garey, M.R. and D.S. Johnson, *Computers and Intractability*, W. H. Freeman, 1979.

[32] Glover, H., J. Huneke and C.S. Wang, 103 graphs that are irreduclble for the projective plane, *J. Comb. Theory*, **B27**(1979), 232–370.

[33] Gross, J.L. and M.L. Furst, Hierarchy of imbedding dirtribution invariants of a graph, *J. Graph Theory*, **11**(1987), 205–220.

[34] Hao, R.X., J.L. Cai and Y.P. Liu, Enumeration of maps on the projective plane, *Acta Math. Appl. Sinica*, English Series **17**(2001), 567–573.

[35] Hao, R.X., W.L. He, Y.P. Liu and E.L. Wei, The genus distributions for a certain type of permutation graphs in orientable surfaces, *Sci. China*, **A50**(2007), 1748–1754.

[36] Hao, R.X. and Y.P. Liu, On chromatic polynomials of some kinds of graphs, *Acta Math. Appl. Sinica*, Eng. Series **20**(2004), 239–246.

[37] Hao, R.X. and Y.P. Liu, The genus polynomials of cross-ladder digraphs in orientable surfaces, *Sci. China*, **A50**(2008), 5: 889–896.

[38] Hao, R.X. and Y.P. Liu, Evaluations of crossing numbers by applying auxiliary graphs, *J. Henan Norn. Univ.* (NS), **30**(2002), 2: 7–13.

[39] He, W.L. and Y.P. Liu, A note on the maximum genus of graph G^3, *J. Syst. Sci. Inf.*, **1**(2003), 615–619.

[40] Heffter, L., Ueber das Problem der Nachbargebiete, *Math. Ann.*, **38** (1891), 477–508.

[41] Hilbert, D. and S. Cohn-Vossen, Anschauliche Geometrie, Julius Springer, 1932. Or *Geometry and Imagination*, Chelsea, New York, 1952.

[42] Huang, Y.Q. and Y.P. Liu, Face size and the maximum genus of a graph, *J. Comb. Theory*, **B80**(2000), 356–370.

[43] Huang, Y.Q. and Y.P. Liu, On the average genus of 3-regular graphs, *Adv. Math.*, **31**(2002), 56–64.

[44] Huang, Y.Q. and Y.P. Liu, Maximum genus, independent number and girth of a graph, *Chin. Ann. Math.*, **B21**(2000), 77–82.

[45] Kempe, A.B., On the geographical problem, *Am. J, Math.*, **2**(1879), 193–200.

[46] Kuratowski, K., Sur le Problem des Courbes Gauches en Topologie, *Fund. Math.*, **15**(1930), 271–283.

[47] Kwak, J. H. and J. Lee, Genus polynomials of dipoles, *Kyungpook Math. J.*, **33**(1993), 115–125.

[48] Lefschetz, S., Planar graphs and related topics, *Proc. Nat. Acad. Sci.*, **54**(1965), 1763–1765.

[49] Li, D.M. and Y.P. Liu, Maximum genus, girth and connectivity, *Eur. J. Comb.*, **21**(2000), 651–657.

[50] Li, D.M. and Y.P. Liu, A tight lower bound on the maximum genus of 3-edge connected loopless graphs, *Acta Math. Appl. Sinica*, **15** (1999), 361–367.

[51] Li, L.F. and Y.P. Liu, The genus distribution of a type of cubic graphs (in Chinese with English abstract), *J. Beijing Jiaotong Univ.*, **28**(2005), 39–42.

[52] Li, Z.X. and Y.P. Liu, Chromatic sums of rooted triangulations on the projective plane, *J. Appl. Math. & Comput.*, **18**(2005), 183–196.

[53] Li, Z.X. and Y.P. Liu, Chromatic sums of 2-edge connected maps on the plane, *Discrete Math.*, **296**(2005), 211–223.

[54] Li, Z.X. and Y.P. Liu, Chromatic sums of biloopless nonseparable near-triangulations on the projective plane, *Acta Math. Appl. Sin.*, **29**(2013), 1: 123–134.

[55] Little, C.H.C., Cubic combinatorial maps, *J. Comb. Theory*, **B44** (1988), 44–63.

[56] Liu, T.Y. and Y.P. Liu, On generators of planar near-triangulations for finding small circuit double covers, *OR Tran.*, **3**(1999), 4: 11–16.

[57] Liu, T.Y. and Y.P. Liu, On the crossing number of circular graphs, *OR Trans.*, **2**(1998), 32–38.

[58] Liu, W.Z. and Y.P. Liu, Enumeration of three kinds of rooted maps on the Klein bottle, *J. Appl. Math. Comput.*, **24**(2007), 411–419.

[59] Liu, W.Z. and Y.P. Liu, 5-essential rooted maps on N_5, *Utilitas Math.*, **72**(2007), 51–64.

[60] Liu, Y.P., The nonorientable maximum genus of a graph (in Chinese with English Abstract), *Sci. Sinica*, Special Issue on Math. **I** (1979), 191–201.

[61] Liu, Y.P., The orientable maximum genus of a graph, *Sci. Sinica*, Special Issue on Math. **II**(1979), 41–55.

[62] Liu, Y.P., A note on the Edmonds surface dual theorem (in Chinese with Englias Abstract), *OR Transactions*, **2**(1983), 1: 62–63.

[63] Liu, Y.P., A polyhedral theory on graphs, *Acta Math. Sinica* **10**(1994), 136–142.

[64] Liu, Y.P., *Theory of Graphic Embeddability* (in Chinese), Science Press, 1994.

[65] Liu, Y.P., *Embeddability in Graphs*, Kluwer, 1995.

[66] Liu, Y.P., Combinatorial invariants on graphs, *Acta Math. Sinica*, **11** (1995), 211–220.

[67] Liu, Y.P., *Enumerative Theory of Maps*. Kluwer, 1999.

[68] Liu, Y.P., *Theory of Counting Combinatorial Maps* (in Chinese), Science Press, 2001.

[69] Liu, Y.P., *Introduction to Combinatorial Maps*, POSTECH, 2002.

[70] Liu, Y.P., Map color theorem and surface embedding of graphs (I–V) (in Chinese), *Math. Theory Prac.*, (1981), 1: 65–78; (1981), 2: 59–63; (1981), 3: 33–44; (1981), 4: 33–41; (1982), 1: 34–44.

[71] Liu, Y.P., New approaches to the double cover conjecture of graphs by circuits, *OR Trans.*, **4**(2000), 50–54.

[72] Liu, Y.P., *Advances in Combinatorial Maps*(in Chinese), Northern Jiaotong University Press, 2003.

[73] Liu, Y.P., *Algebraic Principles of Maps* (in Chinese), Higher Education Press, 2006.

[74] Liu, Y.P., *Theory of Polyhedra*, Science Press, 2008.

[75] Liu, Y.P., A new method for counting rooted trees with vertex partition, *Sci. China*, **A51**(2008), 2000–2004.

[76] Liu, Y.P., *Theory of Counting Combinatorial Maps* (2nd and extended Ed., in Chinese), Science Press, 2008.

[77] Liu, Y.P., *Topological Theory on Graphs*, USTC Press, 2008.

[78] Liu, Y.P., *General Theory of Map Census*, Science Press, 2009.

[79] Liu, Y.P., *Theory of Graphic Embeddability* (2nd and extended Ed., in Chinese), Science Press, 2010.

[80] Liu, Y.P., Advanced graph theory as I have initially realized (XXV)—-Partition by two paths with circuits (in Chinese with English abstract), *Selected Publications of Y.P. Liu*, **Book 23**(2015), Times Literature Press, 11402–11422.

[81] Liu, Y.P., Advanced graph theory as I have initially realized (XXVI)—-Circuit double covers on graphs (in Chinese with English abstract), *Selected Publications of Y.P. Liu*, **Book 23**(2015), Times Literature Press, 11423–11441.

[82] Liu, Y.P., On partition of a travel by circuits, *Selected Publications of Y.P. Liu*, **Book 23**(2015), Times Literature Press, 11442–11464.

[83] Liu, Y.P., Proofs of WEC, CDCC and SEC on graphs, *Selected Publications of Y.P. Liu*, **Book 23**(2015), Times Literature Press, 11465–11481.

[84] Liu, Y.P., On a meson equation of surface type, *J. Kunming Univ. S* & *T*(NS), **37**(2012), 1: 82–89.

[85] Liu, Y.P., Partial differential equations with surfaces (in Chinese with English abstract), *J. Yulin Mormal Univ.*, **33**(2012), 2: 2–9.
[86] Liu, Y.P., A straight difference equation on surface quadrangulations (in Chinese with English abstract), *J. Kunming Univ. S* & *T*(NS), **37**(2012), 3: 78–84.
[87] Liu, Y.P., Partial differential equations in surface quadrangulations (in Chinese with English abstract), *J. Kunming Univ. S* & *T*(NS), **37**(2012), 35: 79–94.
[88] Liu, Y.P., A meson equation of surface type, *J. Jishou Univ.* (NS), **33**(2012), 5: 56–63.
[89] Liu, Y.P., *Lectures on Graph Theory* (two volumes, in Chinese), Graduate School of Academia Sinica, 1979.
[90] Liu, Y.P., *Introductory Map Theory*, Kappar & Omega, 2010.
[91] Liu, Y.P., Algebraic principles on polyhegons, *Orient. J. Math.*, **2**(2010), 2: 83–93.
[92] Liu, Y.P., Surface embeddability of graphs via homology, *Sci. China*, Math. **53**(2010), 11: 2889–2892.
[93] Liu, Y.P., Surface embeddability of graphs via tree-travels, *Int. J. Math. Comb.* **1**(2011), 73–81.
[94] Liu, Y.P., Surface embeddability of graphs via joint trees, *Int. J. Math. Comb.* **2**(2011), 15–20.
[95] Liu, Y.P., Surface embeddability of graphs via reductions, *Int. J. Math. Combin* **3**(2011), 61–67.
[96] Liu, Y.P., Modulo 2 programming and planar embeddings (in Chinese), *Acta Math. Appl.*, **1**(1978), 321–329.
[97] Liu, Y.P., *Theory of Combinatorial Functional Equations* (in Chinese), USTC Press, 2015.
[98] Liu, Y.P., A note on the number of rooted planar triangulations (in Chinese with English absreact), *Acta Math. Appl. Sinica*, **11**(1988), 146–148.
[99] Liu, Y.P., A functional equation of enumerating nonseparable Eulerian planar maps with vertex partition, *Saience Bulletin*, **31**(1986), 941–945.
[100] Liu, Y.P., On the loopless Eulerian vertex partition equation, *Chin. Sci. Bull.*, **36**(1991), 1585–1589.
[101] Liu, Y.P., On the vertex partition equation of loopless Eulerian planar maps, *Acta Math. Appl. Sinica*, English Series **8**(1992), 45–58.
[102] Liu, Y.P., A note on the number of loopless Eulerian planar maps, *J. Math. Res. Expos.*, **12**(1992), 165–170.
[103] Liu, Y.P., A census of rooted petal bundles (in Chinese with English abstract), *J. Xinyang Normal Univ.* (NS), **17**(2004), 249–254.
[104] Liu, Y.P., On the number of Eulerian planar maps, Research Report, Series A, No.24, Dept. Statistics, Uni. Rome "La Sapienza", 1989; Also in *Acta. Math. Sci.*, **12**(1992), 418–423.
[105] Liu, Y.P., Enumeration of rooted separable planar maps, Comb. Optim CORR82-4, Uin. Waterloo, 1982; Also in *Utilitas Math.*, **25**(1984), 77–94.
[106] Liu, Y.P., Counting rooted planar maps, Comb. Optim. CORR 83-26, Uin. Waterloo, 1983; Also in *J. Math. Res. Expos.* **3**(1984), 37–46.
[107] Liu, Y.P., A note on the mumber of cubic planar maps, *Acta Math. Sci*, **12**(1992), 282–285.
[108] Liu, Y.P., A functional equation for enumerating nonseparable planar maps with vertex partition, *KEXUE TONGBAO* (Science Bulletin), **31**(1986), 73–77.
[109] Liu, Y.P., Enumeration of rooted nonseparable bipartite planar maps (in Chinese), *Acta Math. Sci. Sinica*, **9**(1989), 21–28.
[110] Liu, Y.P., Enumeration of rooted vertex non-separable planar maps, Comb. Optim. CORR 83-14, Uni. Waterloo, 1983; Also in *Chin. Ann. Math.* **9B**(1988), 390–403.
[111] Liu, Y.P., Enumerating rooted loopless planar maps, Comb. Optim. CORR 83-4, Uni. Waterloo, 1983; Also in *Acta Math. Appl. Sinica*, English Series **2**(1985), 14–26.
[112] Liu, Y.P., An enumerating equation of rooted loopless planar maps with vertex partition, RUTCOR Research report RRR 36-87, Rutgers Uni., 1987. Main part also in *Acta Math. Appl. Sinica* (Chinese with English absract), **12**(1989), 213–217.

[113] Liu, Y.P., On eumerating equations of planar maps (in Chinese with English abstract), *Adv. Math.*, **18**(1989), 446–460.

[114] Liu, Y.P., Eumeration of rooted outerplanar maps with vertex partition, *Chin. Sci. Bull.*, **32**(1987), 295–299.

[115] Liu, Y.P., On the number of rooted c-nets, Comb. Optim. CORR83-35, Uni. Waterloo, 1983; Also in *J. Comb. Theory*(B) **36**(1984), 118–123.

[116] Liu, Yi. and Y.P. Liu, A characterization of the embeddability of graphs on the surface of given genus, *Chin. Ann. Math.*, **17B**(1996), 457–462.

[117] Lu, J.J., H. Ren and D.J. Ma, The crossing number of $C(m, 3)$ (in Chinese with English abstract), *J. Syst. Sci. & Math. Sci.*, **26**(2006), 4: 504–512.

[118] Lv, S.X. and Y.P. Liu, Up-embeddability and independent number of simple graphs, *Ars Comb.*, **102**(2011), 237–243.

[119] Lv, S.X. and Y.P. Liu, Up-embeddability of graphs with new degree-sum of independent vertices, *J. Math. Res. Appl.*, **32**(2012), 4: 399–406.

[120] Lv, S.X. and Y.P. Liu, Extensions on 2-edge connected 3-regular none up-embeddable graphs, *Adv. Math.* (China), **44**(2015), 1: 55–60.

[121] Lv, S.X. and Y.P. Liu, A lower bound on maximum genus of graphs with girth and minimal degree, *Adv. Math.*, **39**(2010), 6: 729–735.

[122] Lv, S.X. and Y.P. Liu, A new bound on maximum genus of simple graphs, *Ars Comb.*, **98**(2011), 7–14.

[123] MacLane, S., A structural characterization of planar combinatorial graphs, *Duke Math. J.*, **3**(1937), 416–472.

[124] Mao, L.F. and Y.P. Liu, A new approach for enumerating maps on orientable surfaces, *Aust. J. Comb.*, **30**(2004), 247–259.

[125] Mao, L.F., Y.P. Liu and F. Tian, Automorphisms of maps with a given underlying graph and their application to enumeration, *Acta Math. Sinica*, **21**(2005), 225–236.

[126] Mao, L.F., Y.P. Liu and E.L. Wei, The semi-arc automorphism group of a graph with application to map enumeration, *Graphs Comb.*, **22**(2006), 83–101.

[127] Massey, W., *Algebraic Topology*, Harcourt, 1967.

[128] Mullin, R.C. and P.J. Scellenberg, The enumeration of c-nets via quadrangulations, *J. Comb. Theory*, **4**(1963), 259–276.

[129] Ore, O., *The Four Color Problem*, Academic Press, 1967.

[130] Pan, L.Y. and Y.P. Liu, General investigation of loopless maps on surface, *J. Xinyang Normal Univ.* (NS), **26**(2013), 1: 1–6.

[131] Pan L.Y., Y.P. Liu and R.X. Hao, Counting two types of quadrangulations: Rooted near quadrangulations on the disc and nonseparable outerplanar quadrangulations, *Consumer Electronics, Communications and Networks* (CECNet), 2011 International Conference on Date of Conference: 16–18 April 2011, 3998–4001.

[132] Ren, H. and Y.P. Liu, 4-regular maps on the Klein bottle, *J. Combin. Theory*, **B82**(2001), 118–137.

[133] Ren, H. and Y.P. Liu, Enumerating near 4-regular maps on the sphere and the torus, *Discrete Appl. Math.*, **110**(2001), 273–288.

[134] Ren, H. and Y.P. Liu, Bisingular maps on the sphere and the projective plane, *Discrete Math.*, **223**(2000), 275–285.

[135] Ren, H. and Y.P. Liu, On the number of regular maps on the projective plane, *Utilitas Math.*, **56**(1999), 201–209.

[136] Ren, H. and Y.P. Liu, On the number of fair triangulations, *Acta Math. Sinica*, **41**(1998), 1193–1196.

[137] Ringel, G., The toroidal thickness of complete graphs (in German), *Math. Z.*, **87**(1985), 14–26.

[138] Ringel, G., *Farbungsprobleme auf Flachen und Graphen*, VEB Deutscher Verlag der Wis-senschafteng, 1959.
[139] Ringel, G., *Map Color Theorem*, Springer, 1974.
[140] Robertson, N. and P. Seymour, Generalizing Kuratowski's theorem, *Cong. Numer.*, **45**(1984), 129–138.
[141] Shao, Z.L. and Y.P. Liu, Genera of repeated edge amalgamation for two types of graphs (in Chinese), *J. Beijing Jaotong Univ.*, **31**(2007), 3: 58–60.
[142] Shao, Z.L. and Y.P. Liu, On the number of genus embeddings of complete tripartite graph $K_{n,n,l}$, *Ars Comb.*, **104**(2012), 129–142.
[143] Shao, Z.L. and Y.P. Liu, Minimum genera of certain types of graphs, *Utilitas Math.* **81**(2010), 85–98.
[144] Shao, Z.L. and Y.P. Liu, Genus polynomials for three types of graphs, *Ars Comb.*, **96**(2010), 65–73.
[145] Shao, Z.L. and Y.P. Liu, The genus of a type of graph, *Sci. China*, **A53**(2010), 2: 457–464.
[146] Shao, Z.L., Y.P. Liu and Z.G. Li, On the number of genus embeddings of complete bipartite graphs, *Graphs Comb.*, **29**(2013), 1909–1919.
[147] Stahl, S., Generalized embedding schemes, *J. Graph Theory*, **2** (1978), 41–52.
[148] Stahl, S., A combinatorial analog of the Jordan curve theorem, *J. Comb. Theory*, **B35**(1983), 1–21.
[149] Stahl, S., On the zeros of some genus polynomial, *Can. J. Math.*, **49**(1997), 3: 617–640.
[150] Tutte, W.T., Combinatorial oriented maps, *Can. J. Math.*, **31**(1979), 986–1004.
[151] Tutte, W. T., On the spanning trees of self-dual maps, *Ann. New York Acad. Sci.*, **319** (1970), 540–548.
[152] Tutte, W. T., *Graph Theory*, Addison–Wesley, 1984.
[153] Vince, A., Combinatorial maps, *J. Comb. Theory* **34**(1983), 1–21.
[154] Vince, A., Map duality and generalizations, *Ars Comb.*, **39**(1995), 211–229.
[155] Wan, L.X., K.Q. Feng, Y.P. Liu and D.J. Wang, Genus distribution of ladder type and cross type graphs, *Sci. China*, **A52**(2009), 8: 1760–1768.
[156] Wan, L.X., H.J. Lai and Y.P. Liu, Genus distribution for a graph, *Int. J. Math. Comb.*, **4**(2009), 99–109.
[157] Wan, L.X. and Y.P. Liu, The orientable genus distribution of a new type of graphs (in Chinese with English abstract), *J. Beijing Jiaotong Univ.*, **26**(2005), 65–68.
[158] Wan, L.X. and Y.P. Liu, Orientable embedding distributions by genus for certain types of non-planar graphs, *Ars Comb.*, **79**(2006), 97–105.
[159] Wan, L.X. and Y.P. Liu, Orientable embedding genus distribution for certain types of graphs, *J. Comb. Theory*, **B98**(2008), 19–32.
[160] Wan, L.X. and Y.P. Liu, On the embedding distribution of ladders and crosses, *Appl. Math. Lett.*, **21**(2009), 738–742.
[161] Wang, H.Y. and Y.P. Liu, Maximum genus of a 2-connected graph with diameter 3 (in Chinese with English abstract), *J. Xinyang Teacher College* (NS), **17**(2004), 381–384.
[162] Wang, Tao and Y.P. Liu, Implements of some new algorithms for combinatorial maps, *OR Trans.*, **12**(2008), 2: 58–66.
[163] Wawtzynczyk, A., *Group Representations and Spectial Functions*, D. Reidel, 1984.
[164] Wei, E.L. and Y.P. Liu, Strong maximum genus of a type of graphs, *J. Northern Jiaotong Univ.*, **26**(2002), 19–21.
[165] Wei, E.L. and Y.P. Liu, Weak embedding of planar graphs, *J. Appl. Math.* & *Comput.*, **21**(2006), 175–187.
[166] Whitney, H., Planar graphs, *Fund. Math.*, **21** (1933), 73–84.

[167] Xu, Y. and Y.P. Liu, The number of pan-fan maps on the projective plane, *Utilitas Math.*, **72**(2007), 279–286.

[168] Xu, Y. and Y.P. Liu, Counting pan-fan maps on nonorientable surfaces, *Ars Comb.*, **83**(2007), 15–32.

[169] Xu, Y. and Y.P. Liu, A census of petal bundles by genus, *Acta Math. Sinica*, **24**(2008), 6: 937–946.

[170] Xuong, N.H., How to determine the maximum genus of a graph, *J. Comb. Theory*, **B26**(1979), 217–225.

[171] Yan, J.Y. and Y.P. Liu, Asymptotic enumerations of rooted planar and outerplanar maps, *Chin. Sci. Bull.*, **36**(1991), 188–192.

[172] Yang, Y. and Y.P. Liu, Total genus distribution of two classes of 4-regular graphs, *Acta Math. Sinica* (CS), **50**(2007), 1191–1200.

[173] Yang, Y. and Y.P. Liu, Flexibility of embeddings of bouquets of circles on the projective plane and Klein bottle, *Electron. J. Comb.*, **14**(2007), #R80, 1–12.

[174] Yang, Y. and Y.P. Liu, Flexibility of circular graphs $C(2n, 2)$ on the projective plane, *Ars Comb.*, **101**(2011), 75–83.

[175] Yang, Y. and Y.P. Liu, Number of embeddings of wheel graphs on surfaces of small genus, *Ars Comb.*, **101**(2011), 225–249.

[176] Yang, Y. and Y.P. Liu, Embedding of circulant graphs and generalized Petersen graphs on projective plane, *Front Math. China*, **10**(2015), 1: 209–220.

[177] Zeng J.C. and Y.P. Liu, Genus polynomial for similar benzene structure graphs, *Adv. Appl. Discrete Math.*, **11**(2013), 121–133.

[178] Zeng, J.C., Y.P. Liu and R.X. Hao, Counting orientable embeddings by genus for a type of 3-regular graphs, *Graphs and Combin.*, **28**(2012), 1: 133–142.

[179] Zha, S.Y., On minimum-genus embedding, *Discrete Math.*, **149**(1996), 261–278.

[180] Zhang, Y.L. and Y.P. Liu, Enumeration of unicursal maps on surfaces (in Chinese with English abstract), *J. Beijing Jiaotong Univ.* (NS), **34**(2010), 6: 68–70.

[181] Zhang, Y.L., Y.P. Liu and J.L. Cai, An equation for enumerating loopless unicursal maps, *Computer Science and Automation Engineering* (CSAE), 2011 IEEE International Conference on 10–12 June 2011, 1–4.

[182] Zhang, Y.L., Y.P. Liu and R.X. Hao, Enumeration of Eulerian planar near-quadrangulations, *Consumer Electronics, Communications and Networks* (CECNet), 2011 International Conference on Date of Conference: 16–18 April 2011, 4952–4955.

[183] Zhang, Y.S., Y.C, Chen and Y.P. Liu, On average crosscap number of a graph, *Ars Comb.*, **111**(2013), 107–127.

[184] Zhao, X.M. and Y.P. Liu, On the problem of genus polynomials of tree-like graphs (in Chinese with English abstract), *J. Beijing Jiaotong Univ.*, **28**(2004), 7–11.

[185] Zhao, X.M. and Y.P. Liu, Unimodality of genus distribution for tree-like graphs (in Chinese), *J. Shanxi Univ.* (NS), **29**(2006), 242–244.

[186] Zhao, X.M. and Y.P. Liu, Genus distribution for circle-like graphs (in Chinese with English abstract), *Acta Math. Sci.*, **28**(2008), 4: 768–778.

[187] Zhu, Z.L. and Y.P. Liu, The genus distributions of two types of graphs (in Chinese with English abstract), *J. Shenyang Normal Univ.* (NS), **24**(2006), 1–5.

Author Index

DOI 10.1515/9783110481846-023

Subject Index

DOI 10.1515/9783110481846-024

Bei Fragen zur Produktsicherheit wenden Sie sich bitte an:
If you have any questions regarding product safety,
please contact:

Walter de Gruyter GmbH
Genthiner Straße 13
10785 Berlin
productsafety@degruyterbrill.com